Lecture Notes in Computer Science   14487

The series Lecture Notes in Computer Science (LNCS), including its subseries Lecture Notes in Artificial Intelligence (LNAI) and Lecture Notes in Bioinformatics (LNBI), has established itself as a medium for the publication of new developments in computer science and information technology research, teaching, and education.

LNCS enjoys close cooperation with the computer science R & D community, the series counts many renowned academics among its volume editors and paper authors, and collaborates with prestigious societies. Its mission is to serve this international community by providing an invaluable service, mainly focused on the publication of conference and workshop proceedings and postproceedings. LNCS commenced publication in 1973.

Zahir Tari · Keqiu Li · Hongyi Wu
Editors

Algorithms and Architectures for Parallel Processing

23rd International Conference, ICA3PP 2023
Tianjin, China, October 20–22, 2023
Proceedings, Part I

 Springer

Editors
Zahir Tari
Royal Melbourne Institute of Technology
Melbourne, VIC, Australia

Keqiu Li
Tianjin University
Tianjin, China

Hongyi Wu
University of Arizona
Tucson, AZ, USA

ISSN 0302-9743   ISSN 1611-3349 (electronic)
Lecture Notes in Computer Science
ISBN 978-981-97-0833-8   ISBN 978-981-97-0834-5 (eBook)
https://doi.org/10.1007/978-981-97-0834-5

This Springer imprint is published by the registered company Springer Nature Singapore Pte Ltd.
The registered company address is: 152 Beach Road, #21-01/04 Gateway East, Singapore 189721, Singapore

Paper in this product is recyclable.

Preface

On behalf of the Conference Committee, we welcome you to the proceedings of the 2023 International Conference on Algorithms and Architectures for Parallel Processing (ICA3PP 2023), which was held in Tianjin, China from October 20–22, 2023. ICA3PP2023 was the 23rd in this series of conferences (started in 1995) that are devoted to algorithms and architectures for parallel processing. ICA3PP is now recognized as the main regular international event that covers the many dimensions of parallel algorithms and architectures, encompassing fundamental theoretical approaches, practical experimental projects, and commercial components and systems. This conference provides a forum for academics and practitioners from countries around the world to exchange ideas for improving the efficiency, performance, reliability, security, and interoperability of computing systems and applications.

A successful conference would not be possible without the high-quality contributions made by the authors. This year, ICA3PP received a total of 503 submissions from authors in 21 countries and regions. Based on rigorous peer reviews by the Program Committee members and reviewers, 193 high-quality papers were accepted to be included in the conference proceedings and submitted for EI indexing. In addition to the contributed papers, six distinguished scholars, Lixin Gao, Baochun Li, Laurence T. Yang, Kun Tan, Ahmed Louri, and Hai Jin, were invited to give keynote lectures, providing us with the recent developments in diversified areas in algorithms and architectures for parallel processing and applications.

We would like to take this opportunity to express our sincere gratitude to the Program Committee members and 165 reviewers for their dedicated and professional service. We highly appreciate the twelve track chairs, Dezun Dong, Patrick P. C. Lee, Meng Shen, Ruidong Li, Li Chen, Wei Bao, Jun Li, Hang Qiu, Ang Li, Wei Yang, Yu Yang, and Zhibin Yu, for their hard work in promoting this conference and organizing the reviews for the papers submitted to their tracks. We are so grateful to the publication chairs, Heng Qi, Yulei Wu, Deze Zeng, and the publication assistants for their tedious work in editing the conference proceedings. We must also say "thank you" to all the volunteers who helped us at various stages of this conference. Moreover, we were so honored to have many renowned scholars be part of this conference. Finally, we would like to thank all speakers, authors, and participants for their great contribution to and support for the success of ICA3PP 2023!

October 2023

Jean-Luc Gaudiot
Hong Shen
Gudula Rünger
Zahir Tari
Keqiu Li
Hongyi Wu
Tian Wang

Organization

General Chairs

Jean-Luc Gaudiot	University of California, Irvine, USA
Hong Shen	University of Adelaide, Australia
Gudula Rünger	Chemnitz University of Technology, Germany

Program Chairs

Zahir Tari	Royal Melbourne Institute of Technology, Australia
Keqiu Li	Tianjin University, China
Hongyi Wu	University of Arizona, USA

Program Vice-chair

Wenxin Li	Tianjin University, China

Publicity Chairs

Hai Wang	Northwest University, China
Milos Stojmenovic	Singidunum University, Serbia
Chaofeng Zhang	Advanced Institute of Industrial Technology, Japan
Hao Wang	Louisiana State University, USA

Publication Chairs

Heng Qi	Dalian University of Technology, China
Yulei Wu	University of Exeter, UK
Deze Zeng	China University of Geosciences (Wuhan), China

Workshop Chairs

Laiping Zhao	Tianjin University, China
Pengfei Wang	Dalian University of Technology, China

Local Organization Chairs

Xiulong Liu	Tianjin University, China
Yitao Hu	Tianjin University, China

Web Chair

Chen Chen	Shanghai Jiao Tong University, China

Registration Chairs

Xinyu Tong	Tianjin University, China
Chaokun Zhang	Tianjin University, China

Steering Committee Chairs

Yang Xiang (Chair)	Swinburne University of Technology, Australia
Weijia Jia	Beijing Normal University and UIC, China
Yi Pan	Georgia State University, USA
Laurence T. Yang	St. Francis Xavier University, Canada
Wanlei Zhou	City University of Macau, China

Program Committee

Track 1: Parallel and Distributed Architectures

Dezun Dong (Chair)	National University of Defense Technology, China
Chao Wang	University of Science and Technology of China, China
Chentao Wu	Shanghai Jiao Tong University, China

Chi Lin	Dalian University of Technology, China
Deze Zeng	China University of Geosciences, China
En Shao	Institute of Computing Technology, Chinese Academy of Sciences, China
Fei Lei	National University of Defense Technology, China
Haikun Liu	Huazhong University of Science and Technology, China
Hailong Yang	Beihang University, China
Junlong Zhou	Nanjing University of Science and Technology, China
Kejiang Ye	Shenzhen Institute of Advanced Technology, Chinese Academy of Sciences, China
Lei Wang	National University of Defense Technology, China
Massimo Cafaro	University of Salento, Italy
Massimo Torquati	University of Pisa, Italy
Mengying Zhao	Shandong University, China
Roman Wyrzykowski	Czestochowa University of Technology, Poland
Rui Wang	Beihang University, China
Sheng Ma	National University of Defense Technology, China
Songwen Pei	University of Shanghai for Science and Technology, China
Susumu Matsumae	Saga University, Japan
Weihua Zhang	Fudan University, China
Weixing Ji	Beijing Institute of Technology, China
Xiaoli Gong	Nankai University, China
Youyou Lu	Tsinghua University, China
Yu Zhang	Huazhong University of Science and Technology, China
Zichen Xu	Nanchang University, China

Track 2: Software Systems and Programming Models

Patrick P. C. Lee (Chair)	Chinese University of Hong Kong, China
Erci Xu	Ohio State University, USA
Xiaolu Li	Huazhong University of Science and Technology, China
Shujie Han	Peking University, China
Mi Zhang	Institute of Computing Technology, Chinese Academy of Sciences, China

Jing Gong	KTH Royal Institute of Technology, Sweden
Radu Prodan	University of Klagenfurt, Austria
Wei Wang	Beijing Jiaotong University, China
Himansu Das	KIIT Deemed to be University, India
Rong Gu	Nanjing University, China
Yongkun Li	University of Science and Technology of China, China
Ladjel Bellatreche	National Engineering School for Mechanics and Aerotechnics, France

Track 3: Distributed and Network-Based Computing

Meng Shen (Chair)	Beijing Institute of Technology, China
Ruidong Li (Chair)	Kanazawa University, Japan
Bin Wu	Institute of Information Engineering, China
Chao Li	Beijing Jiaotong University, China
Chaokun Zhang	Tianjin University, China
Chuan Zhang	Beijing Institute of Technology, China
Chunpeng Ge	National University of Defense Technology, China
Fuliang Li	Northeastern University, China
Fuyuan Song	Nanjing University of Information Science and Technology, China
Gaopeng Gou	Institute of Information Engineering, China
Guangwu Hu	Shenzhen Institute of Information Technology, China
Guo Chen	Hunan University, China
Guozhu Meng	Chinese Academy of Sciences, China
Han Zhao	Shanghai Jiao Tong University, China
Hai Xue	University of Shanghai for Science and Technology, China
Haiping Huang	Nanjing University of Posts and Telecommunications, China
Hongwei Zhang	Tianjin University of Technology, China
Ioanna Kantzavelou	University of West Attica, Greece
Jiawen Kang	Guangdong University of Technology, China
Jie Li	Northeastern University, China
Jingwei Li	University of Electronic Science and Technology of China, China
Jinwen Xi	Beijing Zhongguancun Laboratory, China
Jun Liu	Tsinghua University, China

Kaiping Xue	University of Science and Technology of China, China
Laurent Lefevre	National Institute for Research in Digital Science and Technology, France
Lanju Kong	Shandong University, China
Lei Zhang	Henan University, China
Li Duan	Beijing Jiaotong University, China
Lin He	Tsinghua University, China
Lingling Wang	Qingdao University of Science and Technology, China
Lingjun Pu	Nankai University, China
Liu Yuling	Institute of Information Engineering, China
Meng Li	Hefei University of Technology, China
Minghui Xu	Shandong University, China
Minyu Feng	Southwest University, China
Ning Hu	Guangzhou University, China
Pengfei Liu	University of Electronic Science and Technology of China, China
Qi Li	Beijing University of Posts and Telecommunications, China
Qian Wang	Beijing University of Technology, China
Raymond Yep	University of Macau, China
Shaojing Fu	National University of Defense Technology, China
Shenglin Zhang	Nankai University, China
Shu Yang	Shenzhen University, China
Shuai Gao	Beijing Jiaotong University, China
Su Yao	Tsinghua University, China
Tao Yin	Beijing Zhongguancun Laboratory, China
Tingwen Liu	Institute of Information Engineering, China
Tong Wu	Beijing Institute of Technology, China
Wei Quan	Beijing Jiaotong University, China
Weihao Cui	Shanghai Jiao Tong University, China
Xiang Zhang	Nanjing University of Information Science and Technology, China
Xiangyu Kong	Dalian University of Technology, China
Xiangyun Tang	Minzu University of China, China
Xiaobo Ma	Xi'an Jiaotong University, China
Xiaofeng Hou	Shanghai Jiao Tong University, China
Xiaoyong Tang	Changsha University of Science and Technology, China
Xuezhou Ye	Dalian University of Technology, China
Yaoling Ding	Beijing Institute of Technology, China

Yi Zhao	Tsinghua University, China
Yifei Zhu	Shanghai Jiao Tong University, China
Yilei Xiao	Dalian University of Technology, China
Yiran Zhang	Beijing University of Posts and Telecommunications, China
Yizhi Zhou	Dalian University of Technology, China
Yongqian Sun	Nankai University, China
Yuchao Zhang	Beijing University of Posts and Telecommunications, China
Zhaoteng Yan	Institute of Information Engineering, China
Zhaoyan Shen	Shandong University, China
Zhen Ling	Southeast University, China
Zhiquan Liu	Jinan University, China
Zijun Li	Shanghai Jiao Tong University, China

Track 4: Big Data and Its Applications

Li Chen (Chair)	University of Louisiana at Lafayette, USA
Alfredo Cuzzocrea	University of Calabria, Italy
Heng Qi	Dalian University of Technology, China
Marc Frincu	Nottingham Trent University, UK
Mingwu Zhang	Hubei University of Technology, China
Qianhong Wu	Beihang University, China
Qiong Huang	South China Agricultural University, China
Rongxing Lu	University of New Brunswick, Canada
Shuo Yu	Dalian University of Technology, China
Weizhi Meng	Technical University of Denmark, Denmark
Wenbin Pei	Dalian University of Technology, China
Xiaoyi Tao	Dalian Maritime University, China
Xin Xie	Tianjin University, China
Yong Yu	Shaanxi Normal University, China
Yuan Cao	Ocean University of China, China
Zhiyang Li	Dalian Maritime University, China

Track 5: Parallel and Distributed Algorithms

Wei Bao (Chair)	University of Sydney, Australia
Jun Li (Chair)	City University of New York, USA
Dong Yuan	University of Sydney, Australia
Francesco Palmieri	University of Salerno, Italy

George Bosilca	University of Tennessee, USA
Humayun Kabir	Microsoft, USA
Jaya Prakash Champati	IMDEA Networks Institute, Spain
Peter Kropf	University of Neuchâtel, Switzerland
Pedro Soto	CUNY Graduate Center, USA
Wenjuan Li	Hong Kong Polytechnic University, China
Xiaojie Zhang	Hunan University of Technology and Business, China
Chuang Hu	Wuhan University, China

Track 6: Applications of Parallel and Distributed Computing

Hang Qiu (Chair)	Waymo, USA
Ang Li (Chair)	Qualcomm, USA
Daniel Andresen	Kansas State University, USA
Di Wu	University of Central Florida, USA
Fawad Ahmad	Rochester Institute of Technology, USA
Haonan Lu	University at Buffalo, USA
Silvio Barra	University of Naples Federico II, Italy
Weitian Tong	Georgia Southern University, USA
Xu Zhang	University of Exeter, UK
Yitao Hu	Tianjin University, China
Zhixin Zhao	Tianjin University, China

Track 7: Service Dependability and Security in Distributed and Parallel Systems

Wei Yang (Chair)	University of Texas at Dallas, USA
Dezhi Ran	Peking University, China
Hanlin Chen	Purdue University, USA
Jun Shao	Zhejiang Gongshang University, China
Jinguang Han	Southeast University, China
Mirazul Haque	University of Texas at Dallas, USA
Simin Chen	University of Texas at Dallas, USA
Wenyu Wang	University of Illinois at Urbana-Champaign, USA
Yitao Hu	Tianjin University, China
Yueming Wu	Nanyang Technological University, Singapore
Zhengkai Wu	University of Illinois at Urbana-Champaign, USA
Zhiqiang Li	University of Nebraska, USA
Zhixin Zhao	Tianjin University, China

| Ze Zhang | University of Michigan/Cruise, USA |
| Ravishka Rathnasuriya | University of Texas at Dallas, USA |

Track 8: Internet of Things and Cyber-Physical-Social Computing

Yu Yang (Chair)	Lehigh University, USA
Qun Song	Delft University of Technology, The Netherlands
Chenhan Xu	University at Buffalo, USA
Mahbubur Rahman	City University of New York, USA
Guang Wang	Florida State University, USA
Houcine Hassan	Universitat Politècnica de València, Spain
Hua Huang	UC Merced, USA
Junlong Zhou	Nanjing University of Science and Technology, China
Letian Zhang	Middle Tennessee State University, USA
Pengfei Wang	Dalian University of Technology, China
Philip Brown	University of Colorado Colorado Springs, USA
Roshan Ayyalasomayajula	University of California San Diego, USA
Shigeng Zhang	Central South University, China
Shuo Yu	Dalian University of Technology, China
Shuxin Zhong	Rutgers University, USA
Xiaoyang Xie	Meta, USA
Yi Ding	Massachusetts Institute of Technology, USA
Yin Zhang	University of Electronic Science and Technology of China, China
Yukun Yuan	University of Tennessee at Chattanooga, USA
Zhengxiong Li	University of Colorado Denver, USA
Zhihan Fang	Meta, USA
Zhou Qin	Rutgers University, USA
Zonghua Gu	Umeå University, Sweden
Geng Sun	Jilin University, China

Track 9: Performance Modeling and Evaluation

Zhibin Yu (Chair)	Shenzhen Institute of Advanced Technology, Chinese Academy of Sciences, China
Chao Li	Shanghai Jiao Tong University, China
Chuntao Jiang	Foshan University, China
Haozhe Wang	University of Exeter, UK
Laurence Muller	University of Greenwich, UK

Lei Liu	Beihang University, China
Lei Liu	Institute of Computing Technology, Chinese Academy of Sciences, China
Jingwen Leng	Shanghai Jiao Tong University, China
Jordan Samhi	University of Luxembourg, Luxembourg
Sa Wang	Institute of Computing Technology, Chinese Academy of Sciences, China
Shoaib Akram	Australian National University, Australia
Shuang Chen	Huawei, China
Tianyi Liu	Huawei, China
Vladimir Voevodin	Lomonosov Moscow State University, Russia
Xueqin Liang	Xidian University, China

Reviewers

Dezun Dong	Xiaolu Li
Chao Wang	Shujie Han
Chentao Wu	Mi Zhang
Chi Lin	Jing Gong
Deze Zeng	Radu Prodan
En Shao	Wei Wang
Fei Lei	Himansu Das
Haikun Liu	Rong Gu
Hailong Yang	Yongkun Li
Junlong Zhou	Ladjel Bellatreche
Kejiang Ye	Meng Shen
Lei Wang	Ruidong Li
Massimo Cafaro	Bin Wu
Massimo Torquati	Chao Li
Mengying Zhao	Chaokun Zhang
Roman Wyrzykowski	Chuan Zhang
Rui Wang	Chunpeng Ge
Sheng Ma	Fuliang Li
Songwen Pei	Fuyuan Song
Susumu Matsumae	Gaopeng Gou
Weihua Zhang	Guangwu Hu
Weixing Ji	Guo Chen
Xiaoli Gong	Guozhu Meng
Youyou Lu	Han Zhao
Yu Zhang	Hai Xue
Zichen Xu	Haiping Huang
Patrick P. C. Lee	Hongwei Zhang
Erci Xu	Ioanna Kantzavelou

Zhixin Zhao
Wei Yang
Dezhi Ran
Hanlin Chen
Jun Shao
Jinguang Han
Mirazul Haque
Simin Chen
Wenyu Wang
Yitao Hu
Yueming Wu
Zhengkai Wu
Zhiqiang Li
Zhixin Zhao
Ze Zhang
Ravishka Rathnasuriya
Yu Yang
Qun Song
Chenhan Xu
Mahbubur Rahman
Guang Wang
Houcine Hassan
Hua Huang
Junlong Zhou
Letian Zhang
Pengfei Wang
Philip Brown
Roshan Ayyalasomayajula

Shigeng Zhang
Shuo Yu
Shuxin Zhong
Xiaoyang Xie
Yi Ding
Yin Zhang
Yukun Yuan
Zhengxiong Li
Zhihan Fang
Zhou Qin
Zonghua Gu
Geng Sun
Zhibin Yu
Chao Li
Chuntao Jiang
Haozhe Wang
Laurence Muller
Lei Liu
Lei Liu
Jingwen Leng
Jordan Samhi
Sa Wang
Shoaib Akram
Shuang Chen
Tianyi Liu
Vladimir Voevodin
Xueqin Liang

Contents – Part I

Short Video Account Influence Evaluation Model Based on Improved
SF-UIR Algorithm 1
    Xiaojun Guo and Zhihao Wu

Deep Hash Learning of Feature-Invariant Representation for Single-Label
and Multi-label Retrieval 17
    Yuan Cao, Xinzheng Shang, Junwei Liu, Chengzhi Qian, and Sheng Chen

Generative Adversarial Network Based Asymmetric Deep Cross-Modal
Unsupervised Hashing 30
    Yuan Cao, Yaru Gao, Na Chen, Jiacheng Lin, and Sheng Chen

CFDM-IME: A Collaborative Fault Diagnosis Method for Intelligent
Manufacturing Equipment 49
    Yue Wang, Tao Zhou, Xiaohu Zhao, and Xiaofei Hu

DFECTS: A Deep Fuzzy Ensemble Clusterer for Time Series 61
    Dechong Wu, Jialun Li, Xuan Mo, and Weigang Wu

Energy-Aware Smart Task Scheduling in Edge Computing Networks
with A3C 81
    Dan Wang, Liang Liu, Binbin Ge, Junjie Qi, and Zehui Zhao

BACTDS: Blockchain-Based Fined-Grained Access Control Scheme
with Traceablity for IoT Data Sharing 97
    Wei Lu, Jiguo Yu, Biwei Yan, Suhui Liu, and Baobao Chai

DeepLat: Achieving Minimum Worst Case Latency for DNN Inference
with Batch-Aware Dispatching 109
    Jiaheng Gao and Yitao Hu

Privacy-Preserving and Reliable Distributed Federated Learning 130
    Yipeng Dong, Lei Zhang, and Lin Xu

Joint Optimization of Request Scheduling and Container Prewarming
in Serverless Computing 150
    Si Chen, Guanghui Li, Chenglong Dai, Wei Li, and Qinglin Zhao

Multi-stage Optimization of Incentive Mechanisms for Mobile Crowd
Sensing Based on Top-Trading Cycles 170
 Jingjie Shang, Haifeng Jiang, Chaogang Tang, Huaming Wu,
 Shuhao Wang, and Shoujun Zhang

A Chained Forwarding Mechanism for Large Messages 187
 Jiaqi Lin, Tao Feng, Nanxin Zhou, Xianming Gao, and Shanqing Jiang

TDC: Pool-Level Object Cache Replacement Algorithm Based
on Temperature Density 204
 HuaCheng Lu, Yong Wang, JunQi Chen, DaHuan Zhang, and ZhiKe Li

Smart DAG Task Scheduling Based on MCTS Method of Multi-strategy
Learning 224
 Lang Shu, Guanyan Pan, Bei Wang, Wenbing Peng, Minhui Fang,
 Yifei Chen, Fanding Huang, Songchen Li, and Yuxia Cheng

CT-Mixer: Exploiting Multiscale Design for Local-Global Representations
Learning 243
 Yin Tang, Xili Wan, Xinjie Guan, and Aichun Zhu

FedQL: Q-Learning Guided Aggregation for Federated Learning 263
 Mei Cao, Mengying Zhao, Tingting Zhang, Nanxiang Yu, and Jianbo Lu

An Adaptive Instruction Set Encoding Automatic Generation Method
for VLIW 283
 Xin Xiao and Zhong Liu

FastDet: Detecting Encrypted Malicious Traffic Faster via Early Exit 301
 Jiakun Sun, Jintian Lu, Yabo Wang, and Shuyuan Jin

Real-EVE: Real-Time Edge-Assist Video Enhancement for Joint
Denoising and Super-Resolution 320
 Liming Ge, Wei Bao, Dong Yuan, and Bing Bing Zhou

Optimizing the Parallelism of Communication and Computation
in Distributed Training Platform 340
 Xiang Hou, Yuan Yuan, Sheng Ma, Rui Xu, Bo Wang, Tiejun Li,
 Wei Jiang, Lizhou Wu, and Jianmin Zhang

FedSC: Compatible Gradient Compression for Communication-Efficient
Federated Learning 360
 Xinlei Yu, Zhipeng Gao, Chen Zhao, and Zijia Mo

Joint Video Transcoding and Representation Selection for Edge-Assisted
Multi-party Video Conferencing 380
   Fanhao Kong, Tuo Cao, Zhuzhong Qian, Xiaoliang Wang, Ming Zhao,
   Liming Wang, and Zhenjie Lin

Performance Comparison of Distributed DNN Training on Optical Versus
Electrical Interconnect Systems 401
   Fei Dai, Yawen Chen, Zhiyi Huang, Haibo Zhang, and Hui Tian

Dynamic Path Planning Based on Traffic Flow Prediction and Traffic
Light Status 419
   Weiyang Chen, Bingyi Liu, Weizhen Han, Gaolei Li, and Bin Song

A Time Series Data Compression Co-processor Based on RISC-V Custom
Instructions 439
   Peiran Du and Zhaohui Cai

MSIN: An Efficient Multi-head Self-attention Framework for Inertial
Navigation 455
   Gaotao Shi, Bingjia Pan, and Yuzhi Ni

Absorb: Deadlock Resolution for 2.5D Modular Chiplet Based Systems 474
   Yi Yang, Tiejun Li, Yi Dai, Bo Wang, Sheng Ma, and Yanqiang Sun

A Multi-server Authentication Scheme Based on Fuzzy Extractor 488
   Wang Cheng, Lin You, and Gengran Hu

Author Index 501

Short Video Account Influence Evaluation Model Based on Improved SF-UIR Algorithm

Xiaojun Guo[1,2,3](✉) and Zhihao Wu[1,2,3]

[1] School of Information Engineering, Xizang Minzu University, Xianyang 712082, Shanxi, China
aikt@xzmu.edu.cn
[2] Key Laboratory of Optical Information Processing and Visualization Technology of Tibet Autonomous Region, Xianyang 712082, Shanxi, China
[3] Cyberspace Governance Research Base of Tibet Autonomous Region, Xianyang 712082, Shanxi, China

Abstract. In view of the lack of numerical value of the influence of the account's own behavior and the lack of the influence of following fans based on the topological structure platform in the evaluation of the influence of short video accounts, this paper proposes a short video platform account influence calculation algorithm FF-SF-UIR (Factor analysis and Fan group chat in SF-UIR) based on the improved SF-UIR (Self and Followers User Influence Rank) algorithm from three perspectives: the influence of the official platform label authentication, the influence of the work, and the influence of the following fans based on the topological structure platform. This method analyzes the rationality of the historicity and periodicity of the influence of the work by using the numerical influence of the official label of the account platform and introducing the factor analysis method. According to the social network topology, the influence of followers based on the topological structure platform is introduced, which makes the evaluation result more reasonable and provides a new way for influence ranking of short video accounts. Taking Douyin (TikTok in China) short video platform as the experimental object, the experimental results show that compared with other existing methods, this method improves the precision by 9.67%, the recall by 23.67%, and the F1-Score by 20.68% in the evaluation index.

Keywords: SF-UIR algorithm · Short video account · Influence · Factor analysis

1 Introduction

The short video platform [1] based on intelligent distribution technology. The short video released on the platform has the characteristics of visualization, fragmentation, mobile lightweight and intuitive image. In recent years, it has gradually become an important channel for the public to obtain information, and the number of users has increased rapidly [2]. According to the 49th Statistical Report on Internet Development in China released by the China Internet Network Information Center, as of December 2021, China's short video users reached 934 million, accounting for 90.5% of the total number of Internet users [3]. Influence is the ability to change the thoughts and actions of others in a way that is acceptable to them. Account influence [4] is an indicator to measure the status and security importance of accounts in social networks. The daily behavior of high-impact accounts in the network may affect or lead others [5]. Risky accounts can also affect the security status of other accounts. The research on the social influence of short video accounts can screen out high-impact accounts to judge and analyze their potential for the platform, to know the influence and the degree of guidance of the account's behavior on other accounts on the short video platform, and provide a reference for the daily safety management and public opinion monitoring of the short video platform [6].

The initial influence evaluation method of short video accounts is relatively simple. For example, the In-Degree algorithm [7] takes an account as a vertex in the social network, and the number of fans of this account is the number of arcs. The Follower algorithm [8] also determines the ranking of account influence according to the number of account fans. In the statistical method, the factor analysis method [9] is widely used in the comprehensive evaluation, which ranks the comprehensive score of factors according to the weight calculation. In terms of dynamic evaluation of influence, Wang Li et al. combined the emergent computation model, proposed a method for evaluating the influence of micro-blog accounts based on the Swarm model [10], which can calculate the influence of accounts at different times. Li Wei [11] proposed a method for dynamically measuring the influence of accounts on social media. Zhao Di [12] proposed a method for measuring the influence of micro-blog topic accounts that supports dynamic update. Qi Jia proposed an improved HITS algorithm [13]. In the network topology, Larry Page proposed PageRank algorithm [14], but in the algorithm, influence of each fan is set to the same value, which does not meet the actual situation of the platform account. Wang Ding et al. [15] proposed the SF-UIR algorithm to evaluate the influence of Weibo account, but due to the certain differences between Weibo and short video platforms, we cannot directly apply it to short video accounts. In terms of short video, Wu Menghan applied the SF-UIR algorithm to the influence evaluation of short video accounts, completely abandoning the influence on account fans and lacking persuasion [16].

Most of the above research solves the problem of Weibo account influence calculation [17], which cannot be directly applied to short video accounts. Moreover, the SF-UIR short video account influence evaluation model only introduces the account's own behavior influence [18], without considering the influence of followers on the platform where the account is located. Based on this, this paper proposes a method for calculating the influence of short video accounts based on the improved SF-UIR algorithm. From the three perspectives of the influence of accounts's tag authentication given by the official platform, the influence of works, and the influence of accounts's following fans [19], the difference between the short video platform and the micro-blog platform is considered, and a new method for calculating the influence of accounts is given.

This article refers to Wang Lifeng's classification of customers into five levels in the optimization of multi-objective cold chain logistics distribution, and every 20% accounts for one level. Since a total of 2008 accounts are used in this chapter, the influence of accounts is divided into three levels, which are high-influence accounts, medium-influence accounts, and low-influence accounts. Therefore, the top 20% of the accounts in the ranking are high-influence accounts. In this chapter, the top 20%–70% of the accounts are medium-influence accounts. The chapter selects accounts ranked 653rd to 1387th, and the influence is between 0.153 and 0.655, which is the account with medium influence; after sorting, about 30% of the accounts are low-impact accounts, and this chapter uses the ranking below 1388th, with an influence of 0.153 The following are low-influence accounts, which provide a new reference method for grading the influence of short video accounts.

We summarize the innovation of this paper as the following three points:

(1) Whether the account is authenticated is no longer given a simple (0, e) influence value. According to the different difficulty of each account authentication category given by the Douyin platform, the platform label authentication influence of the account's own behavior influence is divided into four values [20]: 0, 0.3, 0.5, and 1.
(2) In the influence of the works of the account's own influence, the historical and periodic parallel analysis is carried out in combination with the factor analysis method, which makes the early published works and the recently published works set the importance, abandoning the singleness of considering only the influence of the works for a certain period.
(3) Add the calculation of the influence of accounts's following fans, by combining the topology [21], the number of fans chatting on the short video platform, and the proportion of platform fans chatting and account fans as important factors to calculate.

2 Relevant Technical Basis

2.1 Factor Analysis Method

Factor analysis is to set the KMO (full name) and Bartlett's sphericitytest [22]. When the load matrix estimation method and factor rotation, according to the weight to calculate the new variable Score that factor composite score, then sort. KMO value was used for the KMO value test, and I used p value for the Bartlett test.

2.2 Social Network Account Relationship Directed Graph

A directed graph [23] is a directional graph consisting of a set of vertices and a set of directed edges, each of which is connected by an ordered pair of vertices. n is the number of vertices in Fig. 1, and e is the number of edges or arcs. For a directed graph, the value of e ranges from 1 to n (n-1), and n(n-1) directed edges are directed complete graphs. In the social network topology, it often applied graph structure to the topology of social networks. As shown in Fig. 1

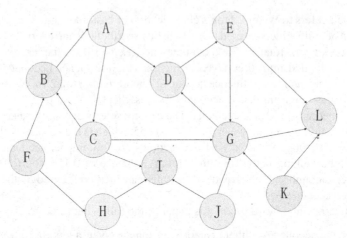

Fig. 1. Account social network diagram Note: The four colors of the nodes represent the four types of accounts with different authentication difficulty levels, and the account relationships are randomly distributed. (color figure online)

2.3 SF-UIR Algorithm

SF-UIR is an algorithm that draws on the PageRank idea. Due to the similarity between the topology of the micro-blog and the network topology, micro-blog users are regarded as a page of the network. Users following the others are compared to pages adding links to other pages. Increasing the in-degree of the page is equal to the user being followed. The SF-UIR algorithm considers both self-factors of users' behavior and factors of followers based on topology. The self-factors of users' behavior include the number of users ' followers, whether users are authenticated, and the dissemination ability of users' micro-blog works. The factors of followers based on the topology structure completely draw on the idea of PageRank, and then add a weight factor [24] on the basis of PageRank to allocate the influence contribution of different users.

3 Short Video Account Influence Calculation Based on Improved SF-UIR Algorithm

3.1 FF-SF-UIR Model Framework

To solve the problem that most of short video account influence algorithms only consider simple numericalization of tag authentication of accounts' own behavior influence and periodic changes of works' influence, and the influence of followers based on the topology is abandoned. Based on SF-UIR algorithm, this paper proposes a FF-SF-UIR algorithm, which combines the accurate quantification of the influence of platform tag authentication, the historical and periodic influence of works, and the influence of platform followers. The structure of the FF-SF-UIR model is shown in Fig. 2.

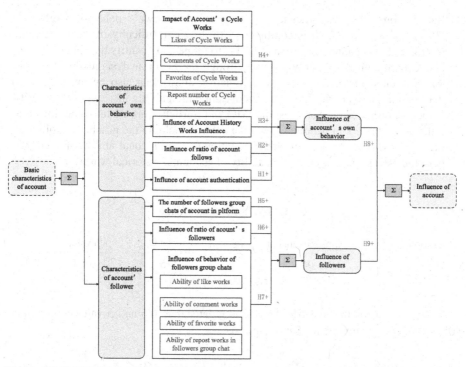

Fig. 2. FF-SF-UIR Model Frame Note: Hn + is related to the positive influence of this factor on the pointing factor, and Hn − is related to the negative influence of this factor on the pointing factor.

3.2 FF-SF-UIR Model Definition and Short Video Account Influence Calculation

The FF-SF-UIR algorithm is divided into two parts. The first part is the influence of accounts ' own behavior. The second part is the influence of fans on the platform that based on the topology. The FF-SF-UIR algorithm is defined as:

$$P_{F-F-SF-UIR}(U_i) = P_{F-F-SF-UIR\_self}(U_i) + P_{F-F-SF-UIR\_fans}(U_i) \qquad (1)$$

$P_{F-F-SF-UIR\_self}(U_i)$ (0 <i < = 2008) is the influence of user i itself, $P_{F-F-SF-UIR\_self}(U_i)$ is the influence of account i based on its own behavior, $P_{F-F-SF-UIR\_fans}(U_i)$ is the followers' influence of account i based on topology platform.

(1) Based on the significance of accounts' own behavior influence $(P_{F-F-SF-UIR\_self}(U_i))$ it is combined with different difficulty of the short video platform account authentication types [19] to refine and quantify the value of different account authentication types, and through the factor analysis method to prove that accounts' historical works dissemination ability can also be introduced, so that the timeliness of accounts' works dissemination ability increases the integrity. Table 1 shows the sphericity test of KMO and Bartlett. This method quantifies four influencing factors of accounts' own behavior, which are the number of followers of accounts, types of account authenttication,disseminational ability of accounts' periodic works, and dissemination ability of accounts' historical works.

Table 1. Test of KMO and Battle.

KMO value	Approximate chi-square	Df	p value
0.731	16942.044	15	0.000

Validation shows that the KMO value is greater than 0.6 and p value is less than 0.05, so the data is suitable for factor analysis.

$$P_{F-F-SF-UIR\_self}(U_i) = \frac{F_{u_i}}{N} + B_{verified}(U_i) + A_{u_i} * T_{u_i} + \frac{H}{h} \quad (2)$$

In formula (2), Fui is the number of followers of account i, N is the number of followers owned by the account with the largest number of followers in the data set, and B is the influence value given by account authentication. According to the difficulty of official authentication, the influence value of account authentication is quantified. Set inf to the user's tag influence. (1)Quality users, quality creators, quality authors in a field, network celebrities, public figures inf = 1. (2) National institutions, media, universities and other well-known institutions inf = 0.5. (3) Professional certification and enterprise certification inf = 0.3. (4) Unauthenticated users inf = 0. Au is released frequency of short videos in a statistical cycle. Tu is the sum of the influence rate of account I's works over the period, and H / h is the total number of likes for account i since registration and the total number of likes for the account with the highest total number of likes in the dataset.

$$A_{u_i} = \frac{M_{u_i}}{T} \quad (3)$$

$$T_{u_i} = \sum_{m_j \in M_{u_i}} (aR_{m_j} + bC_{m_j} + cL_{m_j} + dW_{m_j}) \quad (4)$$

In Eq. (3), Mui is the set of published works of User i in T. T is the statistical period, which is set to 15 days, that is, T = 15.

In Eq. (4), m_j is any work published by user i in T, and R_{m_j} is the praise rate of m_j, that is, the ratio of the number of praises of m_j to the number of followers of user i. C_{m_j} is the comment rate, L_{m_j} is the collection rate, W_{m_j} is the forwarding rate, and a, b, c are their respective weights. A, b, c are their respective weights, here with the relative importance of things table set weights are a = 0.0455, b = 0.1364, c = 0.4091, d = 0.4091.

(2) The significance of the influence of platform followers based on topological structure $(P_{DY-SF-UIR\_fans}(U_i))$ is to excavate the particularity of short video accounts with the platform fans' group chats. All the accounts in the fans' group chats are used as a topological network, and the account i is used as a core. It passed its information to other fans' accounts in the fans' group chats. The fans; accounts have an influence on a series of operations such as likes, comments, collections, and forwarding of its works. Thus quantifies two items of influence of followers of the account, which is the proportion of the influence generated by fans' operation to the number of fans of the account.

$$P_{F-F-SF-UIR\_fans}(U_i) = (1-d) + d * \frac{4 * E_{u_i}}{F_{u_i} + 1} * \frac{F_{u_i}}{N} \qquad (5)$$

The damping coefficient d = 0.85 is set to make the value of the influence of followers on the platform based on the topological structure converge to a certain interval.is the number of followers of the account i, and is to prevent the denominator from being 0.

3.3 FF-SF-UIR Model Algorithm Steps

Algorithm 1: The influence of account i based on the account's own behavior algorithm

Require: $Bui, Fui, N, Rmj, Cmj, Lmj, Wmj\ Mui, T, Aui, Tui, H, h$

Ensure: The influence of account i based on the account's own behavior

    function $F1(N, Bui, Aui, Tui, H/h, Rmj, Cmj, Lmj, Wmj\ Mui, T)$

      $Bui \leftarrow [0, 0.3, 0.5, 1]$ ##The account label is digitized

      $Fui \leftarrow [0...n]$ ##Initialize the assignment of the number of followers

      $N \leftarrow max(Fui) \leftarrow [1....n]$##Take the number of followers owned by the account with the largest number of followers in the data set

      $Rmj \leftarrow [0:1]$ ##praise rate

      $Cmj \leftarrow [0:1]$ ##comment rate

      $Lmj \leftarrow [0:1]$ ## collection rate

      $Wmj \leftarrow [0:1]$ ## forwarding rate

      for $(k=1;(k<=2008);k++)$:**do**

        $Tui \leftarrow 0$

        $G \leftarrow a * Rmj + b * Cmj + c * Lmj + d * Wmj$

        $Tui \leftarrow Tui + G$

    end for ##The sum of the impact rate of works in the calculation period

      $Mui \leftarrow [0....n]$ ## Collection of published works by user i in T

      $T \leftarrow 15$ ##T is the statistical period, and the statistical period is set to 15 days

      $Aui \leftarrow Mui/T$

      $H \leftarrow [0...n]$ ##The total number of likes of account i since registration

      $H \leftarrow max(H)$ ##The total number of likes of the account with the most total likes in the dataset

      $P\_FF\_self\_i \leftarrow Bui + Fui/N + Tui*Aui + H/h$ ##Calculate the influence of account i's own behavior

      return $P\_FF\_self\_i$

    end function

Algorithm 2: Account I's topology-based platform follows fans' influence algorithm

---

Require: d, Eui, Fui, N

Ensure: Account i's topology-based platform follows fans' influence

    function F2(d, Eui, Fui, N)

      $d \leftarrow 0.85$

      $Eui \leftarrow [0...n]$

      $Fui \leftarrow [0...n]$ ##Initialize the assignment of the number of followers

      $N \leftarrow \max(Fui) \leftarrow [1....n]$##Take the number of followers owned by the account with the largest number of followers in the data set

      if(Eui not=0 and Fui not=0) **Then**

        $P\_FF\_fans\_i \leftarrow (1 - d) + (d * 4 * Eui * Fui) / (Fui + 1) * N$

      else

        $P\_FF\_fans\_i \leftarrow 0.15$

      end if

      return $P\_FF\_fans\_i$

    end function

---

4 Account Influence Experiment and Comparative Analysis

4.1 Data Set

The data source used is the Douyin short video platform, by using the crawler software [25], crawling through the specified keyword search, and obtaining 2008 account information from May 28, 2022 to June 12, 2022. The basic information field [21] is shown in Table 2: The user's periodic video information [22] is obtained, as shown in Table 3. After preprocessing, a variety of algorithms and methods were compared.

Table 2. Data set User information data display.

Attribute	Explanation
Name	Username
Account	Douyin ID
Bio	Personal Information
Category	It is the official user label
Followers Group	the number of followers group chats
Products	E-commerce sales show window
Works	Number of Works Published Since Account Registration
Likes	Total number of likes for works since account registration
Following	How many people the user follows
Followers	How much followers this user has received
Website	Click on the link to the user 's personal information interface

Table 3. Video information display of dataset users.

Attribute	Explanation
Link	Click on the link to jump to the video playback interface
Title	Name of video
Account	Name of account that posted or reposted the video
Douyin ID	ID Name of Douyin ID that posted or reposted the video
Video duration	The time playing length of the video
Likes	Likes of other users of this video
Comments	Comments of other users of this video
Repost	the number of repost of other users of this video
Favorites	Favorites of other users of this video
Release time	The time the user posted the video on the Beat platform

4.2 Account Influence Experiment and Comparative Analysis

4.2.1 Experimental Comparison Object

In order to verify and evaluate the effectiveness of the proposed method, the experiment selects the popular or classic user influence algorithm as the comparison of the proposed method in this chapter. The specific methods are as follows.

(1) In-Degree algorithm, to calculate the influence of an account with an intuitive number of followers.
(2) Factor analysis, mathematical statistics to obtain the proportion of different characteristics, weight calculation account influence.

(3) Original short video platform SF-UIR algorithm, improved on the basis of PageRank algorithm, mainly through the user's own behavior to calculate influence of accounts.

4.2.2 Specific experimental design

The accuracy of experimental results was evaluated by precision, recall and F1-score. To ensure the objectivity of the experimental results and avoid the influence of human factors on the experimental results, it is used in the experiment that the Top-k influence account set of four methods to construct the standard set [23]. Build as shown in the formula.

$$
\begin{aligned}
B^k = \left(I_1^k \cap I_2^k\right) \cup \left(I_1^k \cap I_3^k\right) \cup \left(I_1^k \cap I_4^k\right) \cup \left(I_2^k \cap I_3^k\right) \\
\cup \left(I_2^k \cap I_4^k\right) \cup \left(I_3^k \cap I_4^k\right)
\end{aligned}
\tag{6}
$$

$I_1^k, I_2^k, I_3^k, I_4^k$ denotes the user set of Top-k for four computing methods. According to the standard set Bk, calculate precision($precison_i^k$), recall($recall_i^k$), F1-score($F1 - Score_i^k$), of Top-k in each method, calculation method is shown in the formula.

$$
precison_i^k = \frac{|B^k \cap I_i^k|}{|I_i^k|}
\tag{7}
$$

$$
recall_i^k = \frac{|B^k \cap I_i^k|}{|B^k|}
\tag{8}
$$

$$
F1_i^k = \frac{2 * precison_i^k * recall_i^k}{precison_i^k + recall_i^k}
\tag{9}
$$

(2) Compare the list of accounts with or without the influence ranking of followers based on the topology platform. To verify the account regardless of the level of influence, the necessity of introducing the influence of followers based on the topology platform. On the data set, two adjacent accounts, with and without followers' group chats with an influence interval of about 500, are selected for comparison. The difference between the ranking of the FF-SF-UIR algorithm that deletes the influence of followers and the ranking of the complete FF-SF-UIR algorithm is analyzed to illustrate the necessity of introducing the topological structure platform to calculate the influence of followers.

4.3 Experimental Result Analysis

(1) Comparison of Precision, Recall and F1-Score for Each Algorithm

Through the four algorithms of FF-SF-UIR algorithm, In-Degree algorithm, factor analysis and SF-UIR algorithm, the ranking of accounts' influence is obtained respectively, and then the precision, recall and F1-Score [24] of the ranked top-k accounts are compared, respectively. It showed the experimental results in the figure. The results show that the FF-SF-UIR algorithm is higher than the user in-degree algorithm, factor analysis method and SF-UIR algorithm in precision, recall and F1-Score. This is because the FF-SF-UIR algorithm makes a reasonable refinement and quantification of the account authentication types through the official authentication difficulty of the short video platform, and takes the historicity and periodicity of the influence of works through the factor analysis into account, and also introduces the influence characteristics brought by account fans through the topology of social networks [25] (Figs. 6, 7 and 8).

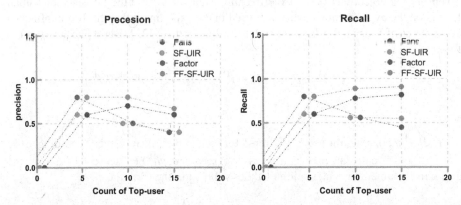

Fig. 3. Top-user precision algorithm comparison

Fig. 4. Top-user recall algorithm comparison

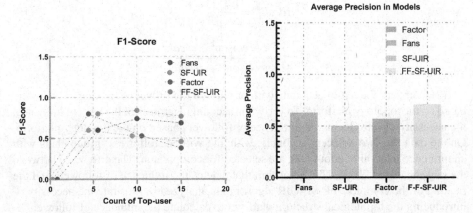

Fig. 5. Top-user F1-Score algorithm comparison

Fig. 6. Average Precision in Models

From Fig. 3, Fig. 4 and Fig. 5, it is not difficult to see that the improved method is superior to the other three influence calculation methods in precision, recall and F1-Score. Among them, precision is 21% higher than SF-UIR algorithm, 14% higher than the factor analysis method, and 8% higher than the IN-Degree algorithm; the recall is 30% higher than the SF-UIR algorithm, 17% higher than the factor analysis method, and 14% higher than the IN-Degree algorithm. F1-Score is 27% higher than the SF-UIR algorithm, 23% higher than the factor analysis method, and 12% higher than the IN-Degree algorithm. This is because the improved method not only digitizes the influence

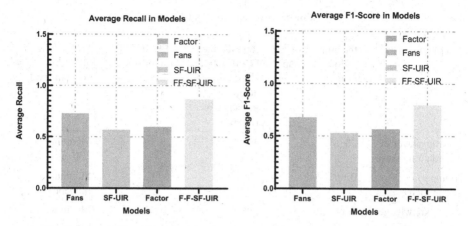

Fig. 7. Average Recall in Models   **Fig. 8.** Average F1-Score in Models

of the official label of the account platform, but also integrates the historical and periodic influence of works and introduces influence of followers on the platform based on the topology structure. The accuracy and applicability of the improved method are higher, and the advantages are more obvious.

(2) Compare the list of accounts with or without the influence ranking of followers based on the topology platform.

From Table 4, it is proved that the ranking of high-impact accounts is almost unchanged, because influence value of the high-impact accounts based on the accounts' own behavior is very high, and fluence relative value of the short video platform's follower is small, which can hardly change the ranking. After the introduction of influence of followers in middle-impact accounts, the large number of followers' group chats makes the relative improvement of the influence ranking, which helps the short video platform to increase its user stickiness. Low-impact users because the number of followers is generally too small and no followers' group chats, so the two algorithms ranking have not changed users' ranking. The improved method inherits the SF-UIR algorithm and considers the particularity of the chat stickiness of the short video platform, so as to better describe the influence of accounts in the short video platform and improve the accuracy of calculation.

Table 4. User ranking comparison.

Username	the number of followers group chats	Ranking the FF-SF-UIR algorithm that deletes the influence of followers	FF-SF-UIR algorithm ranking	FF-SF-UIR algorithm account level
Forever power of fashion tomorrow	26	199	195	Medium influence
Younger E-commerce	0	500	514	Medium influence
Game knowledge sister	8000	998	1005	Medium influence
Gamer A Guan	0	999	1118	Medium influence
E-commerce Li Yifeng	0	1501	1531	Low impact
E-commerce people with one lamp	96	1502	1400	Low impact
People's Daily	0	1	1	High impact
CCTV news	0	2	2	High impact
Blessings 00	0	2004	2004	Low impact
Niceni	0	2005	2005	Low impact

5 Conclusions

This paper analyzes real accounts in a short video platform. Considering the different difficulty of account authentication, the different cycles of works' influence and the influence of short video platform followers' group chats, an improved account influence evaluation model based on the SF-UIR algorithm is proposed. The algorithm in this paper not only references the influence algorithm of micro-blog accounts but also combines the particularity of the short video platform. From the numerical influence of the official label of the platform in the influence of the accounts' own behavior, the historical and periodic analysis of the influence of works is carried out in combination with the factor analysis method. Considering the number of followers' group chats in the account, and combining the topological structure, the influence of followers based on the topological structure is introduced. It can not only obtain the calculation results of the influence of short video accounts but also improve accuracy of the calculation results compared with other traditional influence calculation methods. The research is of great significance to understand the importance of accounts in the short video platform and the security of platform management.

There are also some shortcomings in this paper. In terms of influence evaluation, the evaluation indicators used in this paper are relatively few. In terms of calculating the influence of followers based on the topological structure platform, this paper only analyzes the followers' group chats of the short video platform accounts, without considering the other influence of followers. In terms of data, the time complexity of crawling information is too high and imperfect. The follow-up work will address the above issues and explore how to improve the SF-UIR method on other social platforms.

References

1. Liu, Yi., Zhang, Y., Haidong, Hu., Liu, X., Zhang, L., Liu, R.: An extended text combination classification model for short video based on albert. J. Sens. **2021**, 1–7 (2021)
2. Chen, X., Valdovinos, K.D.B., Zeng, J.: # Positive Energy Douyin: constructing "playful patriotism" in a Chinese short-video application. Chin. J. Commun. **14**(1), 97–117 (2021)
3. Ge, J., Sui, Y., Zhou, X.: Effect of short video ads on sales through social media: the role of advertisement content generators. Int. J. Advert. **40**(6), 870–896 (2021)
4. Klug, D., Qin, Y., Evans, M.: Trick and please: a mixed-method study on user assumptions about the TikTok algorithm. In: 13th ACM Web Science Conference 2021, pp. 84–92 (2021)
5. Liang, X., Tao, X., Wang, Y.: Impact analysis of short video on users behavior: users behavior factors of short video evidence from users data of Tik Tok. In: 2021 7th International Conference on E-Business and Applications, pp. 18–24 (2021)
6. Purba, K.R., Asirvatham, D., Murugesan, R.K.: Classification of Instagram fake users using supervised machine learning algorithms. Int. J. Electr. Comput. Eng. **10**(3), 2763 (2020)
7. Fortunato, S., Boguñá, M., Flammini, A., Menczer, F.: Approximating PageRank from indegree. In: Aiello, W., Broder, A., Janssen, J., Milios, E. (eds.) WAW 2006. LNCS, vol. 4936, pp. 59–71. Springer, Heidelberg (2008). https://doi.org/10.1007/978-3-540-78808-9_6
8. Shah, D., Zaman, T.: Community detection in networks: the leader-follower algorithm. Statistics **1050**(2) (2010)
9. Watkins, M.W.: Exploratory factor analysis: a guide to best practice. J. Black Psychol. **44**(3), 219–246 (2018)
10. Batoo, K.M., Pandiaraj, S., Muthuramamoorthy, M.: Behavior-based swarm model using fuzzy controller for route planning and E-waste collection. Environ. Sci. Pollut. Res. **29**(14), 19940–19954 (2022)
11. Agres, K., Abdallah, S., Pearce, M.: Information-theoretic properties of auditory sequences dynamically influence expectation and memory. Cogn. Sci. **42**(1), 43–76 (2018)
12. Wang, Z., Liu, H., Liu, W.: Understanding the power of opinion leaders' influence on the diffusion process of popular mobile games: travel frog on Sina Weibo. Comput. Hum. Behav. **109**, 106354 (2020)
13. Kanathey, K., Thakur, R.S., Jaloree, S.: Ranking of web pages using aggregation of page rank and hits algorithm. Int. J. Adv. Stud. Comput. Sci. Eng. **7**(2), 17–22 (2018)
14. Yeh, W.C., Zhu, W., Huang, C.L.: A new BAT and PageRank algorithm for propagation probability in social networks. Appl. Sci. **12**(14), 6858 (2022)
15. Wang, D., Xu, J., Duan, C.: Improved algorithm for user influence evaluation based on PageRank. J. Harbin Inst. Technol. **50**(5), 60–67 (2018)
16. Wu, M.: Research and implementation of user analysis model based on short video platform. Jilin University (2020)
17. McCartney, G., Pao, C., Pek, R.: An examination of Sina Weibo travel blogs' influence on sentiment towards hotel accommodation in Macao. J. Chin. Tourism Res. **14**(2), 146–157 (2018)

18. Thielmann, I., Spadaro, G., Balliet, D.: Personality and prosocial behavior: a theoretical framework and meta-analysis. Psychol. Bull. **146**(1), 30 (2020)
19. Li, Y., Fan, J., Wang, Y.: Influence maximization on social graphs: a survey. IEEE Trans. Knowl. Data Eng. **30**(10), 1852–1872 (2018)
20. de Morais, W.E.A., Alfinito, S., Barbirato, L.L.: Certification label and fresh organic produce category in an emerging country: an experimental study on consumer trust and purchase intention. British Food J. **123**, 2258–2271 (2021)
21. Tóth, G., Wachs, J., Di, C.R.: Inequality is rising where social network segregation interacts with urban topology. Nat. Commun. **12**(1), 1–9 (2021)
22. Thao, N.T.P., Van Tan, N., Tuyet, M.T.A.: KMO and bartlett's test for components of workers' working motivation and loyalty at enterprises in Dong Nai province of Vietnam. Int. Trans. J. Eng. Manage. Appl. Sci. Technol. **13**(10), 1–13 (2022)
23. Shi, L., Zhang,Y., Cheng, J.: Skeleton-based action recognition with directed graph neural networks. In: Proceedings of the IEEE/CVF Conference on Computer Vision and Pattern Recognition, pp. 7912–7921 (2019)
24. Zachary, Z., Brianna, F., Brianna, L.: Self-quarantine and weight gain related risk factors during the COVID-19 pandemic. Obes. Res. Clin. Pract. **14**(3), 210–216 (2020)
25. Wang, J., Li, Q., Zhou, J.: News crawling based on Python crawler. J. Phys. Conf. Ser. **1757**(1), 012117 (2021)

Deep Hash Learning of Feature-Invariant Representation for Single-Label and Multi-label Retrieval

Yuan Cao[1]($\boxtimes$), Xinzheng Shang[1], Junwei Liu[1], Chengzhi Qian[1], and Sheng Chen[2]

[1] Ocean University of China, Qingdao, China
cy8661@ouc.edu.cn
[2] Tianjin University, Tianjin, China
chensheng@tju.edu.cn

Abstract. In large-scale retrieval, hash learning is favored by people owing to its fast speed. Nowadays, many hashing methods based on deep learning are proposed, because they have better performance than traditional feature representation methods. Both in supervised hash learning and unsupervised hash learning, similarity matrix is used in the objective function. In the similarity matrix, if two images share at least one label, the similarity is "1", otherwise it is "0". However, this kind of similarity can not reflect the similarity ranking of multi-label images well, which is vulnerable to pixel interference. Therefore, in order to improve the retrieval accuracy of multi-label data, we improve the traditional deep learning hashing method by dividing the multi-label images into "strong similarity" and "weak similarity". In addition, although the deep neural network can judge the label of the images directly through the pixels, it does not understand the high-level semantics of the images. Hence, we take the feature invariance of the images into consideration, which means that the transformed image should have the same feature representation with the original image. In this way, we propose a novel Deep Hash learning method based on Feature-Invariant representation (FIDH), which focuses on deep understanding rather than deep learning. Experiments on common single-label and multi-label datasets show that our method obtains better performance than state-of-the-art methods in large-scale image retrieval.

Keywords: Large-scale retrieval · Hash learning · Multi-label similarity

1 Introduction

With the rapid development of multimedia big data, the number of images is explosive growth, which requires fast and accurate retrieval methods. Accurate nearest neighbor retrieval (KNN) [1] consumes a lot of time and is not suitable for big data retrieval, while approximate nearest neighbor retrieval (ANN) [2–5] is more popular since it balances time and efficiency. Generally, the existing

Z. Tari et al. (Eds.): ICA3PP 2023, LNCS 14487, pp. 17–29, 2024.
https://doi.org/10.1007/978-981-97-0834-5_2

hashing methods can be divided into supervised hashing [6–10] and unsupervised hashing [11–15]. Supervised hashing uses manually labeled tags as supervision signals, and usually constructs semantic similarity matrix based on labels [16–19]. The disadvantage of the supervised hashing method is that it costs a lot of manpower and material resources to obtain the tag, so the unsupervised hash method is proposed. The unsupervised hashing method maps the images from the original space to the hashing space and sets a series of loss functions to maintain the similarity. Usually, a pre-training model is used to get the feature vectors of the images, and the semantic similarity matrix is obtained after ranking by the distance [20].

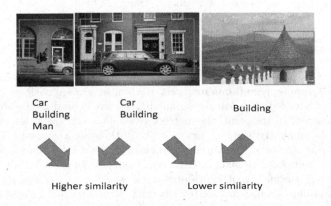

Car
Building
Man

Car
Building

Building

Higher similarity   Lower similarity

Fig. 1. The similarity between the middle and left images should be greater than that between the middle and right images.

Gidaris et al. [21] proposed a self-monitoring method based on image rotation. However, this leads to covariance between feature representation and image transformation. Misra et al. [22] solved this problem. However, they do not map the similarity matrix of similar images in the original space to the feature space. However, whether it is supervised hash learning or unsupervised hash learning, most of the existing methods set the value of semantic similarity matrix to 1 or 0 (If at least one label is the same, the value is 1, If no label is the same, the value is 0). For multi-label images, this method can not well reflect the image similarity ranking [23]. As shown in Fig. 1, the similarity between the middle and left images should be greater than that between the middle and right images. Recent studies have shown that the deep neural network can judge the image category according to the pixels of a small part of the region, so it is easy to be affected by pixels in the process of training. [24–26].

In order to solve the above problems, we propose a novel Deep Hash learning method based on Feature-Invariant representation (FIDH). We improve the traditional method of constructing semantic similarity, add the weight to the semantic similarity matrix, and set the loss function to reduce the impact of pixels on the training process, and propose an image retrieval method suitable for single-label and multi-label. Specifically, we divide image similarity into"strong similarity" and "weak similarity". "Strong similarity" means that all labels are the same or none of them are the same. "Weak similarity" means that at least one group of labels is the same and at least one group of labels is different. We evaluate our method on popular single-label and multi-label datasets, and the results show that our method has better performance than traditional methods. The contributions of our proposed method are illustrated as follows.

- We improve the training method on the multi-label data sets by using "strong similarity" and "weak similarity" instead of the traditional "similarity" and "dissimilarity".
- We set the loss function to reduce the interference of the image pixels to the neural network, so that the feature vectors of the images are as similar as the original images with any transformation.
- Experiments show that our method obtains advanced performance both in single-label and multi-label image retrieval. At the same time, it is an attempt to train network from deep learning to deep understanding.

2   Related Work

With the rapid growth of multimedia data, accurate nearest neighbor [1] search has been unable to meet people's needs, since it needs to consume a lot of retrieval time. In recent years, approximate nearest neighbor search [2–4] represented by hash learning has been gradually favored by people. According to whether tags are used, hash learning can be divided into supervised hash learning [6–10] and unsupervised hash learning [11–15].

Supervised Hash Learning. Supervised learning is a common technique for training neural networks [16–19] and decision trees. These two technologies highly rely on the information given by the pre-determined classification system. For neural networks, the classification system uses the information to judge the network errors, and then constantly adjusts the network parameters. For the decision tree, the classification system uses it to determine which attributes provide the most information. Representative methods based on pairwise tags are minimum loss hash (MLH) [27] and supervised hash with kernel (KSH) [28].

Supervised hash learning uses artificially labeled tags as supervised information to learn hash functions, this method is usually better than the unsupervised method. In recent years, deep learning [16,17] has become popular, and a variety of deep hash learning methods have been proposed.

For example, Lin et al. [28] proposed an unsupervised deep learning method DeepBit [15] , which imposed three standards on binary code (i.e. minimum loss quantization, evenly distributed codes and uncorrelated bit) to learn compact binary descriptors, so as to achieve efficient visual object matching. The ITQ method proposed by Gong et al. [29] maximizes the variance of each binary bit and minimizes the loss of binarization, thus obtaining higher image retrieval performance. Liong et al. [10] proposed using deep neural network to learn hash codes, and achieved three goals through optimization: (1) minimizing the loss between real valued feature descriptors and learned binary codes. (2) The binary code is evenly distributed on each bit. (3) Different bits are as independent as possible.

Unsupervised Hash Learning. Using unlabeled data to learn the distribution of data or the relationship between data is called unsupervised learning. Unsupervised learning is an algorithm of artificial intelligence network. Its purpose is to classify the original data in order to understand the internal structure of the data. LSH [30] is one of the representatives of unsupervised hash learning, which maps similar items to the same bucket. Different from supervised learning network, unsupervised learning network does not know whether its classification result is correct or not, that is, it is not enhanced by supervision (tell it what kind of learning is correct). In recent years, many unsupervised hash algorithms [11–15] have been proposed, because supervised hash learning algorithms need a lot of labels labeled manually, which costs a lot of manpower and material resources.

As the new method of unsupervised hash learning, self-supervised hash learning is popular in the field of deep learning. And many methods based on "pretext task" have been proposed [21,22,25,31,32]. But these methods rely on the pre-training model, and the accuracy is obviously lower than the supervised hash learning method.

3   Deep Hash Learning Based on Feature-Invariant Representation

3.1   Overview

Here, we give an overview of the proposed Deep Hash learning method based on Feature-Invariant representation (FIDH). Given a training set of n images $I = \{I_1, I_2, ..., I_n\}$. Firstly, the similarity matrix S is calculated by labels. The traditional calculation method is that if I_i and I_j have any of the same labels, then $s_{ij} = 1$, otherwise $s_{ij} = 0$. We follow the method in [33] and use the percentage to calculate . The formula is as follows.

$$s_{ij} = \frac{< l_i, l_j >}{||l_i||||l_j||}, \tag{1}$$

where l_i and l_j denote the label vectors of image I_i and I_j. $< l_i, l_j >$ denotes the inner product of l_i and l_j. According to the formula (1), We classify images into two categories: strong similarity and weak similarity. Hard similarity is divided into completely similarity and completely dissimilarity. Figure 2 shows we propose an improved method, which inputs the images into the convolutional neural network in pairs. By setting the loss function, CNN can learn the high-level semantic information of the image, and hash the image in the last layer.

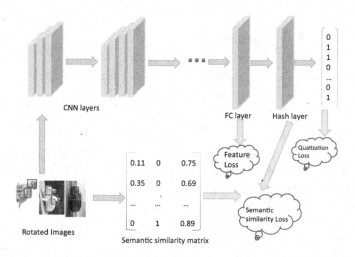

Fig. 2. The framework of FIDH. First, the image is rotated. Then, it is input into the convolutional neural network in pairs. After network adjustment, the binary code is obtained. The upper part shows the structure of the neural network and the lower part shows that the pairwise similarity of hash codes is guaranteed by the proposed semantic similarity loss, feature loss and quantization loss.

3.2   Neural Network Structure

Many hash learning methods use VGG19 as their network structure and prove its availability. VGG19 contains 19 hidden layers (16 convoluted layers and 3 fully connected layers). VGGNet structure is very simple, the whole network uses the same size of convolution core size (3×3) and maximum pool size (2×2). We replace the $fc8$ layer with hash layer. And our method can also be extended to other models, such as AlexNet and GoogLeNet.

3.3   Objective Function

We set a series of loss functions to ensure the pairwise similarity of images.

Semantic Similarity Loss. The value of traditional semantic similarity matrix can only have 1 and 0. Given the hash code B and semantic similarity matrix $S = s_{ij}$ of all images, The conditional probability $p(s_{ij}|B)$ can be expressed as below.

$$p(s_{ij}|B) = \begin{cases} \sigma(\Omega), & s_{ij} = 1, \\ 1 - \sigma(\Omega), & s_{ij} = 0. \end{cases} \tag{2}$$

where $\sigma(x) = \frac{1}{1+e^{-x}}$ denotes the sigmoid function. The inner product can well represent the Hamming distance, so we construct $\Omega = b_i^T b_j$. Then we use the negative log-likelihood as the loss function.

$$L_1 = - \sum_{s_{ij} \in S} log(p(s_{ij}|B))$$

$$= - \sum_{s_{ij} \in S} (s_{ij}log(\sigma(\Omega_{ij})) + (1 - s_{ij})log(1 - \sigma(\Omega_{ij}))) \qquad (3)$$

$$= \sum_{s_{ij} \in S} (log(1 + e^{\Omega_{ij}}) - s_{ij}\Omega_{ij}).$$

We use formula (3) to calculate the loss of images with strong similarity. For images with partial similarity (weak similarity), we use the following formula to calculate the loss below.

$$L_2 = \sum_{s_{ij} \in S} (\frac{<b_i, b_j> + q}{2} - s_{ij} \cdot q)^2. \qquad (4)$$

Combining formulas (3) and (4), we use W_{ij} to mark the two cases, that is, $W_{ij} = 1$ means that the two pictures are strongly similar, $W_{ij} = 0$ means that the two pictures are weakly similar. Therefore, the objective function can be written as follows.

$$L_s = \sum_{s_{ij} \in S} [W_{ij}(log(1 + e^{\Omega_{ij}}) - s_{ij}\Omega_{ij})$$

$$+ \gamma \cdot (1 - W_{ij})(\frac{<b_i, b_j> + q}{2} - s_{ij} \cdot q)^2], \qquad (5)$$

where γ denotes a weight parameter.

Quantization Loss. Since binary code B is discrete, direct optimization will lead to the disappearance of gradient in the process of back propagation. Based on the previous work, we use the continuous relaxation method to solve this problem [17,34]. We use continuous u instead of discrete b, which will cause quantization loss. Therefore, we set the objective function to reduce the loss and encourage the network to input accurate binary code.

$$L_q = \sum_{i,j \in N} (|||u_i| - 1||_1 + |||u_j| - 1||_1), \qquad (6)$$

where $||\cdot||_1$ denotes the L1-norm of the vector and $|\cdot|$ denotes the absolute value operation.

Feature Loss. In order to better represent the high-level semantic information of the image, we set the feature loss to adjust the network parameters, and input the image into the network after rotation. We think that the pixel information changes after image rotation, that is, the network will not determine the image label according to a small part of pixels. Our goal is to make the network understand the high-level semantic information of similar images as much as possible, and approach deep understanding rather than deep learning. We use cosine distance to measure the loss of paired features.

$$L_f = \sum_{i,j \in N} \frac{f_i \cdot f_j}{|f_i||f_j|}, \tag{7}$$

where f denotes the output feature vector of fc7 layer of neural network. Combining formula (7) with formula (5), the following objective function is obtained.

$$L_{sf} = \sum_{s_{ij} \in S} [W_{ij}(log(1 + e^{\Omega_{ij}}) - s_{ij}\Omega_{ij} - \frac{f_i \cdot f_j}{|f_i||f_j|}) \\ + \gamma \cdot (1 - W_{ij})(\frac{< b_i, b_j > +q}{2} - s_{ij} \cdot q)^2]. \tag{8}$$

Combining the quantitative loss and semantic loss, the final objective function is as follows.

$$L = L_{sf} + \lambda L_q, \tag{9}$$

where λ denotes the parameter to control the quantitative loss.

3.4 Optimization

We use standard back propagation and gradient descent method to optimize the objective function. By replacing b with u, the objective function is rewritten as follows.

$$L_{sf} = \sum_{s_{ij} \in S} [W_{ij}(log(1 + e^{\Omega_{ij}}) - s_{ij}\Omega_{ij} - \frac{f_i \cdot f_j}{|f_i||f_j|}) \\ + \gamma \cdot (1 - W_{ij})(\frac{< u_i, u_j > +q}{2} - s_{ij} \cdot q)^2] \\ + \lambda \cdot (|||u_i| - 1||_1 + |||u_j| - 1||_1). \tag{10}$$

After the learning process, we get the approximate hash code whose value is in $(-1,1)$. In order to evaluate the effectiveness of our proposed method, we use the following formula to get the accurate binary code (Table 1).

$$b = sgn(u). \tag{11}$$

Through our series of formulas, we can learn the hash code in an end-to-end way.

4   Experiments

4.1   Datasets

Flickr. In order to verify the effectiveness of the method, we have carried out experiments on the widely used dataset Flickr, and compared with other advanced methods. The Flickr is a data set which contains 25000 images, and each image has at least one label [35].

Cifar-10. Cifar-10 is a color image data set closer to universal objects. There are 50000 training images and 10000 test images in the dataset.

4.2 Evaluation Metrics

The evaluation index Mean Average Precision (mAP) and Average Precision (AP) is used to evaluate our method [41]. For each query, the Average Precision (AP) is the average of the first k results, and the Mean Average Precision (mAP) is the average of all queries.

$$AP = \frac{1}{N} \sum_{k=1}^{K} P(k)\delta(k), \tag{12}$$

where N is the number of instances related to the ground truth in the database used for query. $P(k)$ is the precision of the first k instances. When the $k-$th instance is related to the query(hey have at least one identical label), $\delta(k) = 1$, otherwise $\delta(k) = 0$.

5 Results

Table 2 shows the mAP results on 12-bit, 24-bit, 36-bit and 48-bit on Flickr. It can be seen that the map results on 12-bit, 24-bit, 36-bit and 48-bit are improved by 2.06%, 1.62%, 2.22% and 3.3% respectively. The experimental results on higher bit numbers are the best, which shows that our high bit number hash code can better represent the high-level semantic information of the image. In order to better verify our scheme, we rotate the image 90 °C, 180 °C and 270 °C respectively, and compare the experimental results in Fig. 3. In order to extend our method to the single-label data set, we have carried out experiments in the single-label data set Cifar-10, for the sake of fairness, we combine several methods of comparison with VGG19 model, the results are shown in Table 3. Compared with other deep unsupervised hash learning methods, our supervised method will get better results.

Table 1. The training and testing time of DH, BGAN and FIDH on Flickr.

Methods	Training time	Testing time
DH [10]	1 h	0.5 ms
BGAN [25]	5 h	3 ms
FIDH	0.5 h	0.5 ms

Table 2. The mAP results after image rotation of 90 °C on Flickr.

Methods	12-bit	24-bit	36-bit	48-bit
LSH [30]	0.5968	0.6068	0.6265	0.6369
ITQ [29]	0.6845	0.6950	0.6973	0.6978
SH [36]	0.6451	0.6512	0.6505	0.6463
MLH [27]	0.7033	0.7073	0.7163	0.7103
KSH [28]	0.7907	0.8070	0.8141	0.8181
DLBHC [37]	0.7236	0.7566	0.7573	0.7761
HashNet [38]	0.7909	0.8262	0.8414	0.8483
DMSSPH [23]	0.7800	0.8080	0.8096	0.8159
DHN [34]	0.8227	0.8393	0.8446	0.8471
DQN [39]	0.8092	0.8227	0.8298	0.8270
IDHN [33]	<u>0.8746</u>	<u>0.8830</u>	<u>0.8834</u>	<u>0.8843</u>
FIDH	**0.8952**	**0.8992**	**0.9056**	**0.9173**

Table 3. The mAP results after image rotation of 90 °C on Cifar-10.

Method	12-bit	24-bit	32-bit	48-bit
ITQ [29]	0.162	0.169	0.172	0.175
SH [36]	0.131	0.135	0.133	0.130
LSH [30]	0.121	0.126	0.120	0.120
Spherical [40]	0.138	0.141	0.146	0.150
ITQ+VGG	0.196	0.246	0.289	0.301
SH+VGG	0.174	0.205	0.220	0.232
LSH+VGG	0.101	0.128	0.132	0.169
Spherical+VGG	0.212	0.247	0.256	0.281
DeepBit [15]	0.185	0.218	0.248	0.263
DH [10]	0.160	0.164	0.166	0.168
BGAN [25]	<u>0.401</u>	<u>0.512</u>	<u>0.531</u>	<u>0.558</u>
FIDH	**0.713**	**0.727**	**0.744**	**0.757**

We compare the training time and test time of several hash methods (DH, BGAN and FIDH) [10,25], as shown in Table 1. The results show that our method has advantages in training time and test time. Figure 4 shows the comparison of training time for binary adversarial hashing (BGAN), efficient similarity hashing (DHN), deep quantization hashing (DQN), soft similarity hashing (IDHN), and supervised multi label hash learning (FIDH). The results show that the method in this paper has a slight advantage in training time.

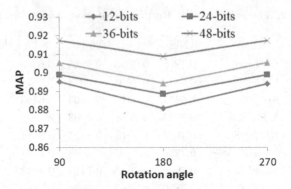

Fig. 3. Comparison of mAP results of different rotation angles on Flickr.

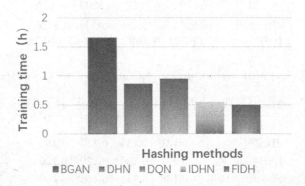

Fig. 4. Comparison of training time between FIDH model and other methods in FLICKR-25K dataset.

6   Conclusion

This paper improves the traditional deep hash learning method for multi-label retrieval by adopting the method of similarity classification and the idea of invariant feature representation. In our model the similarity between two images is divided into "strong similarity" and "weak similarity". Furthermore, the invariance of image rotation feature is added to the training process and the quantization loss is minimized to ensure the quality of hash codes. Compared with the traditional hash learning method, the hash codes obtained by our model can represent the high-level semantic information of the image better. Experimental results show that our method performs better than state-of-the-art methods on multi-label datasets. Meanwhile, it can be extended to single-label datasets easily.

Acknowledgment. This work is supported by the NSFC Grant No. 62202438; the Natural Science Foundation of Shandong Province Grant No. ZR2020QF041; the 22th batch of ISN Open Fund Grant No. ISN22-21.

References

1. Yi, J., Jiang, D., Sun, H., Zhao, X.: The research on nearest neighbor search algorithm based on vantage point tree. In: IEEE International Conference on Software Engineering and Service Science (2017)
2. Figueroa, K., Paredes, R.: Approximate direct and reverse nearest neighbor queries, and the k-nearest neighbor graph. In: International Workshop on Similarity Search and Applications, vol. 33, pp. 91–98 (2009)
3. Ke, Q., Ge, T., He, K., Sun, J.: Optimized product quantization for approximate nearest neighbor search. In: IEEE Conference on Computer Vision and Pattern Recognition, pp. 2946–2953. IEEE (2013)
4. Guan,T., Ai, L., Yu, J., He, Y.: Efficient approximate nearest neighbor search by optimized residual vector quantization. In: International Workshop on Content-Based Multimedia Indexing, pp. 1–4 (2014)
5. Cao, Y., Qi, H., Gui, J., Li, K., Tang, Y.Y., Kwok, J.T.: Learning to hash with dimension analysis based quantizer for image retrieval. **23**, 3907–3918 (2021)
6. Lin, K., Yang, H., Chen, C.: Supervised learning of semantics-preserving hash via deep convolutional neural networks. IEEE Trans. Pattern Anal. Mach. Intell. **40**, 437–451 (2018)
7. Li, Z., Xiao, F., Qi, G., Jin, L., Li, K., Tang, J.: Deep semantic-preserving ordinal hashing for cross-modal similarity search. IEEE Trans. Neural Netw. Learn. Syst. **30**, 1429–1440 (2019)
8. Luo, X., Nie, L., Li, C., Yan, T., Xu, X.: Supervised robust discrete multimodal hashing for cross-media retrieval. IEEE Trans. Multimedia **21**, 2863–2877 (2019)
9. Cui, X., Jiang, Q., Li, W.: Deep discrete supervised hashing. IEEE Trans. Multimedia **27**, 5996–6009 (2018)
10. Wang, G., Moulin, P., Liong, V.E., Lu, J., Zhou, J.: Deep hashing for compact binary codes learning. In: IEEE Conference on Computer Vision and Pattern Recognition, pp. 2475–2483 (2015)
11. Zhang, D., Lang, B., Liu, X., Mu, Y., Li, X.: Large-scale unsupervised hashing with shared structure learning. IEEE Trans. Cybern. **45**, 1811–1822 (2015)
12. Liu, T., Li, J., Liu, W., Deng, C., Yang, E., Tao, D.: Unsupervised semantic-preserving adversarial hashing for image search. IEEE Trans. Image Process. **28**, 4032–4044 (2019)
13. Peng, J., Xia, Z., Feng, X., Hadid, A.: Unsupervised deep hashing for large-scale visual search. In: International Conference on Image Processing Theory, Tools and Applications, pp. 1–5 (2016)
14. Li, Y., Zhu, Y., Wang, S.: Unsupervised deep hashing with adaptive feature learning for image retrieval. IEEE Signal Process. Lett. **26**, 395–399 (2019)
15. Chen, C., Lin, K., Lu, J., Zhou, J.: Learning compact binary descriptors with unsupervised deep neural networks. In: IEEE Conference on Computer Vision and Pattern Recognition, pp. 1183–1192 (2016)
16. Yang, P., Wang, S., Zhong, G., Xu, H., Dong, J.: Deep hashing learning networks. In: International Joint Conference on Neural Networks, pp. 2236–2243 (2016)
17. Hsiao, J., Lin, K., Yang, H., Chen, C.: Deep learning of binary hash codes for fast image retrieval. In: IEEE Conference on Computer Vision and Pattern Recognition Workshops, pp. 27–35 (2015)
18. Moreno, P.J., Carneiro, G., Chan, A.B., Vasconcelos, N.: Supervised learning of semantic classes for image annotation and retrieval. IEEE Trans. Pattern Anal. Mach. Intell. **29**, 394–410 (2007)

19. Wang, S., Li, W.W.: Feature learning based deep supervised hashing with pairwise labels (2015)
20. Liu, T., Wei, L., Yang, E., Deng, C., Tao, D.: Semantic structure-based unsupervised deep hashing. In: Twenty-Seventh International Joint Conference on Artificial Intelligence (2018)
21. Singh, P., Gidaris, S., Komodakis, N.: Unsupervised representation learning by predicting image rotations (2018)
22. Xu, C., Feng, Z., Tao, D.: Self-supervised representation learning by rotation feature decoupling. In: IEEE/CVF Conference on Computer Vision and Pattern Recognition, pp. 10356–10366 (2019)
23. Li, B., Ye, M., Wu, D., Lin, Z., Wang, W.: Deep supervised hashing for multi-label and large-scale image retrieval. In: International Conference on Multimedia Retrieval, pp. 150–158 (2017)
24. Goodfellow, I., et al.: Generative adversarial networks. In: Advances in Neural Information Processing Systems, pp. 2672–2680 (2014)
25. Song, J.: Binary generative adversarial networks for image retrieval. Int. J. Comput. Vision **32**, 1–22 (2020)
26. Misra, I., van der Maaten, L.: Self-supervised learning of pretext-invariant representations. In: IEEE/CVF Conference on Computer Vision and Pattern Recognition, pp. 6706–6716 (2020)
27. Norouzi, M., Blei, D.M.: Minimal loss hashing for compact binary codes. In: International Conference on Machine Learning, pp. 353–360 (2011)
28. Ji, R., Jiang, Y.-G., Liu, W., Wang, J., Chang, S.-F.: Improved deep hashing with soft pairwise similarity for multi-label image retrieval. In: IEEE Conference on Computer Vision and Pattern Recognition, pp. 2074–2081 (2012)
29. Gong, Y., Lazebnik, S.: Iterative quantization: a procrustean approach to learning binary codes. In: CVPR, pp. 817–824 (2011)
30. Datar, M.: Locality-sensitive hashing scheme based on p-stable distributions (2004)
31. Noroozi, M., Favaro, P.: Unsupervised learning of visual representations by solving jigsaw puzzles (2016)
32. Gupta, A., Doersch, C., Efros, A.A.: Unsupervised visual representation learning by context prediction. In: IEEE International Conference on Computer Vision, pp. 1422–1430 (2015)
33. Zhang, Z., Zou, Q., Lin, Y., Chen, L., Wang, S.: Improved deep hashing with soft pairwise similarity for multi-label image retrieval. IEEE Trans. Multimedia **22**, 540–553 (2020)
34. Wang, J., Zhu, H., Long, M., Cao, Y.: Deep hashing network for efficient similarity retrieval. In: AAAI Conference on Artificial Intelligence, pp. 2415–2421 (2016)
35. Huiskes, M.J., Lew, M.S.: The MIR Flickr retrieval evaluation. In: International Conference on Multimedia Retrieval, pp. 39–43 (2008)
36. Torralba, A., Weiss, Y., Fergus, R.: Spectral hashing. In: Advances in Neural Information Processing Systems, pp. 1753–1760 (2009)
37. Hsia, J.H., Lin, K., Yang, H.F., Chen, C. S.: deep learning of binary hash codes for fast image retrieval. In: Computer Vision and Pattern Recognition Workshops, pp. 27–35 (2015)
38. Wang, J., Cao, Z., Long, M., Yu, P. S.: HashNet: deep learning to hash by continuation. In: IEEE International Conference on Computer Vision, pp. 5609–5618 (2017)
39. Wang, J., Zhu, H., Cao, Y., Long, M., Wen, Q.: Deep quantization network for efficient image retrieval. In: AAAI Conference on Artificial Intelligence, pp. 3457–3463 (2016)

40. Ribeiro-Neto, B., Baeza-Yates, R.: Modern Information Retrieval. ACM Press, New York, 463 p. (1999)
41. He, J., Chang, S., Heo, J., Lee, Y., Yoon, S.: Spherical hashing: binary code embedding with hyperspheres. IEEE Trans. Pattern Anal. Mach. Intell. **11**, 2304–2316 (2015)

Generative Adversarial Network Based Asymmetric Deep Cross-Modal Unsupervised Hashing

Yuan Cao[1(✉)], Yaru Gao[1], Na Chen[1], Jiacheng Lin[1], and Sheng Chen[2]

[1] Ocean University of China, Qingdao, China
cy8661@ouc.edu.cn
[2] Tianjin University, Tianjin, China
chensheng@tju.edu.cn

Abstract. With the explosive growth of internet information, cross-modal retrieval has become an important and valuable frontier hotspot. Due to its low storage consumption and high search speed, deep hashing has achieved significant success in cross-modal retrieval. Current research on unsupervised cross-modal hashing algorithms mainly focuses on two aspects: extracting high-level semantic information from given instances' raw data and designing network structures suitable for unsupervised learning. However, despite the abundance of unsupervised method research found in the literature, many current studies overlook the fact that the data distributions of different modalities are highly distinct. In fact, asymmetric network structures are more in line with cross-modal data learning. Therefore, this paper proposes an asymmetric deep cross-modal unsupervised hashing method based on generative adversarial networks (referred to as UDCMH-GAN algorithm). This method utilizes the image network channel as the reconstruction network to learn more valuable high-level semantic information, while the text network is set to a conventional network structure. The introduction of generative and adversarial mechanisms aims to achieve better modality fusion and bridge the semantic gap. The proposed method is validated on widely used datasets, and the results demonstrate that asymmetric learning methods are indeed more reasonable and accurate for different modalities.

Keywords: Cross-modal Retrieval · Hash Learning · Generative Adversarial Network

1 Introduction

The rapid development of science and technology has brought convenience to people's daily life, but it also puts forward higher requirements for the continuous development of science and technology. People share every aspect of their lives on social media, and the data on the network includes not only text, pictures, but also sound, video and so on. These data contain rich information and

Z. Tari et al. (Eds.): ICA3PP 2023, LNCS 14487, pp. 30–48, 2024.
https://doi.org/10.1007/978-981-97-0834-5_3

are of great significance to social development. However, these massive amounts of data also pose a huge challenge to technological development: how to retrieve and store data quickly, efficiently, and with low consumption while making full use of the data. To address this challenge, cross-modal retrieval methods based on hash learning have emerged. The cross-modal retrieval method is mainly used to solve the semantic gap caused by the inconsistency of data distribution between modalities. The hash learning method is used to solve the problems of high storage space cost and slow retrieval speed in the context of massive data. Therefore, cross-modal retrieval methods based on hash learning have attracted more and more attention from researchers in the field of computer vision. The main problem of cross-modal retrieval is to bridge the data semantic gap between different modalities. Its goal is to achieve mutual retrieval by learning to eliminate data inconsistencies between modalities. Hashing learning methods aim to map high-dimensional semantic representations to low-dimensional Hamming spaces, while still maintaining the similarity relationship of the original semantic space in Hamming spaces. Compared to real-valued representations, hash codes greatly reduce computational cost and storage space. Deep learning trains a neural network on sample data to learn its internal rules and semantic representation. Its ultimate goal is to enable machines to analyze and learn like humans, recognizing questions, pictures, sounds and other data. Deep learning mainly uses the powerful fitting ability of deep neural network to achieve superior visual relationship learning and accurate reasoning.

Depending on whether labels are used, deep cross-modal hashing can be classified into supervised and unsupervised methods. Supervised methods involve using labels to train a supervised network. While strong supervision utilizing labels can improve retrieval accuracy, manual labeling of all data is highly resource intensive, especially in the context of multi-modal data. Therefore, unsupervised methods are now a more promising direction, which is also the method adopted in this paper. In unsupervised cross-modal hashing algorithms, current research mainly focuses on two aspects. One is to improve methods for extracting high-level semantic information from raw data for a given instance. The other is to design a network structure for unsupervised learning. To design a network structure that implements unsupervised learning, let's take Generative Adversarial Networks as an example. The network achieves a balance between generative and discriminative modules, and learns high-level semantics for different modalities. Currently, Generative Adversarial Networks have shown excellent performance in most computer domains, thus, they are widely used in cross-modal retrieval models. However, recent literature studies show that most research on unsupervised methods focuses on setting similar structures for image and text channels, without considering their differences in real-world data distributions. In fact, asymmetric network structures may be more suitable for cross-modal learning. Therefore, this paper proposes an asymmetric deep cross-modal unsupervised hashing method based on generative adversarial networks, referred to as UDCMH-GAN algorithm. In this method, the picture network channel is set as a reconstruction network to learn more important high-level semantic information, while the text network adopts a conventional network structure. Furthermore, generative and adversarial mechanisms are introduced to enhance modality fusion and eliminate semantic gaps. The proposed method is validated on widely used datasets, and the results show that asymmetric learning methods are indeed better suited for different modalities. Therefore, the main innovations of this paper are as follows:

1. In this paper, an asymmetric feature extraction method is proposed. Image modality adopts reconstruction method to extract high-level semantic features, and text modality adopts conventional feature extraction method. The misalignment method is used to extract more semantic information in a way that is more consistent with the distribution of different modalities.
2. In this paper, we propose an asymmetric countermeasure that mitigates the attack using reconstructed features in images and deep features in text. To ensure a fair and efficient comparison, we use two independent discriminators to learn common features of different modalities separately. This approach can provide more accurate and robust countermeasures against attacks on cross-modal data.
3. Our proposed method has been validated on commonly used datasets, and our results have demonstrated a noticeable improvement in retrieval accuracy, as well as overall performance.

2 Related Work

The unsupervised method is different from the supervised method, that is, it does not use the manually marked label information as the supervised information, but uses the original characteristic information of the given instance data to learn the hash code and the corresponding hash function. This kind of method also aims to maintain the same similarity between the hash code and the original instance space. The following will briefly introduce the representative unsupervised methods in recent years.

The compound correlation quantization method [14] for efficient multi-mode retrieval published in 2016, which converts multi-modal data into correlated maximal mapping of isomorphic potential space, and converts this feature into compact hash code. In 2017, a cross-modal hash code learning method based on fusion similarity was proposed [11]. The method first constructs the similarity matrix of each modality, combines them, then uses an asymmetric graph to learn a hash function, and finally optimizes the whole learning process using an alternating update optimization algorithm. The results show that this method outperforms other unsupervised methods in different data sets. In 2019, a deep joint semantic reconstruction hash for large-scale unsupervised cross-modal retrieval was proposed [18]. This method is the first to construct a joint semantic similarity matrix based on multi-modal input instances. This matrix carefully integrates raw neighborhood relations from different modalities to capture intrinsic semantic relations. In 2020, an unsupervised depth estimation hash method [1] was proposed for local cross-modal retrieval. This method fully expands the originally limited and incomplete multi-modal data, and then conducts information estimation through the constructed weighted graph. Supervised neural network learning with weighted triple loss, showing better retrieval performance on multiple datasets. Joint Modal Distribution Based Large-Scale Unsupervised Deep Cross-Modal Retrieval Hashing [12] improves on the previous method and proposes a new distribution-based similarity decision weighting method. By setting two similarity thresholds, similar instances are closer and dissimilar instances are farther apart. The hash codes generated by this method are more discriminative. Another representative unsupervised knowledge distillation method is called cross-modal hashing [5], which uses the output of unsupervised methods to guide supervised training, that is, uses the teacher-student optimization strategy to learn the semantic information of instance data, Significant performance gains on different datasets.

In 2021, an attention-directed semantic hashing method [16] for unsupervised cross-modal retrieval was proposed. This method constructs an attention-aware semantic fusion matrix, which combines the original similarity matrix with the learned self-attention matrix to ensure the cross-modal invariance and semantic meaning of features. In 2022, the application method of deep adaptive enhanced hashing based on discriminant similarity guidance in unsupervised cross-modal retrieval was proposed [17]. This method uses discriminant similarity guidance and adaptive enhanced optimization strategy to learn hash functions, which also shows very superior performance. In the same year, [4] proposed a new method called Multi-modal Mutual Information Maximization: A New method for Unsupervised Deep Cross-mode hashing, which uses the strategy of maximizing mutual information to solve the unsupervised learning problem of hash codes. Another unsupervised cross-modal retrieval graph convolutional hashing method based on aggregation [24] proposes an efficient aggregation strategy for multiple metrics. The structural information of various modalities is mined from different perspectives to generate a joint similarity matrix. Based on this, a graph convolutional network is utilized to learn the similarity between cross-modal data. In 2023, the unsupervised contrastive cross-modal hashing method [6] was proposed, which overcomes two important current challenges. The first is to propose a new momentum optimizer to overcome the performance degradation caused by binary optimization, and the second is to propose a cross-modal ranking learning loss function to solve the adverse effects caused by false negative paired data, both of which can greatly improve performance.

3 Method

This chapter focuses on the details of our asymmetric deep cross-modal unsupervised hash method based on generative adversarial network, including the symbolic definition of the method, the design of the network framework, the objective function, the optimization process and the analysis of the experimental results.

3.1   Definition

In order to better describe the method in this chapter, the first step is to describe the symbols used in the method interpretation. Like most existing popular methods, this method uses image modality and text modality as representatives of different modalities for training and learning. Formula $O = \{(x_i, y_i)_{i=1}^n\}$ is used to represent the cross-modal dataset, where n represents the number of datasets, x_i represents the i-th instance of the image modality, and y_i represents the i-th instance of the text modality. For the image modality, $b_i = f(x_i; \theta_{Enc}) \in \mathbb{R}^n$ is used to represent the hash function of learning, and $x_i = f(b_i; \theta_{Dec}) \in \mathbb{R}^n$ is used to represent the decoding process corresponding to it. For the text modality, $b_t = g(y_i; \theta_T) \in \mathbb{R}^n$ is used to represent the hash function of learning, where n represents the dimension where the data is mapped to the common space, x represents the sample data of image modality, and θ_I represents the network parameters involved in the picture network model. Similarly, y represents the sample data of text modality, and θ_T represents the network parameters involved in the text network model. The method covered in this chapter are unsupervised, like the joint semantic methods, is expressed in $S^{n \times n}$. The hash code learned in picture network model is represented by B_I, and the hash code learned in text network model is represented by B_T.

3.2   Framework Introduction

This section will introduce the network structure based on the asymmetric deep cross-mode unsupervised hashing algorithm generative adversarial network, as shown in Fig. 1 below. The method framework proposed in this section is mainly composed of four parts: image modal channel, text modal channel, image discriminator and text discriminator.

1. Image Modal Channel

   For the image channel, the pre-training feature of the image is obtained by the pre-training model VGGNet19 first, which is represented by $M = \{m_1, m_2, ..., m_k\}$, and the original of the image is represented by the three-channel pixel. The obtained pre-trained feature M is taken as the input of the structure of the automatic encoder, and the output is the common feature that the encoder learns to represent the modality of the picture, expressed as $F^{Enc} = \{f_i^{Enc}\}_{i=1}^k$. The encoder network structure uses the common two layer fully connected MLP structure. The output of the encoder structure is the input of the hash layer, that is, the hash layer is connected at the back of the encoder. The output of the hash layer is the hash code, represented by the B_I descriptor. The decoder part of the automatic encoder is connected behind the hash layer, which is symmetrically similar to the encoder, that is, the network structure of the decoder is also a two layer fully connected MLP structure, whose function is to decode the hash code into features similar to the pre-training features, represented by $F^{Dec} = \{f_i^{Dec}\}_{i=1}^k$.

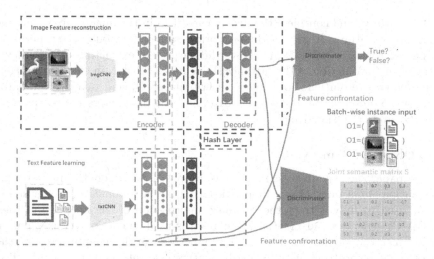

Fig. 1. Framework of UDCMH-GAN

2. Text Modal Channel

   For the text modal channel, The BOW features means that all the words that have appeared in the text dataset are counted to form the corresponding word dictionary. In the dictionary composed of each word, N word with the highest frequency is selected to form the final dictionary. Each text data is represented by a 0–1 vector of dimension N, represented by an identifier $G = \{g_1, g_2, ..., g_k\}$. The BOW feature G is input into the text network to obtain the common features, which is represented by $F^{Txt} = \{f_i^{Txt}\}_{i=1}^k$. The text network is composed of two layers of fully connected MLP structure. Similar to the image network, After the text network is connected to the hash layer, the output hash code is expressed as B_T.

3. Image Discriminator

   For the image discriminator, it is represented by D_x, the network structure is fully connected, the dimension of the first layer is the same as the characteristic dimension of the image decoder, and the number of neurons in the last layer is two. The output result represents the authenticity result judged by the discriminator. The decoder features of the image network are regarded as the real image features, and the features in the text network are regarded as false image features. Finally, the network parameters are learned by generating the antagonistic process, so as to reduce the semantic gap.

4. Text Discriminator

   For the text discriminator, it is represented by D_y, which is similar to the image discriminator. The dimension of the first layer is the same as the feature dimension of the text, and the number of neurons in the last layer is two. The output result represents the authenticity result judged by the discriminator. The text discriminator takes the image network as the generator of text features, takes the features learned from the text network as the real text features, and considers the decoder features in the image network as the false text features. It adjusts the network parameters through the generation and confrontation process to achieve the purpose of extracting the high-level semantic information of the instance. In order to achieve an efficient discriminator with resolution, both image discriminator D_x and text discriminator D_y are represented by three-layer MLP, and the parameters are represented by Θ_x and Θ_y respectively.

3.3   Objective Function

In this paper, an asymmetric method is proposed based on the inconsistent distribution of data in different modalities, that is, the automatic encoder structure is used to fully extract the semantic information of the image, and then the GAN's generative adversarial mechanism is used to eliminate the semantic gap between the modalities. Therefore, the methods in this chapter propose the loss of image reconstruction, inter-modal feature loss, inter-modal similarity discrimination loss, quantization loss and the most critical loss of generation and adversarial for asymmetric structures. The specific meaning and formula of each loss function will be described in detail below.

The unsupervised method is adopted in this paper, so the joint semantic matrix is selected as the inter-category supervision information to join the training of supervision network parameters. The specific method of constructing joint semantic matrix in the research is shown in the following formula 1 :

$$S = \alpha \frac{\hat{S}\hat{S}^T}{k} + (1 - \alpha)\hat{S} \tag{1}$$

where, $\hat{S}$ represents the fusion similarity matrix of image modal similarity matrix and text modal similarity matrix, expressed as $\hat{S} = S_I + lS_T$. It is worth noting that $S_I = MM^T \in R^{k \times k}$, $S_T = GG^T \in R^{k \times k}$, where features represent standardized operations before calculation. The above formula 1 also conforms to the diffusion theory and can effectively express the spatial distribution information of the original data of different modes, so it can also effectively supervise the parameter training of the network.

The following will carefully introduce the loss functions applied to the whole network part, including automatic encoder reconstruction loss, common space mode fusion loss, hash code generation and quantization loss, and generative adversarial loss, etc.

Firstly, for image network channel, this study is to learn the high-level semantic information of potential information through the encoding and decoding process of automatic encoder. At the same time, the output result of the encoder is taken as the input of the hash layer, and the hash code is learned and expressed as $B_I = sgn(U)$, $U = u_1, u_2..., u_k$, where u_i represents the unquantized continuous hash code. On the contrary, the decoder decodes the learned hash code into a depth feature, expressed as $m_i = Dec(u_i, \theta_i^d)$. Therefore, for the encoding and decoding process of the automatic encoder, this study sets the image reconstruction loss, that is, the depth features obtained by decoding should theoretically be similar to the features before encoding, in order to learn more potential semantic information of the image through the encoding and decoding process. The loss function reconstructed according to the above analysis picture is shown in the following Eq. 2 :

$$L_1 = \|M - F^{Dec}\|_2 \tag{2}$$

where, $\|A\|_2$ represents the 2-Norm of the matrix A, which is used to measure the distance between two matrices in space.

Secondly, the above analysis shows that the common features learned in the common space represent the potential semantic information of different modalities based on the same instance. Therefore, for the common features of the same instance, the distance in space should be very close, that is, the common features of the two modes should be similar or the same. Therefore, we set the modality fusion loss function of the common space as shown in Formula 3 :

$$L_2 = \|F^T - F^{Enc}\|_2 \tag{3}$$

According to theoretical analysis, the above formula 3 can effectively learn the common features between different modalities of the same instance, but the discriminant between different instances and different classes of instances is not taken into account. Therefore, it is necessary to set the discriminant loss of similarity between modalities in this study. The purpose is to make different instances of different modalities and different classes of instances learn more discriminant hash code values. The specific inter-modal similarity discrimination loss function is shown in Formula 4 :

$$L_3 = \|kS - UV\|_F^2 + \|kS - VU\|_F^2 \tag{4}$$

Among them, S is the unsupervised joint semantic matrix mentioned above, and U and V represent the hash code values of the two modalities without discrete operation. Therefore, in order to retain higher accuracy of hash code in the discretization process, quantization loss is set in this study to ensure that information loss is reduced in the discretization process. The specific quantified loss function is shown in the following formula 5 :

$$L_4 = \|U - B_I\|_F^2 + \|V - B_T\|_F^2 \tag{5}$$

wherein, $\|A\|_F^2$ mentioned in the formula 4 and 5 refers to the square of the F-Norm of the matrix A, which is another standard to measure the size of the distribution distance of the matrix in the space.

At the same time, both discriminators set in this paper need to gradually acquire the ability to distinguish truth from falseness during the learning process. The role of the image network for the text discriminator is thought to be to generate features to obfuscate its judgment, whereas the role of the text network for the image discriminator is thought to generate features to obfuscate the judgment of the image discriminator. Therefore, in this study, the corresponding two discriminators D_x and D_y are set to generate antagonistic losses.

The specific loss functions are expressed as L_{D_x} and L_{D_y} respectively, as shown in the following formula 6 :

$$
\begin{aligned}
L_5 &= L_{D_x} + L_{D_y} \\
&= -\frac{1}{k}[\sum_{i=1}^{k}(log(D_x(f_i^{Dec};\Theta_x)) + log(1 - D_x(f_i^{Txt};\Theta_x)) \\
&= +\sum_{i=1}^{k}(log(D_y(f_i^{Txt};\Theta_y)) + log(1 - D_y(f_i^{Dec};\Theta_y)))]
\end{aligned}
\tag{6}
$$

where, f_i^{Dec} represents the feature information output by the image decoder, and f_i^{Txt} represents the common feature learned by the text channel. It is very important for feature learning and hash code learning to choose decoding feature and image feature to oppose. This antagonistic approach can promote the image channel to learn more rich high-level semantic information, and then promote the text channel to learn the same or similar common semantic features as the image channel. At the same time, combining with the constraint of inter-modal loss function, text channel and image channel can learn more rich semantic information. Compared with the symmetric structure, the proposed method can learn the potential semantic information of the two modalities more comprehensively to generate accurate and compact hash code.

To sum up, in order to obtain more abundant potential semantic information between modes, a compact and discriminant hash code is generated. This method sets loss functions that are conducive to mining potential information, including automatic encoder reconstruction loss, common space mode fusion loss, hash code generation and quantization loss, and generation and adversarial loss, etc. Specific total objective function is shown in Formula 7.

$$
L = L_1 + \gamma_1 L_2 + \gamma_2 L_3 + \gamma_3 L_4 + \gamma_4 L_5
\tag{7}
$$

Among them, γ_1, γ_2, γ_3, γ_4 are hyperparameters, used to adjust the proportion of loss function of each part.

3.4   Optimization

For the optimization process of network parameters, the method proposed in this chapter adopts the iterative updating method. In the first step, the whole network parameters are initialized, and the training data and the hyperparameters determined according to the training results are given in the training process. The second step is to find the hash codes B_I and B_T with fixed initialization parameters. The third step updates the network parameters θ_I, θ_T, Θ_x and Θ_y according to the fixed hash code. When updating network parameters according to hash code, this method uses U and V instead of discrete hash code to participate in network update. The specific algorithm process and optimization process are shown in the table below.

Algorithm 1. Learning and optimization of UDCMH-GAN

Require: $O_k = \{x_i, y_i\}_{i=1}^k$; The original features M and G corresponding to the image modality and text modality; Enter the dimension of the hash code c; The amount of data contained in the batch input training network; The learning rate of network learning parameter l_r; The number of rounds of the required training network n_t; The hyperparameters γ_1, γ_2, γ_3, γ_4 involved in all objective functions.

Ensure: The hash function of image modal $b_i = f(x_i; \theta_I) \in R^n$, text modal hash function $b_t = g(y_i; \theta_T) \in R^n$; Network parameters learned by the image channel and text channel θ^{Enc}, θ^{Dec}, θ_y; Hash codes B_I and B_T are learned by different modalities for different instances.

1: The encoder and decoder network parameters, text network parameters θ^{Enc}, θ^{Dec}, θ_y, and two discriminant network parameters Θ_x and Θ_y are randomly initialized.

2: *for* $i = 1, 2, ..., n_t$ *do*;

3:     *for* $i = 1, 2, ..., b_s$ *do*;

4:         According to the original data distribution of the two modalities, the similarity matrix S needed by the training network is calculated.

5:         B_I and B_T are calculated based on the initialization parameters.

6:         Fixed parameter θ^{Enc}, θ^{Dec}, θ_y According to formula 6, the discriminator parameters Θ_x and Θ_y are updated by gradient descent;

7:         With fixed parameters Θ_x and Θ_y, according to formula 6, gradient descent method is used to update generators for two discriminators, namely text generation network for image discriminator and image generation network for text discriminator;

8:         Fixed θ_y, Θ_x and Θ_y, updating auto encoders θ^{Enc} and θ^{Dec} with gradient descent according to reconstruction loss formula 2;

9:         θ^{Enc}, Θ_x, Θ_y and θ^{Dec} are fixed. According to the remaining inter-modal losses and quantization losses, the network parameters θ^{enc}, Θ_y are updated to generate hash codes in image modal and text modal.

10:     *end for*

11:     Update the network parameters of the auto encoder, image channel and two modal discriminators;

12: *end for*

---

4 Experiment

4.1 Specific Experiment Setting

1. Datasets

   MIRFlickr25K [7] : It is a cross-modal data set widely used for cross-modal hash retrieval. The data set consists of 25,000 image text pairs collected from the Flickr website. Each instance is annotated with one or more tags selected from 24 categories. This method uses the cleaned data set, and there are 20015 text image pairs left after cleaned. Examples containing at least 20 text descriptions were selected in the experiment. The text modal of the examples was represented as a 1386-dimensional bag of words vector. This method randomly selects 2000 image text pairs as query data set, and the rest data as training database, and then randomly selects 10,000 instances for training during the training process.

NUS-Wide [2] : The dataset consists of 269,498 network images with text labels containing one or more labels in 81 conceptual categories. In the method experiment of this chapter, 20000 image text pairs are selected as the database, which belong to the 21 most common category concepts. Each instance text instance feature is represented by a 1000-dimensional bag vector. In the experiment, 2000 image text pairs were randomly selected as query data set, and 10,000 data were randomly selected as training sets.

2. Evaluation

When evaluating retrieval performance, the method proposed in this chapter performs two different cross-modal retrieval tasks: one is to query text modal data using image modal data, and the other is to query image modal data using text modal data. Set adjacent definitions related to real data to at least share the same semantic category. In the experiment, the widely used Hamming sort and hash search were used as retrieval protocols. Meanwhile, the widely used evaluation standard Mean Average Precision (mAP), which represents the average of the average precision (AP) of each query instance, is used to weigh the exact value of Hamming ranking results, and the retrieval precision and the ranking of the returned results are also taken into account. This paper also uses the Precision-Recall Curve as the performance evaluation standard, aiming to reflect the accuracy of different recall rates. According to the results of many experiments, the search performance is the best when the super parameter of the total objective function is set as $\gamma_1 = 1.2$, $\gamma_2 = 0.5$, $\gamma_3 = 0.5$, $\gamma_4 = 0.8$.

3. Baseline

In this method, 15 most representative cross-modal hash methods are selected as baseline standards in the experiment. Six supervised cross-modal hash methods (including DLFH, MTFH, FOMH, DCH, DCHU, and SCAHN) and nine unsupervised cross-modal hash methods (including CVH, LSSH, CMFH, FSH, UGACH, DJSRH, JDSH, UCH, DGCPN) were selected. Among the baseline comparison methods mentioned above, DCHUC, SCAHN, UGACH, DJARH, JDSH, UCH, and DGCPN are deep cross-modal hash methods, while other methods are not.

DLFH [8] : This method belongs to the category of supervised methods. A new discrete potential factor model is proposed to model tag information containing supervision information, and can directly learn hash codes without relaxation.

MTFH [13] : The method uses an efficient objective function to flexibly learn hash codes of different lengths for a particular modality. The method semantically correlates different hash representations of heterogeneous data, showing good retrieval performance.

FOMH [15] : This method is an online cross-modal hash method, which adaptively fuses heterogeneous modalities and flexibly learns new data to identify hash codes.

DCH [21] : The method generates a unified binary code based on class labels and a modality-specific hash function. This method also proposes an efficient discrete optimization algorithm for jointly learning modality-specific hash functions and unified binary codes.

DCGUC [19] : The method adopts an iterative optimization algorithm to learn a unified hash code for image-text pairs, and realizes an end-to-end deep cross-modal hash learning process.

SCAHN [20] : This method proposes a sooner or later feature fusion attention mechanism, which can set the length of the opcode according to the importance of each bit. This mechanism can make each bit of the hash code unequally weighted, so that no additional hashing is required. Model training.

CVH [9] : The method learns a hash function for each modality of a multimodal training instance. Hash functions map similar objects to similar hash codes in modals to preserve the distance-preserving nature of hashing.

LSSH [25] : This method uses sparse coding and matrix decomposition for the two modalities to capture the salient features of the image and the semantic features of the text, and maps the modal features to the common abstract space to obtain a unified binary code.

CMFH [3] : The key idea of the method is to learn a common hash code for different modalities in a shared semantic space. Thus, different modalities can be effectively linked in the public space.

FSH [11] : Inspired by diffusion fusion, this method constructs an undirected asymmetric graph to model the fusion similarity between different modalities. Based on this, an alternative optimization graph is introduced to learn the fusion similarity.

UGACH [23] : This method makes full use of the unsupervised feature learning ability of generating adversarial models to develop the underlying data structure of cross-modal data so as to learn more representative hash codes.

DJSRH [18] : The method builds a novel joint semantic matrix that carefully integrates the original structures from different modalities to generate a similarity matrix of multimodal instances. This matrix is then used as supervision information to generate compact hash codes.

JDSH [12] : In this method, a joint-modal similarity matrix is constructed to fully preserve the cross-modal semantic correlation between instances. Based on this, a sampling and weighting method is applied. Finally, a more discriminative hash code is learned.

UCH [10] : This method is an unsupervised cyclic generation adversarial hash method. In this method, the outer periodic network is used to learn a powerful common representation, and the inner periodic network is used to generate a reliable hash code.

DGCPN [22] : This method is a graphical model-based method to explore semantic consistency. Specifically, it refers to ensuring sufficient and effective supervision information through graph neighbor consistency, coexistence similarity, and modal similarity.

4.2 Results

1. Results on Datasets

As shown in the Table 1, it is the mAP comparison of this method on the MIRFlickr25k dataset. It is found from the experimental results that the mAP performance of the method proposed in this chapter is superior to that of other methods. According to the longitudinal comparison, the method UDCMH-GAN proposed in this chapter outperforms the method with the best performance compared with all the benchmark methods. Especially in the process of image text retrieval, the average accuracy is about 0.06 higher than the existing standard method. In the process of text image retrieval, the method proposed in this chapter has also been significantly improved, with the average accuracy improved by about 0.02 in different bits of length. According to horizontal comparison, the results of this study and most of the baseline methods show that the retrieval accuracy increases with the increase of hash code length. Specifically, for the same research method, 16-bit hash code has the lowest retrieval accuracy, while 128-bit hash code has the highest retrieval accuracy. This reflects that although the hash method consumes less storage space and the retrieval speed is fast, it also sacrifices some retrieval precision to some extent. Throughout the experimental results of various methods, compared with the sacrifice of retrieval accuracy, hash method is more meaningful in the improvement of resource consumption and retrieval speed. 3 represents the result of PR curve with hash code length of 128 in data set MIRFlickr25k, which also reflects the superiority of the method proposed in this paper. According to the experimental results of different methods, the performance of the same method in two different retrieval tasks is also different, but this study has improved the performance of the two retrieval tasks. In summary, the performance of this study on the MIRFlickr dataset is superior, which also shows that further research on cross-modal hashing methods is needed in the future.

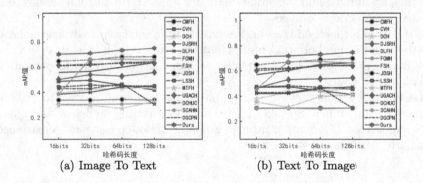

(a) Image To Text   (b) Text To Image

Fig. 2. mAP results on the NUS-Wide dataset

Table 1. mAP

	MIRFlickr				NUS-Wide			
	16 bits	32 bits	64 bits	128 bits	16 bits	32 bits	64 bits	128 bits
CVH	0.620	0.608	0.594	0.583	0.487	0.495	0.456	0.414
LSSH	0.597	0.609	0.606	0.605	0.442	0.457	0.45	0.451
CMFH	0.557	0.557	0.556	0.557	0.339	0.338	0.343	0.339
FSH	0.581	0.612	0.635	0.662	0.557	0.565	0.598	0.635
DLFH	0.638	0.658	0.677	0.684	0.385	0.399	0.443	0.445
MTFH	0.507	0.512	0.558	0.554	0.297	0.297	0.272	0.328
FOMH	0.575	0.640	0.691	0.659	0.305	0.305	0.306	0.314
DCH	0.596	0.602	0.626	0.636	0.329	0.422	0.43	0.436
UGACH	0.685	0.693	0.704	0.702	0.613	0.623	0.628	0.631
DJSRH	0.652	0.697	0.700	0.716	0.502	0.538	0.527	0.556
DCHUC	0.585	0.593	0.626	0.630	0.429	0.429	0.467	0.306
JDSH	0.724	0.734	0.741	0.745	0.647	0.656	0.679	0.68
SCAHN	0.558	0.605	0.603	0.612	0.385	0.661	0.661	0.656
DGCPN	0.711	0.723	0.737	0.748	0.61	0.614	0.635	0.641
UCH	0.654	0.669	0.6702	/	/	/	/	/
Ours	**0.726**	**0.737**	**0.741**	**0.760**	**0.701**	**0.708**	**0.732**	**0.749**
CVH	0.629	0.615	0.599	0.587	0.47	0.475	0.444	0.412
LSSH	0.602	0.598	0.598	0.597	0.473	0.482	0.471	0.457
CMFH	0.553	0.553	0.553	0.553	0.306	0.306	0.306	0.306
FSH	0.576	0.607	0.635	0.660	0.569	0.604	0.651	0.666
DLFH	0.675	0.700	0.718	0.725	0.421	0.421	0.462	0.474
MTFH	0.514	0.524	0.518	0.581	0.353	0.314	0.399	0.41
FOMH	0.585	0.648	0.719	0.688	0.302	0.304	0.3	0.306
DCH	0.612	0.623	0.653	0.665	0.379	0.423	0.444	0.459
UGACH	0.673	0.676	0.686	0.690	0.603	0.614	0.64	0.641
DJSRH	0.662	0.691	0.683	0.695	0.456	0.532	0.538	0.545
DCHUC	0.614	0.629	0.553	0.553	0.429	0.429	0.467	0.306
JDSH	0.710	0.720	0.733	0.720	0.649	0.669	0.689	0.699
SCAHN	0.590	0.635	0.627	0.632	0.456	0.674	0.674	0.664
DGCPN	0.695	0.707	0.725	0.731	0.617	0.621	0.642	0.647
UCH	0.661	0.667	0.668	/	/	/	/	/
Ours	**0.721**	**0.723**	**0.733**	**0.740**	**0.711**	**0.718**	**0.73**	**0.747**

In order to have a clearer understanding of performance improvement, this study draws the mAP results on the NUS-Wide data set into a broken line graph for display, as shown in the following Fig. 2.

Figure 2(a) represents the results of the image query text task. It can be seen that the retrieval accuracy of the methods proposed in this chapter from 16-

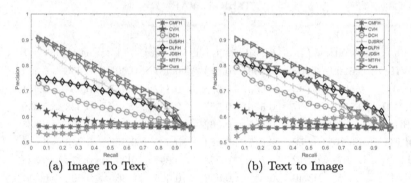

(a) Image To Text   (b) Text to Image

Fig. 3. PR curve on MIRFlickr dataset

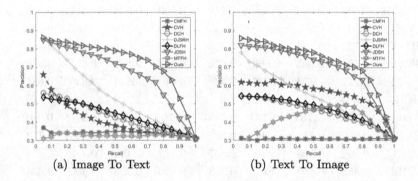

(a) Image To Text   (b) Text To Image

Fig. 4. PR curve on the NUS-Wide dataset

bit hash code to 128-bit hash code is higher than that of various baseline methods. According to the longitudinal comparison, it can be clearly seen that the method proposed in this chapter is folded above all the baseline methods, which indicates that the retrieval accuracy of the method proposed in this chapter is higher than other methods. According to the horizontal comparison, it is found that the discounting of the method in this study shows an upward trend, which indicates that the retrieval accuracy will increase with the increase of the length of the hash code, which is also consistent with the characteristics of the hash method. 2(b) represents the retrieval results of text query images. It can be seen that mAP accuracy values of almost all methods show an upward trend from 16 bits with fewer hash code points to 128 bits with more hash code points. According to the longitudinal comparison, it can be concluded that the mAP value in this study is about 0.04 higher than that of the most advanced method at that time in the text modal task of image query. 4 represents the PR curve with hash code length of 128 bits on this data set. It can be seen from the figure that this research achieves high precision retrieval before other methods, which reflects the scientific nature of this research method. In summary, the method proposed in this chapter shows superior performance in the deep cross-modal hashing method.

2. Ablation Experiment

In order to further investigate the methods proposed in this chapter, some ablation experiments will be performed in this section to prove that the design of this study is reasonable and efficient in various aspects. The data set used in the ablation experiment was the widely used MIRFlickr25k, and the mean average accuracy was selected as the evaluation criteria. The ablation experiment is mainly divided into three parts. UDCMH-GAN1 represents the use of a common network without the auto-encoder structure and the generative adversarial model. Then the method proposed in this chapter degenerates into an unsupervised cross-modal hash method similar to the DJSRH method, as shown in Fig. 5(a). UDCMH-GAN2 represents a model with only one discriminator and an auto encoder structure added on the basis of UDCMH-GAN1, as shown in Fig. 5(b). Based on UDCMH-GAN1, UDCMH-GAN3 uses both an automatic encoder and a generative adversarial model with two discriminators. The mAP value training results of the three methods are shown in Table 2.

According to the results, the performance of UDCMH-GAN1 without automatic encoder and antagonistic model is the worst for both image retrieval text task and text retrieval image task. On the basis of UDCMH-GAN1, the performance of UDCMH-GAN2 is improved, and the accuracy is improved by about 0.04, which indicates that the automatic encoder is of great benefit to the feature extraction of image modal, and the generative adversarial network is also of great significance to eliminate the semantic gap between modalities. UDCMH-GAN3 is the method proposed in this chapter, and the experimental results show that the accuracy of this study is improved by about 0.05 on the basis of UDCMH-GAN2. The significant performance improvement indicates that it is more scientific to use two discriminators of generate adversarial networks. In conclusion, the model designed in this study is reasonable and the design of loss function is superior (Figs. 3 and 4).

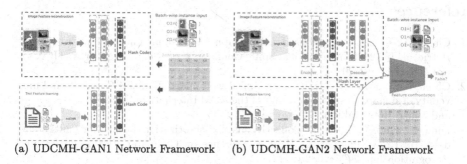

(a) UDCMH-GAN1 Network Framework   (b) UDCMH-GAN2 Network Framework

Fig. 5. Different Frameworks of Ablation Experiment

Table 2. Ablation Results

Methods	Image2Text				Text2Image			
	16 bits	32 bits	64 bits	128 bits	16 bits	32 bits	64 bits	128 bits
UDCMH-GAN1	0.625	0.697	0.700	0.716	0.662	0.691	0.683	0.695
UDCMH-GAN2	0.682	0.699	0.711	0.725	0.672	0.697	0.697	0.706
UDCMH-GAN3	0.726	0.737	0.741	0.760	0.721	0.723	0.733	0.740

5   Conclusion

This chapter mainly describes the specific framework and algorithm process of asymmetric deep cross-modal unsupervised hashing algorithm based on generative adversarial model. This method is an unsupervised cross-modal hash retrieval method, which integrates an automatic encoder to obtain more potential information of an image and generate a confrontation model. The main contribution of the method in this chapter can be summarized into three aspects. The first is to use the double discriminator model to train image features and text features. In the process of confrontation, asymmetric features are also used as input, decoding features are used for image, and common features are used for text, which is conducive to learning more potential semantic information and eliminating inter-modal heterogeneity. The second is to propose an asymmetric feature extraction method. The image channel uses the decoding process of the auto encoder to learn more potential semantic information, while the text uses the conventional feature extraction network, which is due to the inconsistent spatial distribution of image data and text data. Finally, this method is also verified on the widely used datasets MIRFlikr25k and NUS-Wide, showing strong competitiveness.

Acknowledgment. This work is supported by the NSFC Grant No. 62202438; the Natural Science Foundation of Shandong Province Grant No. ZR2020QF041; the 22th batch of ISN Open Fund Grant No. ISN22-21.

References

1. Chen, D., et al.: Unsupervised Deep Imputed Hashing for Partial Crossmodal Retrieval. In: 2020 International Joint Conference on Neural Networks (IJCNN), pp. 1–8. IEEE (2020)
2. Chua, T.-S., et al.: Nus-wide: a real-world web image database from national university of singapore. In: Proceedings of the ACM International Conference on Image and Video Retrieval, pp. 1–9 (2009)
3. Ding, G., Guo, Y., Zhou, J.: Collective matrix factorization hashing for multimodal data. In: Proceedings of the IEEE Conference on Computer Vision and Pattern Recognition, pp. 2075–2082 (2014)
4. Hoang, T., et al.: Multimodal mutual information maximization: a novel approach for unsupervised deep cross-modal hashing. IEEE Trans. Neural Netw. Learn. Syst. (2022)

5. Hu, H., et al: Creating something from nothing: unsupervised knowledge distillation for cross-modal hashing. In: Proceedings of the IEEE/CVF Conference on Computer Vision and Pattern Recognition, pp. 3123–3132 (2020)
6. Hu, P., et al.: Unsupervised contrastive cross-modal hashing. IEEE Trans. Pattern Anal. Mach. Intell. **45**, 3877–3889 (2022)
7. Huiskes, M.J., Lew, M.S.: The MIR Flickr retrieval evaluation. In: Proceedings of the 1st ACM International Conference on Multimedia Information Retrieval, pp. 39–43 (2008)
8. Jiang, Q.-Y., Li, W.-J.: Discrete latent factor model for cross-modal hashing. IEEE Trans. Image Process. **28**(7), 3490–3501 (2019)
9. Kumar, S., Udupa, R.: Learning hash functions for cross-view similarity search. In: Twenty-second International Joint Conference on Artificial Intelligence (2011)
10. Li, C., et al.: Coupled cyclegan: unsupervised hashing network for cross-modal retrieval. In: Proceedings of the AAAI Conference on Artificial Intelligence, vol. 33, no. 01, pp. 176–183 (2019)
11. Liu, H., et al.: Cross-modality binary code learning via fusion similarity hashing. In: Proceedings of the IEEE Conference on Computer Vision and Pattern Recognition, pp. 7380–7388 (2017)
12. Liu, S., et al.: Joint-modal distribution-based similarity hashing for largescale unsupervised deep cross-modal retrieval. In: Proceedings of the 43rd International ACM SIGIR Conference on Research and Development in Information Retrieval, pp. 1379–1388 (2020)
13. Liu, X., et al.: MTFH: a matrix tri-factorization hashing framework for efficient cross-modal retrieval. IEEE Trans. Pattern Anal. Mach. Intell. **43**(3), 964–981 (2019)
14. Long, M., et al.: Composite correlation quantization for efficient multimodal retrieval. In: Proceedings of the 39th International ACM SIGIR Conference on Research and Development in Information Retrieval, pp. 579–588 (2016)
15. Lu, X., et al.: Flexible online multi-modal hashing for large-scale multimedia retrieval. In: Proceedings of the 27th ACM International Conference On Multimedia, pp. 1129–1137 (2019)
16. Shen, X., et al.: Attention-guided semantic hashing for unsupervised cross-modal retrieval. In: 2021 IEEE International Conference on Multimedia and Expo (ICME), pp. 1–6. IEEE (2021)
17. Shi, Y., et al.: Deep adaptively-enhanced hashing with discriminative similarity guidance for unsupervised cross-modal retrieval. IEEE Trans. Circ. Syst. Video Technol. **32**(10), 7255–7268 (2022)
18. Su, S., Zhong, Z., Zhang, C.: Deep joint-semantics reconstructing hashing for large-scale unsupervised cross-modal retrieval. In: Proceedings of the IEEE/CVF International Conference on Computer Vision, pp. 3027–3035 (2019)
19. Tu, R.-C., et al.: Deep cross-modal hashing with hashing functions and unified hash codes jointly learning. IEEE Trans. Knowl. Data Eng. **34**(2), 560–572 (2020)
20. Wang, X., et al.: Self-constraining and attention-based hashing network for bit-scalable cross-modal retrieval. Neurocomputing **400**, 255–271 (2020)
21. Xu, X., et al.: Learning discriminative binary codes for large-scale cross-modal retrieval. IEEE Trans. Image Process. **26**(5), 2494–2507 (2017)
22. Yu, J., et al.: Deep graph-neighbor coherence preserving network for unsupervised cross-modal hashing. In: Proceedings of the AAAI Conference on Artificial Intelligence, vol. 35, no. 5, pp. 4626–4634 (2021)

23. Zhang, J., Peng, Y., Yuan, M.: Unsupervised generative adversarial cross-modal hashing. In: Proceedings of the AAAI Conference on Artificial Intelligence, vol. 32, no. 1 (2018)
24. Zhang, P.-F., et al.: Aggregation-based graph convolutional hashing for unsupervised cross-modal retrieval. IEEE Trans. Multimedia **24**, 466–479 (2022)
25. Zhou, J., Ding, G., Guo, Y.: Latent semantic sparse hashing for cross-modal similarity search. In: Proceedings of the 37th International ACM SIGIR Conference on Research & Development in Information Retrieval, pp. 415–424 (2014)

CFDM-IME: A Collaborative Fault Diagnosis Method for Intelligent Manufacturing Equipment

Yue Wang[1,2], Tao Zhou[1,2], Xiaohu Zhao[1,2(✉)], and Xiaofei Hu[3]

[1] National and Local Joint Engineering Laboratory of Internet Applied Technology on Mines, China University of Mining and Technology, Xuzhou 221008, China
[2] School of Information and Control Engineering, China University of Mining and Technology, Xuzhou 221116, China
xiaohuzhaoxz@163.com
[3] School of Information and Business Management, Chengdu Neusoft University, Chengdu 611844, China
huxiaofei@nsu.edu.cn

Abstract. The stability of the intelligent manufacturing industry will directly affect the development of the social economy. The privacy of data among different smart factories (SF) leads to a lack of generalization of existing deep learning-based fault diagnosis methods. In order to solve the problems existing in the fault diagnosis method, this paper combines blockchain, federated learning, and deep learning to propose a collaborative fault diagnosis method for intelligent manufacturing equipment (CFDM-IME). Specifically, firstly, a fault diagnosis model based on LSTM is proposed to realize local model training of local intelligent manufacturing equipment. Then, a domain parameter aggregation method based on a federated average is proposed to realize the aggregation of internal model parameters of smart factories. Then, the parameter collaborative optimization smart contract is designed and implemented to achieve the aggregation of global parameters. Finally, we conduct simulation experiments on the proposed method. Theoretical and simulation experiments prove that our proposed architecture is feasible.

Keywords: Predictive maintenance · Fault diagnosis · Blockchain · Deep learning · Federated learning

1 Introduction

In recent years, with the continuous development of artificial intelligence, big data, blockchain, and other technologies, intelligent manufacturing has moved from industry 4.0 to industry 5.0 [1]. Industry 4.0 focuses on the realization of intelligent manufacturing based on cyber-physical systems (CPS), while the emerging Industry 5.0 surpasses CPS and covers the entire manufacturing value chain [2]. Intelligent manufacturing equipment (IME) failure as a tricky problem will run through all stages of the intelligent manufacturing industry [3,4].

Z. Tari et al. (Eds.): ICA3PP 2023, LNCS 14487, pp. 49–60, 2024.
https://doi.org/10.1007/978-981-97-0834-5_4

The maintenance of IME can effectively reduce the probability of failure, thereby reducing the economic loss caused by equipment failure in smart factories. The existing maintenance solutions for IME can be mainly divided into three categories: breakdown maintenance [5], preventive maintenance [6], and predictive maintenance [7].

Breakdown maintenance will cost the most to maintain [8]. In the practical application of preventive maintenance, there are two prominent contradictions between excessive maintenance and insufficient maintenance [9]. Predictive maintenance avoids unnecessary downtime caused by excessive intervention as well as cascading failures caused by failure to take action in a timely manner [10]. Predictive maintenance is widely used in the industrial Internet of Things due to its excellent advantages [11].

As one of the important links of predictive maintenance, fault diagnosis will directly affect its effect [12]. In the real intelligent manufacturing industry, the same type of IME in different smart factories can produce different kinds of products according to the task demand. This will not only affect the accuracy of the fault diagnosis model of IME but also lead to the failure of predictive maintenance methods in severe cases. By updating the fault diagnosis models of IME in different smart factories, a fault diagnosis model with generalization can be obtained. However, when IME shares model parameters, attackers can obtain personal privacy information through data mining technology, threatening customer security.

In order to ensure the safe, smooth, and efficient operation of IME. This paper combines blockchain, federated learning, and deep learning technologies and proposes a collaborative fault diagnosis method for intelligent manufacturing equipment (CFDM-IME) to solve the problem of generalization of fault diagnosis models for IME among multiple smart factories. The main contributions of this paper are as follows:

(1) We combine blockchain and federated learning technology to propose a collaborative fault diagnosis architecture for IME. This architecture can realize the collaborative fault diagnosis of IME among smart factories under the premise of ensuring the privacy of each SF.
(2) We combine LSTM and Softmax to propose a fault diagnosis model based on LSTM. The IME that deployed this model can not only realize fault diagnosis but also realize the identification of multiple fault types, providing an important basis for fault maintenance.
(3) We implement a parameter collaborative optimization method for the fault diagnosis model. Specifically, the in-domain parameter aggregation method based on federated average realizes the aggregation and optimization of in-domain parameters of smart factories. The parameter synergy optimizes smart contracts (PSOSC) realizes the aggregation of model parameters among smart factories.

The rest of this article is organized as follows: Sect. 2 summarizes the work related to equipment fault diagnosis. Section 3 introduces the CFDM-IME proposed in this article in detail. Section 4 introduces the principles of CFDM-IME.

Section 5 conducts simulation experiments and analyzes the experimental results. Section 6 summarizes the full text.

2 Related Work

The recent research on deep learning-based fault diagnosis and collaborative fault diagnosis will be summarized and analyzed in this section.

A bearing fault diagnosis method for flywheel energy storage systems based on parameter optimization of variational mode decomposition energy entropy is proposed in the literature [13] to solve the problem that the complex nonlinear and non-stationary characteristics of the vibration signal of the bearing bring difficulties to the fault diagnosis of the bearing, but the method ignores the problem of an insufficient sample. A new variational generative adversarial network model based on multi-source signal fusion is proposed in the literature [14] to solve the problem of low fault diagnosis accuracy due to insufficient samples and imbalance.

A new collaborative deep learning framework is proposed in the literature [15] to solve the problem of data sharing and deep learning structure application in distributed complex systems based on deep learning schemes, but the collaborative learning in this method is relatively limited, and the communication cost needs to be further optimized. A new method for collaborative diagnosis of maintenance faults based on multi-sensory fusion is proposed in the literature [16] to solve the obstacles of multi-perception data fusion and maintenance strategy fusion in collaboration efficiency based on deep learning schemes, but they ignore the data privacy issues among different domains. A cloud-edge collaborative adaptive fault diagnosis method with expanded label sampling space in cloud manufacturing environments is proposed in the literature [17] to solve the problem that existing methods need fault labels and a large number of target fault data to solve the lack and imbalance of fault data.

3 CFDM-IME in Detail

In this section, CFDM-IME will be introduced in detail from the two aspects of the overall architecture of the method and the method flow.

3.1 Overall Architecture of CFDM-IME

The overall architecture of CFDM-IME is shown in Fig. 1. The architecture consists of a terminal layer, an edge layer, and a blockchain layer. The detailed description of each layer is shown below.

Terminal Layer: The terminal layer consists of IME in a smart factory (SF). The IME in the same SF plays the role of federation in federated learning, training its own fault diagnosis model and uploading local optimal parameters.

Edge Layer: The edge layer consists of servers in numerous smart factories. SF plays the role of aggregator in federated learning, aggregating the local parameters of all its subordinate IME to obtain in-domain model parameters.

Blockchain Layer: The blockchain layer is built by the SF that deploys the Ethereum blockchain client through P2P networking. The PSOSC is deployed in the blockchain layer to obtain global model parameters by secondary aggregation of parameters aggregated by SF.

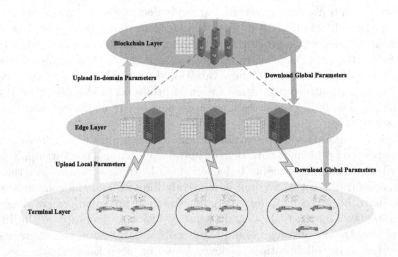

Fig. 1. The overall architecture of CFDM-IME.

3.2   Flow of CFDM-IME

The CFDM-IME method is implemented by IME at the terminal layer, SF at the edge layer, and PSOSC at the blockchain layer. The flow of CFDM-IME is shown in Fig. 2.

The specific process is as follows:

(1) The PSOSC sends global parameters to the SF.
(2) The SF sends global parameters to the IME.
(3) All IME updates the local fault diagnosis model based on the parameters received.
(4) The IME trains the local fault diagnosis model and uploads the latest parameters to the superior SF.
(5) After receives all the parameters from the IME, the SF aggregates the parameters in the domain and uploads the aggregated parameters to the PSOSC.
(6) After the PSOSC receives all the parameters in the domain aggregated by the SF, the global parameter aggregation is carried out.
(7) Iterate over the above steps until the model is optimal.

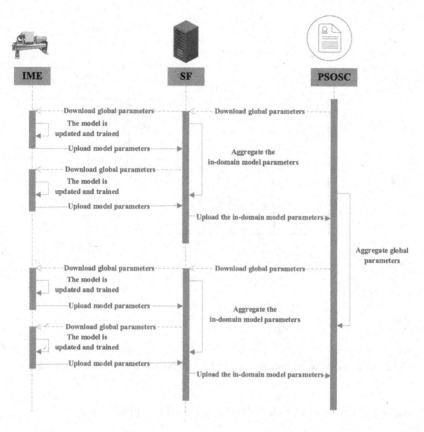

Fig. 2. The flow of CFDM-IME.

4   CFDM-IME Implementation

In this section, the principles of an LSTM-based fault diagnosis model, a federated average-based in-domain parameter aggregation method, and a PSOSC will be introduced.

4.1   LSTM-Based Fault Diagnosis Model

The overall structure of the LSTM-based fault diagnosis model is shown in Fig. 3.

The LSTM cell updates unit state $c_{(t)}$ at each point in time based on the forget, update, and output gates, which determine the value of $\alpha_{(t)}$.

The forgotten gate uses Eq. (1) to calculate whether the previous state in memory is forgotten.

$$f_{(t)} = \sigma \left(W_f \left[\alpha_{(t-1)}, x_{(t)} \right] + b_f \right) \tag{1}$$

If $f_{(t)} = 0$, LSTM ignores the previous state. If $f_{(t)} = 1$, LSTM remains in its previous state.

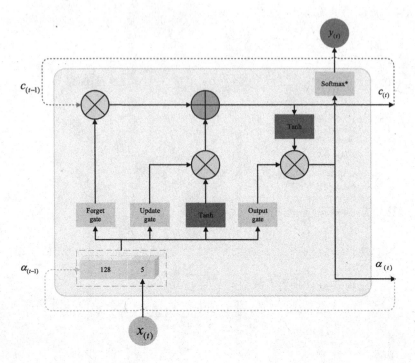

Fig. 3. The overall structure of the LSTM-based fault diagnosis model.

Equation (2) is used to calculate the value of the update gate.

$$i_{(t)} = \sigma \left(W_i \cdot \left[\alpha_{(t-1)}, x_{(t)} \right] + b_i \right) \qquad (2)$$

Equation (3) is used to calculate the information that may be stored in the current cell state at the current point in time.

$$\tilde{C}_{(t)} = \tanh \left(W_C \cdot \left[\alpha_{(t-1)}, x_{(t)} \right] + b_C \right) \qquad (3)$$

According to the value $i_{(t)}$ of the update gate, the value $f_{(t)}$ of the forgotten gate, the current cell state $\tilde{C}_{(t)}$, and the previous moment cell state $c_{(t-1)}$, the current cell state $c_{(t)}$ is calculated using Eq. (4).

$$C_{(t)} = f_{(t)} * C_{(t-1)} + i_{(t)} * \tilde{C}_{(t)} \qquad (4)$$

Equation (5) is used by the output gate to control the output at the current moment.

$$o_{(t)} = \sigma \left(W_o \left[\alpha_{(t-1)}, x_{(t)} \right] + b_o \right) \qquad (5)$$

Equation (6) is used to calculate the hidden state at the current moment.

$$\alpha_{(t)} = o_{(t)} * \tanh \left(C_{(t)} \right) \qquad (6)$$

Softmax is used for the classification of faults, namely:

$$z_{(t)} = W_y \alpha_{(t)} + b_y \tag{7}$$

$$y_{(t)}^{pred} = soft \max(z_{(t)}) \tag{8}$$

4.2   Federated Average-Based In-Domain Parameter Aggregation Method

Assume that there are N IME $\{C_i, i = 1, 2, \ldots, N\}$ trained collaboratively under the coordination of the SF. Each IME C_i has its corresponding data set D_i, then the total data set of all IME is $D = \{D_1, D_2, \ldots, D_N\}$. The data set of each IME obeys the $\Im_i$ distribution, and the data set size is $|D_i|$. Any term of each dataset is (x, y), where $x \in R^d$ is the input feature, $y \in R^t$ is the corresponding label.

Assuming that the model parameter corresponding to each IME is ω_i, then the predicted value of the model is $\hat{y} = f(\omega_i; x)$. Defining the loss function for each IME as $L_i(\hat{y}, y) : R^t \to R$, the optimization goal for each IME is:

$$\arg \min_{\omega_i} L_i(\omega_i) = E_{D_i \sim \varsigma_i} [L_i(\hat{y}, y)] \tag{9}$$

Substituting $\hat{y} = f(\omega_i; x)$ into Eq. (9) yields:

$$\arg \min_{\omega_i} L_i(\omega_i) = E_{D_i \sim \varsigma_i} [L_i(f(\omega_i, x), y)] \approx \frac{1}{|D_i|} \sum_j^{|D_i|} L_i(f(\omega_i, x_j), y_j) \tag{10}$$

The cross-entropy loss function is used for optimization, i.e.,

$$\arg \min_{\omega_i} L_i(\omega_i) = E_{D_i \sim \varsigma_i} [L_i(f(\omega_i, x), y)] \approx -\frac{1}{|D_i|} \sum_j^{|D_i|} y_j \log(f(\omega_i, x_j)) \tag{11}$$

The domain parameters of each IME are aggregated and optimized according to Eq. (12) to obtain the final fault diagnosis model.

$$\arg \min_{\omega} L(\omega) = \sum_i^N E_{D_i \sim \varsigma_i} [L_i(f(\omega, x_i), y_j)] \approx \sum_i^N \frac{1}{|D_i|} \sum_j^{|D_i|} L_i(f(\omega, x_j), y_j) \tag{12}$$

4.3   Parameter Synergy Optimizes Smart Contracts

Suppose there are M smart factories $\{S_i, i = 1, 2, \ldots, M\}$ coordinated by PSOSC for synergistic aggregation. Each SF S_i has a corresponding parameter dataset P_i, and the total global parameter of all smart factories is $P = \{P_1, P_2, \ldots, P_M\}$.

PSOSC randomly selects S_t among M smart factories and sends the global parameter P_t of the current round (the global parameter of the first round is P_0) to all smart factories.

Each selected SF updates the global parameters issued by PSOSC to its own local and sends them to the IME. The IME trains the LSTM-Based Fault Diagnosis Model and sends updated parameters to the SF. Federated Average-based In-domain Parameter Aggregation Method is used by SF to aggregate parameters and get p_{t+1}^m, which sends p_{t+1}^m to the PSOSC.

After receiving the global parameters sent by each SF, PSOSC aggregates them according to Eq. (13).

$$p_{t+1} = \sum_{m=1}^{M} \frac{n_m}{n} p_{t+1}^m \tag{13}$$

where n_m is the amount of data per smart factory, $n = \sum_{m=1}^{M} n_m$.

The PSOSC sends the latest global parameters to the SF.

4.4 Intelligent Manufacturing Equipment Collaborative Fault Diagnosis Algorithm

The intelligent manufacturing equipment collaborative fault diagnosis algorithm is divided into three parts: parameter synergy optimizes smart contracts, smart factory and intelligent manufacturing equipment. This is shown in Algorithm 1.

5 Experimental Simulation

We use 4 Maxtang microcomputers produced by Shenzhen Datang Computer Co., Ltd. to simulate smart factories (SF A, SF B, SF C and SF). Deploy the Ethereum blockchain client in the smart factory and build a blockchain ecosystem within the local area network through the P2P networking method provided. At the same time, the PySyft framework is used to build federated learning environments.

Based on the above experimental environment, we use the predictive maintenance dataset (AI4I2020) provided by Torcianti et al. to simulate the collaborative fault diagnosis method for intelligent manufacturing equipment proposed in this paper [18].

5.1 Experimental Results and Analysis

Table 1 shows the model parameter settings of the CFDM-IME.

In order to explore the influence of the number of IME participating in federated training and the number of local model updates on the accuracy of the entire model, the other parameters in Table 1 are left unchanged, and only the values of *epochs* and ρ_i are changed. Set the local update round *epochs* to $\{1, 5, 10, 20, 40\}$, and the value of ρ_i to $\{0.1, 0.2, 0.4, 0.6, 0.8, 1.0\}$, that is, $\{4, 8, 16, 24, 32, 40\}$ intelligent manufacturing devices are randomly selected to participate in federation training in each round. The experimental results are shown in Fig. 4.

Algorithm 1: Intelligent Manufacturing Equipment Collaborative Fault Diagnosis Algorithm (IMECFDA)

---

1   *ParameterSynergyOptimizesSmartContracts():*
2   PSOSC initializes the global parameter P_0 and sends the original global parameter P_0 to all smart factories.
3   **for** *Each global parameter update turn $t = 1, 2, ..., T$* **do**
4   The PSOSC to determine S_t.
5   **foreach** $s \in S_t$ **do**
6         $p_{t+1}^{(s)} \leftarrow$ *SmartFactory(s, $\bar{p}_t$)* //The interdomain parameters are updated.
7   Send the updated parameter $P_{t+1}^{(s)}$ to PSOSC.
8   **end**
9      $\bar{p}_{t+1} = \sum\limits_{m=1}^{M} \frac{n_m}{n} p_{t+1}^m$ is used by PSOSC to aggregate the received parameters.
10   PSOSC sends the aggregated global parameter $\bar{p}_{t+1}$ to all smart factories.
11   **end**
12   *SmartFactory(s, $\bar{p}_t$):*
13   Get the latest parameters from PSOSC, i.e., $\omega_t = \bar{p}_t$.
14   **foreach** $i \in s$ **do**
15      $\omega_{t+1}^{(i)} \leftarrow$ *InteligentManufacturingEquipment(i, $\bar{\omega}_t$)* //The in-domain parameters are updated.
16   Send the updated parameter $\omega_{t+1}^{(i)}$ to SF.
17      $\bar{\omega}_{t+1} = \sum\limits_{i=1}^{I} \frac{n_i}{n} \omega_{t+1}^i$ is used by SF for the aggregation of parameters.
18   The SF checks whether the maximum number of training rounds has been reached.
19   SF sends the aggregated parameters $\bar{\omega}_{t+1}$ to all IMEs.
20   **end**
21   *InteligentManufacturingEquipment(i, $\bar{\omega}_t$)*
22   Obtain the latest parameters from SF, i.e, $\omega_{1,1}^i = \bar{\omega}_t$.
23   **for** *Each iteration ite from 1 to the number of iterations epochs* **do**
24   *batches* $\leftarrow D_i$ //Divide the dataset D_i into batches the size of R.
25   Obtain the parameter $\omega_{1,ite}^{(i)} = \omega_{B,ite-1}^{(i)}$ from the previous iteration.
26   **for** *Number b from batch 1 to lot $B = \frac{n_i}{R}$* **do**
27         $g_i^{(b)} = \nabla L_i^{(b)}$ //The gradient is calculated.
28         $\omega_{b+1,ite}^{(i)} \leftarrow \omega_{b,ite}^{(i)} - \eta g_i^{(b)}$ //Local parameters are updated.
29   **end**
30   **end**
31   Get the latest local parameter $\omega_{t+1}^{(i)} = \omega_{B,epochs}^{(i)}$.

As can be seen from Fig. 4, the accuracy of the model becomes higher and higher as the number of IME participating in federal training increases. When all IME participants participate in federated training, the accuracy of the model is at its highest. In addition to this, when *epochs* $= 1$, the model's accuracy is lower.

Table 1. Parameter settings for model training process.

Parameter	Description	Parameter value
epochs	Local update round	5
lr	Learning rate	0.01
R	Local training batch size	64
ρ_s	The proportion of SF selected in each round	1.0
ρ_i	The proportion of IME selected in each round	1.0
T	Global update rounds	50
M	Total number of smart factories	4
N	Total number of intelligent manufacturing equipment	40

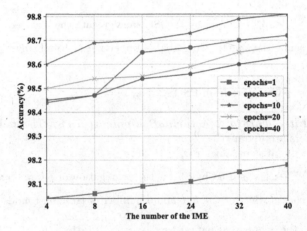

Fig. 4. The accuracy of different IME and local update rounds.

This is because local training is not enough, and the model does not get enough information from the local dataset. When $epochs = 10$, the accuracy reaches its highest, and as $epochs$ increases, the accuracy gradually decreases because the local update round is too large. Therefore, the more IME that participate in the federation, the higher the accuracy of the model obtained with the appropriate number of iterations ($epochs = 10$), and the better the generalization.

In order to explore the influence of the number of smart factories participating in global parameter optimization on the accuracy of the entire model, leave the other parameters in Table 1 unchanged and only change the value of ρ_s. Set the value of ρ_s to $\{0.25, 0.5, 0.75, 0.1\}$, that is, $\{1, 2, 3, 4\}$ smart factories are randomly selected to participate in global parameter optimization in each round. The experimental results are shown in Fig. 5.

As can be seen from Fig. 5, the accuracy of the model becomes higher and higher as the number of smart factories participating in parameter synergy optimization increases. When all smart factories participate in parameter synergy optimization, the accuracy of the model is at its highest. Therefore, the more

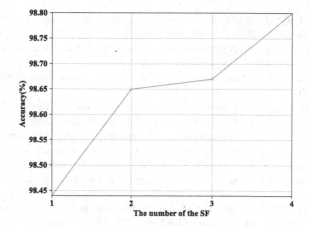

Fig. 5. The accuracy of different numbers of smart factory.

smart factories that participate in parameter synergy optimization, the higher the accuracy of the obtained model and the better the generalization.

6   Conclusion

This paper proposes a collaborative fault diagnosis method for intelligent manufacturing equipment to solve the generalization problem of existing fault diagnosis models. This method trains the local fault diagnosis model in the intelligent manufacturing equipment, aggregates the parameters in the domain in the smart factory, and optimizes the global parameters in the parameter synergy optimizes smart contracts, so as to finally ensure that the intelligent manufacturing equipment realizes collaborative fault diagnosis under the premise of private data. In our future work, we will carry out our work in depth from the two aspects of communication delay between intelligent manufacturing equipment and smart factories and a lightweight fault diagnosis model.

Acknowledgements. This work was supported by the Fundamental Research Funds for the Central Universities under Grant 2020ZDPY0223.

References

1. Maddikunta, P.K.R., et al.: Industry 5.0: a survey on enabling technologies and potential applications. J. Ind. Inf. Integr. **26**, 100257 (2022)
2. Xu, X., Lu, Y., Vogel-Heuser, B., Wang, L.: Industry 4.0 and industry 5.0—inception, conception and perception. J. Manuf. Syst. **61**, 530–535 (2021)
3. Chi, Y., Dong, Y., Wang, Z.J., Yu, F.R., Leung, V.C.: Knowledge-based fault diagnosis in industrial internet of things: a survey. IEEE Internet Things J. **9**(15), 12886–12900 (2022)

4. Sinche, S., et al.: A survey of IoT management protocols and frameworks. IEEE Commun. Surv. Tut. **22**(2), 1168–1190 (2019)
5. Sheut, C., Krajewski, L.: A decision model for corrective maintenance management. Int. J. Prod. Res. **32**(6), 1365–1382 (1994)
6. Barlow, R., Hunter, L.: Optimum preventive maintenance policies. Oper. Res. **8**(1), 90–100 (1960)
7. Mobley, R.K.: An Introduction to Predictive Maintenance. Elsevier (2002)
8. Amelia, M., Aspiranti, T.: Analisis pemeliharaan mesin conveyor menggunakan metode preventive dan breakdown maintenance untuk meminimumkan biaya pemeliharaan mesin pada pt x. Jurnal Riset Manajemen dan Bisnis **1**, 1–9 (2021)
9. Dui, H., Zhang, C., Tian, T., Wu, S.: Different costs-informed component preventive maintenance with system lifetime changes. Reliab. Eng. Syst. Saf. **228**, 108755 (2022)
10. Zhang, W., Yang, D., Wang, H.: Data-driven methods for predictive maintenance of industrial equipment: a survey. IEEE Syst. J. **13**(3), 2213–2227 (2019)
11. Ong, K.S.H., Wang, W., Hieu, N.Q., Niyato, D., Friedrichs, T.: Predictive maintenance model for IIoT-based manufacturing: a transferable deep reinforcement learning approach. IEEE Internet Things J. **9**(17), 15725–15741 (2022)
12. Liu, Z., Fang, L., Jiang, D., Qu, R.: A machine-learning-based fault diagnosis method with adaptive secondary sampling for multiphase drive systems. IEEE Trans. Power Electron. **37**(8), 8767–8772 (2022)
13. He, D., Liu, C., Jin, Z., Ma, R., Chen, Y., Shan, S.: Fault diagnosis of flywheel bearing based on parameter optimization variational mode decomposition energy entropy and deep learning. Energy **239**, 122108 (2022)
14. Zhang, L., Zhang, H., Cai, G.: The multiclass fault diagnosis of wind turbine bearing based on multisource signal fusion and deep learning generative model. IEEE Trans. Instrum. Meas. **71**, 1–12 (2022)
15. Wang, H., Liu, C., Jiang, D., Jiang, Z.: Collaborative deep learning framework for fault diagnosis in distributed complex systems. Mech. Syst. Sig. Process. **156**, 107650 (2021)
16. Shao, H., Lin, J., Zhang, L., Galar, D., Kumar, U.: A novel approach of multi-sensory fusion to collaborative fault diagnosis in maintenance. Inf. Fus. **74**, 65–76 (2021)
17. Ren, L., Jia, Z., Wang, T., Ma, Y., Wang, L.: LM-CNN: a cloud-edge collaborative method for adaptive fault diagnosis with label sampling space enlarging. IEEE Trans. Industr. Inf. **18**(12), 9057–9067 (2022)
18. Torcianti, A., Matzka, S.: Explainable artificial intelligence for predictive maintenance applications using a local surrogate model. In: 2021 4th International Conference on Artificial Intelligence for Industries (AI4I), pp. 86–88. IEEE (2021)

DFECTS: A Deep Fuzzy Ensemble Clusterer for Time Series

Dechong Wu, Jialun Li[ID], Xuan Mo, and Weigang Wu[✉][ID]

School of Computer Science and Engineering, Sun Yat-sen University, Guangzhou, China
wudch3@mail3.sysu.edu.cn, {lijlun3,moxuan}@mail2.sysu.edu.cn,
wuweig@mail.sysu.edu.cn

Abstract. Time series clustering plays an important role in various fields such as anomaly detection and resource scheduling. With the increase of complexity and scale of time series datasets, many deep-learning-based time series clustering methods have emerged and achieved great success. These methods mostly use a deep neural network to extract features which are fed into classic clustering methods. The focus of such methods is to enhance the neural network to extract more meaningful features. Differently, in this paper, we propose the first ensemble clusterer, which uses time series splitting method to construct different sub-series representing various aspects of the original time series. For each group of sub-series, a clusterer is generated, which consists of an autoencoder and a fuzzy C-means algorithm. After collecting the results from each clusterer, a consensus is reached by voting. However, traditional voting algorithm widely-used in ensemble classifier does not work in ensemble clusterer because labels mismatch among clusterers. To deal with this problem, we propose an efficient label aligning algorithm. Then, a weighted voting algorithm is applied and a new mechanism with fitness matrix is utilized to further improve the performance. Extensive experiments show that our method outperforms seven representative algorithms under three metrics.

Keywords: Fuzzy clustering · Ensemble clusterer · Time series · Fuzzy C-means · Deep learning

1 Introduction

The technique to cluster time series has been widely-used in many data mining based scenarios, including anomaly detection [17], resource scheduling [29], etc. The main task of time series clustering is to discover meaningful patterns from complex datasets [2] and extracts valuable information for data analysts. Although general clustering methods, including hierarchical, density-based, spectral, partitional [8, 23] clustering can be used for time series, these methods ignore

This work is partially supported by the Key-Area Research and Development Program of Guangdong Province (No. 2020B0101090005).

Z. Tari et al. (Eds.): ICA3PP 2023, LNCS 14487, pp. 61–80, 2024.
https://doi.org/10.1007/978-981-97-0834-5_5

the temporality of time series. Therefore, many clustering algorithms specifically for time series have been proposed. Some of them try to convert time series into discrete data with characteristic of temporality so that the existing algorithms for clustering can be directly used [18]. A more popular approach is to design specific distance measures for time series clustering [6,7]. Besides, shapelet recognition is widely used in time series clustering [27]. Nevertheless, the above methods usually perform poorly when outliers and noise occur.

With the increase of complexity and scale of datasets [25] in the last decade, deep learning has been adopted in clustering, which typically consists of three phases. 1) A neural network is used to extract features from datasets. 2) A clustering algorithm is utilized to cluster these features. 3) Several loss functions are used to update the parameters of the neural network. Different deep clustering methods may employ neural networks with different architectures. One of the most popular is autoencoder (AE) [26] due to its simple but efficient structure. The autoencoder aims at reconstructing data points for unsupervised data representation. Usually, deep clustering employs a variety of traditional clustering loss functions like K-means [11,15,32,34], fuzzy C-means (FCM) [38], spectral clustering [35] and other tricks [10,36] to enhance the ability of the autoencoder for a better cluster result.

Deep learning models have also been employed in time series clustering. DTC [24] depicts a model using recurrent-network-based autoencoder to obtain feature representations. However, it exploits Kullback-Leibler divergence as clustering loss function, which leads to instability. DTCR [20] uses an auxiliary classification task to enhance the ability of autoencoder. But it has a large time and space complexity. Roughly, all these deep time series clustering methods focus on modifying the structure and promoting the ability of autoencoder specially for time series.

Unlike existing approaches of deep time series clustering, in this paper, we propose the first ensemble clusterer with deep learning to capture the characteristics of time series. Currently, several types of ensemble clusterer methods are proposed, such as feature extracting [1], label aligning [39], graph generating [3] and so on. Besides, integrating deep learning into these methods becomes popular due to its powerful data processing capability. Normally, an ensemble clusterer, combines basic clusterers with various clustering algorithms, can greatly improve performance and robustness compared to single clustering methods. However, deep ensemble clusterer specially for time series has not been studied.

We present a novel deep fuzzy ensemble clusterer for time series clustering (DFECTS). DFECTS first splits time series into multiple sub-series which represent different aspects of data to each clusterer. By splitting, the volume of data to be handled by DFECTS can be reduced efficiently. The clusterer then exploits a GRU-based [12] autoencoder and FCM algorithm to cluster these sub-series. Specifically, GRU has the ability to capture temporality of time series and FCM is a fast clustering algorithm to accelerate the convergence of our model. To handle the problem of label mismatch occurring only in ensemble clusterer, a label aligning method is proposed, which fully utilizes the rich information offered by FCM. Most existing ensemble learning methods exploit weighted voting algorithms that make use of true labels to reach a consensus among clusterers. However, there are no true labels in unsupervised learning. To cope this problem, we present a novel algorithm that realizes weighted voting based on fitness matrix for ensemble clusterer which handles the bias of different clusterers. Our contributions can be summarized as follows:

- We propose a novel deep ensemble clusterer model for time series clustering, which has a lower time complexity compared to many other ensemble clusterer models. To the best of our knowledge, this is the first work using deep ensemble clusterer for time series.
- We propose a label-aligning method for soft clustering algorithm to capture more information between data and clusters.
- We propose a weighted voting strategy based on our fitness matrix for unsupervised clustering. The fitness matrix guides clusterers to make a more accurate judgement, which further improves the performance.
- We conduct extensive experiment, which show that the DFECTS achieves promising performance by several metrics in many datasets.

The rest of the paper is organized as follows. In Sect. 2, recent advances related to our work are introduced. Section 3 details the DFECTS algorithm. In Sect. 4, we present our experimental evaluation. Finally, we conclude our work in Sect. 5.

2   Related Works

2.1   Time Series Clustering

Depending on whether explicit feature extraction mechanism is adopted, we broadly classify existing time series clustering methods into two categories: raw-time-series-based methods and feature-extracting-based methods.

For raw-time-series-based methods, the key point is the design of distance measures that fit time series characteristics. DTW [7] is a well-known distance measure which exploits dynamic programming to align the time series and obtains amplitude invariance and offset invariance. Batista et al. [6] design a complexity-invariant distance measure (CID) which focuses on the complexity of shape of the time series. They then incorporate it into hierarchical clustering algorithm. Barragan et al. [5] propose a distance measure basing on wavelet transform and principal component analysis. Time series are first converted to space-time-frequency domain and principal component analysis is applied to get the distance between different time series afterward. The distance measure is incorporated into FCM algorithm to obtain cluster results. Another popular approach is making use of shapelets. A shapelet is a subsequence of the time series. Paparrizos and Gravano [27] try to figure out the most common shapelets among time series as cluster centroids. Zhang et al. [37] learn shapelets by minimizing the distance between shapelets and time series. Based on these learned shapelets, they convert the original time series into a shapelet-based space. Pseudo-class labels and spectral analysis are applied to train the model and get the results later.

Feature-extracting-based methods extract features from raw time series using approaches like independent component analysis, spectral analysis and deep neural network. Deep neural network models have been recently used for time series feature extracting and achieve gréat success. The key point lies in enhancing the ability of deep neural network to extract more meaningful features. Trosten et al. [31] propose to improve the ability of GRU by aggregating three loss function about clustering. Among others, autoencoder is a very popular model. Madiraju et al. [24] employ CNN and LSTM as the autoencoder and utilize a clustering loss function with distance measures specially for time series to achieve high performance. Zheng et al. [20] employ the bidirectional multi-layer Dilated RNN as autoencoder. They propose a fake-sample generation strategy and auxiliary classification task. By back propagating the gradient of loss function of this task, the ability of autoencoder can be further improved.

Our DFECTS method also employs autoencoder to extract features. Rather than enhancing the autoencoder itself, the novelty of our work lies in mechanisms to deep ensemble clustering.

2.2   Ensemble Clusterer

Ensemble clusterer, composed of basic clusterers and an ensemble strategy, attempts to exploit the diversity among clusterers to achieve high robustness and performance. There are several heuristics to handle diversity. 1) Making datasets multifarious for each clusterers. Xue et al. [33] create multiple views of the same dataset for clusterers. Fischer and Buhmann [14] apply a resampling scheme with similarity to bootstrap method and obtain some subsets. Goh et al. [19] apply an oversampling method which generates synthetic data. 2) Personalizing the clusterers themselves. Affeldt et al. [1] use deep autoencoder for each clusterers with parameters initialized randomly. Ma et al. [21] adopt different clustering algorithm and use several combining strategies. Kuncheva and Vetrov [16] use K-means with a random target number of clusters as its basic clusterer.

Our model DFECTS combines these two heuristics to construct diverse and robust clusterers.

The other important part for ensemble clusterer is ensemble strategy. Unlike supervised learning, in unsupervised learning like clustering, the labels (the clusters to which the samples belong) given by the model have no practical meaning. Labels can only indicate whether the samples are similar to each other or not. Hence, for ensemble clusterer, the widely-used voting algorithms in ensemble learning do not work. To cope with this difficulty, researchers propose various mechanisms. Bai et al. [4] generate a weighted graph and graph-based algorithm is applied to obtain the results. Tao et al. [30] get basic clusters from each clusterer and seek for a common latent subspace. These strategies are usually time consuming to get results. Zhou and Tang [39] propose a fast algorithm to align

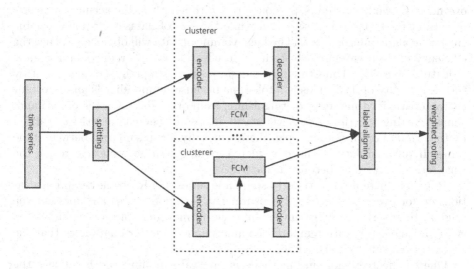

Fig. 1. The overall architecture of DFECTS. We firstly split time series into multiple sub-series. Then, these sub-series are fed into multiple clusterers respectively. We then apply a label aligning algorithm to correct the label mismatch among clusterers. Finally, we employ a weighted voting algorithm to obtain the results.

labels and apply a voting algorithm afterward. The underlying idea is that similar clusters should contain similar samples. Given two clusterers, the algorithm counts the number of overlapped samples between each pair of clusters from different clusterers. Then, the pair of clusters with the largest number of overlapped samples are matched by denoting them using the same label. Such a process is repeated until all the clusters are matched. The algorithm is utilized for hard clustering which assigns data to a cluster without probability. Inspired by this, we propose a new label aligning method for soft clustering and fully exploit the assignment (probability) values.

3 Methodology

The overall structure of DFECTS is illustrated in Fig. 1. Time series are firstly split into multiple sub-series. Then, these sub-series are fed into multiple clusterers, respectively. A label aligning algorithm is applied to correct the label mismatch among clusterers. Finally, a weighted voting algorithm is employed to get the results. In the following subsections, we describe the details of all these parts.

3.1 Time Series Splitting

Many ensemble clusterer algorithms use all features for each clusterer. However, for deep ensemble learning, such approach will significantly cause large training overhead. Given T clusterers, N time series with length L, the accumulative size of dataset for training is $T \times N \times L$ while the size of dataset for non-ensemble clustering algorithms is $N \times L$. The huge volume of data will obviously reduce the efficiency of deep ensemble clustering. An effective way to reduce data volume is feature sampling. Unfortunately, classic feature sampling strategies destroy the temporality of data. This is intolerable because temporality is an important characteristic for time series distinguishing from other data. On the other hand, simply cutting the time series will result in loss of key characteristics carried by the original data. For example, there is usually local similarity between time series. If such similarity is not considered in time series cutting, clustering the time series blocks may become ineffective.

Roughly, there are two rules to split time series: (1) Keep the temporal relationship for every sub-series. (2) Ensure that each sub-series can represent the original time series to a certain extent. The former guarantees each sub-series is still a time series with temporal dynamics and the latter guarantees that the effect of each sub-series is not too bad.

One of effective sampling methods is interval sampling which satisfies the above two rules. Let $S = \{s_1, s_2, ..., s_L\}$ be a time series of length L, and t be the interval, this method samples time series into T sub-series where $T = t + 1$: $S^p = \{s_p, s_{p+t}, ...s_{p+it}, ...\}, 1 \leq p \leq T, p + it \leq L$. The length of each sub-series is $\lfloor \frac{L}{T} \rfloor$ except the first sub-series whose length is $\lceil \frac{L}{T} \rceil$. We take the same interval sampling for all time series.

With interval sampling, each sub-series reflects the original time series from different aspects, which can enhance diversity of clusterers. After splitting time series, T datasets $X^p, 1 \leq p \leq T$ are generated. The model thus employs T clusterers to cluster these datasets. It is obviously that the size of dataset for our model is reduced to $N \times L$.

3.2 Single Clusterer

As mentioned above, every single clusterer of DFECTS uses an autoencoder-based deep clustering model. This deep clustering model contains two parts: autoencoder and clustering algorithm.

For autoencoder, a bidirectional GRU and fully connected layers are used as shown in Fig. 2.

More precisely, the autoencoder includes an encoder composed of 1 bi-GRU and 4 fully connected layers and a decoder composed of 4 fully connected layers. The decoder does not use GRU because it increases time consumption without improving the performance according to our experiments. Encoder encodes data layer by layer into feature space: $Z^p = e(X^p)$, where function $e(\cdot)$ denotes the encoding process. Decoder decodes features and gets simulated data: $\overline{X}^p = d(Z^p)$, where function $d(\cdot)$ denotes the decoding process. The task for autoencoder is to make $\overline{X}^p$ as similar as possible to X^p. The loss function is

$$L^p_{reconstruction} = ||X^p - \overline{X}^p||_2^2 \tag{1}$$

To make the model more robust, Gaussian noise is added to time series just as many existing algorithms do [15,20]. Then the reconstruction loss function is changed to

$$L_{reconstruction}^p = ||X - d(e(X^p + X_{noise}^p))||_2^2 \tag{2}$$

For clustering module, FCM [8] is used as a basic clustering algorithm for its fast convergence ability. Given assignment matrix U, its objective function is describe as

$$\min_{u_{ij}, y_k} \sum_{i=1}^{N} \sum_{j=1}^{K} (u_{ij})^m ||x_i - c_j||_2^2 \tag{3}$$

$$s.t. \quad \sum_{j=1}^{K} u_{ij} = 1, 0 < u_{ij} < 1$$

where u_{ij} is the i-th row and the j-th column of U indicating the probability that data x_i belongs to the j-th cluster with centroid c_j, and $m \geq 1$ is a weighted exponent. According to Eq. 3 with features Z^p provided by encoder and m set to 2, the loss function of FCM in our model is

$$L_{cluster}^p = \sum_{i=1}^{N} \sum_{j=1}^{K} (u_{ij}^p)^2 ||z_i^p - c_j^p||_2^2 \tag{4}$$

$$s.t. \sum_{j=1}^{K} u_{ij}^p = 1, i = 1, ..., N$$

where z_i^p denotes Z_{i*}^p indicating the features of the i-th time series, c_j^p denotes C_{j*}^p indicating the centroid of the j-th cluster, N is the size of dataset and K is the number of clusters. By using the Lagrange multiplier procedure, u_{ij}^p is solved as

$$u_{ij}^p = \frac{1}{||z_i^p - c_j^p||_2^2 \sum_{j=1}^{K} \frac{1}{||z_i^p - c_j^p||_2^2}} \tag{5}$$

and c_j^p is solved as

$$c_j^p = \frac{\sum_{i=1}^{N} (u_{ij}^p)^2 z_i^p}{\sum_{i=1}^{N} (u_{ij}^p)^2} \tag{6}$$

At the beginning, the cluster centroid matrix C^p is initialized by K-means. The assignment matrix U^p and cluster centroid matrix C^p are updated iteratively during training. Finally, the clusterer uses assignment matrix U^p to calculate the labels of time series

$$s_i^p = argmax_{1 \leq j \leq K}(u_{ij}^p) \tag{7}$$

3.3   Ensemble Clusterer

Given T clusterers, by combining all clusterers, the reconstruction loss function of ensemble clusterer is

$$L_{reconstruction} = \sum_{p=1}^{T} L_{reconstruction}^{p} \tag{8}$$

and the clustering loss function of ensemble clusterer is

$$L_{cluster} = \sum_{i=1}^{N} \sum_{j=1}^{K} \sum_{p=1}^{T} (u_{ij}^{p})^2 ||z_i^p - c_j^p||_2^2 \tag{9}$$

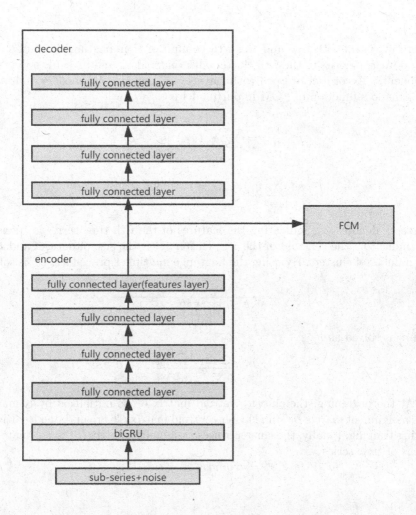

Fig. 2. Autoencoder of single clusterer

Just like many other deep clustering algorithms, we jointly utilize the reconstruction loss function and clustering loss function. So, the objective function is

$$L = \beta L_{reconstruction} + \gamma L_{cluster} \tag{10}$$

where β and γ are hyperparameters.

So far, the model simply adds up all clusterers. To achieve better performance, we introduce the fitness matrix.

Fitness Matrix. Normally, each clusterer learns a different feature space because of the diversity. That is, some clusterers may have a high probability of correct label to some time series, but a high probability of wrong label to other time series. It can be described as the different fitness of the clusterer to different time series. Clusterer should not be deeply involved in those time series that have poor fitness. Therefore a fitness matrix $\Theta \in \mathbb{R}^{N \times (T)}$ is introduced. Depending on this matrix, the clustering loss function is modified to

$$L_{cluster} = \sum_{i=1}^{N} \sum_{j=1}^{K} \sum_{p=1}^{T} \theta_{ip}(u_{ij}^p)^2 ||z_i^p - c_j^p||_2^2$$

$$s.t. \sum_{j=1}^{K} u_{ij}^p = 1, i = 1, ..., N \tag{11}$$

where θ_{ip} denoting the i-th row and the p-th column of Θ is the fitness between the p-th clusterer and the i-th time series.

Using similar derivation, u_{ij}^p is solved as

$$u_{ij}^p = \frac{1}{||z_i^p - c_j^p||_2^2 \sum_{j=1}^{K} \frac{1}{||z_i^p - c_j^p||_2^2}} \tag{12}$$

and c_j^p is solved as

$$c_j^p = \frac{\sum_{i=1}^{N} \theta_{ip}(u_{ij}^p)^2 z_i^p}{\sum_{i=1}^{N} \theta_{ip}(u_{ij}^p)^2} \tag{13}$$

The parameters of autoencoder W are updated as

$$W^* = W - \alpha \nabla_W L \tag{14}$$

where $\alpha > 0$ is a diminishing learning rate and L is the objective function in our model.

The detailed mechanism to calculate the fitness matrix will be introduced later in Sect. 3.3. The training process for single clusterer is showed as Algorithm 1.

Algorithm 1. the training of single clusterer

Input: Input data matrix $\{X^1, X^2, ..., X^K\}$, Centroid matrix of cluster $\{C^1, C^2, ..., C^K\}$, Parameters $W = \{W^1, W^2, ..., W^K\}$
  1: Using K-means to initialize C^p
  2: Pretrain autoencoder
  3: **repeat**
  4:    Update u_{ij}^p by Eq. 12
  5:    Update W by Eq. 14
  6:    Update c_j^p by Eq. 13
  7: **until** exceed the maximum iterations
Output: clustering assignment matrix $U = \{U^1, U^2, ..., U^K\}$

Label Aligning. Zhou and Tang [39] provide a label aligning algorithm only for hard clustering, and cannot be used in our clustering design, which applies soft clustering. Inspired by their algorithm, we propose a new label aligning algorithm specially for soft clustering. Generally, compared to hard clustering with 0-1 matrix, soft clustering can offer more details about relationships between clusters and time series using assignment matrix U. To make full use of assignment matrix U and fitness matrix Θ, a correlation score $Corr$ is designed in our model to represent the similarity between clusters from different clusterers. Given two clusters C_n^a and C_m^b, the correlation score $Corr$ is

$$Corr(C_n^a, C_m^b) = \sum_{i=1}^{N} min(\theta_{ia} u_{in}^a, \theta_{ib} u_{im}^b) \tag{15}$$

where C_j^i denotes the j-th cluster of the i-th clusterer and $Corr \in [0,1]$. The closer the $Corr$ is to 1, the better the two clusters match. If $min(\theta_{ip} u_{ia}^a, \theta_{ip} u_{ib}^b)$ is small, either θ is small, which means clusterer should not be deeply involved in this time series or u is small, which means this time series is not shared by C_n^a and C_m^b. The correlation scores between each pair of clusters of different clusterers are computed. Then, the pair of clusters with the largest correlation score are matched. Such a process is conducted repeatedly until all the clusters are matched. The algorithm is showed as Algorithm 2.

Note that, in the following, we suppose to use the first clusterer as a benchmark and have aligned labels for all clusterers. After label aligning, the label of the i-th time series s_i^* can be calculated by

$$q_{ij} = \sum_{p=1}^{T} \theta_{ip} u_{ij}^p \tag{16}$$

$$s_i^* = argmax_{1 \le j \le K}(q_{ij})$$

where q_{ij} is the score of the i-th time series to the j-th cluster.

The Mechanism of Θ. According to Eq. 16, given the i-th time series, the voting result s_i^* denoted as j^* and the corresponding assignment value $u_{ij^*}^p$ $(1 \le$

Algorithm 2. label aligning for fuzzy C-means

1: **for** $i = 1 \rightarrow K, j = 1 \rightarrow K$ **do**
2:     $Corr_{ij} = \sum_{i=1}^{N} min(\theta_{ia} u_{in}^a, \theta_{ib} u_{im}^b)$
3: **end for**
4:   $\Gamma = \Phi$
5: **while** $\Gamma \neq \{C_1^b, C_2^b, ..., C_K^b\}$ **do**
6:     $(u, v) = arg(max(Corr_{ij}))$
7:     Match(C_u^a, C_v^b) /*C_u^a is matched to C_v^b*/
8:     Delete $Corr_{u*}$
9:     Delete $Corr_{*v}$
10:    $\Gamma = \Gamma \cup \{C_v^b\}$
11: **end while**

$p \leq T$), clusterers having small $u_{ij*}^p{}^*$ are considered to have poorer fitness to the i-th time series compared to other clusterers having high $u_{ij*}^p{}^*$. We can design a strategy to decrease the former's fitness value and increase the latter's fitness value without changing the voting result under single training epoch. According to Eq. 13, time series having small fitness value cannot affect the computing of the offset of cluster centroid too much. Therefore, by using fitness matrix, clusterers just focus on time series which has large fitness value and generate different cluster centroids with high reliability. Because clusterers have good capability (large $u_{ij*}^p{}^*$) to cluster time series with large fitness values, by combining all clusterers, the model will be further improved.

The strategy is shown below. We first use a naive version of voting algorithm

$$g_{ij} = \sum_{p=1}^{T} u_{ij}^p$$

$$s_i' = argmax_{1 \leq j \leq K}(g_{ij})$$

(17)

Different from q_{ij}, g_{ij} is independent of θ because we believe that in this step every clusterer is equal. Given the i-th time series, the voting result s_i' denoted as j' and the corresponding assignment value $u_{ij'}^p$, to find those clusterers with larger $u_{ij'}^p$, $u_{ij'}^p$ is compared to $\frac{g_{ij}}{T}$ for $1 \leq p \leq T$. That is, if $u_{ij'}^p > \frac{g_{ij}}{T}$, $u_{ij'}^p$ is considered to be larger and the p-th clusterer has a good fitness to the i-th time series. In fact, $\frac{g_{ij}}{T}$ is an average score. In extreme cases, either there are no clusterers having poor fitness or there is only one clusterer having good fitness. θ_{ip} is updated as

$$\theta_{ip}^* = \begin{cases} 1 - (1 - \theta_{ip}) * \left(1 - max_{1 \leq j \leq K}(u_{ij}^p)\right), if \ u_{ij'}^p > \dfrac{g_{ij}}{T} \\ \theta_{ip} * max_{1 \leq j \leq K}(u_{ij}^p), if \ u_{ij'}^p < \dfrac{g_{ij}}{T} \end{cases}$$

(18)

where $max_{ij}^p = max_{1 \leq j \leq K}(u_{ij}^p)$ indicates the ability of the clusterer dealing with this time series. If p-th clusterer has poor ability to tackle i-th time series, the u_{ij}^p

---

Algorithm 3. the DFECTS method

Input: Input data matrix X, Centroid matrix of cluster $\{C^1, ..., C^p, ..., C^T\}$, Parameters $W = \{W^1, W^2, ..., W^T\}$

  1: split X into $\{X^1, X^2, ..., X^T\}$
  2: Using K-means to initialize C^p respectively
  3: Pretrain autoencoder for all clusterers
  4: Set $\Theta = \mathbb{1}$
  5: **repeat**
  6:     Update each clusterer by Algorithm 1
  7:     **if** time to update Θ **then**/*Update Θ*/
  8:         align labels by Algorithm 2
  9:         **for** each clusterer p and each x_i **do**
10:             Update θ_{ip}^* by Eq. 18
11:         **end for**
12:     **end if**
13: **until** exceed the maximum iterations
14: Get s^* by Eq. 16
Output: labels s^*

---

will degenerate to $\frac{1}{K}$ for $1 \leq j \leq K$ since $\sum_{j=1}^{K} u_{ij}^p = 1$, and max_{ij}^p will be equal to $\frac{1}{K}$ which is small. So we can conclude that, the larger max_{ij}^p is, the better ability the clusterer has. For the update of θ, we are more willing to accelerate the increase for the first condition or slow the decrease for the second condition if the clusterer has a good ability. The matrix Θ is set to 1 as default. If the first condition always triggers, θ will converge to 1. Otherwise, if the second condition triggers frequently, θ will converge to 0. After updating the fitxness matrix, the voting result remains unchanged under single training epoch, but it guides the next epoch of training. The whole process of DFECTS is showed as Algorithm 3.

4   Experiment

In this section, we apply the proposed DFECTS model to time series clustering and evaluate the performance on ten datasets with three widely-used metrics. Our experiments consist of three parts: evaluation of clustering performance and comparisons with existing methods, hyperparameter analysis, and visualization analysis.

4.1   Experimental Settings

In consideration of the number of clusterers and the efficiency of our model, we adopt the following splitting settings: 1) If the length of time series is greater than 1000, taking into account the efficiency, we split it into ten sub-series. 2) If the length of time series is greater than 300, we split it into several sub-series with a length close to 100. 3) Otherwise, we split the time series with interval t set to 2 or 3. The protocol may be fine-tuned toward different datasets.

As mentioned above, we employ a bi-GRU layer and 4 fully connected layers as the encoder and 4 fully connected layers as the decoder. In our experiments, we set the units per fully connected layer of the encoder to $[400, 200, 100, 50]$ and $[100, 200, 400, l]$ to decoder where l is the length of sub-series. The last layer of encoder is feature layer. DFECTS performs well under this skeleton which is unchanged for all experiments. The epoch of pretraining is set to 20 as default and the epoch of fine tuning is set to 100 as default. The β and γ of Eq. 10 are set to 1 and 0.1 respectively. With further tuning, the performance could be improved. Considering that the update of Θ may bring fluctuations in other parameters, we set the update frequency to 5 epochs according to some experiments. The experiments are conducted on a workstation equipped with a NVIDIA GTX 2070 GPU. To reduce the impact of random initialization, we ran each experiment five times and report the mean of results.

Table 1. The details of UCR time series datasets, including the size of datasets, the number of clusters, and the length of time series.

datasets	size	clusters	length
ACSF1	100	10	1460
Beef	30	5	470
DistalPhalanxTW	139	6	80
FaceFour	88	4	350
Fungi	186	18	201
HandOutlines	370	2	2709
Lightning7	73	7	319
Mallat	2345	8	1024
PowerCons	180	2	144
ProximalPhalanxOutlineCorrect	291	2	80

4.2   Datasets

We conduct the experiments on ten UCR [13] time series datasets. The size of datasets, the number of clusters, and the length of time series are listed in Table 1. Since our experiments are based on unsupervised learning, we omit the training set and utilize test set in this study. These datasets cover different sizes and different lengths of time series, which can effectively verify the performance of each model.

4.3   Performance Evaluation and Comparison

We compare DFECTS with seven clustering methods, including one traditional clustering algorithm (K-means [23]), one non-deep clustering algorithm for time

Table 2. Performance comparison of baseline methods and DFECTS under NMI on ten datasets.

Dataset	K-means	DCN	DEC	IDEC	k-Shape	DTC	DTCR	d-FCM	DFECTS
ACSF1	0.3292	0.3490	0.4429	0.4412	0.4082	0.4413	0.3450	0.4304	*0.4936*
Beef	0.2673	0.2599	0.2390	0.2509	0.2182	0.2742	0.3092	*0.3297*	0.3265
DistalPhalanxTW	0.4644	0.4794	0.4349	0.4247	0.4808	0.4184	0.4482	0.5195	*0.5224*
FaceFour	0.4522	0.4275	0.4487	0.4066	0.3698	0.4723	0.4939	0.4733	*0.4952*
Fungi	0.8989	0.8930	0.9131	0.8989	0.6395	0.7220	0.8796	0.9002	*0.9267*
HandOutlines	0.1846	0.1801	0.1935	0.1609	0.2386	0.1838	0.0877	0.1875	*0.2765*
Lightning7	0.4809	0.4576	0.5395	0.4587	*0.5739*	0.4060	0.5269	0.5225	0.5522
MALLAT	0.9170	0.9368	0.8072	0.9285	0.8702	0.8947	0.5686	0.9151	*0.9601*
PowerCons	0.6400	0.5751	0.6056	0.5658	0.1346	0.4388	0.4087	0.6723	*0.6965*
ProximalPhalanxOutlineCorrect	0.1407	0.1407	0.1156	0.1081	0.1407	0.1425	*0.1601*	0.1501	0.1501

Table 3. Performance comparison of baseline methods and DFECTS under ACC on ten datasets.

Dataset	K-means	DCN	DEC	IDEC	k-Shape	DTC	DTCR	d-FCM	DFECTS
ACSF1	0.2500	0.2300	0.2900	0.3240	0.3080	0.3500	0.2480	0.3340	*0.3920*
Beef	0.3333	0.3000	0.3133	0.3333	0.3400	0.3400	*0.4067*	0.3667	0.4000
DistalPhalanxTW	0.5612	0.5827	0.5396	0.5137	0.5180	0.5424	0.5655	0.5813	*0.6115*
FaceFour	0.5909	0.5796	0.6250	0.5886	0.5614	0.6000	*0.6477*	0.6318	0.6364
Fungi	0.8065	0.8118	0.8194	0.8129	0.1914	0.5441	0.7516	0.8226	*0.8548*
HandOutlines	0.7310	0.7254	0.7398	0.7142	0.7950	0.7670	0.7054	0.6541	*0.8081*
Lightning7	0.5205	0.5343	0.5479	0.5014	0.3890	0.4575	0.5370	0.5342	*0.5562*
MALLAT	0.8222	0.8239	0.7619	0.8239	0.5096	0.8000	0.5568	0.9190	*0.9765*
PowerCons	0.9278	0.9111	0.8944	0.9056	0.6889	0.8500	0.8500	0.9356	*0.9444*
ProximalPhalanxOutlineCorrect	0.6460	0.6461	0.6357	0.6323	0.6460	0.6357	*0.6509*	0.6495	0.6495

Table 4. Performance comparison of baseline methods and DFECTS under RI on ten datasets.

Dataset	K-means	DCN	DEC	IDEC	k-Shape	DTC	DTCR	d-FCM	DFECTS
ACSF1	0.6061	0.5731	0.7325	0.7591	0.7056	0.7480	0.5858	0.7428	*0.7834*
Beef	0.6506	0.6069	0.6198	0.6575	0.5926	0.6437	*0.6708*	0.6138	0.6690
DistalPhalanxTW	0.8238	0.8240	0.7971	0.7914	0.8151	0.6683	0.8145	0.8392	*0.8406*
FaceFour	0.7226	0.7304	0.7608	0.7519	0.7134	0.7518	0.7568	0.7617	*0.7696*
Fungi	0.9722	0.9709	0.9704	0.9722	0.7188	0.8462	0.9672	0.9700	*0.9774*
HandOutlines	0.6063	0.6015	0.6182	0.5960	0.6637	0.6422	0.5833	0.5462	*0.6890*
Lightning7	0.8143	0.7911	0.8203	0.8050	0.8074	0.7520	0.8014	0.8265	*0.8303*
MALLAT	0.9593	0.9615	0.9270	0.9608	0.9156	0.9420	0.8639	0.9692	*0.9889*
PowerCons	0.8652	0.8371	0.8101	0.8280	0.5716	0.7455	0.7436	0.8813	*0.8945*
ProximalPhalanxOutlineCorrect	0.5411	0.5411	0.5353	0.5334	0.5411	0.5353	*0.5431*	*0.5431*	*0.5431*

series (k-Shape [27]), three non-time-series deep clustering methods (DCN [34], DEC [32] and IDEC [15]) and two deep clustering methods for time series (DTC [24], DTCR [20]). Besides, deep FCM (d-FCM) algorithm utilized in single clusterer is added to our experiments.

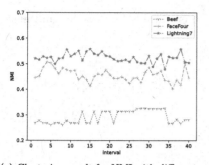
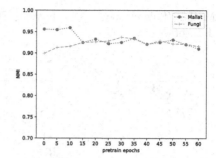

(a) Clustering result for NMI with different sampling interval.

(b) Clustering result for NMI with different pretraining epoch.

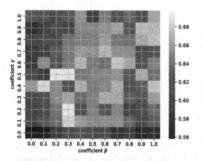
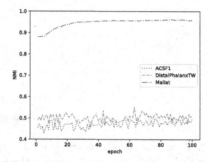

(c) Clustering result for NMI with different β and γ on dataset PowerCons.

(d) Clustering result for NMI during fine tuning.

Fig. 3. Hyperparameter analysis for sampling interval, pretraining epoch, *beta* and *gamma* using NMI as criterion.

To evaluate the performance, we employ three widely-used external metrics: Rand Index (RI) [28], Normalized Mutual Information (NMI) [9] and unsupervised clustering accuracy (ACC). Table 2, Table 3, Table 4 present NMI, ACC and RI of nine methods on ten benchmark datasets, respectively. The best clustering result is highlighted as boldface.

As showed in these tables, DFECTS obviously outperforms the other methods on all three clustering metrics. More precisely, among all ten datasets, DFECTS achieves seven best results for NMI, seven best results for ACC, and nine best results for RI.

Furthermore, several insights can be observed. Firstly, the performance of deep-learning-based methods is superior to other methods. This implies that feature extracting by autoencoder achieves great success. Secondly, DTC and DTCR have poor performance on some datasets. This should be caused by over-fitting due to the complex neural network. Thirdly, compared to d-FCM algorithm of single clusterer, DFECTS has a great improvement in every aspect, which confirms the effectiveness of the mechanism in DFECTS. Lastly, more superiority is achieved by DFECTS on large-scale datasets like HandOutlines and Mallat. This clearly shows that, DFECTS is suitable for complex and massive datasets.

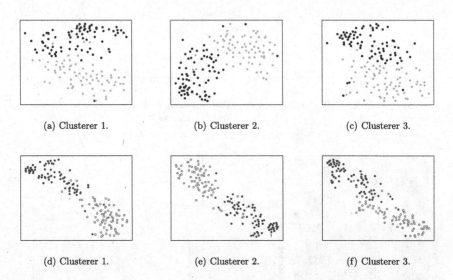

<table>
<tr><td>(a) Clusterer 1.</td><td>(b) Clusterer 2.</td><td>(c) Clusterer 3.</td></tr>
<tr><td>(d) Clusterer 1.</td><td>(e) Clusterer 2.</td><td>(f) Clusterer 3.</td></tr>
</table>

Fig. 4. Visualization of clusters. (a)–(c) shows the initial state of the cluster distribution in three clusterers. (d)–(e) shows the final state of the cluster distribution in the same clusterers. (Color figure online)

4.4 Hyperparameter Analysis

Sampling Interval Analysis. To investigate the effect of sampling interval t, we measure the clustering results of NMI on dataset Beef, FaceFour and Lightning7 with varied value of t in Fig. 3(a). Grid search is applied in this experiment and t is searched in grid of $[1, 40]$ with step of 1. According to these curves, peaks can be observed from dataset FaceFour and Lightning7 when t is small. This benefits from the important role of autoencoder. Generally, the smaller the t is, the longer length the sub-series has. And sub-series contain more time series information with longer length, which autoencoder can capture excellently. Note that the size of dataset Beef is small which is unfriendly to autoencoder and causes poor results in small t. Besides, peaks also occur when t is large on three datasets. A possible reason is that pattern is found by time series splitting.

Pretraining Epoch Analysis. We conduct an experiment to study the performance of DFECTS on different pretraining epochs. The parameter is searched in grid of $[0, 60]$ with step of 1. Figure 3(b) shows the clustering results of NMI on dataset Mallat and Fungi. Actually, DFECTS does not rely heavily on pretraining. Dataset Mallat achieves a high score even without pretraining. This is because we jointly utilize reconstruction loss function and clustering loss function in fine tuning phase and the objective function works well in feature extracting. As the result keeps unchanged when the parameter increases, we set the pre-training epoch to 5–50 for all datasets.

Parameter β and γ Analysis. In order to study the effect of reconstruction loss function and clustering loss function, we perform an additional experiment on the parameter β and γ which are searched in grid of $[0, 1]$ with step 0.1, respectively. Figure 3(c) presents the NMI score with varied value of β and γ on dataset PowerCons. By focusing on the row where γ is 0.0, we can observe that the score is poor regardless of the value of β. This demonstrates the effectiveness of clustering loss function. In addition, by observing the column with β equal to 0.0, we can conclude that reconstruction loss function also plays an important role. NMI reaches the highest value combining reconstruction loss function and clustering loss function with $\beta = 0.3$ and $\gamma = 0.2$. Note that we set $\beta = 1$ and $\gamma = 0.1$ for other experiments without fine tuning.

Fine Tuning Epoch Analysis. Figure 3(d) reports the convergence of our model on three datasets ACSF1, DistalPhalanxTW and Mallat in fine tuning phase. The whole phase lasts for 100 epochs and after a few iterations, we can observe the NMI score becomes stable with slight fluctuation. This indicates the effect of fast convergence of FCM.

4.5 Visualization Analysis

Besides numerical measurement, we also evaluate the performance via visualization analysis. We conduct the experiment on dataset PowerCons with two clusters. The technique t-SNE [22] is utilized to map the learned representations into two-dimensional data. Figure 4 illustrates the visualization of these two clusters with three clusterers. The first row shows the distribution at beginning and the second row shows the distribution at final state. At the beginning, these two clusters are discrete and have many data crossed together without a clear demarcation line. At the final state, the data are clearly grouped into two clusters. Additionally, a red outline is added to these points with fitness value less than 0.5. Most points with red outline appear on the border of clusters where the model is easy to make mistakes. This exactly demonstrates the effectiveness of fitness matrix. Figure 4(f) also shows that, clusterer 3 performs poorly on the border without using fitness matrix. However, by exploiting the fitness matrix, it can ignore these points appearing on the border in part and distinguishes these two clusters successfully. Note that if one clusterer has small fitness value to one point, there must be another clusterer having large fitness value to this point. Then, the model will have great performance by integrating these clusterers.

5 Conclusion

In this paper, we propose a novel deep fuzzy ensemble clusterer model called DFECTS to cluster time series. We integrate the autoencoder and fuzzy C-means algorithm into the basic clusterer, which enables the model to encode time series data, and construct a data distribution that easily to be clustered. To handle the problem of label mismatch, we propose a label aligning method to overcome such inherent defect of ensemble clusterer. Moreover, we propose the fitness matrix for weighted voting algorithm to deal with the complex and massive time series. In comparison with existing approaches, our model achieves superior performance on ten datasets with three metrics, as shown in extensive experiments.

To the best of our knowledge, this should be the first deep ensemble clusterer for time series, and it can be further improve in many directions. For example, we can investigate more time series splitting approaches. Additionally, studying small patterns described in the experiment of sampling interval analysis is also an interesting direction.

References

1. Affeldt, S., Labiod, L., Nadif, M.: Spectral clustering via ensemble deep autoencoder learning (SC-EDAE). Pattern Recogn. **108**, 107522 (2020)
2. Aghabozorgi, S., Shirkhorshidi, A.S., Wah, T.Y.: Time-series clustering - a decade review. Inf. Syst. **53**, 16–38 (2015)
3. Bagherinia, A., Minaei-Bidgoli, B., Hosseinzadeh, M., Parvin, H.: Reliability-based fuzzy clustering ensemble. Fuzzy Sets Syst. **413**, 1–28 (2021)
4. Bai, L., Liang, J., Guo, Y.: An ensemble clusterer of multiple fuzzy k-means clusterings to recognize arbitrarily shaped clusters. IEEE Trans. Fuzzy Syst. **26**(6), 3524–3533 (2018)
5. Barragan, J.F., Fontes, C.H., Embiruçu, M.: A wavelet-based clustering of multivariate time series using a multiscale SPCA approach. Comput. Ind. Eng. **95**, 144–155 (2016)
6. Batista, G.E., Keogh, E.J., Tataw, O.M., De Souza, V.M.: CID: an efficient complexity-invariant distance for time series. Data Min. Knowl. Disc. **28**(3), 634–669 (2014). https://doi.org/10.1007/s10618-013-0312-3
7. Berndt, D.J., Clifford, J.: Using dynamic time warping to find patterns in time series. In: KDD Workshop, Seattle, WA, USA, vol. 10, pp. 359–370 (1994)
8. Bezdek, J.C.: Pattern Recognition with Fuzzy Objective Function Algorithms. Springer, New York (2013). https://doi.org/10.1007/978-1-4757-0450-1
9. Cai, D., He, X., Han, J.: Document clustering using locality preserving indexing. IEEE Trans. Knowl. Data Eng. **17**(12), 1624–1637 (2005)
10. Chang, J., Wang, L., Meng, G., Xiang, S., Pan, C.: Deep adaptive image clustering. In: Proceedings of the IEEE International Conference on Computer Vision, pp. 5879–5887 (2017)
11. Chen, D., Lv, J., Zhang, Y.: Unsupervised multi-manifold clustering by learning deep representation. In: Workshops at the Thirty-First AAAI Conference on Artificial Intelligence (2017)
12. Cho, K., et al.: Learning phrase representations using RNN encoder-decoder for statistical machine translation. In: Proceedings of the Empirical Methods in Natural Language Processing, pp. 1724–1734. ACL (2014)
13. Dau, H.A., et al.: Hexagon-ML: the UCR time series classification archive, October 2018. https://www.cs.ucr.edu/~eamonn/time_series_data_2018/

14. Fischer, B., Buhmann, J.M.: Bagging for path-based clustering. IEEE Trans. Pattern Anal. Mach. Intell. **25**(11), 1411–1415 (2003)
15. Guo, X., Gao, L., Liu, X., Yin, J.: Improved deep embedded clustering with local structure preservation. In: IJCAI, pp. 1753–1759 (2017)
16. Kuncheva, L.I., Vetrov, D.P.: Evaluation of stability of k-means cluster ensembles with respect to random initialization. IEEE Trans. Pattern Anal. Mach. Intell. **28**(11), 1798–1808 (2006)
17. Li, J., Izakian, H., Pedrycz, W., Jamal, I.: Clustering-based anomaly detection in multivariate time series data. Appl. Soft Comput. **100**, 106919 (2021)
18. Liao, T.W.: Clustering of time series data - a survey. Pattern Recogn. **38**(11), 1857–1874 (2005)
19. Lim, P., Goh, C.K., Tan, K.C.: Evolutionary cluster-based synthetic oversampling ensemble (eco-ensemble) for imbalance learning. IEEE Trans. Cybern. **47**(9), 2850–2861 (2016)
20. Ma, Q., Zheng, J., Li, S., Cottrell, G.W.: Learning representations for time series clustering. In: Advances in Neural Information Processing Systems, vol. 32, pp. 3781–3791 (2019)
21. Ma, T., et al.: Multiple clustering and selecting algorithms with combining strategy for selective clustering ensemble. Soft. Comput. **24**(20), 15129–15141 (2020)
22. Van der Maaten, L., Hinton, G.: Visualizing data using t-SNE. J. Mach. Learn. Res. **9**(11), 2579–2605 (2008)
23. MacQueen, J., et al.: Some methods for classification and analysis of multivariate observations. In: Proceedings of the Fifth Berkeley Symposium on Mathematical Statistics and Probability, Oakland, CA, USA, vol. 1, pp. 281–297 (1967)
24. Madiraju, N.S., Sadat, S.M., Fisher, D., Karimabadi, H.: Deep temporal clustering: fully unsupervised learning of time-domain features. arXiv preprint arXiv:1802.01059 (2018)
25. Min, E., Guo, X., Liu, Q., Zhang, G., Cui, J., Long, J.: A survey of clustering with deep learning: from the perspective of network architecture. IEEE Access **6**, 39501–39514 (2018)
26. Ng, A., et al.: Sparse autoencoder. CS294A Lecture Notes **72**(2011), 1–19 (2011)
27. Paparrizos, J., Gravano, L.: k-shape: efficient and accurate clustering of time series. In: Proceedings of the 2015 ACM SIGMOD International Conference on Management of Data, pp. 1855–1870 (2015)
28. Rand, W.M.: Objective criteria for the evaluation of clustering methods. J. Am. Stat. Assoc. **66**(336), 846–850 (1971)
29. Sarma, B., Kumar, R., Tuithung, T.: Optimised fuzzy clustering-based resource scheduling and dynamic load balancing algorithm for fog computing environment. Int. J. Comput. Sci. Eng. **24**(4), 343–353 (2021)
30. Tao, Z., Liu, H., Li, S., Ding, Z., Fu, Y.: Marginalized multiview ensemble clustering. IEEE Trans. Neural Netw. Learn. Syst. **31**(2), 600–611 (2019)
31. Trosten, D.J., Strauman, A.S., Kampffmeyer, M., Jenssen, R.: Recurrent deep divergence-based clustering for simultaneous feature learning and clustering of variable length time series. In: 2019 IEEE International Conference on Acoustics, Speech and Signal Processing (ICASSP), ICASSP 2019, pp. 3257–3261. IEEE (2019)
32. Xie, J., Girshick, R., Farhadi, A.: Unsupervised deep embedding for clustering analysis. In: International Conference on Machine Learning, pp. 478–487. PMLR (2016)
33. Xue, Z., Du, J., Du, D., Lyu, S.: Deep low-rank subspace ensemble for multi-view clustering. Inf. Sci. **482**, 210–227 (2019)

34. Yang, B., Fu, X., Sidiropoulos, N.D., Hong, M.: Towards k-means-friendly spaces: simultaneous deep learning and clustering. In: International Conference on Machine Learning, pp. 3861–3870. PMLR (2017)

35. Yang, X., Deng, C., Zheng, F., Yan, J., Liu, W.: Deep spectral clustering using dual autoencoder network. In: Proceedings of the IEEE/CVF Conference on Computer Vision and Pattern Recognition, pp. 4066–4075 (2019)

36. Zhang, J., et al.: Self-supervised convolutional subspace clustering network. In: Proceedings of the IEEE/CVF Conference on Computer Vision and Pattern Recognition, pp. 5473–5482 (2019)

37. Zhang, Q., Wu, J., Zhang, P., Long, G., Zhang, C.: Salient subsequence learning for time series clustering. IEEE Trans. Pattern Anal. Mach. Intell. **41**(9), 2193–2207 (2018)

38. Zhang, R., Li, X., Zhang, H., Nie, F.: Deep fuzzy k-means with adaptive loss and entropy regularization. IEEE Trans. Fuzzy Syst. **28**(11), 2814–2824 (2019)

39. Zhou, Z.H., Tang, W.: Clusterer ensemble. Knowl. Based Syst. **19**(1), 77–83 (2006)

Energy-Aware Smart Task Scheduling in Edge Computing Networks with A3C

Dan Wang[1], Liang Liu[1(✉)], Binbin Ge[1], Junjie Qi[2], and Zehui Zhao[2]

[1] Zhongguancun Laboratory, Beijing, People's Republic of China
{wangdan,liul,gebinbin}@zgclab.edu.cn
[2] Beijing University of Posts and Telecommunications, Beijing,
People's Republic of China
{qijunjie,zehuizhao}@bupt.edu.cn

Abstract. With the rapid development of the Internet of Things, a huge amount of data is generated from mobile users. Mobile Edge Computing (MEC) extends cloud computing capabilities to the edge of the network which enables real-time and low-latency services for mobile users. However, how to intelligently schedule tasks to save energy consumption in an edge computing environment remains a challenge. In this paper, we formulate a dynamic offloading optimization problem to find out how to minimize system energy consumption based on a two-layer heterogeneous edge cloud cellular architecture. In addition, we model the problem as a Markov Decision Process (MDP) and employ the asynchronous advantage actor-critic (A3C) algorithm to propose a suitable scheduling strategy. The simulation results verify the effectiveness of our proposed method.

Keywords: task scheduling · Asynchronous Advantage Actor-critic (A3C) · Mobile Edge Computing (MEC)

1 Introduction

With the rapid development of mobile communication technology and the Internet of Things, terminal devices and data are increasing explosively. The energy consumption of the system has significantly increased. However, energy consumption is a key optimization goal for task scheduling decisions [1–3]. On the one hand, offloading tasks to the Mobile Edge Computing (MEC) server can save device battery power because computing does not have to be done locally. On the other hand, the device needs to consume energy to offload data to the server [4]. When the energy consumed for local execution is less than the energy required for offloading to MEC, the task will be processed locally. Otherwise, the device will offload the task to MEC [5].

Wireless heterogeneous networks play an important role in 5G high transmission capacity. And wireless backhaul link is a cost-effective and feasible solution.

Supported by organization x.

In [6], they consider a Mobile User (MU) can offload its computing tasks to a Small Base Station (SBS) MEC server, and the SBS is connected to a Macro Base Station (MBS) through a wireless backhaul link. The network architecture minimizes the system computing overhead. [7] considers a multi-access edge computing system with forward and backward constraints and proposes an energy consumption optimization problem for joint radio resource allocation and offloading decisions. [8] studies the power-delay tradeoff problem of MEC in backhaul-constrained wireless networks and proposes an online intelligent computing and communication resource allocation algorithm. In [9], they study the relationship between server deployment and user offloading in wireless edge networks and wireless backhaul links. And they propose a joint optimization based on genetic algorithms.

Many articles study energy consumption optimization in edge cloud systems. In [10], an energy optimization task offloading strategy based on differential evolution optimization is proposed. By analyzing the total number of computing bits and total energy consumption of the system, differential evolution optimization is performed to achieve low energy consumption and high computing bits. [11] proposes a cloud-based multi-computing resource collaborative task offloading model, designs an adaptive computing strategy, and proposes a particle swarm optimization algorithm for energy consumption. [12] proposes a centralized algorithm based on average sample approximation to solve the random optimization problem of minimizing the cost of the MEC network system. It designs a distributed algorithm for practical implementation to find an almost optimal strategy to minimize information exchange between MBS and SBS. In [13], a jointly allocating computing resources and bandwidth resources algorithm is proposed to optimize energy consumption. And an approximate algorithm with energy discretization is proposed to reduce the complexity of the problem. However, most of the above articles make task-scheduling decisions offline. Offline decisions can generally achieve nearly global optimal optimization results. However, in practical applications, it is unrealistic. Although online decisions can only get approximate optimal results, decision results can be obtained in real-time, which is consistent with actual production applications.

To achieve online dynamic scheduling, online learning is used to get task scheduling strategies. In [14], an adaptive application-aware task scheduling algorithm is proposed in the heterogeneous edge cloud network. The proposed scheduling algorithm provides QoS for applications and improves overall scheduling performance and resource utilization. [15] combines deep reinforcement learning and online learning to estimate network conditions and server load in real-time, allocates tasks dynamically to edge servers and combines circular mechanisms with deep reinforcement learning to allocate resources based on the time sensitivity of the task.

In this paper, an energy-aware dynamic task scheduling strategy is proposed to optimize the long-term average energy consumption of the system. Jointly considering delay constraints and resource allocation, a multi-constraint optimization problem is formulated. And an algorithm based on deep reinforcement

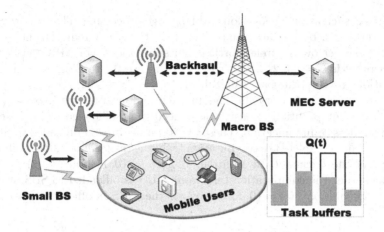

Fig. 1. Architecture of the two-tier small cell network.

learning is proposed to solve this problem. The main contributions of this work are summarized as follows:

- Proposing a two-layer heterogeneous edge cloud cellular architecture with backhaul links, and proposing a dynamic offloading optimization problem to optimize the long-term energy consumption of the system.
- Proposing a dynamic task scheduling algorithm based on Asynchronous Advantage Actor-Critic (A3C) to solve this problem. The algorithm achieves higher exploration efficiency and optimization effects through asynchronous work of each edge node, interactive updates with the central agent, and interactive feedback training between the actor network and the critic network.

Simulation results show that the proposed scheduling strategy can optimize the system's long-term average energy consumption while satisfying the task's delay constraints.

2 System Model and Problem Formulation

2.1 System Architecture

In this paper, we consider a two-tier small cell network (TSCN) architecture, as illustrated in Fig. 1. The TSCN consists of a MBS denoted by n, multiple SBSs denoted by $\mathbf{S} = \{1, 2, ..., m, ..., M\}$, and a set of MUs denoted by $\mathbf{U} = \{1, 2, ..., i, ..., I\}$. Between a SBS and the MBS, there is a backhaul link, which relays the transmission from the SBS to the MBS. We deploy multiple MEC servers with powerful computation capacity at the edge of base stations (BSs). We assume that the index set of the time slots is $\mathbf{T} = \{1, 2, ..., t, ..., T\}$, and the length of each time slot is Δt. We further assume that intensive and indivisible tasks arrive at the MUs independently and identically distributed (i.i.d.), which include tasks such as image location and language processing. The task from MU i can be described by its data size as $d_i(t)$ and its latency requirement as $\tau_i(t)$. Each MEC server owns wireless orthogonal subcarriers. For MEC servers, the subcarriers set is $\mathbf{K} = \{1, 2, ..., k, ..., K\}$. To facilitate the reading, we summarize the notations used in this paper in Table 1.

Due to the short battery life and limited computation resource of mobile devices, some computation-intensive tasks need to be offloaded to MEC servers through wireless communication. These tasks can be directly offloaded to SBS. If the SBS can not complete the computation task, the task will be further offloaded to MBS via the backhaul. In addition, $x_{i,m}^k(t) = \{0,1\}$ and $x_{i,n}^k(t) = \{0,1\}$ denotes whether the computation task from MU i is offloaded to SBS or MBS. For the backhaul, $y_{m,n}^k(t) = \{0,1\}$ means whether SBS m offload task to MBS.

Table 1. Summary of key notations

Notation	Description
B	The bandwidth of all the subcarriers
$d_i(t)$	Size of task data from MU i
$\tau_i(t)$	Tolerable latency for task from MU i
$p_i^{tr}(t)$	Transmission power of MU i
$p_m^{tr}(t)$	Transmission power of SBS m
$\eta^i(t)$	CPU cycles required for processing 1 bit of task on MU i
$\eta^m(t)$	CPU cycles required for processing 1 bit of task on SBS m
$\eta^n(t)$	CPU cycles required for processing 1 bit of task on MBS
F_m	Total computation capacity of the SBS m
F_n	Total computation capacity of the MBS
$f_i(t)$	computation capacity allocated to task from MU i
N_0	Density of thermal noise

2.2 Transmission Model

We first introduce the transmission model. For task offloaded from MU i to MEC s via one hop, the transmission rate is denoted as Eq. (1), where $\alpha_k(t)$ is the bandwidth of subcarrier k. Power gain between MU i and MEC servers s is denoted by $g_{i,s}(t) = \Xi \cdot [d(i,s)]^{-\gamma}$. Ξ is a constant related to the path loss, and γ is the path-loss exponent, $d(i,s)$ refers to the euclidean distance between MU i and MEC servers s. According to the transmission model proposed before, the time and energy consumption for transmitting the task via one hop can be computed respectively as Eqs. (2) and (3).

$$r_{i,s}^k(t) = \alpha_k(t)log_2(1 + \frac{p_i^{tr}(t)g_{i,s}(t)}{N_0}) \tag{1}$$

$$T_{i,s}^{tr}(t) = \frac{d_i(t)}{r_{i,s}^k(t)} \tag{2}$$

$$E_{i,s}^{tr}(t) = p_i^{tr}(t)T_{i,s}^{tr}(t) \tag{3}$$

If the SBS m is constrained by its computation capacity, it will offload the task to the MBS. The transmission rate, required time, and energy cost of transmission from SBS m to the MBS via backhaul link are as follows.

$$r_{m,n}^k(t) = \alpha_k(t)log_2(1 + \frac{p_m^{tr}(t)g_{m,n}(t)}{N_0}) \tag{4}$$

$$T_{m,n}^{tr}(t) = \frac{d_i(t)}{r_{m,n}^k(t)} \tag{5}$$

$$E_{m,n}^{tr}(t) = p_m^{tr}(t)T_{m,n}^{tr}(t) \tag{6}$$

Considering that MEC servers have different computing capacities, some tasks that arrived but have not yet been executed will be queued in the task buffer at each MEC server. In this paper, rather than consider the queue lengths of the task buffers, we introduce transmission rate control parameters θ_m and θ_n to control the transmission rate of MEC servers. The task that reaches the MEC server cannot exceed its computational capacity.

$$\sum_{i \in U} \sum_{k \in K} x_{i,m}^k(t)r_{i,s}^k(t) \leq \sum_{k \in K'} y_{m,n}^k(t)r_{m,n}^k(t) + \frac{F_m}{\theta_m \eta^m} \tag{7}$$

$$\sum_{i \in U} \sum_{k \in K} x_{i,n}^k(t)r_{i,s}^k(t) + \sum_{m \in S} \sum_{k \in K} y_{m,n}^k(t)r_{m,n}^k(t) \leq \frac{F_n}{\theta_n \eta^n} \tag{8}$$

Equation (7) guarantees the total bits of tasks transmitted to SBS can not exceed the sum of the maximum computing rate on SBS and tasks transmitted from SBS to MBS. Equation (8) guarantees the total bits of tasks transmitted from MUs and SBSs to MBS cannot exceed the maximum computation rate of the MBS. Then we express the value of $y_{m,n}^k(t)$ as

$$y_{m,n}^k(t) = \begin{cases} 0, if \ \sum_{i \in U} \sum_{k \in K} x_{i,m}^k(t)r_{i,s}^k(t) \leq \frac{F_m}{\theta_m \eta^m} \\ 1, if \ \sum_{i \in U} \sum_{k \in K} x_{i,m}^k(t)r_{i,s}^k(t) > \frac{F_m}{\theta_m \eta^m} \end{cases} \tag{9}$$

2.3   Computing Model

In this section, we consider the task computing process locally and on edge. We assume that MU i computes only one task at the same time, so the local computing delay of the task from MU i is computed as

$$T_i^{com}(t) = \frac{d_i(t)\eta^i}{F_i(t)} \tag{10}$$

and the energy cost in time slot t is given by

$$E_i^{com}(t) = p_i^{com}(t)T_i^{com}(t) \tag{11}$$

where $p_i^{com}(t)$ is the CPU power of user device MU i, $p_i^{com}(t) = w_i^{com}F_i^3$, w_i^{com} is the effective switched capacitance of the CPU at MU i.

For edge computing, the difference depends on the computation resource allocated to the task from MU i. The delay and energy cost are denoted as Eqs. (12) and (13) respectively.

$$T_{i,s}^{com}(t) = \frac{d_i(t)\eta^s}{f_i(t)} \tag{12}$$

$$E_{i,s}^{com}(t) = p_s^{com}(t)T_{i,s}^{com}(t) \tag{13}$$

where $p_s^{com}(t)$ is the CPU power of MEC s, $p_s^{com}(t) = w_s^{com}f_s^3$, w_s^{com} is the effective switched capacitance of the CPU at MEC s.

2.4   Problem Formulation

In summary, task processing can be divided into three categories, local calculating, offloading directly to SBS or MBS, and offloading from SBS to MBS. The total delay and energy cost in time slot t for the task from MU i can be expressed as Eqs. (14) and (15).

$$T_i(t) = \begin{cases} T_i^{com}(t), & \text{if } x_{i,m}^k(t) = 0, x_{i,n}^k(t) = 0 \\ T_{i,s}^{tr}(t) + T_{m,n}^{tr}(t) + T_{i,s}^{com}(t), & \text{if } y_{m,n}^k(t) = 1 \\ T_{i,s}^{tr}(t) + T_{i,s}^{com}(t), & \text{otherwise} \end{cases} \tag{14}$$

$$E_i(t) = \begin{cases} E_i^{com}(t), & \text{if } x_{i,m}^k(t) = 0, x_{i,n}^k(t) = 0 \\ E_{i,s}^{tr}(t) + E_{m,n}^{tr}(t) + E_{i,s}^{com}(t), & \text{if } y_{m,n}^k(t) = 1 \\ E_{i,s}^{tr}(t) + E_{i,s}^{com}(t), & \text{otherwise} \end{cases} \tag{15}$$

To minimize the averaged sum energy consumption, we formulate the problem as a constrained optimization. The task offloading decision is restricted by constraint $C1$, which allows each task to choose only one approach to complete computation: locally, remotely on MBS or SBS. Bandwidth allocation is constrained by $C2$, while $C3$ ensures that each task is completed within an acceptable latency. During task offloading, there are constraints of computation resource allocation. As $C4$ and $C5$, the computation resources allocated to tasks cannot exceed the total resources. Additionally, we have already explained constraints $C6$ and $C7$ in the above system model and will not provide further explanations in this section.

$$\min \frac{1}{T} \sum_{t=0}^{T-1} \sum_{i \in \mathbf{U}} E_i(t)$$

s.t.

$$C1: \ x_{i,m}^k(t), x_{i,n}^k(t), y_{m,n}^k(t) \in \{0,1\}$$

$$\sum_{m \in \mathbf{S}} \sum_{k \in \mathbf{K}} x_{i,m}^k(t) \cdot \sum_{k \in \mathbf{K}} x_{i,n}^k(t) = 0, \forall i \in \mathbf{U}$$

$$C2: \ \sum_{k \in \mathbf{K}} (x_{i,m}^k(t) + x_{i,n}^k(t)) \alpha_k(t) + \sum_{k \in \mathbf{K}} y_{m,n}^k(t) \alpha_k(t) \le B$$

$$C3: \ T_i(t) \le \tau_i(t), \forall t \in \mathbf{T} \qquad\qquad (16)$$

$$C4: \ \sum_{i \in \mathbf{U}} \sum_{k \in \mathbf{K}} x_{i,m}^k(t)(1 - y_{m,n}^k(t)) f_s(t) \le F_m, \forall m \in \mathbf{S}$$

$$C5: \ \sum_{i \in \mathbf{U}} \sum_{k \in \mathbf{K}} x_{i,n}^k(t) f_s(t) + \sum_{m \in \mathbf{S}} \sum_{k \in \mathbf{K}} y_{m,k}(t)^n f_{m,n}(t) \le F_n$$

$$C6: \ \sum_{i \in \mathbf{U}} \sum_{k \in \mathbf{K}} x_{i,m}^k(t) r_{i,s}^k(t) \le \sum_{k \in \mathbf{K}'} y_{m,n}^k(t) r_{m,n}^k(t) + \frac{F_m}{\theta_m \eta^m}$$

$$C7: \ \sum_{i \in \mathbf{U}} \sum_{k \in \mathbf{K}} x_{i,n}^k(t) r_{i,s}^k(t) + \sum_{m \in \mathbf{S}} \sum_{k \in \mathbf{K}} y_{m,n}^k(t) r_{m,n}^k(t) \le \frac{F_n}{\theta_n \eta^n}$$

3 Asynchronous Deep Reinforcement Learning Based Offloading Algorithm

In this section, we propose an Asynchronous Deep Reinforcement Learning Offloading (ADRLO) algorithm to solve the optimization problem with constraints in a dynamic and complex environment. Our offloading algorithm employs model-free Deep Reinforcement Learning (DRL) because of its ability to handle complex environments. DRL deals with problems by placing learning agents in the environment to achieve goals.

3.1 Markov Decision Process

Because the state of MEC in the current time slot is influenced only by the offloading decision and resource allocation in the previous time slot, we define the problem as a Markov Decision Process (MDP). This MDP process is represented as $M = \{S, A, P, r\}$, where S denotes the state space observed from the environment, A represents the action space in which the agent can choose an action and interact with the environment, P is the state transition probability, and r is the reward function.

State Space. The state space consists of three parts as $S_t = \{\mathbf{D}(t), \mathbf{T}(t), \mathbf{F}(t)\}$. $\mathbf{D}(t)$ and $\mathbf{T}(t)$ represent the data size and tolerable latency of tasks at time slot t. $\mathbf{F}(t)$ represents the computation resources of the MEC servers at time slot t.

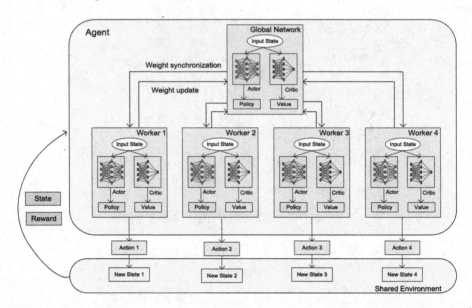

Fig. 2. Architecture of A3C.

Action Space. We focus on the offloading decision and resource allocation. Therefore, the action space $A(t)$ at time slot t is represented as $A_t = \{\mathbf{X(t)}, \mathbf{Y(t)}, \mathbf{f(t)}\}$. $\mathbf{X(t)}$ and $\mathbf{Y(t)}$ denotes the offloading decision for the first and second hops. $\mathbf{f(t)}$ means the computation resources allocated to the task when it is computed at the edge node.

Transition Probability. $P(S_{t+1}|S_t, A_t)$ The state $S_t \in S$ at moment t takes action with a certain probability $A_t \in A$ and then the state of moves to $S_{t+1} \in S$ at next moment $t+1$.

Reward Function. r_t represents the immediate reward, and the agent receives a reward based on the action taken. To minimize the system energy cost, we set the reward negatively related to the objective function. To promote satisfying constraints, when the constraint is not satisfied, the reward is smaller than otherwise. So we define the immediate reward as

$$r_t = \begin{cases} -E_i(t), & \text{if constraints are satisfied} \\ -2\max E_i(t), & \text{otherwise} \end{cases} \tag{17}$$

3.2 Task Scheduling Algorithm Based on A3C

There are several DRL algorithms, including both value-based and policy-based algorithms. Value-based algorithms, such as DQN, optimize based on value functions. However, when dealing with large state and action spaces, it will lead to high complexity. Policy-based algorithms, on the other hand, are more effective for high-dimensional and continuous action spaces, but the evaluation process can be inefficient and may lead to local optimum solutions. Due to the high-dimensional state and large action space in TSCN, we have chosen the A3C algorithm, which combines the advantages of value-based and policy-based algorithms. In addition, it utilizes a multi-threading strategy to improve efficiency.

Algorithm 1. A3C based Offloading Decision Algorithm

1: **Initialization:**
   Assume global shared parameter vectors θ_a and θ_c
   Assume current thread parameter vector θ'_a and θ'_c
   Set global shared counter $T = 0$ and thread step counter $t = 1$
2: **repeat**
3:    Reset gradients: $d\theta_a \leftarrow 0$ and $d\theta_c \leftarrow 0$
4:    Synchronize thread-specific parameters $\theta'_a = \theta_a$ and $\theta'_c = \theta_c$
5:    Set $t_{start} = t$ and Get start state S_t
6:    **repeat**
7:       Perform at according to policy $\pi(A_t|S_t; \theta'_a)$
8:       Receive reward r_t and new state S_{t+1}
9:       $t \leftarrow t + 1$
10:      $T \leftarrow T + 1$
11:   **until** terminal S_t or $t - t_{start} == t_{max}$
12:   **if** S_t is terminal state **then**
13:      $R = 0$
14:   **else**
15:      $R = V(S_t, \theta'_c)$
16:   **end if**
17:   **for** $i \in t - 1, ..., t_{start}$ **do**
18:      $R \leftarrow r_i + \gamma R$
19:      Cumulative Actor's local gradient update
20:      Cumulative Critic's local gradient update
21:   **end for**
22:   Perform asynchronous update of θ_a using $d\theta_a$
23:   Perform asynchronous update of θ_c using $d\theta_c$
24: **until** $T > T_{max}$

Figure 2 illustrates the A3C network model that we have proposed. A3C employs a multi-threading technique consisting of a global network and several workers. They are composed of the actor and critic networks. The actor network selects an action based on the policy, while the critic network generates value estimations for the actions. The actor network utilizes the value to optimize the policy, which aims to maximize the future discounted reward over a long time running. The global network and workers have identical structures but different parameters. Workers operate concurrently in various threadings and explore each thread at random to compute strategy gradients and update parameters.

Regularly, each worker asynchronously updates parameters to the global network, which promptly distributes them to all workers. The parameters of the global network and each thread continuously synchronize and update until the network converges.

In A3C, actor network is used to approximate the policy $\pi(A_t|S_t; \theta'_a)$ and critic network maintains the value function $V(S_t; \theta'_c)$. A3C updates parameters according to discounted returns:

$$R = \sum_{i=0}^{k-1} \gamma^i r_{t+i} + \gamma^k V(S_{t+k}; \theta'_c) \tag{18}$$

where k varies from state to state and has an upper bound of t_{max}. The discount factor γ is in the range of $(0, 1]$, and r_{t+i} is the immediate reward. To reduce the estimation's variance, we use a baseline that is only related to the state. The advantage function is the difference between discounted return and the state-value function, which is defined as

$$A(S_t, A_t) = R - V(S_t; \theta'_c) \tag{19}$$

where k varies from state to state and has an upper bound of t_{max}. The discount factor γ is in the range of $(0, 1]$, and r_{t+i} is the immediate reward. To reduce the estimation's variance, we use a baseline that is only related to the state. The advantage function is the difference between discounted return and the state-value function, which is defined as Eq. (21).

$$L_\pi(\theta'_a) = log\pi(A_t|S_t; \theta'_a)A(S_t, A_t) \tag{20}$$

$$L_v(\theta'_c) = (R - V(S_t; \theta'_c))^2 \tag{21}$$

where k varies from state to state and has an upper bound of t_{max}. The discount factor γ is in the range of $(0, 1]$, and r_{t+i} is the immediate reward. To reduce the estimation's variance, we use a baseline that is only related to the state. The advantage function is the difference between discounted return and the state-value function, which is defined as

$$d\theta_a \leftarrow d\theta_a + \nabla_{\theta'_a} log\pi(A_t|S_t; \theta'_a)A(S_t, A_t) \tag{22}$$

$$d\theta_c \leftarrow d\theta_c + \frac{\partial(R - V(S_t; \theta'_c))^2}{\partial\theta'_c} \tag{23}$$

Finally, asynchronous updates for θ_a and θ_c are performed as shown in Eqs. (24) and (25), where α and β are the learning rates.

$$\theta_a = \theta_a + \alpha d\theta_a \tag{24}$$

$$\theta_c = \theta_c - \beta d\theta_c \tag{25}$$

In summary, each worker interacts with the environment simultaneously, allowing them to acquire independent experiences and obtain different exploration policies. After each worker completes a training episode, it asynchronously updates the parameters to the global network. This architecture solves the problem of empirical timing correlation and efficiently utilizes computational resources to improve training efficiency. The details are shown in Algorithm 1.

4 Simulation

In this section, we simulate the proposed algorithm and prove its accuracy and effectiveness by analyzing the results and comparing it with other algorithms.

Table 2. Parameters

Parameter	Value
The number of SBSs	5
The number of MUs	20
The computation capacity of MBS server	128×10^8 cycle/s
The computation capacity of SBS server	$[16,32] \times 10^8$ cycle/s
The computation capacity of MU	$[4,8] \times 10^8$ cycle/s
The memory resources of edge server R_i^m	$[256\text{--}1024]$ Gb
The bandwidth of the channel	$[10\text{--}30]$ MHz
The processing complexity of task	$[300\text{--}1000]$ cycle/byte
The size of task data	$[2\text{--}5]$ Mb
Signal-to-noise rate $\frac{s}{\sigma^2}$	100

4.1 Simulation Parameters Setting

We consider a scenario with 5 SBSs and 20 MUs, and the total number of time slots is 100. The parameter setting is given in Table 2. The neural network settings are as follows. The learning rate of the actor network is set to 0.0001, and the learning rate of the critic network is set to 0.001. After every five episodes of training, the agent's network parameters are updated. The probability of random exploration is gradually decreasing from 1 to 0.01. Meanwhile, we use RMSProp optimizer.

4.2 Accuracy Analysis

We evaluate the convergence of the proposed algorithm under different neural network hyperparameters and algorithm settings. Figure 3 and Fig. 4 show the proposed algorithm's convergence under different learning rates, which is the step size in each iteration for moving towards the minimum of the loss function. As we can see in Fig. 3, fixed critic learning rate, when the actor learning rate is too large, such as 0.0005, the convergence result can not reach the optimal value. When the actor learning rate is too low such as 0.00005, the convergence speed will become very slow. When the learning rate of the actor network is too high, it may complete learning before the critic network provides accurate value estimates, leading to convergence to a suboptimal strategy. Therefore, we

Fig. 3. Accuracy performance under different actor network learning rate

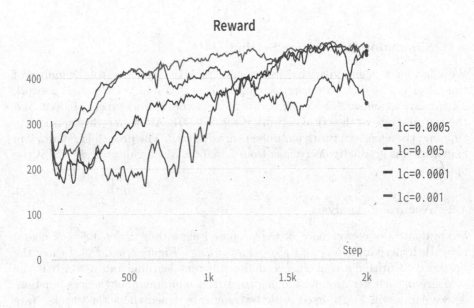

Fig. 4. Accuracy performance under different critic network learning rate

set the learning rate of the actor network to 0.0001, which can have the fastest convergence speed and ensure the convergence effect.

Figure 4 shows the convergence of the proposed algorithm under different values of the critic learning rate, when the learning rate of the actor network is fixed, the convergence speed of the algorithm increases with the increase of the learning rate of the critic network. However, when the critic learning rate is too large, such as 0.005, the algorithm convergence is not optimal. When the critic network learns too quickly, it may not be able to keep up with the changing policies of the actor network, resulting in learning instability and shocks. Therefore, the learning rate of the critic network is set to 0.001 to ensure the convergence results while maximizing the algorithm's convergence.

4.3   Effectiveness Analysis

We compare our proposed algorithm with several baseline methods, including the random offloading algorithm (denoted by RandOff), the offloading algorithm without backhaul links (denoted by NoBackhaul), and the Proximal Policy Optimization (PPO) algorithm based on the AC framework. The random offloading algorithm makes random decisions and randomly selects the computing resources allocated to the task within a range. The offloading algorithm without backhaul links refers to the algorithm proposed not including backhaul links. The AC-PPO algorithm introduces the clipping mechanism of the PPO algorithm into the objective function based on the AC algorithm, limiting policy updates to a certain range, preventing excessive policy updates, and stabilizing the learning process.

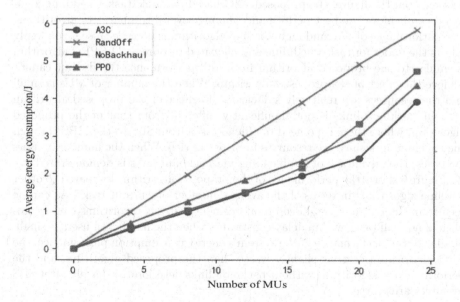

Fig. 5. System energy consumption performance under different numbers of MUs

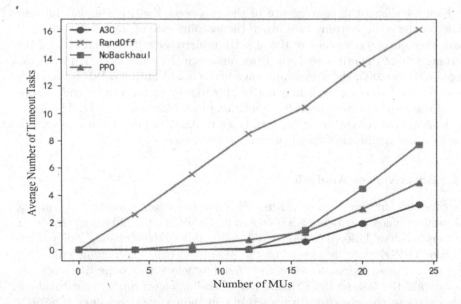

Fig. 6. Number of timeout tasks performance under different number of MUs

Figure 5 shows the performance of the proposed algorithm compared to other baseline algorithms in terms of the average system energy consumption. As can be seen from the figure, the proposed A3C-based dynamic task scheduling algorithm can achieve the best performance among several comparative algorithms. The performance of the random unloading algorithm is poor because it randomly selects the nodes for task scheduling and allocated resources. The PPO algorithm is similar to the proposed algorithm in overall performance trends, but cannot achieve the effect of the proposed algorithm. When the number of MUs is small and the resources are relatively sufficient, the effect of the proposed algorithm without backhaul links is not significantly different from that of the proposed algorithm, where there is no need to offload tasks from SBS to the MBS through the backhaul link since the resources in SBSs are rich. When the number of users increases, the advantage of re-offloading via backhaul links is demonstrated.

Figure 6 shows the performance of the proposed algorithm compared to other baseline algorithms in terms of the average number of timeout tasks. As can be seen from the figure, several deep reinforcement learning algorithms can accomplish almost all tasks within delay constraints when the number of users is small. Similar to the performance of the system's energy consumption performance, the PPO algorithm performs slightly worse than the proposed algorithm, and the performance of algorithms without backhaul links deteriorates sharply after SBS resources are scarce.

5 Conclusion

In this paper, we proposed a task scheduling scheme in the two-tier small cell network based on reinforcement learning. Moreover, We took advantage of A3C to manage the task offloading process. We considered the energy consumption under constrained conditions to improve the quality of service. The simulation results show that the proposed scheduling strategy can get greater long-term returns while satisfying the task's delay limitations.

References

1. Fan, R., Li, F., Jin, S., Wang, G., Jiang, H., Wu, S.: Energy-efficient mobile-edge computation offloading over multiple fading blocks. In: 2019 IEEE Global Communications Conference (GLOBECOM), pp. 1–6 (2019). https://doi.org/10.1109/GLOBECOM38437.2019.9014035
2. Ali, Z., Jiao, L., Baker, T., Abbas, G., Abbas, Z.H., Khaf, S.: A deep learning approach for energy efficient computational offloading in mobile edge computing. IEEE Access 7, 149623–149633 (2019). https://doi.org/10.1109/ACCESS.2019.2947053
3. Guo, M., Wang, W., Huang, X., Chen, Y., Zhang, L., Chen, L.: Lyapunov-based partial computation offloading for multiple mobile devices enabled by harvested energy in MEC. IEEE Internet Things J. 9(11), 9025–9035 (2022). https://doi.org/10.1109/JIOT.2021.3118016
4. Yang, Y., Chen, X., Chen, Y., Li, Z.: Green-oriented offloading and resource allocation by reinforcement learning in MEC. In: 2019 IEEE International Conference on Smart Internet of Things (SmartIoT), pp. 378–382 (2019). https://doi.org/10.1109/SmartIoT.2019.00066
5. Mach, P., Becvar, Z.: Mobile edge computing: a survey on architecture and computation offloading. IEEE Commun. Surv. Tut. 19(3), 1628–1656 (2017). https://doi.org/10.1109/COMST.2017.2682318
6. Pham, Q.V., Le, L.B., Chung, S.H., Hwang, W.J.: Mobile edge computing with wireless backhaul: joint task offloading and resource allocation. IEEE Access 7, 16444–16459 (2019). https://doi.org/10.1109/ACCESS.2018.2883692
7. Chen, J., Guo, X., Chang, Z., Hämäläinen, T.: Resource allocation for multi-access edge computing with fronthaul and backhaul constraints. In: 2021 17th International Symposium on Wireless Communication Systems (ISWCS), pp. 1–6 (2021). https://doi.org/10.1109/ISWCS49558.2021.9562171
8. Sun, Y., Yang, G., Zhou, X.S.: Online intelligent resource management for power-delay tradeoff in backhaul-limited mobile edge computing systems. In: 2019 IEEE SmartWorld, Ubiquitous Intelligence Computing, Advanced Trusted Computing, Scalable Computing Communications, Cloud Big Data Computing, Internet of People and Smart City Innovation (SmartWorld/SCALCOM/UIC/ATC/CBDCom/IOP/SCI), pp. 1019–1024 (2019). https://doi.org/10.1109/SmartWorld-UIC-ATC-SCALCOM-IOP-SCI.2019.00198
9. Song, H., Gu, B., Son, K., Choi, W.: Joint optimization of edge computing server deployment and user offloading associations in wireless edge network via a genetic algorithm. IEEE Trans. Netw. Sci. Eng. 9(4), 2535–2548 (2022). https://doi.org/10.1109/TNSE.2022.3165372

10. Sun, Y., Song, C., Yu, S., Liu, Y., Pan, H., Zeng, P.: Energy-efficient task offloading based on differential evolution in edge computing system with energy harvesting. IEEE Access **9**, 16383–16391 (2021). https://doi.org/10.1109/ACCESS.2021.3052901

11. Aiwen, Z., Leyuan, L.: Energy-optimal task offloading algorithm of resources cooperation in mobile edge computing. In: 2021 4th International Conference on Advanced Electronic Materials, Computers and Software Engineering,(AEMCSE), pp. 707–710 (2021). https://doi.org/10.1109/AEMCSE51986.2021.00146

12. Liu, Y., Xie, S., Yang, Q., Zhang, Y.: Joint computation offloading and demand response management in mobile edge network with renewable energy sources. IEEE Trans. Veh. Technol. **69**(12), 15720–15730 (2020). https://doi.org/10.1109/TVT.2020.3033160

13. Zhao, T., Zhou, S., Song, L., Jiang, Z., Guo, X., Niu, Z.: Energy-optimal and delay-bounded computation offloading in mobile edge computing with heterogeneous clouds. China Commun. **17**(5), 191–210 (2020). https://doi.org/10.23919/JCC.2020.05.015

14. Oo, T., Ko, Y.B.: Application-aware task scheduling in heterogeneous edge cloud. In: 2019 International Conference on Information and Communication Technology Convergence (ICTC), pp. 1316–1320 (2019). https://doi.org/10.1109/ICTC46691.2019.8939927

15. Yuan, H., Tang, G., Li, X., Guo, D., Luo, L., Luo, X.: Online dispatching and fair scheduling of edge computing tasks: a learning-based approach. IEEE Internet Things J. **8**(19), 14985–14998 (2021). https://doi.org/10.1109/JIOT.2021.3073034

BACTDS: Blockchain-Based Fined-Grained Access Control Scheme with Traceablity for IoT Data Sharing

Wei Lu[1], Jiguo Yu[2(✉)], Biwei Yan[1], Suhui Liu[3], and Baobao Chai[4]

[1] School of Computer Science and Technology, Qilu University of Technology, Jinan 250353, People's Republic of China
[2] Big Data Institute, Qilu University of Technology, Jinan 250353, People's Republic of China
jiguoyu@sina.com
[3] School of Cyber Science and Engineering, Southeast University, Nanjing 211102, People's Republic of China
suhuiliu@seu.edu.cn
[4] School of Computer Science and Engineering, Shandong University of Science and Technology, Qingdao 266590, People's Republic of China
bbchai_915@sdust.edu.cn

Abstract. Due to the growing prevalence of Internet of Things (IoT) services, there is an opportunity to leverage the data generated by IoT devices to create greater value. However, how individuals can ensure user privacy and data security when sharing their private data remains a major challenge. To address the above problems, we propose a trackable and fine-grained private data sharing access control scheme (BACTDS) based on blockchain. First, to ensure data security, in our proposed system model, data users can interact with the cloud to obtain ciphertext information only if they have permission information and can decrypt the obtained ciphertext to obtain plaintext information only if they meet access policies. Second, we store the user's identity information in the probability polynomial. With the help of the traceability algorithm, our proposed scheme supports the tracking of malicious users who have leaked private keys. The malicious user's identity is recovered through lagrangian interpolation polynomials and then traced to malicious users, this prevents illegal sharing and misuse of keys. Finally, we provide a formal security proof of our scheme under the assumption of the random oracle model, which demonstrates its provable security. Moreover, experimental results and performance analysis indicate that our scheme (BACTDS) is time-efficient and cost-effective.

Keywords: Internet of Things (IoT) · Traceability · Blockchain · Data Sharing · Access Control

This work was supported in part by the Major Program of Shandong Provincial Natural Science Foundation for the Fundamental Research (ZR2022ZD03), the NSF of China under Grants 62272256 and 61832012, and the Piloting Fundamental Research Program for the Integration of Scientific Research, Education and Industry of Qilu University of Technology (Shandong Academy of Sciences) under Grant 2022XD001.

1   Introduction

The Internet of Things (IoT) technology has a significant impact on our daily lives and is extensively used in various fields such as transportation [1], smart factory [2], and wireless health monitoring [3]. Personal health data generated by wearable IoT devices, for example, might provide individuals with particular benefits as personal digital assets. Encouraging individuals to share their data can benefit society as a whole.

At present, most organizations or individuals use cloud service providers (CSPs) to store tons of data at lower costs, but the premise of this approach is full trust in cloud service providers. However, this assumption exposes outsourced data to substantial safety and privacy risks.

In response to the challenges of data privacy and tracking malicious users in data sharing, we introduce a novel solution called BACTDS, which supports blockchain-based traceability and fine-grained access control. This paper presents several key contributions, including

1) We introduce a novel access control scheme based on blockchain and ABE that provides traceability and privacy preservation for IoT data sharing. Our proposed scheme encourages the sharing of private data to a certain extent, and users who share personal data should be rewarded.
2) In our proposed system model, data users can interact with the cloud to obtain ciphertext information only if they have permission information and can decrypt the obtained ciphertext to obtain plaintext information only if they meet access policies.
3) Our proposed scheme makes it possible to trace tracking malicious users who have leaked private keys. Our scheme puts user information into polynomials, recovers malicious user identities through lagrangian interpolation polynomials, and then traces malicious users, preventing unlawful key sharing and illegal use.
4) The rigorous security analysis of the BACTDS is under the random oracle model, and the performance analysis and experimental comparisons show the superiority and effectiveness of our scheme over existing solutions.

2   Related Works

CP-ABE [4] is an effective solution for secure data sharing in scenarios where users may have different levels of clearance or authority. Therefore, many papers applied CP-ABE to the blockchain. In particular, Yu et al. [5] developed a lightweight hybrid attribute-based signcryption scheme (LH-ABSC) to address the security issues associated with outsourced data storage, such as data leakage and unauthorized access. Liu et al. [6] introduced a novel multi-authority traceable ABE scheme. The proposed scheme aims to enhance security and efficiency by offloading most of the decryption operations to a cloud server, resulting in reduced computational and equipment overhead. To enable multiple users to share the same ciphertext data without exposing the identity and data information of the data participants, Liang et al. [7] use proxy re-encryption technology for the storage of big data.

Some works exploited the blockchain to realize more secure access control for IoT. Tan et al. [8] proposed a scheme that uses blockchain technology in combination with CP-ABE to enable secure and private sharing of electronic medical records (EMRs) with the ability to track and direct revoke access to COVID-19 EMRs. In a similar vein, Zuan et al. [9] aimed at tracing maliciously modified data by storing raw data and transaction data on separate blockchains. Meanwhile, Liu et al. [10] replaced the traditional search server with a blockchain to improve searchable attribute encryption with efficient revocation and decryption.

In summary, while the existing work above explores the potential of applying blockchain technology and some privacy-preserving techniques to contribute data, there are still some challenges in blockchain-based data sharing. These challenges are how to solve the problem of tracing malicious users and how to reduce storage space costs. To solve these challenges, this scheme adopts a Lagrange interpolation polynomial to store user identities, which greatly reduces the cost of storage space. In addition, the solution also discusses privacy protection and traceability issues such as data collection and access control.

3   System Model

The BACTDS system model comprises various components, including the personal health data owner (DO), wearables, data user (DU), cloud storage (CS), authority center (AC), and blockchain (BC), as illustrated in Fig. 1.

1) Authoritative center (AC): The authoritative center is responsible for generating the system's public parameters, denoted as PP, and the master private key, denoted as MSK. The PP is then made public, while the MSK is kept confidential by the authoritative center.
2) Personal health data owner (DO): The personal health data owner integrates the information collected from the wearable device to obtain the information m. Next, they use a symmetric key to encrypt the information m, resulting in the CT_m, which is stored in the CS for future access. Then encrypt the symmetric key k with the access policy (M, ρ) made by yourself to obtain the ciphertext CT. CT uploads blockchain storage. At the same time, DO formulates sales rules on the blockchain.
3) Blockchain (BC): The blockchain stores the encrypted ciphertext CT of k and the sales rules made by the data's owner.
4) Cloud storage (CS): The cloud platform stores the ciphertext CT_m. The restriction on the cloud platform is that only users with tokens can obtain the ciphertext CT_m from it.
5) Data user (DU): Data users are able to obtain encrypted keys and access policies from the blockchain via CT. If the attribute set S possessed by the data user satisfies the access policy of ciphertext CT, then the symmetric key

k can be obtained by decrypting CT. The user can then access the ciphertext CT_m from its corresponding location in the CS and decrypt it using the symmetric key k to obtain their personal medical and health data m. Data is an important digital asset, and DU is willing to pay for private data on DO.

6) Wearables: Wearables are responsible for collecting records of personal information, which are source authenticated by the BLS signature algorithm and then send to the data owner for aggregation.

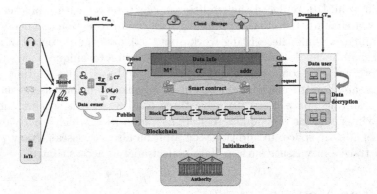

Fig. 1. System model.

4   Design of Our Scheme

The entity interactions of the proposed BACTDS scheme is illustrated in Fig. 2. Firstly, the scheme is initialized to upload the PP to the BC. Then, the symmetric encryption algorithm is employed to encrypt the personal information health data and generate the corresponding ciphertext CT_m. The ciphertext CT_m is uploaded and stored in the CS. Next, the scheme employs ABE that utilizes a ciphertext policy to encrypt the symmetric secret key k. This encryption generates the corresponding ciphertext CT, which is then uploaded and stored on the BC. DU initiates a purchase request to the BC, the BC allows the request and generates the corresponding request permission token. The requested permission information is then transmitted to the CS. The DU sends his token information to the CS, and searches for the license information corresponding to the token, subsequently the CS transfers the ciphertext to the DU.

The detailed description of our proposed blockchain-based fine-grained access control traceability and reward personal health and medical data sharing scheme is as follows, where the symbols used in this scheme are listed in Table 1.

- $Setup(1^\lambda) \rightarrow (PP, MSK)$: The system is initialized with a security parameter λ, where λ determines the system's level of security. We define $(\mathbb{G}_1, +)$

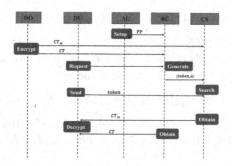

Fig. 2. Interactions among entities.

Table 1. Notations.

Notation	Meaning
$\mathbb{G}_1, \mathbb{G}_2$	Two multiplication cyclic groups
$\mathbb{A}$	A is Access Structure
(M, ρ)	Matrix and map function
PP	System public parameters
MSK	System master key
S	User's attribute set
UID	User's identity
$SK_{UID,S}$	The private key of the user UID
λ	The security parameter
U	The system attribute universe
A, C	A is a adversary, C is challenger
CT_m	Ciphertext corresponding to message

and $(\mathbb{G}_2, *)$. The bilinear pairing function is denoted by $e : \mathbb{G}_1 \times \mathbb{G}_1 \to \mathbb{G}_2$. Additionally, we define two safe hash functions: $H : \{0,1\}^* \to \mathbb{Z}_p$ and $H_1 : \mathbb{G}_1 \to \mathbb{Z}_p$. The attribute universe U is mapped to the group $\mathbb{Z}_p$ by a mapping function denoted as $\Phi : U \to \mathbb{Z}_p$. Randomly chooses $g, w, \mu, \beta, \eta \in \mathbb{G}_1$, $\alpha, a \in \mathbb{Z}_p$, and a probabilistic encryption algorithm (Enc, Dec) with two distinct secret keys k_1 and k_2 (where $k_1, k_2 \in \mathbb{Z}_p$). In addition, it selects the symmetric key encryption/decryption algorithm $(E_{\text{sym}}, D_{sym})$ with secret key k (where $k \in G_2$). Then it initializes the Shamir's threshold scheme $STS_{(k,n)}$ by selecting an order $k-1$ polynomial $f(x)$ and $k-1$ distinct points $\{(x_1, y_1), (x_2, y_2), (x_3, y_3)..., (x_{k-1}, y_{k-1})\}$ (where $f(x_i) = y_i, i \in [1, k-1]$) of the polynomial f(x) for secret storage.

The system's public parameters PP (publicized and stored on the blockchain) and the master key MSK are:

$$PP = (\mathbb{G}_1, \mathbb{G}_2, q, P, g, \mu, \beta, w, \eta, e(g,g)^\alpha, g^a, H, H_1),$$

$$MSK = (\alpha, a, b, k_1, k_2).$$

- $KeyGen(PP, MSK, UID, S = \{attr_i, i \in [1, n] \subseteq \mathbb{Z}_p\}) \rightarrow SK_{UID,S}$: Upon receiving the identity UID and corresponding attribute set S from the data user, the AC calculates the secret key of the user:

$$x = Enc_{k_1}(UID), y = f(x), \zeta = Enc_{k_2}(x||y).$$

And it randomly chooses $b_1, b_2, ..., b_n \in Z_p$, and calculates the decryption key of user $SK_{UID,S}$ as follow: $SK_{UID,S} = \{K_0, K', K_1, K_1', \{K_{\Gamma,1}, K_{\Gamma,2}\}, \Gamma \in [n]\}$ where

$$K_0 = g^{\alpha/(a+\zeta)}w^b, \quad K' = \zeta, \quad K_1 = g^b, \quad K_1' = g^{ab},$$
$$K_{\Gamma,1} = g^{br}, \quad K_{\Gamma,2} = (\mu^{attr_\Gamma}\beta)^{br}\eta^{-(a+\zeta)b},$$

and randomly select a number $x_l \in \mathbb{Z}_p$, and calculate its public key $PK_{UID} = x_l P$. Therefore, the key pair of DU is used by (x_l, PK_{UID}) for the generation of the license token. To enhance computational efficiency, we map the user's UID to the cyclic group $\mathbb{Z}_p$.

- $Encrypt(PP, UID, m, (M, \rho)) \rightarrow (CT_m, CT)$: First, the DO selects the symmetric key encryption/ decryption algorithm (E_{sym}, D_{sym}) with secret key k to encrypt personal health and medical data records m, and calculates the ciphertext as $CT_m = E_{sym_k}(m)$. After encrypting the data, the system calculates $M^* = H(CT_m)$, which is a hash of ciphertext, and uploads it to the BC. The purpose of this is to provide an additional layer of security and prevent tampering with the data by adversaries. Next, the ciphertext CT_m and M^* are sent to the CS and stored. We assume the existence of a matrix M that can generate shares. This matrix has l rows and n columns. For $i = 1, 2, .., l$, a function ρ is defined to label row i of M with an attribute from the S. In the LSSS-based expression, the access policy is defined as (M, ρ), and $s, y_2, ..., y_n \in \mathbb{Z}_P$ are randomly selected to form the secret sharing matrix $v = (s, y_2, y_3, ..., y_n)^T$. It gets the vector of the shares $\lambda = (\lambda_1, \lambda_2, ..., \lambda_l)$ by computing the inner product $\lambda_i = M_i v$, where M_i is the i-th row of M. When a specific policy is used to encrypt k, DO provides access control to the data and allows only data users who meet the policy requirements to access the data. DO can complete encryption by itself. To encrypt k, DO first choose a random element $s \in \mathbb{G}_1$, and then computes $C = ke(g, g)^{\alpha s}$. Finally, get the ciphertext $CT = (CT_1, CT_2)$, where CT_1 contains $\{C, C_0, C_0'\}$, CT_2 contains $\{(M, \rho), C_{i,1}, C_{i,2}, C_{i,3}\}$. The detailed calculation is as follows:

$$C_0 = g^s, \quad C_0' = g^{as}, \quad C_{i,1} = w^{\lambda_i}\eta^{t_i}, \quad C_{i,2} = (\mu^{\rho(i)}\beta)^{-t_i}, \quad C_{i,3} = g^{t_i},$$

and t_i is randomly chosen on $t_i \in Z_p$, $(i \in [1, l])$. Finally, the CT is stored on the BC.

- $TokenGen(PP, UID, address, time) \rightarrow Token$: The token generation algorithm inputs the public parameters PP, user UID, account address, and time, and outputs the license token $Token$. In this phase, smart contracts are utilized to facilitate payments between DO and DU. Once the payment has been successfully made, the smart contract grants a license token to the CS.

Suppose a data user DU is interested in $DO's$ personal health and medical data and accepts the sales rules published by DO on the blockchain, DU initiates a purchase request and then records it on the BC. The data user who made the purchase is denoted as DU_L. When the BC receives the purchase request of DU_L, the BC performs the following steps:

1) The BC verifies whether the balance of $DU_L's$ account address meets the pricing of the data owner's sales rules. If the balance meets the pricing, the next step is performed; Otherwise, the data user's request is rejected.
2) Generate the licence $Token$ for DU_L, where $Token = \{$spe$_L, M^*, PK_{UID}, t_s, t'_s\}$.
   - spe$_L$ is the unique identification of $Token$.
   - M^* is the hash value of personal health medical data.
   - PK_{UID} is the public key of DU_L.
   - t_s is the time stamp of $Token$.
   - t'_s is the expiration time of the token.
3) Compute $\tilde{m} = H(Token)$, encrypt $Token$ with DU'_L public key PK_{UID} as $cl = PK_{UID}(Token)$. Then, send $(cl, \tilde{m})$ to DU_L and send $(Token, \tilde{m})$ to CS.
   - $Decrypt(PP, SK_{UID,S}, CT) \rightarrow k \ or \ \perp$: First, the DU sends his token information $Token$ to the cloud to search the corresponding license information, and then the CS transmits the respective ciphertext CT_m to the DU. Next, the DU obtains CT from the BC. Then the data user calculates $I = i : \rho(i) \in S$ to check if the attribute set S satisfies the access policy. If yes, a set of constants $\{\omega_i \in \mathbb{Z}_p\}_{i \in I}$ such that $\sum_{i \in I} \omega_i M_i = (1, 0, 0, ...0)$ can be gained, where M_i is the i-th row of matrix M. Note that $\sum_{i \in I} \omega_i \lambda_i = s$ if the attribute set S is authorized. Otherwise, the decryption algorithm outputs $\perp$. The following is the calculation process:

$$X = e(K_0, C_0^{K'} C'_0) = e(g, g)^{\alpha s} e(g, w)^{s(a+\varsigma)b},$$
$$Y = \prod_{i \in I} (e(K_1^{K'} K'_1, C_{i,1}) e(K_{\Gamma,1}, C_{i,2}) e(K_{\Gamma,2}, C_{i,3}))^{w_i}$$
$$= e(g, w)^{sb(a+\varsigma)},$$
$$Z = X/Y = e(g, g)^{\alpha s}.$$

Then output the symmetric key to get the encrypted message $k = C/Z$. Finally, the DU decrypts the CT_m obtained from the CS through the obtained symmetric key k, thereby obtaining the desired plaintext information m, where $m = D_{sysmk}(CT_m)$. Verify whether $M^* = H(E_{symk}(m))$ is equal to determine whether the information has been tampered with by the adversary. If the verification equation is satisfied, the data owner will accept the information m. Otherwise, the message m is rejected.

- $KeyforCheck(PP, MSK, SK_{UID,S}) \rightarrow 1 \ or \ 0$: Checks whether the form of the key $SK_{UID,S}$ is intact, and outputs 1 if the key form is intact. Execute the key tracking algorithm, otherwise, the algorithm outputs 0, which indicates that the check failed. The key form check is performed by the authority center AC with the following checks:

$$(1)\ K' \in Z_p^*, K_0, K_1, K_1', K_{\Gamma,1}, K_{\Gamma,2} \in \mathbb{G}_1,$$
$$(2)\ e(K_1', g) = e(K_1, g^a),$$
$$(3)\ e(K_0, g^a g^{K'}) = e(K_1' K_1^{K'}, w)e(g, g)^\alpha.$$

If and only if formulas 1, 2, and 3 are true, the key form can be determined to be intact.

- $Trace(PP, STS(k, n), MSK, SK_{UID,S}) \to UID$ or $\perp$: Once the key-form check for $SK_{UID,S}$ is successfully completed, the algorithm executes the following steps to find the user's UID to determine the real identity of the data consumer.

1) The tracking algorithm starts by extracting the relevant parameters K' from $SK_{UID,S}$, and then uses K' to decrypt and obtain $x||y$, where $x||y = Dec_{k_2}(K')$.
2) Assuming that let x' $= x, y' = y$, if $(x', y') \in \{(x_1, y_1), (x_2, y_2), (x_3, y_3)..., (x_{k-1}, y_{k-1})\}$, the algorithm identifies a malicious user by calculating $UID = Dec_{k_1}(x')$ to get the user's UID.
3) If $(x', y') \notin \{(x_1, y_1), (x_2, y_2), (x_3, y_3)..., (x_{k-1}, y_{k-1})\}$, the algorithm combines the other points $\{(x_2, y_2), (x_3, y_3)..., (x_{k-1}, y_{k-1})\}$ and (x', y') to recover the secret value a_0^* of STS(k,n) by means of Lagrangian interpolation. Determine whether the equation $a_0^* = f(0)$ holds, and if the equation holds, find the malicious user by calculating $UID = Dec_{k_1}(x')$ to obtain the user's UID. Otherwise, the algorithm terminates.

5   Security Analysis

5.1   CPA-Security

Theory: No PPT adversary can selectively defeat our scheme against chosen-plaintext attacks (CPA) given the $q - type$ assumption.

Proof: If there exists a PPT adversary A that can break our scheme with a non-negligible advantage ϵ assuming the security of choice, then the adversary A can use this advantage to construct a challenger C that can solve the q-type problem with a non-negligible advantage ϵ.

5.2   Data Authenticity

IoT wearables are considered untrustworthy, and some unauthorized devices may falsify data records. An intuitive way to solve this issue is using signatures. Specifically, each wearable device uses its own private key si to sign its own transmitted information mi, and the DO use the wearable device's public key

pi to verify the signature σ. However, in IoT applications, there may be tons of devices generating data all the time. Thus, in our scheme, we use BLS aggregate signature algorithm to realize the authentication of source information from IoT wearable devices and improve the verification efficiency of DO. Each wearable device signs the generated data before sending it to the data owner. Then, the DO can perform batch verification. In this way, the source authenticity of the personal medical record is guaranteed efficiently.

5.3 Anonymity

To assure the users' privacy, the system needs to guarantee anonymity of users. Therefore, our scheme anonymized the user's private information in two aspects. On one hand, during the key generation phase, the identity information of the user identity UID is encrypted using a probabilistic encryption algorithm. This type of encryption uses random algorithms, which produce different ciphertexts for the same information. Since probabilistic encryption offers polynomial security, it is virtually impossible for adversaries to derive any information about the original plaintext from the ciphertext, even if they perform calculations or attempt to obtain experimental plaintexts. In other words, the ciphertext completely masks any information about the user's identity, thereby ensuring their anonymity.

On the other hand, it is anonymity protection for the privacy information of data users: in the licensing token generation stage, the DU's public key is generated by randomly selecting a value x_l from $\mathbb{Z}_p$ and calculated $PK_{UID} = x_l P$, the random number $x_l \in \mathbb{Z}_p$ is in the hands of the data users, and spe_L is used to represent the unique identity of the token, which is designed to protect the privacy of DU by not revealing their actual identity information.

6 Experiment Analysis

To establish a clear understanding of proposed scheme's performance, we use the following notations to represent computational costs. C_a, C_m, C_p, and C_{pr} are used to denote the computational overheads of performing point addition operation on $\mathbb{G}_1$ group, point multiplication operation on $\mathbb{G}_1$ group, power operation, and bilinear pairing operation. C_s, C_{po}, and C_h denote the calculation cost of the symmetric encryption and decryption algorithm, the probabilistic encryption and decryption algorithm, and mapping the hash value to $\mathbb{Z}_p$. Other computational costs are ignored. The calculation cost of each stage is shown in Table 2.

To quantify the computational cost of our scheme, we break it down into several phases and analyze the complexity of each phase separately. In the initialization phase, the computation overheads are $2C_p + C_{pr}$. In the key generation phase, the computation overheads are $C_a + (3n + 5)C_p + (2n+4)C_m + 2C_{po}$. In the encryption phase, the computation overheads are $C_h + C_s + (5n + 2)C_p + (2n + 5)C_m$. In the decryption phase, the computation overheads are

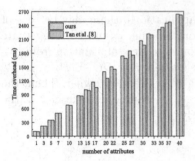

Fig. 3. Encryption time cost comparison.

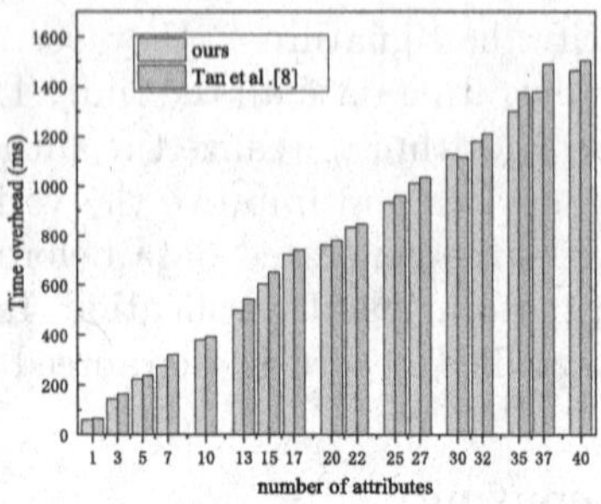

Fig. 4. Decryption time cost comparison.

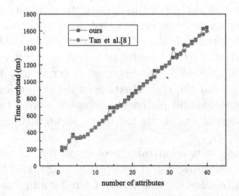

Fig. 5. Key generation time for different numbers of attribute.

Table 2. Computational Overhead.

Scheme	*Computational* Overheads
[8]	$3C_p + C_{pr}$
	$2C_a + (3+n)C_p + (4+n)C_m + nC_h$
	$C_{pr} + (4+n)C_p + (3+n)C_m + nC_h$
	$(3r+6)C_p + (7r+7)C_m + (3r+4)C_{pr}$
	$(2n+1)C_{pr} + 2C_a + (3n+2)C_m + (2n+2)C_p$
Ours	$2C_p + C_{pr}$
	$C_a + (3n+5)C_p + (2n+4)C_m + 2C_{po}$
	$C_h + C_s + (5n+2)C_p + (2n+5)C_m$
	$(2n+2)C_p + C_s + (n+2)C_m + (n+2)C_{pr}$
	$5C_{pr} + 3C_m + 2C_{po}$

$(2n + 2)C_{pr} + C_s + (n + 2)C_m + (n + 2)C_p$, and in the tracing phase, the computation overheads are $5C_{pr} + 3C_m + 2C_{po}$.

Through performance comparison, the calculation cost of our scheme is obviously less than that of Tan et al. [8] in the tracking stage. This is because our scheme converts the user information into a point conforming to the $f(x)$ polynomial through calculation and recovers it by Lagrange interpolation polynomial, while the Tan et al. [8] scheme needs to query whether the user's information is in the registration list before tracing. In the verification phase of tracking, our scheme only needs to check the integrity of the secret key form, which requires a small amount of computation, while the scheme of Tan et al. [8] needs to input the public parameters and intermediate parameters of the user's attribute set for verification, which requires a large amount of computation.

To measure the effectiveness of our proposed scheme, we conducted a simulation experiment. The hardware setup for this experiment included an Intel(R) Core(TM) i5 CPU @2.2 GHz, 8 GB RAM, and the Windows 10 (64b) operating system. For simulating encryption and decryption operations, we used the Java pair-based cryptography library. To perform our experiments, we opted for a supersingular elliptic curve with a base field consisting of 512-bit elements and a group order size of 160 bits. We illustrate the link between key generation time and the number of attributes in Fig. 5. Both schemes show a linear increase in key generation time as the number of attributes increases. The comparison of encryption time cost is illustrated in Fig. 3. As the number of attributes increased, we observed a gradual increase in the encryption time for both schemes. When the number of attributes is 1, 3, 5, and 7, the data encryption time of the two is basically the same. However we discover the BACTDS scheme spends a slightly longer than Tan et al. [8] scheme when the number of attributes is increasing.

Figure 4 presents a comparison of the decryption time cost. It can be observed that both schemes experience a gradual increase in decryption time as the number of attributes increases. However, it is worth noting that the decryption time of the BACTDS scheme is lower compared to that of the Tan et al. [8] scheme. The reason is that in Tan et al. [8] scheme, the cloud service provider performs the decryption delegation function, providing the patients with new ciphertext and decryption parameters. Only users who satisfy the attributes and possess decryption parameters during decryption can decrypt the ciphertext. In addition, the BACTDS scheme stores the user identity in a probability polynomial, which greatly reduces the storage space cost, while Tan et al. [8] the scheme stores the user identity information in the revocation list, which greatly increases the storage space.

7   Discussions and Conclusion

Our paper introduces a novel fine-grained access control scheme called BACTDS, which utilizes blockchain technology to trace data access and control from the data source. To uphold the security of data, in BACTDS, the data user must satisfy both the attribute set of the access policy and possess the token to obtain the ciphertext information from the cloud and decrypt it to obtain the plaintext information. With the help of the traceability algorithm, the BACTDS recovers the identity of malicious users through lagrangian interpolation polynomials and then traces the malicious users, which can prevent illegal data sharing and misuse of secret keys. Our paper outlines the system and security models for the BACTDS scheme and provides formal security proof under the random oracle model. This proof validates the provable security of our proposed scheme. Through experiments and performance analysis, it is proved that our scheme (BACTDS) is more advantageous in terms of storage space and time consumption, and the BACTDS scheme is effective and attractive.

Distributed data-sharing systems are increasingly being recognized as valuable tools in big data applications. In our future work, we plan to conduct a penalty study to investigate the malicious behaviors of various users and ensure the security of data sharing.

References

1. Zhu, F., Lv, Y., Chen, Y., Wang, X., Xiong, G., Wang, F.Y.: Parallel transportation systems: toward IoT-enabled smart urban traffic control and management. IEEE Trans. Intell. Transp. Syst. **21**(10), 4063–4071 (2019)
2. Liu, Y., Dillon, T., Yu, W., Rahayu, W., Mostafa, F.: Noise removal in the presence of significant anomalies for industrial IoT sensor data in manufacturing. IEEE Internet Things J. **7**(8), 7084–7096 (2020)
3. Liu, S., Chen, L., Wang, H., Fu, S., Shi, L.: O³HSC: outsourced online/offline hybrid signcryption for wireless body area networks. IEEE Trans. Netw. Serv. Manage. **19**(3), 2421–2433 (2022)
4. Bethencourt, J., Sahai, A., Waters, B.: Ciphertext-policy attribute-based encryption, pp. 321–334 (2007)
5. Yu, J., Liu, S., Wang, S., Xiao, Y., Yan, B.: LH-ABSC: a lightweight hybrid attribute-based signcryption scheme for cloud-fog-assisted IoT. IEEE Internet Things J. **7**(9), 7949–7966 (2020)
6. Liu, S., Yu, J., Hu, C., Li, M.: Traceable multiauthority attribute-based encryption with outsourced decryption and hidden policy for CIoT. Wirel. Commun. Mob. Comput. **2021**, 1–16 (2021)
7. Liang, K., Susilo, W., Liu, J.K.: Privacy-preserving ciphertext multi-sharing control for big data storage. IEEE Trans. Inf. Forensics Secur. **10**(8), 1578–1589 (2015)
8. Tan, L., Yu, K., Shi, N., Yang, C., Wei, W., Lu, H.: Towards secure and privacy-preserving data sharing for Covid-19 medical records: a blockchain-empowered approach. IEEE Trans. Netw. Sci. Eng. **9**(1), 271–281 (2021)
9. Wang, Z., Tian, Y., Zhu, J.: Data sharing and tracing scheme based on blockchain. In: 2018 8th International Conference on Logistics, Informatics and Service Sciences (LISS), pp. 1–6 (2018)
10. Liu, S., Yu, J., Xiao, Y., Wan, Z., Wang, S., Yan, B.: BC-SABE: blockchain-aided searchable attribute-based encryption for cloud-IoT. IEEE Internet Things J. **7**(9), 7851–7867 (2020)

DeepLat: Achieving Minimum Worst Case Latency for DNN Inference with Batch-Aware Dispatching

Jiaheng Gao and Yitao Hu$^{(\boxtimes)}$

College of Intelligence and Computing, Tianjin University, Tianjin, China
{gjhagonsle,yitao}@tju.edu.cn

Abstract. Deep neural network (DNN) has achieved the state-of-the-art results in multiple fields, and has been widely used to build latency sensitive applications for its high performance. When dispatching requests among GPU machines for DNN execution, the inference system hosted in the cloud needs to guarantee that the maximum latency of all requests, denoted as the worst case latency, is within the latency objectives of the clients. In this paper, we design and implement a request dispatch system, called DeepLat, which distributes client requests among the GPU machines efficiently to minimize the worst case latency of the DNN-based application. DeepLat uses batch-aware dispatch policy to minimize the batch collecting time, proposes duration-based algorithm to reduce the average latency and supports partial-batch dispatching to minimize the waiting time for bottleneck machines. Evaluation shows that compared to existing request dispatch systems, DeepLat can reduce the worst case latency by 37.7% on average without using extra computing resources. Besides, DeepLat achieves the theoretical lower bound for the worst case latency for over 48% workload. With the capability to minimize the worst case latency, DeepLat reduces the total cost of DNN serving system by 43.2% on average.

1 Introduction

In recent year, massive datasets, advanced algorithms and powerful hardware contribute to the rapid development of deep neural network (DNN), which has achieved the state-of-the-art results in multiple complex tasks, such as object detection and product recommendation [1–6]. To leverage the advances in DNN, more and more applications are using DNN models for its high performance [7–12]. For example, when user uploads photos, the social network will run object detection model to detect human faces in each photo [13].

These DNN-based applications are usually deployed in the inference system hosted in the cloud using the classic client-server architecture, where the client sends the requests to the inference system for DNN model execution. The data volume for popular DNN-based applications is usually beyond the processing capability of any individual machine. Therefore, the inference system auto-scales the amount of computing resources (*e.g.*, hundreds of GPU machines) according to the real-time workload [14,15]. Once receiving the requests from the clients, the inference system is responsible to dispatch the request among the allocated machines for model execution.

The DNN inference is often associated with a latency constraint [7–9,12], where the client wants to receive the result within a predefined latency budget. Existing systems define the maximum latency of all requests as the *worst case latency* for the entire workload. Therefore, when dispatching the requests, the inference system should distribute the requests in a proper way to guarantee that the worst case latency of the entire workload is smaller than the latency objective, which is the primary goal of this paper.

To achieve the minimum worst case latency, we design and implement a request dispatch system, called DeepLat, which distributes client requests among computing machines efficiently to minimize the worst case latency for the entire workload.

First, DeepLat distributes the request among GPU machines for model execution in a batch-aware manner. Existing systems [7,9,10] distribute individual request to each GPU machine in rotation and generate batched request at the GPU machine, which unnecessarily increases the batch collecting time. DeepLat takes the batch information of each GPU machine into account and designs a duration-based heuristic to distribute batched request to GPU machine directly, which largely reduces the batch collecting time.

Second, DeepLat supports partial-batch dispatching to further reduce the worst case latency. Being batch-aware reduces the batch collecting time, while allocating batched requests evenly minimizes the average latency. However, this heuristic can *not* achieve the minimum worst case latency, because the batch boundary makes the pipelining between the batch collecting process and the model execution process challenging. DeepLat proposes a threshold-based heuristic to switch a group of requests at batch boundary in the request list to reduce the waiting time of bottleneck machines at the cost of a larger batch collecting time of other machines, which reduces the worst case latency for the entire workload when the latency reduction of bottleneck machines is larger.

Evaluation shows that DeepLat achieves a smaller worst case latency than the state-of-the-art inference system [7] for *all* workload. Specifically, DeepLat reduces the worst case latency by 37.7% on average without using extra computing resources. For certain workload, the worst case latency of existing systems is more than 15 times of DeepLat's. Besides, DeepLat achieves the lower bound for the worst case latency for at least 48.4% workload. For the remaining 51.6% workload, DeepLat might still achieve the lower bound, but can *not* be verified in polynomial time. Ablation study shows that batch-awareness, optimization during the starting phase and partial-batch is necessary to minimize the worst case latency. Besides, DeepLat supports multiple model configurations, while existing systems only support a maximum of two model configurations. With the capability to minimize the worst case latency, DeepLat can reduce 43.2% of the total cost for serving system on average by choosing machine configurations with large throughput, while existing systems have to choose ones with small throughput due to latency constraints.

2   Background and Motivation

In this section, we start with the request dispatch policy of existing systems, followed by three insights that motivate the design of DeepLat: batch-awareness, reducing average latency and partial-batch dispatch capability.

How Existing Systems Dispatch Request? Batching is widely used in inference system to increase hardware throughput [7–12]. It combines multiple client requests into one inference job to invoke a larger GPU kernel, which suits the parallel computation architecture of the underlying GPU. For example, as shown in Table 1, when the batch size is 2, the model throughput is relatively low. When we double the batch size, the runtime duration is *not* doubled, leading to a larger model throughput. Requests in the batch are processed together on the GPU, and the results of requests in the same batch are returned at the same time. Inference system defines the *worst case latency* as the maximum latency of all requests, and schedules the DNN workload to guarantee that the worst case latency is smaller than the latency objectives.

Table 1. The execution duration and model throughput of a given DNN model under various batch sizes, and machines that have used the corresponding configuration in Sect. 2.

Batch size	Execution duration (s)	Model throughput	Machine
2	0.25	8	D
4	0.4	10	E
6	0.5	12	A, B, C
12	0.6	20	F

Therefore, batching, though increasing the model throughput, also increases the worst case latency for the request because of the time taken to collect enough requests to form the batch. Given the request rate for a given DNN model, the scheduler for inference system needs to deal with two problems while using batching. First, the scheduler needs to choose the model configuration (*e.g.,* batch size) for each GPU machine to handle the workload. Second, the scheduler needs to decide the request dispatch policy to distribute requests among the GPU machines.

Existing systems [7–10] have studied the first problem extensively, while paying little attention to the second one. In this paper, we mainly focus on the second problem. Namely, given the request rate information and the model configuration of each GPU machine, how to distribute the requests among these GPU machines to minimize the worst case latency.

In most cases, the underlying GPU could be busy executing the previous batched request when the current batched request is ready, leading to a waiting time for the current one. To deal with the waiting time, each machine maintains a local request queue to store requests, and only starts model executing after the underlying GPU is idle. Therefore, the worst case latency L_{wc} consists of three parts: the model execution time d, the batch collecting time l_c and the potential waiting time l_w which contains scheduling waiting time and gRPC delay.

$$L_{wc} = d + l_c + l_w \tag{1}$$

The model execution time d is related to the batch size and can be derived from offline profiling [7,8,12]. The batch collecting time l_c and the waiting time l_w is decided by the dispatch policy of the inference system. To satisfy the latency objective, the scheduler should choose a proper dispatch policy to reduce the batch collecting time or the waiting time, so as to minimize the worst case latency for the entire workload.

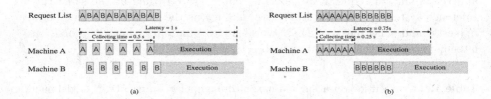

(a)   (b)

Fig. 1. (a) Rotation distribution; (b) Batch-aware distribution.

DeepLat Must be Batch-Aware. When dispatching requests among machines, existing systems [7,8] distribute the request as individual request in rotation between the GPU machines, and generate batched requests at the GPU machine locally. The frontend of the inference system will distribute requests in the rate of machine's maximum throughput to guarantee line-rate on the GPU machine, leading to a minimum waiting time ($l_w = 0$). Meanwhile, the requests are collected by GPU machine's local request queue in rate of maximum throughput, leading to a batch collecting time of $l_c = \frac{b}{t} = \frac{b}{b/d} = d$, where b is the batch size, d is the execution duration and t is the model throughput for the given GPU machine. Existing systems use two times the execution duration as the worst case latency, since $L_{wc} = d + d + 0 = 2d$. For example, Fig. 1(a) illustrates the dispatching result of existing systems for a workload of 24 req/s. The frontend distributes requests among two machines A and B with batch size of 6 and execution time of 0.5 s in rotation. The batch collecting rate of each machine equals its model throughput of 12 req/s, which results in a batch collecting time of $6/12 = 0.5$ s and a waiting time of 0 s. Therefore, under the rotation dispatch policy of existing systems, the worst case latency is $0.5 + 0.5 + 0 = 1.0$ s.

However, we argue that the dispatch policy of existing systems can *not* minimize the worst case latency. The batch collecting time can be reduced if we dispatch requests as batched request from the frontend directly, as opposed to dispatching requests as individual ones in existing systems. For example, as shown in Fig. 1(b), instead of sending individual requests between machines in rotation, the frontend sends a consecutive of 6 requests to machine A to match its batch size before sending the next 6 consecutive requests to machine B. The equivalent batch collecting rate is the total workload (*e.g.*, 24 req/s) instead of the maximum throughput of each machine (*e.g.*, 12 req/s), leading to a worst case latency of $0.5 + 6/24 + 0 = 0.75$ s. Therefore, dispatching batched request can reduce the worst case latency.

Takeaway. Existing systems dispatch requests as individual requests among machines in rotation and generate batched requests at the GPU machines, which increases the batch collecting time, leading to a larger worst case latency. DeepLat must be batch-aware and generate batched requests at the frontend directly to minimize the worst case latency of the entire workload.

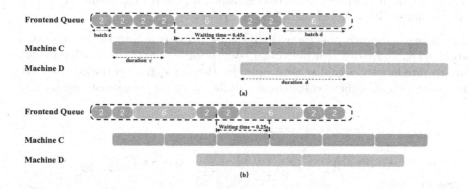

Fig. 2. Two batch-aware dispatch policies with the same worst case latency: (a) dispatch policy 1; (b) dispatch policy 2.

DeepLat Must Take the Average Latency Into Account. The scheduler should dispatch requests among GPU machines with various batch sizes properly to guarantee that the *worst case latency* of the entire workload is smaller than the latency objectives. Meanwhile, each machine from the workload has its *individual latency*. Dispatching policies with the same worst case latency could have various individual latency. For example, for a workload of 20 req/s, the scheduler choose two machines, where machine C has batch size of 6 and execution duration of 0.5 s and machine D has batch size of 2 and execution duration of 0.25 s. Figure 2 shows two batch-aware dispatch policies. Table 2 shows the worst case latency for the entire workload, as well as the individual latency for two machines under the corresponding dispatch policy. Both dispatch policies are batch-aware and have the same worst case latency of 0.8 s. However, the individual latency of machine C in dispatch policy 1 is 0.725 s, while the one in dispatch policy 2 is only 0.425 s.

Table 2. The worst case latency for the entire workload, machine C and D under two dispatch policies in Fig. 2.

Policy ID	Entire workload	Machine C	Machine D
1	0.80	0.725	0.80
2	0.80	0.425	0.80

This is because during the starting phase, dispatch policy 1 sends *four* consecutive batched requests in batch of 2 to machine C before sending batched request to machine B. Though it will *not* affect the worst case latency of the entire workload, the consecutive batched requests increase the waiting time (*e.g.*, 0.45 s) of machine C's following batches. In response to this, dispatch policy 2 only sends *two* consecutive batched requests to machine C before sending batched request to machine D at the starting phase, leading to a smaller waiting time (*e.g.*, 0.25 s) on machine C. Therefore, dispatch policy 2 has a smaller average latency among its assigned machines.

Takeaway. Dispatch policies with the same worst case latency for the entire workload can have different individual latency for each machine, leading to various average latency. DeepLat must optimize the dispatch order during the starting phase to reduce the average latency while minimizing the worst case latency.

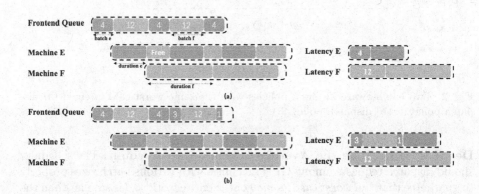

Fig. 3. Partial-batch dispatching example: (a) before using partial-batch; (b) after using partial-batch.

DeepLat Must Support Partial-Batch Dispatching. As discussed above, dispatching request in batch at the frontend can reduce the batch collecting time, leading to a smaller worst case latency. However, dispatching request in full batch across all machine might *not* guarantee the minimum worst case latency. For certain workload, breaking the boundary of batch and dispatching request in partial batch can further reduce the worst case latency.

For example, as shown in Fig. 3, assuming a workload of 30 req/s, the scheduler allocated two machines to deal with the workload, where machine E has batch size of 4 and execution duration of 0.4 s and machine F has batch size of 12 and execution duration of 0.6. Dispatch policy 3 in Fig. 3(a) dispatches requests between machines in full batch, leading to a worst case latency of 1.07 s. Dispatch policy 4 in Fig. 3(b) splits the 5th batch into two parts, and puts the first half in front of the 4th batch, leading to a worst case latency of only 1.0 s.

The intuition is that by switching a certain number of requests at the boundary, we will be able to reduce the waiting time of the bottleneck machine (*e.g.*, machine F) in the cost of a larger batch collecting time of the other machines (*e.g.*, machine E). As long as the latency reduction for the bottleneck machine is larger than the latency increase for the other machines, partial-batch dispatching can reduce the worst case latency for the entire workload. For example, in the above example, partial batch can decrease machine F's individual latency from 1.07 to 1.0 s, while increasing machine E's individual latency from 0.800 to 0.934 s, leading to a lower worst case latency for the entire workload. However, finding the proper requests to switch is challenging, since doing it naively (*e.g.*, switching one request at the boundary between the 3rd and 4th batch in the request list) will only increase the worst case latency.

Takeaway. Switching a certain number of requests at batch boundary to support partial-batch dispatching can reduce the waiting time for the bottleneck machines. DeepLat should support partial-batch dispatching to minimize the worst case latency for the entire workload.

DeepLat is Cost-Effective with Minimum Worst Case Latency. As discussed above, DeepLat dispatch requests in batches, take the average latency into account and support partial-batch dispatching. By doing so, for *each* given input machine configuration, DeepLat can achieve a much lower worst case latency than existing systems due to a smaller batch collecting time and waiting time. DeepLat can potentially use machine configurations with greater throughput, while existing systems have to choose machine configurations with lower throughput to satisfy the latency constraints. By using machines with larger throughput, DeepLat can reduce the total cost.

Takeaway. Reducing the worst case latency for each candidate machine configuration enables DeepLat to choose machine configurations with larger throughput, which reduces the total cost. With such capability, DeepLat is cost-effective.

3   Design

In this section, we start with an overview of DeepLat, followed by a detailed description of its components.

3.1   System Overview

As shown in Fig. 4, DeepLat has three key components: frontend, model repository and server management. Given the request rate for the DNN model, as

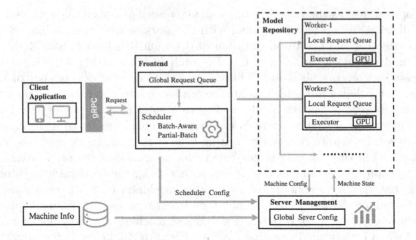

Fig. 4. DeepLat runtime overview.

well as the GPU machines to deal with the workload, the scheduler in frontend generates the dispatch policy to minimize the worst case latency for the entire workload, while the frontend distributes requests to GPU machines for model execution according to the dispatch policy. Clients send requests to the global request queue in the frontend, which then distributes requests to each machine's local request queue. Server Management globally controls the state of the model instance by detecting the system config at frontend.

DeepLat minimizes worst case latency as follows:

▶ It generates batch-aware dispatch policy to minimize the batch collecting time. To do so, DeepLat proposes a duration-based heuristic to decide the dispatch order. Following the dispatch policy, DeepLat distributes batched requests from the global request queue to machine's local request queue directly to reduce the batch collecting time which reduces the average latency of machines serving the workload, while minimizing the worst case latency.

▶ It supports partial-batch in the dispatch policy to minimize the worst case latency. To do so, DeepLat proposes a threshold-based heuristic to opportunistically switch a group of request order at the batch boundary. Though the partial-batch might increase the batch collecting time for certain machines, it can reduce the waiting time of the bottleneck machines, leading to a lower worst case latency for the entire workload.

3.2   Batch-Aware Request Dispatching

To reduce the worst case latency, DeepLat must decide the dispatch policy to distribute requests received at the frontend to GPU machines for model execution, leading to two subtasks. First, when distributing requests to machines, DeepLat needs to choose the request dispatch granularity at the frontend (*e.g.,* distributing requests as individual ones to generate batches at the GPU machine or distributing batched requests directly at the frontend). Second, Among the machines for model execution, DeepLat needs to decide the dispatch order among machines assigned for the workload.

Table 3. The model configuration for three machines.

Machine	Batch size	Execution duration (s)	Model throughput
A	2	0.25	8
B	6	0.50	12
C	6	0.50	12

For the first subtask of request dispatch granularity, DeepLat chooses batched request to reduce the batch collecting time in the worst case latency. Once receiving requests from the clients, the frontend will store requests in the global request queue. Existing systems [7,8] ignore the batch information and distributes the requests to machine's local request queue as individual request in rotation. When enabling batching, each machine can fetch batches of requests at once thus reducing request collection time and communication consumption.

For the second subtask of dispatch order, deciding the optimal dispatch order is known to be NP-hard [7], so an intuitive idea is to use a greedy-based heuristic. The key idea is to calculate the expected latency of a given request on all available GPU machines and then dispatch the request to the GPU machine with the minimum worst case latency. However, as discussed in Sect. 2, the greedy-based heuristic will unnecessarily repeat the machine with smaller execution duration during the starting phase, leading to a larger average latency. For example, assuming a workload of 32 req/s, the scheduler assigns three machines to deal with the workload with configuration shown in Table 3. For the first 16 requests, greedy-based algorithm will allocate req 1–4 as two batches of 2 to machine A, req 5–10 to machine B and req 11–16 to machine C. It will repeat the above order for every 16 requests. The bottleneck of the average latency comes from machine A's two consecutive batches, which unnecessarily increases the waiting time of the second batch on machine A, since it can *not* start execution until the first batch on machine A has finished.

To reduce the average latency while minimizing the worst case latency for the entire workload, DeepLat proposes a duration-based heuristic Algorithm 1 to optimize the starting phase during request dispatching. The key intuition is to distribute the batched requests among machines *evenly*, so the waiting time for machines with small duration is reduced. Suppose the set of machines M_i is S, for the scheduling $order_g$ obtained by the greedy algorithm, Algorithm 1 checks whether the current one keeps the batched requests evenly on each machine (Line 9). If the dispatch order is *not* even, Algorithm 1 will update the dispatch order to make it even (Line 10–11).

Algorithm 1. Batch-Aware Request Dispatching

1: **DistributionScheduling**($order_g, S$)
2: $item, order_d, exec, state \leftarrow [\ ], [\ ], [\ ], [\ ]$
3: $Clock, wcl_d \leftarrow 0, 0$
4: **for** $M_i =< b_i, d_i > in\ S$ **do**
5:    $exec[M_i] = \frac{b_i}{d_i}$
6: **for** o in $order_g$ **do**
7:    $M_i \leftarrow min(item)$
8:    $Clock \leftarrow time_{now}$
9:    **if** $item(M_i)! = item(o)$ **then**
10:       $item[M_i] \leftarrow item[M_i] + d_i$
11:       $order_d.push(M_i)$
12:    **else**
13:       $order_d.push(o)$
14:       $item[o] \leftarrow item[o] + d_o$
15:    $state[M_i] \leftarrow UpdateState(state[M_i], exec[M_i], d_i)$
16:    $wcl_d = max(wcl_d, state[M_i] - Clock)$
17: **return** $order_d, wcl_d$
18:
19: **UpdateState**($state, exec, d$)
20: $Clock \leftarrow time_{now}$
21: **if** $Clock + exec + d < state$ **then**
22:    $state = clock + exec + d$
23: **else**
24:    $state = state + d$
25: **return** $state$

For machine k among a workload containing N machines, Algorithm 1 will use the execution duration of machine k to calculate its *dispatch step* as $\frac{\sum_{i=1}^{N} \frac{1}{d_i}}{\frac{1}{d_k}}$ to update the dispatch order when the dispatch order is *not* even. If the position is occupied, Algorithm 1 will choose the next available one. For example, in the above example of workload 32 req/s, the dispatch step of machine A should be $\frac{1/0.25+1/0.5+1/0.5}{1/0.25} = 2$, and the dispatch step of machine B and C should be $\frac{1/0.25+1/0.5+1/0.5}{1/0.5} = 4$. Therefore, Algorithm 1 will update the dispatch order to ensure that machine A takes dispatch position $0, 2, 4, \ldots$ in the request list. Since position $0, 4, 8, \ldots$ is occupied, machine B will take position $1, 5, 9, \ldots$ and machine C will take position $3, 7, 11, \ldots$ instead. Specifically, for the first 16 requests, Algorithm 1 will allocate req 1–2 to machine A, req 3–8 to machine B, req 9–10 to machine A and req 11–16 to machine C. Different from the dispatch order from greedy-based algorithm, the one from Algorithm 1 separates the allocated batches on machine A to reduce the waiting time on it, leading to a lower average latency among the machines. By doing so, DeepLat reduces the average latency while keeping the worst case latency unchanged, which makes the dispatch order a better candidate for future worst case latency optimization (Sect. 4).

Algorithm 2. Supporting partial-batch

1: **PartialScheduling**$(Order_d, S)$
2: $state, order_p \leftarrow [\ ], [\ ]$
3: $wcl_p \leftarrow 0$
4: **for** $M_i = < b_i, d_i > in\ S$ **do**
5:    $exec[M_i] \leftarrow \frac{b_i}{T}$
6:    $wcl_p \leftarrow max(wcl_p, exec[M_i] + d_i)$
7: **for** o in $Order_d$ **do**
8:    **if** $predict_{wcl}(state[o]) > wcl_p$ **then**
9:      **while** $predict_{wcl}(state[o]) > wcl_p$ **do**
10:        $o_{next} \leftarrow next(o)$
11:        $split(o, o_{next})$
12:        **if** $predict_{wcl}(state[o_{next}]) < predict_{wcl}(state[o])$ **then**
13:          $order_p.push(split(o, o_{next}))$
14:          $wcl_p \leftarrow max(wcl_p, predict(o_{next}))$
15:          $o_{next} \leftarrow next(o_{next})$
16:          $state[o] \leftarrow UpdateState(state[o_{next}], exec[o_{next}], d_{next})$
17:        **else**
18:          $wcl_p \leftarrow predict_{wcl}(state[o])$
19:          $order_p.push(o)$
20:          break
21:    **else**
22:      $order_p.push(o)$
23: **return** $order_p, wcl_p$

3.3 Supporting Partial-Batch

The dispatch policy derived from Sect. 3.2 minimizes the worst case latency for the entire workload by being batch-aware and reduces the average latency for each GPU machine by optimizing the dispatch order during the starting phase. However, as we discussed in Sect. 2, support for partial-batch is required to further reduce the worst case latency. To do so, we propose a threshold-based heuristic to switch requests at the batch boundary.

Given a workload of T req/s with N GPU machines, for GPU machine i with configuration of batch size b_i and execution duration d_i, the lower bound for its individual latency should be $min(L_{wc}) = min(d_i + l_c + l_w) = d_i + b_i/T$, since the minimum batch collecting time is b_i/T and the minimum waiting time is 0. We use $max_{i=1}^{N}(d_i + b_i/T)$ as the latency threshold wcl_p for the entire workload. The intuition behind the threshold-based heuristic is to compare the current worst case latency for the entire workload against the latency threshold. If the current worst case latency is larger than the latency threshold, DeepLat will try to switch a group of requests at the batch boundary for partial-batch to reduce the worst case latency.

Given the above intuition, DeepLat uses Algorithm 2 to implement the threshold-based heuristic to switch requests at the batch boundary. It first calculate the latency threshold wcl_p(Line 6) as we described. Then for each request order derived from Algorithm 1, Algorithm 2 checks whether the current latency of the given request is larger than the latency threshold (Line 8). If it is larger than the latency threshold, Algorithm 2 generates a potential splitting strategy (Line 11). If splitting the batch can indeed reduce the worst case latency, Algorithm 2 will update the request order accordingly (Line 13–15).

DeepLat considers four cases for potential splitting strategies. In the first one, the top request list in Fig. 5(a) shows the dispatch order without partial-batch. The bottleneck of the worst case latency comes from the 5th batch (the second orange one), where the GPU machine is still running the 1st batch (the first orange one) when the 5th batch is collected, leading to a relatively long waiting time. Therefore, DeepLat will split the 6th batch (the fourth blue one) to postpone the collecting time for the 5th batch as the bottom request list in Fig. 5(a). By doing so, though the batch collecting time for the 6th batch is increased, the waiting time for the 5th batch is reduced, leading to a lower worst case latency. Namely, in the first case, DeepLat splits one batch to put another batch in between to reduce the worst case latency for the entire workload.

In the second case (Fig. 5(b)), DeepLat splits one batch to put multiple batches in between. In the third case (Fig. 5(c)), DeepLat splits multiple batches to put one batch in between. In the fourth case (Fig. 5(d)), DeepLat splits multiple batches to put multiple batches in between. Note that for the last three cases, when DeepLat considers multiple batches, these batches should have the model configuration with the same batch size. According to our experience, any batch splitting strategies to minimize the worst case latency can be transformed into one of the four cases.

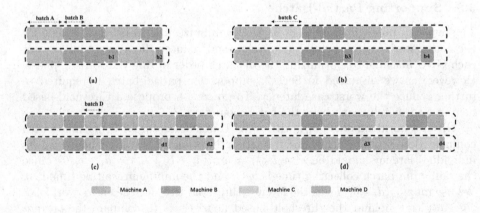

Fig. 5. Four cases to use partial-batch to further reduce the worst case latency.

Lemma 1. *If there is an optimal dispatch policy with partial-batch, then there must exist at least one optimal dispatch policy where the splitted batch is filled with batched request from other machines at full batch.*

Proof. We use proof of contradiction, where for the optimal dispatch policy with partial-batch, the splitted batch must be filled with batched request from other machines as *partial-batch* as well. Without loss of generality, we assume that machine A has the splitted batch filled with batched request from machine B. There will be two cases.

In the first case, the request list for machine A and B starts by requests for machine A and ends by requests for machine B. Namely, the request list is $A_1 B_1 A_2 B_2$, where A_i and B_i represents certain amount of individual requests that the dispatch policy will distribute to machine A or B, where $\sum len(A_i)$ and $\sum len(B_i)$ equal the batch size for machine A and B. Now we switch the request order of B_1 and A_2, leading to a request list of $A_1 A_2 B_1 B_2$. We keep the model execution of machine A unchanged. Namely, after collecting requests from A_2, machine A will not start model execution until B_1 has been collected at machine B. For the updated request list, machine A's latency is unchanged, while machine B's latency will be reduced, since requests from $B1$ will have a smaller batch collecting time. Therefore, the updated request list, which fills splitted batch with batched request at full batch (*e.g.*, $B_1 B_2$), has a worst case latency no larger than the optimal, which contradicts the assumption.

In the second case, the reqeust list starts and ends by machine A. Namely, the request list is $A_1 B_1 A_2 B_2 A_3$. Similar to the first case, we can switch $B1$ and A_2 for a new request list, which fills splitted batch with batched request at full batch. The new request list has a worst case latency no larger than the optimal, which contradicts the assumption.

Theorem 1. *Given a workload of T req/s with two configuration types $c_A = \langle b_A, d_A \rangle$ and $c_B = \langle b_B, d_B \rangle$[1] with n_A and n_B machines respectively, if the following two conditions are satisfied, then Algorithm 2 can guarantee the lower bound for the worst case latency.*

$$\frac{b_A}{T} + \lceil \frac{d_B}{d_A} \rceil d_A - (\lceil \frac{d_B}{d_A} \rceil - 1) \frac{d_A}{T} \leq \frac{b_B}{T} + d_B \qquad (2)$$

$$\frac{b_A + b_B}{T} + d_A \leq \frac{b_B}{T} + d_B \qquad (3)$$

Proof. Algorithm 2 takes the duration-based dispatch policy as a baseline. According to the definition, machine with c_A has dispatch step $\frac{n_A/d_A + n_B/d_B}{1/d_A}$, while machine with c_B has dispatch step $\frac{n_A/d_A + n_B/d_B}{1/d_B}$. Therefore, the average request frequency between machine with c_A and machine with c_B should be $\frac{d_B}{d_A}$. Since $\frac{d_B}{d_A}$ is not necessary an integer. The actual request frequency is either $\lceil \frac{d_B}{d_A} \rceil$ or $\lceil \frac{d_B}{d_A} \rceil - 1$, leading to the following two cases.

In the first case, the actual request frequency is $\lceil \frac{d_B}{d_A} \rceil$, where we have $\lceil \frac{d_B}{d_A} \rceil$ consecutive batches with c_A followed by one batch with c_B. To ensure the lower bound for the worst case latency, we need to guarantee that the latency for

---

[1] Without loss of generality, we assume $b_A < b_B$.

executing a consecutive of $\lceil \frac{d_B}{d_A} \rceil$ batched request on machines with c_A is smaller than the latency for executing one batched request on a machine with $_B$, which is equivalent to Eq. 2.

In the second case, the actual request frequency is $\lceil \frac{d_B}{d_A} \rceil - 1$. DeepLat will use partial-batch to reduce the worst case latency. According to Lemma 1, DeepLat will split the first batch with c_A in the next iteration and put the batch with c_B as a full batch in between. To ensure the lower bound for the worst case latency, DeepLat will guarantee that the updated latency for machine with c_A is smaller than the latency for machine with c_B, which is equivalent to Eq. 3.

Theorem 1 provides the *sufficient condition* to achieve the lower bound for the worst case latency. Note that if the input to DeepLat can *not* satisfy the conditions for Theorem 1, it might still achieve the lower bound for the worst case latency. By supporting partial-batch, DeepLat further reduces the worst case latency by minimizing the waiting time of the bottleneck machines (Sect. 4).

4   Evaluation

We compare DeepLat against the state-of-the-art inference system Nexus [7], then perform an ablation study that quantifies the importance of DeepLat's design choices.

4.1   Methodology

We have implemented all the features of DeepLat as described in Sect. 3. Each GPU machine runs TF-Serving [16] as the computation engine for DNN model execution. The input to DeepLat includes (1) the model configurations for each GPU machine, which is chosen among the model profiling derived on a commercial cloud, and (2) the request rate for each DNN model, which is synthesized to verify DeepLat's capability to minimize the worst case latency under various workload. The goal of DeepLat is to dispatch the requests among the GPU machines properly to minimize the worst case latency for the entire workload.

We compare DeepLat against the state-of-the-art DNN inference system Nexus [7], which distributes individual requests among GPU machines in rotation as we described in Sect. 2. For an input with N GPU machines, we define $max_{i=1}^{N}(d_i + b_i/T)$ as the *potential* lower bound for the worst case latency, denoted as L_{pot}, where T is the request rate for a given DNN model, b_i and d_i is the batch size and execution duration of the given DNN model on machine i. Note that L_{pot} is not necessarily reachable, while deriving the *actual* lower bound for the worst case latency is known to be NP-hard. Therefore, we use L_{pot} as an indicator of DeepLat's worst case latency minimization capability. To verify DeepLat's capability to minimize the worst case latency, we used tensorflow as the backend in a real cluster and generate a variety of workloads with request rate ranging from 130 to 3000 req/s and latency SLO ranging from 7 to 37 ms, using model profiles of three popular DNN models ResNet [5], Inception [17] and Mobilenet [18].

For most experiments, we mainly focus on the worst case latency for the entire workload, which is the maximum latency of all requests in a given workload. An ideal request dispatch system should have the minimum worst case latency. In Sect. 4.3, we also measure the average latency to show that duration-based heuristic is a better candidate to use partial-batch for worst case latency reduction.

4.2   Comparison Results

Figure 6a shows the average normalized worst case latency for DeepLat and Nexus, as well as the normalized potential lower bound L_{pot}, while Fig. 6b shows the cumulative distribution function (CDF) of normalized worst case latency of Nexus and L_{pot} over DeepLat. For *all* workload, DeepLat achieves a smaller worst case latency than Nexus, while Nexus requires an average of 60.5% extra worst case than DeepLat's. Namely, DeepLat achieves a 37.7% worst case latency reduction. For 5.1% of the workload, Nexus's worst case latency is more than 2 times of DeepLat's with a maximum up to 15.5 times. We have the following observations.

First, DeepLat outperforms Nexus in the worst case latency for *all* workload. As discussed in Sect. 3, Nexus dispatches requests from the frontend to machine's local request queue as individual request in rotation and generates batched request at each machine independently, leading to an equivalent batch collecting rate of model throughput on an individual machine. Meanwhile, DeepLat dispatches requests from the frontend to machine's local request queue as batched request directly, leading to an equivalent batch collecting rate of the total workload. Therefore, DeepLat has a lower batch collecting time, leading to a smaller worst case latency. For example, for the workload used in the experiment, the batch collecting time of Nexus is 6.248 times of DeepLat's on average.

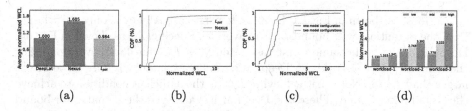

(a)   (b)   (c)   (d)

Fig. 6. (a) Average normalized worst case latency for DeepLat, Nexus and L_{pot}; (b) CDF of normalized worst case latency of Nexus over DeepLat; (c) Normalized worst case latency of Nexus over DeepLat under 1 or 2 model configurations; (d) Normalized worst case latency of Nexus over DeepLat under low, mid and high request rate for three workload.

Second, the advantage of DeepLat over Nexus on worst case latency reduction is related to the number of model configurations. Figure 6c shows the average

normalized worst case latency of Nexus over DeepLat under 1 or 2 model configurations. When there is 1 model configuration, the normalized worst case latency
of DeepLat equals the one of L_{pot}. Namely, DeepLat achieves the lower bound
for the worst case latency for *all* workload with 1 model configuration. Besides,
the advantage of DeepLat over Nexus is larger with 1 model configuration. Note
that for all workload in Sect. 4.2, we use a maximum of 2 model configurations
for fair comparison, since Nexus only supports at most 2 model configurations.
DeepLat has no limit on the number of model configurations, so we remove the
limitation in Sect. 4.3.

Third, the advantage of DeepLat over Nexus is larger when the number
of GPU machines is larger. Figure 6d shows the ratio of Nexus's worst case
latency over DeepLat's under three workload. For each workload, we provide
three request rate (*e.g.*, low, mid and high) as input. For all workload, the worst
case latency increases with the request rate, and the *high* request rate achieves
the largest worst case latency ratio. This is because with the increase in request
rate, the benefit of batch-aware dispatching is larger with higher batch collecting
rate over Nexus's rotation-based dispatching.

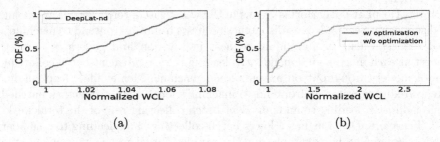

(a)   (b)

Fig. 7. (a) CDF of normalized worst case latency of DeepLat-nd over DeepLat; (b)
Average latency with and without duration-based optimization.

Besides, as shown in Fig. 6a, DeepLat's worst case latency is only 1.6% larger
than the potential lower bound L_{pot} on average. Specifically, as shown in Fig. 6b,
DeepLat achieves the same worst case latency as L_{pot} for 48.4% workload. For
the remaining 51.6% workload, DeepLat's worst case latency is only up to 11.0%
larger than L_{pot}'s. Note that achieving L_{pot} is the sufficient condition to achieve
the actual lower bound. Therefore, DeepLat has achieved the actual lower bound
for the worst case latency for *at least* 48.4% workload. For the remaining workload, DeepLat might still achieve the actual lower bound, but can *not* be verified
in polynomial time. According to our experience, the verification time for each
workload takes more than 24 h, leading to a total verification time over a year
for the all workload.

4.3   Ablation Study

The Importance of Duration-Based Optimization. DeepLat uses
duration-based heuristic to optimize the dispatch order derived from the greedy-

based heuristic, so as to minimize the average latency, which is critical for partial-dispatch. To evaluate the benefit of it, we compare DeepLat against DeepLat-nd, which disables the duration-based heuristic.

Figure 7a shows the CDF of normalized worst case latency of DeepLat-nd over DeepLat, where DeepLat-nd requires an extra of 2.7% worst case latency on average than DeepLat and a maximum of 6.7% extra worst case latency. This is because without duration-based heuristic to optimize the dispatch order during the staring phase, though the worst case latency of the input to Algorithm 2 remains the same, the average latency is much larger, leaving minimum room for Algorithm 2 to use partial-batch to reduce the worst case latency. To verify it, Fig. 7b shows the average latency with and without duration-based optimization, where the average latency without duration-based optimization can be up to 2.69 times larger. Therefore, the support for duration-based optimization is critical to achieve the minimum worst case latency.

The Importance of Partial-Batch Dispatching. DeepLat supports partial-batch to further reduce the worst case latency for the entire workload by switching the dispatch order of a group of requests at the batch boundary. To quantify the importance of it, we compare DeepLat against DeepLat-np, which disables partial-batch dispatching. Namely, we compare the worst case latency derived from Algorithm 2 against the one derived from Algorithm 1.

Figure 8 shows the average normalized worst case latency of DeepLat-np over DeepLat, where DeepLat-np requires an average of 3.8% extra worst case latency and a maximum of 13.4% extra worst case latency. Therefore, DeepLat can further reduce the worst case latency by switching a group of requests at the batch boundary.

Worst Case Latency Reduction with Multiple Model Configurations. As discussed in Sect. 4.2, DeepLat has no limit on the number of model configurations, as opposed to Nexus's maximum of 2 model configuration limitation. To verify DeepLat's capability to reduce the worst case latency, we compare the worst case latency of DeepLat against two times the execution duration L_{2d} with various multiple model configurations. We can *not* use Nexus directly due to its limitation on the number of model configurations.

Figure 9 shows the average normalized worst case latency of L_{2d} over DeepLat with various number of model configurations, where the rotation-based L_{2d}

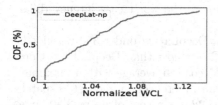

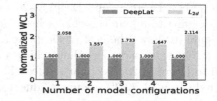

Fig. 8. CDF of normalized worst case latency of DeepLat-np over DeepLat.

Fig. 9. Average normalized worst case latency of L_{2d} over DeepLat.

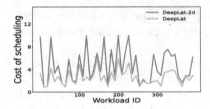
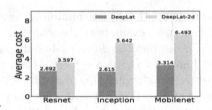

Fig. 10. Cost of DeepLat and DeepLat-2*d* under 350 workloads.

Fig. 11. Average cost of DeepLat and DeepLat-2*d* for three DNN models.

requires an extra of 55.7% to 111.4% worst case latency than DeepLat on average. Therefore, DeepLat can reduce the worst case latency under various number of model configurations.

4.4 Further Analysizing DeepLat's Performance

DeepLat Largely Reduces the Cost. As discussed, in contrast to existing systems [7,8] which uses two times the model duration 2*d* as the worst case latency, DeepLat uses batch-aware strategies to reduce the worst case latency to $d + \frac{b}{T}$, where d is the model duration, b is the batch size of the given machine configuration and T is the total request rate for the DNN model. By reducing the worst case latency, DeepLat is able to select machine configurations with larger throughput, which leads to a lower total cost when serving the DNN model.

To verify DeepLat's cost reduction capability with minimum worst case latency, we compare DeepLat against DeepLat-2*d*, which uses 2*d* as the expected worst case latency, similar to existing systems. We generated a total of 350 workloads associated with various request rates and latency SLOs using ResNet, Inception or Mobilenet. For each workload with a fixed request rate and latency SLO of a given DNN model, DeepLat and DeepLat-2*d* calculate the cost[2] of scheduling in two steps. First, both of them calculate the worst case latency for all candidate machine configurations, where DeepLat uses various optimizations as described in Sect. 3 to minimize the worst case latency and DeepLat-2*d* uses 2*d*. Second, both of them greedily select the machine configuration which has the largest throughput while its estimated worst case latency is less than the latency SLO. DeepLat and DeepLat-2*d* estimate the worst case latency differently, leading to different cost of scheduling.

Figure 10 shows the cost of DeepLat and DeepLat-2*d* under 350 workloads. For *all 350* workloads, DeepLat achieves a lower cost than DeepLat-2*d*. Compared to DeepLat-2*d*, DeepLat reduces 43.2% cost on average with a maximum cost reduction of 73.7%. Namely, due to the differences in request dispatching, batch-unaware systems such as DeepLat-2*d* has a cost of up to 3.80 times

---

[2] When calculating the cost, we consider homogeneous hardware, so the cost is the number of machines required to serve the workload. We leave the case of heterogeneous hardware to future work.

of DeepLat's. To understand the model sensitivity of DeepLat's cost reduction capability, Fig. 11 shows the average cost for DeepLat and DeepLat-2d on three DNN models, where DeepLat reduces an average of 25.2%, 53.7% and 49% for ResNet, Inception and Mobilenet respectively. DeepLat achives a larger cost reduction on smaller DNN models (*e.g.*, Inception and Mobilenet) because they provides more machine configurations to satisfy the latency SLOs, leaving larger search space for DeepLat to optimize.

Sensitivity of Cost Reduction to Request Rate and Latency SLO. Each workload has unique combination of request rate and latency SLO. The corresponding cost of scheduling for each workload is influenced by the request rate and latency SLO. To understand the sensitivity of cost reduction to them, we compare the cost of DeepLat and DeepLat-2d on varying latency SLOs or request rates.

Figure 12 shows the cost of DeepLat and DeepLat-2d under varying latency SLOs while keeping the request rate fixed. When the latency SLO is small (*e.g.*, ≤ 9.6 ms), DeepLat and DeepLat-2d has the same cost. This is because when the latency SLO is small, there is only one machine configuration (*e.g.*, batch size of 1) that can satisfy the latency SLO. So both systems have to choose the same machine configuration, leading to the same cost. When the latency SLO is larger (*e.g.*, 9.6–32.2 ms, which is most common in practice [7]), more machine configurations can satisfy the latency SLO, where DeepLat always chooses machine configurations with larger throughput by leveraging its efficient dispatching policies, leading to a lower cost. When the latency SLO is large (*e.g.*, ≥ 32.2 ms), the benefit of minimizing worst case latency on cost reduction is diminishing, since the latency SLO is large enough for all candidate machine configurations to satisfy it, regardless of the dispatching policy.

Figure 13 shows the cost of DeepLat and DeepLat-2d under varying request rates while keeping the latency SLO fixed. Similar to Fig. 12, when the request rate is small, DeepLat and DeepLat-2d have comparable performance. When the request rate is large, DeepLat largely reduces the cost. Therefore, by dispatching requests properly for minimum worst case latency, DeepLat can further reduce the cost of scheduling under a variety of latency SLOs and request rates.

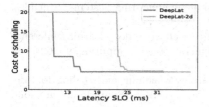

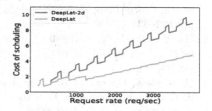

Fig. 12. Cost of DeepLat and DeepLat-2d with the same request rate but different latency SLOs.

Fig. 13. Cost of DeepLat and DeepLat-2d with the same SLO but different request rates.

5   Conclusion

In this paper, we present a request dispatch system, called DeepLat, which distributes requests for DNN-based applications among the GPU machines efficiently to reduce the worst case latency for the entire workload. DeepLat achieves the minimum worst case latency goal with a three-round optimization. It uses batch-aware dispatching to reduce the batch collecting time. It reorders the dispatch order during the starting phase to achieve low average latency. It supports partial-batch to further reduce the waiting time for bottleneck machines. Evaluation shows that DeepLat reduces the worst case latency by 37.7% than the state-of-the-art on average. Besides, DeepLat achieves the lower bound for the worst case latency for more than 48% workload.

Acknowledgements. This work was funded by National Key Research and Development Program of China (2022YFB4500223) and National Natural Science Foundation of China under Grant Nos.62202328. This work was partially supported by Open Subjects of State Key Laboratory of Computer Architecture (CARCHA202116) and Tianjin University Independent Innovation Fund - Qiming Programme (2023XQM-0005).

References

1. LeCun, Y., Bengio, Y., Hinton, G.: Deep learning. Nature **521**(7553), 436–444 (2015)
2. Goodfellow, I., Bengio, Y., Courville, A.: Deep Learning. MIT Press (2016)
3. Krizhevsky, A., Sutskever, I., Hinton, G.E.: Imagenet classification with deep convolutional neural networks. Adv. Neural Inf. Process. Syst. **25** (2012)
4. Hirschberg, J., Manning, C.D.: Advances in natural language processing. Science **349**(6245), 261–266 (2015)
5. He, K., Zhang, X., Ren, S., Sun, J.: Deep residual learning for image recognition. In: Proceedings of the IEEE Conference on Computer Vision and Pattern Recognition, pp. 770–778 (2016)
6. Covington, P., Adams, J., Sargin, E.: Deep neural networks for youtube recommendations. In: Proceedings of the 10th ACM Conference on Recommender Systems, pp. 191–198 (2016)
7. Shen, H., et al.: Nexus: a gpu cluster engine for accelerating dnn-based video analysis. In: Proceedings of the 27th ACM Symposium on Operating Systems Principles, pp. 322–337 (2019)
8. Hu, Y., Ghosh, R., Govindan, R.: Scrooge: a cost-effective deep learning inference system. In: Proceedings of the ACM Symposium on Cloud Computing, pp. 624–638 (2021)
9. Crankshaw, D., et al.: Inferline: latency-aware provisioning and scaling for prediction serving pipelines. In: Proceedings of the 11th ACM Symposium on Cloud Computing, pp. 477–491 (2020)
10. Crankshaw, D., Wang, X., Zhou, G., Franklin, M.J., Gonzalez, J.E., Stoica, I.: Clipper: a {Low-Latency} online prediction serving system. In: 14th USENIX Symposium on Networked Systems Design and Implementation (NSDI 17), pp. 613–627 (2017)

11. Romero, F., Li, Q., Yadwadkar, N.J., Kozyrakis, C.: {INFaaS}: automated model-less inference serving. In: 2021 USENIX Annual Technical Conference (USENIX ATC 21), pp. 397–411 (2021)
12. Hu, Y., et al.: Rim: offloading inference to the edge. In: Proceedings of the International Conference on Internet-of-Things Design and Implementation, pp. 80–92 (2021)
13. Girshick, R., Radosavovic, I., Gkioxari, G., Dollár, P., He, K.: Detectron. https://github.com/facebookresearch/detectron
14. Dean, J., André Barroso, L.: The tail at scale. Commun. ACM 56(2), 74–80 (2013)
15. Gandhi, A., Harchol-Balter, M., Raghunathan, R., Kozuch, M.A.: Autoscale: dynamic, robust capacity management for multi-tier data centers. ACM Trans. Comput. Syst. (TOCS), 30(4), 1–26 (2012)
16. https://github.com/tensorflow/serving (2022)
17. Szegedy, C., Vanhoucke, V., Ioffe, S., Shlens, J., Wojna, Z.: Rethinking the inception architecture for computer vision. In: Proceedings of the IEEE Conference on Computer Vision and Pattern Recognition, pp. 2818–2826 (2016)
18. Howard, A.G., et al.: Mobilenets: efficient convolutional neural networks for mobile vision applications. arXiv preprint arXiv:1704.04861 (2017)

Privacy-Preserving and Reliable Distributed Federated Learning

Yipeng Dong[1,2,3], Lei Zhang[1,2,3](✉), and Lin Xu[1,2,3]

[1] Shanghai Key Laboratory of Trustworthy Computing, Software Engineering Institute, East China Normal University, Shanghai 200062, China
{51215902087,52265902009}@stu.ecnu.edu.cn, leizhang@sei.ecnu.edu.cn
[2] Guangxi Key Laboratory of Cryptography and Information Security, Guilin, Guangxi 541000, China
[3] Engineering Research Center of Software/Hardware Co-Design Technology and Application, Ministry of Education, East China Normal University, Shanghai 200062, China

Abstract. Federated learning enables collaborative training of the global model by participants with diverse data sources while preserving data privacy. However, the traditional federated learning architecture faces some challenges, including single-point of server failure and privacy disclosure. To address these challenges, this paper proposes a distributed federated learning scheme based on multi-key homomorphic encryption, which fundamentally solves the problems of server single-point failure and malicious behavior, while effectively protecting the data privacy of participants. The trusted execution environment (TEE) is used to detect the quality of the models and to prevent some malicious participants from executing malicious behavior. Furthermore, an incentive mechanism is designed to encourage participants to actively and honestly perform training tasks. Our scheme satisfies privacy, robustness, and fairness criteria, as demonstrated in our analysis.

Keywords: Federated learning · Data privacy · Multi-key homomorphic encryption · Intel SGX

1 Introduction

Deep learning based on neural networks has made great progress in various tasks, such as speech recognition [32], image recognition [27], and automatic driving [26]. To achieve higher accuracy, it is necessary to provide a large amount of training data to the deep learning model. However, with the continuous improvement of data privacy awareness, it has become difficult to centralize all the participants' training data. To solve this problem, many privacy protection deep learning frameworks have emerged [5,12,29]. Of these frameworks, federated learning (FL) [21] is a widely used system. It allows users to participate in collaborative training without uploading local data. The traditional federated learning architecture is a centralized structure, which includes a central server and multiple participants. Participants use local data for model training and upload the obtained model updates to the central server. After the central server obtains the updates of all participants, it aggregates them and returns the aggregated global model. The participants adjust their local models based on the global model returned by the central server. This process iterates over several rounds until the global model converges.

Although federated learning enables participants to perform collaborative training without disclosing local data, it still faces some challenges. The single point of failure and malicious behavior of the central server are some of the problems that do great harm to the system [33]. The central server undertakes the task of model aggregation. If it is disconnected or damaged, the entire federated learning process will be stopped. In the FL task, existing studies assume that the central server is honest or semi-honest. However, part of the research did not protect the global model or aggregation. Whether it is safe or not remains to be verified. After the central server obtains the unprotected global model, an attacker can use the model for profit. What's more, once the central server is controlled by the attacker, they can use the server to destroy the training task arbitrarily. Even if the central server performs tasks normally, it can infer a lot of sensitive information from the gradient of model updates through privacy inference attacks [15,24]. Moreover, in some scenarios (for example, the participants are edge mobile devices), sending data to the central server each time will incur a large cost.

Therefore, to solve the above problems, serverless federated learning is an effective solution [37]. Instead of using the central server as an aggregator, participants are used for performing aggregation tasks. This method fundamentally solves the problem of server failure and evil. However, in this framework, there are still some other challenges to be addressed. This paper focuses on privacy, model quality, and incentive fairness. Preventing privacy leakage is the most basic goal that the paper needs to focus on. Attackers can use eavesdropping and other means to obtain unprotected model updates to launch attacks. Therefore, a reasonable protection method needs to be proposed so that the attacker cannot obtain any additional information.

As for model quality detection, to make the final global model better, we must detect the model quality to eliminate low-quality models. We need to encourage participants to use more and better data for training to provide more effective models. A non-IID dataset of a participant may result in low-quality updates, weakening the generalization ability of the final global model. In the distributed framework, the participants perform the model quality detection, unlike traditional federated learning where the server performs this task. To gain more benefits, some participants may conceal the detection results, thereby affecting the performance of the global model. This paper focuses on the detection of low-quality models and evaluates the detection results to avoid the impact of low-quality models on the global model.

Incentive fairness is another problem that needs to be solved. The existing FL scheme assumes that all participants unconditionally contribute their resources. However, during global model training, a large number of participants' computing and communication resources will be consumed [18]. Assuming that participants will complete the work without compensation is overly optimistic. In reality, without a reasonable incentive system, it is impossible to mobilize the enthusiasm of participants. However, [36] proposed a horizontal federal learning incentive mechanism based on reputation and reverse auction. It is aimed at the centralized federal learning framework, and the incentive strategy does not apply to the decentralized framework. Therefore, it is very important to design a fair incentive mechanism.

Our Contributions: To solve the single point of failure problem of the central server in the existing privacy-protected federated learning scheme, we first construct a privacy-protected distributed federated learning scheme. Firstly, to protect the participant's local model from being obtained by other participants, we use a multi-key homomorphic encryption scheme to encrypt the local model update. Then, we designed a fair incentive mechanism that enables participants to participate honestly and actively throughout the FL task. Finally, we check the quality of the participant's model to ensure the performance of the global model. Trusted execution environments are used to verify the model quality evaluation results and to prevent malicious participants from tampering with or uploading false models.

2 Related Work

In research on distributed federated learning frameworks, it is necessary to ensure data privacy, model quality, and incentive fairness. [25] proposed a linear model of distributed federated learning for fully distributed data. It is called gossip learning. Hegedűs *et al.* [14] used gossip learning as an alternative to federated learning. They found that the overall performance of gossip learning was similar to that of federated learning when the training data were evenly distributed on the nodes. However, their work did not consider privacy protection, which could lead to the disclosure of the model. [30] proposed a decentralized federated learning method that uses blockchain to protect local gradient privacy. It protects participants' gradient privacy to a certain extent. Gao *et al.* [11] proposed another method, which uses differential privacy to protect the model and can detect low-quality models. However, it faces the threat of differential privacy attacks [16] and cannot resist free-rider attacks [10]. Research on decentralized FL structure is ongoing, and this is also our concern.

For the gradient privacy problem of model updating, many different research directions are proposed. Abdelmoniem *et al.* [1] described a statistically-based gradient compression technology for distributed training systems, which can directly reduce the degree of FL privacy leakage because they reduce the information source for privacy inference. However, this method is not effective against all attacks, such as model inversion attacks. The gradient perturbation algorithm can effectively mask the gradient of the model. Differential privacy (DP) protects data privacy by adding random noise to sensitive information. Shokri *et al.* [28] used DP to protect gradients, aiming to achieve a trade-off between data privacy and training accuracy. However Hitaj *et al.* [15] points out that Shokri's work fails to adequately protect data privacy. An attacker can learn private data by generating adversarial networks (GAN). Using encryption algorithms to encrypt FL model updates is an effective means. Homomorphic encryption (HE) [8] and secure multi-party computing (MPC) [19] are the most commonly used algorithms. HE allows data to be encrypted and homomorphic calculation, and the decryption result is equivalent to performing operations on the original data. Homomorphic encryption will not change the original information. Therefore, it can theoretically ensure that the model will not suffer performance loss in the convergence process [34]. Aono *et al.* [2] used HE to protect the training model. The disadvantage of their scheme is that it assumes the participants are honest. If the participant is curious, this may lead to the disclosure of the honest participant's model. To solve the problem of the leakage of the private models of the participants, [4] proposed a secure and practical aggregation method based on MPC. However, this will require a large amount of computing and communication overhead, and it is necessary to ensure that participants are online in real-time during the implementation of the protocol.

The quality of model updates uploaded by participants will have a great impact on the final global model. Most of the existing model quality detection schemes [20,22] evaluate model quality by calculating similarity. To use this method, the server needs to obtain updates from each participant to calculate the similarity. However, in the serverless scenario, because participants cannot access the model information in the unencrypted state of other participants, the method of model quality checking with the help of servers in the traditional federated learning framework is no longer applicable. Another method is to use DP, but it will affect the accuracy of the model. It is an effective method to eliminate the impact of low-quality updates by using the idea of "reject on negative impact" (RONI) [3,9]. The idea is that if the update significantly reduces the accuracy of the global model, the update will be rejected. However, in the decentralized framework, each participant connet access any unencrypted information about other participants, and therefore, the existing solutions based on the traditional federated learning framework cannot effectively solve the challenges we face.

To achieve incentive fairness, the most widely used methods are based on reputation or contracts. In the reputation-based method, participants with high reputation value can choose and reward [35]. In the serverless framework, the calculation method of the reputation value of participants will not be applicable. The contract-based approach uses contracts to limit the behavior of participants [17,31]. However, it is often necessary to allocate rewards based on the resources that participants have. Malicious participants are likely to claim that they have more resources to obtain higher rewards. Therefore, we need to design a fair and reasonable incentive mechanism for serverless federated learning.

3   Background

3.1   System Architecture

In a traditional federated learning architecture [23], the server plays the role of model aggregation. However, the harm caused by single-point failure and the malicious server is always one of the key problems to be solved in federated learning research. Therefore, we study the distributed federated learning framework.

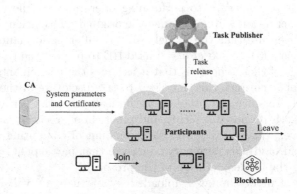

Fig. 1. System architecture.

Figure 1 shows the architecture of our system, which consists of a certificate authority(CA), a task publisher, and participants.

- CA: This is a trusted third party. It is responsible for generating system parameters and issuing certificates to entities in the system.
- Task publisher: This is an individual or an organization. It is responsible for publishing federated training tasks to the system and providing compensation. And it partitions the task area. The task publisher needs to generate a test dataset to test the effectiveness of the model.
- Participants: They have local datasets. After initializing a federated training task, participants can join the training task to train the global model together. In each round of the federated training task, participants train local models and generate local gradients. During the training process, they will upload the information that needs to be synchronized to the blockchain. Each participant has a local TEE for model quality detection to prevent malicious behavior from participants and external opponents. Participants are divided into groups based on geographic location to reduce overhead.

3.2   Threat Model and Goals

In contrast to traditional federated learning architectures, no servers are used in this framework. This can fundamentally solve the problem of a single point of failure of the server and the problem of malicious server stealing participants' private data information. A distributed federated learning framework can only be threatened by external attackers and malicious participants. We assume that the task publisher and most of the participants in the system are semi-trusted (honest but curious). But some of the participants may be malicious, they can perform some malicious operations, such as uploading fake models, launching active attacks, etc. Moreover, the number of participants in each group needs to be greater than two, otherwise, the privacy of the local model cannot be guaranteed. Therefore, our framework needs to guarantee the following properties:

Privacy: It ensures that participants' local data and models are not known to attackers.

High Model Quality: It requires that the solution resistant to low-quality and false models. Moreover, in the process of detecting model quality, it can resist attack methods such as poisoning attacks and free-rider attacks.

Server Freeness: There is no server to perform the aggregation of local gradients. The aggregation task is left to the participants.

Incentive Fairness: The process of model training, aggregation, and evaluation will consume participants' resources, and participants are willing to calculate and upload reliable model updates only when they get enough compensation. This can be ensured by appropriate incentives.

3.3 Multi-Key Homomorphic Encryption

Multi-key homomorphic encryption (MKHE) is a cryptographic system. It allows homomorphic computation of information encrypted with different keys. We choose to use Multi-key CKKS [6] as the encryption method and use homomorphic addition calculation in the scheme. Let M be the message space with an arithmetic structure. MKHE consists of five polynomial algorithms (Setup, KeyGen, Enc, Eval, Dec). We assume that each participant P_i has an index id_i and a public-private key pair (pk_i, sk_i). Multi-key ciphertext implicitly contains an ordered set of associated indexes. For example, a fresh ciphertext $ct \leftarrow MKHE.Enc(m; pk_i)$ corresponds to a single-element set $S = id_i$. As homomorphism is computed between ciphertexts from different parties, the size of the index set will increase. It consists of the following algorithms:

- $pp \leftarrow MKHE.Setup(1^\lambda)$. Given the security parameter λ, generate the public parameters pp. We assume that all the other algorithms implicitly take pp as an input.
- $(pk, sk) \leftarrow MKHE.KeyGen(pp)$. Each participant P_i generates a public-private key pair (pk_i, sk_i). Each participant P_i is able to perform this process in parallel.
- $ct \leftarrow MKHE.Enc(m; pk)$. Encrypts a plaintext message $m \in M$ by pk and outputs a ciphertext ct.
- $\overline{ct} \leftarrow MKHE.Eval(C, (\overline{ct_1}, ..., \overline{ct_l}), (pk_1, ..., pk_K)$. l is the number of ciphertexts to be computed, and K is the number of participants in the process of generating these ciphertexts. Given a boolean circuit C, a tuple of multi-key

ciphertexts $(\overline{ct_1}, ..., \overline{ct_l})$ and the corresponding set of public keys $(pk_1, ..., pk_K)$. After the *Eval* algorithm is executed for the ciphertexts tuple, a ciphertext result $\overline{ct}$ is generated and output. In the training process, only the addition operation in the evaluation algorithm is required, Therefore, the addition evaluation is described: $\overline{ct} \leftarrow MKHE.Add((\overline{ct_1}, \overline{ct_2}),$
$(pk_1, pk_2))$. Given two ciphertexts $\overline{ct_i}$, return the ciphertext $\overline{ct} = \overline{ct_1} + \overline{ct_2}$.

- $m \leftarrow MKHE.Dec(\overline{ct}; sk_1, ..., sk_K)$. Decrypts a ciphertext $\overline{ct}$ with a corresponding sequence of secret keys and outputs a plaintext m. Let $\overline{ct}$ be a ciphertext associated to K participants and $sk_1, ..., sk_K$ be their secret keys.

In the classic definition of MKHE primitive, decrypting multi-key ciphertext requires all private keys of related parties. However, in our scenario, the private key of the participant cannot be disclosed. So we use the idea of threshold decryption. $MKHE.Dec$ includes two algorithms: partial decryption and merging. The algorithms are described as follows:

- $\rho_i \leftarrow MKHE.PartDec(\overline{ct}, (pk_1, ..., pk_K), i, sk_i)$. When the extended ciphertext $\overline{ct}$ is input under the K key sequence and the i private key, the output part decrypts ρ_i.
- $m \leftarrow MKHE.FinDec(\rho_1, ..., \rho_K)$. On input K partial decryption, output the plaintext μ.

We use $(MKHE.Setup, MKHE.KeyGen, MKHE.Enc, MKHE.Eval,$ $MKHE.Dec)$ to define an MKHE scheme used in this paper.

4   The Scheme

4.1   High-level Description

Based on Sect. 3.1, we study distributed FL. Our scheme can be described in four stages: system setup, device initialization, task publishing, training and aggregation. The sketches of these stages are as follows.

- System Setup: The CA generates certificates and system parameters. The CA publishes them to the public channel.
- Device Initialization: Participants generate public and secret key pairs according to the system parameters and generate their index values. And each participant uses the authentication service to authenticate the local TEE to establish a secure channel.
- Task Publishing: The task publisher initializes the global model and selects a region as the task area that is divided into several subareas. It will broadcast training tasks into the task area. Participants receiving the training task were grouped according to their geographic location within the task partition.
- Training and Aggregation: At this stage, each participant trains the local model and uploads it to the local TEE through a secure channel. TEE encrypts the model and signs the encryption result. The participant downloads packaged the encrypted model and signature information from TEE

and uploads them to the blockchain. At the same time, participants download models and signature information from other participants in the group from the blockchain to achieve synchronization. Each participant aggregates the downloaded ciphertext model and decrypts it to obtain a partially decrypted model. Upload the partially decrypted model to the blockchain, download the decrypted models of the same group of participants, and aggregate them to obtain the plaintext sub-global model. After generating the global model in each round, the task publisher evaluates the quality of the participants' local models using the participants' TEEs and selects participants who can participate in the next round of training based on the results. Update the credibility score of participants according to their overall performance in the task, and give rewards or punishments. It is worth noting that before the start of each round of training tasks, the participants will be re-grouped according to their positions.

4.2 System Setup

At this stage, the CA selects a security parameter λ and executes $MKHE.Setup$ (1^λ) to generate the system parameter pp.

4.3 Device Initialization

At this stage, each participant in the system executes $MKHE.KeyGen(pp)$ to generate its public-private key pair. For the i-th mobile device $\mathcal{P}_i$, we assume its public-private key pair is (pk_i, sk_i). We also assume $\mathcal{P}_i$ has a local dataset D_i with dataset size s_i and a local TEE TEE_i. $\mathcal{P}_i$ has to obtain a reputation certificate $RC_i = (id_i, pk_i, rv_i, au_i, \sigma_i^{CA})$ from the CA, where id_i is $\mathcal{P}_i$'s identity, rv_i is $\mathcal{P}_i$'s reputation value that is managed by CA, au_i is any auxiliary information of $\mathcal{P}_i$, σ_i^{CA} is CA's signature on (id_i, pk_i, rv_i, au_i). Each participant authenticates the identity of the local TEE and establishes a secure channel.

4.4 Task Publishing

A task publisher initializes a global model ω_0 and selects an area that is divided into several subareas. We assume there are K subareas. It broadcasts the task $(location, tt, \sigma^T)$ to the mobile devices in the area, where $location$ is the task area and the information about how the area is divided into K subareas, tt is the task information which includes an initial global model ω_0, the maximum number of training rounds T, loss function $F(\cdot)$, etc., σ^T is the task publisher's signature on $(location, tt)$.

Any mobile device that is interested in the task can respond to this request by sending a response message. Suppose device $\mathcal{P}_i$ is interested in the task. It sends a response message $RM_i = (RC_i, location_i, \sigma_i^P)$ to the task publisher, where $location_i$ is the position of $\mathcal{P}_i$, σ_i^P is $\mathcal{P}_i$'s signature on $(RC_i, location_i)$.

Suppose the response messages received by the task publisher are $RM_1, ...,$ RM_N. It selects the proper mobile devices based on the locations and reputation values of the devices. Further, based on the locations of the devices, the devices are separated into K subareas. For instance, the device located in the l-th sub-area, then the device is assigned to the l-th group. Suppose $\mathcal{P}_1, ..., \mathcal{P}_n$ are selected as the participtants. Finally, the message $\{(\mathcal{P}_i, l_i)\}_{i \in \{1,...,n\}}$ is published, where l_i denotes that $\mathcal{P}_i$ belongs to the l_i-th subarea. And the task publisher remotely authenticates the participants' TEEs and establishes secure channels. Then he pre-stored the test dataset to the TEEs through secure channels.

Grouping participants has the following advantages over direct training:

- When performing threshold decryption on the encrypted sub-global model, the decryption cost will increase significantly with the number of participants. Therefore, when the participants are grouped, the decryption cost of the sub-global models will be greatly reduced compared to that of the global model.
- During task execution, participants may be disconnected. The advantage of grouping is that if a participant is offline, it will not affect the generation of sub-global models of other groups. Moreover, even if some groups that may have frequent participant disconnections, resulting in a longer time to generate sub-global models, the sub-global models generated by other groups can be aggregated as the final global model.

For the group q, assume that the time required for the participants in the group to perform the task is T_q. If a participant does not return a decryption result within time T_q, we consider him or her offline and require online participants in the group to re-aggregate the received models and decrypt the sub-global model.

There are certain restrictions when decrypting and grouping. It is worth noting that when decrypting the sub-global model, it is necessary to ensure that the participants who contribute to the model are online. Otherwise, partial decryption will fail. In addition, the number of aggregated models at any time must be greater than or equal to three, because if there is only an aggregation of bilateral models, the decrypted aggregated model cannot protect the information of participants. We need to ensure that at least two participants in each group do not collude with each other. Since there are not few members in each group in the real world, this is easy to be guaranteed.

4.5   Training and Aggregation

This stage involves multiple training rounds, and Fig. 2 shows one round of the training process. Each training round consists of three steps, i.e., model generation, model aggregation, and model evaluation. Suppose the n participants are divided into Q groups, and the q-th group has a participants $\{\mathcal{P}_{q_1}, ..., \mathcal{P}_{q_a}\}$. Suppose it is the t-th training round. Next, we show each step in detail.

In the model generation step, each participant selected by the task publisher first trains a local model based on its local dataset, and uploads the local model and his public key to the local TEE through a secure channel. TEE encrypts the model using MKHE.Enc to generate an encrypted model and use its own internal private key to sign the encrypted results. Each participant downloads

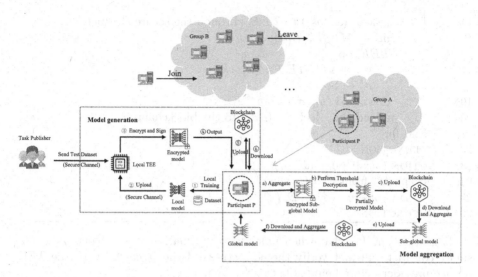

Fig. 2. Training and aggregation.

the packaged encrypted model and signature information from the local TEE and uploads it to the blockchain. The detailed processes are shown in Algorithm 1.

Algorithm 1. Local Training and Encryption.

Require: Participants $\mathcal{P} = \{\mathcal{P}_1, ..., \mathcal{P}_n\}$, where $\mathcal{P}_i \in \mathcal{P}$ holds his/her own dataset D_i, public-private key pair (pk_i, sk_i), id_i, a local TEE TEE_i; s_i is the dataset size of D_i, Acc_{t-1}^i is the accuracy of $\mathcal{P}_i$'s local model at round $t-1$-th; ω_{t-1} is the global model corresponding to the $t-1$-th round and is initially set to be a random value; η is the learning rate; $F(\cdot)$ is the loss function; E is the maximum number of epochs that the local model training is allowed in a training round.

Ensure: The encrypted models $m_t^1, ..., m_t^n$ corresponding to $\mathcal{P}_1, ...,$ $\mathcal{P}_n$ respectively.

1: FUNCTION
2:   For $\mathcal{P}_i \in \mathcal{P}$ do
3:     $\omega_t^i \leftarrow LocalUpdate(\omega_{t-1}, D_i)$
4:     If $t == 1$:
5:       $\omega_t^i \leftarrow s_i \cdot \omega_t^i$
6:       upload (ω_t^i, pk_i) to TEE_i through the secure channel
7:       TEE_i do
8:         $m_t^i \leftarrow MKHE.Enc(\omega_t^i; pk_i)$
9:         $\sigma_t^i \leftarrow Sign(m_t^i)$
10:        download (m_t^i, σ_t^i) from TEE_i
11:        upload $(m_t^i, s_i, id_i, \sigma_t^i)$ to the blockchain
12:      Else:
13:        $\omega_t^i \leftarrow Acc_{t-1}^i \cdot \omega_t^i$

14:                upload (ω_t^i, pk_i) to TEE_i through the secure channel
15:                $m_t^i \leftarrow MKHE.Enc(\omega_t^i; pk_i)$
16:                TEE_i do
17:                    $m_t^i \leftarrow MKHE.Enc(\omega_t^i; pk_i)$
18:                    $\sigma_t^i \leftarrow Sign(m_t^i)$
19:                download (m_t^i, σ_t^i) from TEE_i
20:                upload $(m_t^i, Acc_{t-1}^i, id_i, \sigma_t^i)$ to the blockchain
21:        End for
22: END FUNCTION
23: FUNCTION LocalUpdate(ω, D_i):
24:        If epoch i from 1 to E then
25:            $\omega \leftarrow \omega - \eta \cdot \nabla F(\omega, D_i)$
26: END FUNCTION

It is worth noting that when uploading information to the blockchain, the smart contract will first verify the correctness of the signature with the TEE service provider before being able to go online.

In the model aggregation step, for the q-th group with participants $\{\mathcal{P}_{q_1}, ..., \mathcal{P}_{q_a}\}$, a participant P_{q_i} downloads the information $\{(m_t^{q_1}, Acc_{t-1}^{q_1}, id_{q_1}, \sigma_t^{q_1}), ..., (m_t^{q_{i-1}}, Acc_{t-1}^{q_{i-1}}, id_{q_{i-1}}, \sigma_t^{q_{i-1}}), (m_t^{q_{i+1}}, Acc_{t-1}^{q_{i+1}}, id_{q_{i+1}}, \sigma_t^{q_{i+1}}), ..., (m_t^{q_a}, Acc_{t-1}^{q_a}, id_{q_a}, \sigma_t^{q_a})\}$ of other participants in the same group from the blockchain. Then the participants in the q-th group run the following Algorithm 2 to generate a sub-global model. Finally, each participant downloads the sub-global models corresponding to the other groups and computes the updated global model $\omega_t = \sum \omega_t^{q_1 q_a}$.

Algorithm 2. Model Aggregation.

Require: Participants $\mathcal{P}_q = \{\mathcal{P}_{q_1}, ..., \mathcal{P}_{q_a}\}$, where $\mathcal{P}_{q_i} \in \mathcal{P}_q$ holds his/her public-private key pair (pk_{q_i}, sk_{q_i}), id_{q_i}; $m_t^{q_1}, ..., m_t^{q_a}$ are the encrypted local models generated by the participants in $\mathcal{P}_q$ corresponding to the t-th round.
Ensure: The sub-global model $\omega_t^{q_1 q_a}$ of t-th round.
   1: FUNCTION
   2:     For $\mathcal{P}_{q_i} \in \mathcal{P}_q$ do
   3:         Computes $m_t^{q_1 q_a} = MKHE.Add((m_t^{q_1}, ..., m_t^{q_a}), (pk_{q_1}, ..., pk_{q_a}))$
   4:         Computes $\rho_{q_i} = MKHE.PartDec(m_t^{q_1 q_a}, (pk_{q_1}, ...pk_{q_a}), q_i, sk_{q_i})$, and uploads ρ_{q_i} to the blockchain.
   5:         Downloads $\{\rho_{q_1}, ..., \rho_{q_{i-1}}, \rho_{q_{i+1}}, ..., \rho_{q_a}\}$
   6:     End for
   7:     Participants in the group q compute the sub-global model $\omega_t^{q_1 q_a} = MKHE.FinDec(\rho_{q_1}, ..., \rho_{q_a})$ and uploads $\omega_t^{q_1 q_a}$ to the blockchain through consensus.
   8: END FUNCTION

It is worth noting that each participant will obtain a partially decrypted model during collaborative decryption, and merging them will generate a plaintext sub-global model. If the plaintext model cannot be generated, there is an error in the aggregation of some participants, preventing malicious aggregation behavior.

In the model evaluation step, the task publisher needs to evaluate the quality of each participant's local model. In this step, to prevent malicious participants and external enemies from engaging in malicious behavior, all processes are implemented based on TEE. After each round of model aggregation task, the task publisher uploads the global model and threshold to the TEEs through secure channels. Run the model evaluation algorithm in TEEs to evaluate the local models uploaded by participants. And return the evaluation results to the task publisher through a secure channel. The task publisher filters participants based on the results. The detailed processes are shown in Algorithm 3.

Algorithm 3. Model Evaluation.

Require: Participants $\mathcal{P} = \{\mathcal{P}_1, ..., \mathcal{P}_n\}$, Where $\mathcal{P}_i$ holds a local TEE TEE_i; ω_t is the global model corresponding to the t-th round, ω_t^i is the local model of $\mathcal{P}_i$ corresponding to the t-th round, rv_i is the reputation value of participant $\mathcal{P}_i$, Acc_i is the sub-global model accuracy excluding participant $\mathcal{P}_i$, ϵ is the threshold for low quality models.

1: FUNCTION EVALUATION:
2:   Task publisher do
3:     For $\mathcal{P}_i \in \mathcal{P}$
4:       remote authentication of TEE_i and establishment of a secure channel.
5:       download the global model ω_t from the blockchain and upload threshold ϵ and ω_t to TEE_i through the secure channel.
6:       TEE_i do
7:         $\overline{\omega}_t^i = \omega_t - \omega_t^i$
8:         $Acc_i = ACC(\overline{\omega}_t^i)$
9:         $Acc = ACC(\omega_t)$
10:        If $\frac{Acc_i}{Acc} - 1 > \epsilon$
11:          $rv_{\mathcal{P}_i} = rv_{\mathcal{P}_i} - (\frac{Acc_i}{Acc} - 1)$
12:        End if
13:        output evaluation results (Acc_i, Acc) and reputation value rv_i to the task publisher.
14:      End for
15:   The task publisher broadcasts the evaluation results and selects participants for the next round of training based on the evaluation results.
16: END FUNCTION

For participants to participate in the training task actively and honestly, we need a fair incentive mechanism to encourage users. The reputation value of the participants is the fundamental dependence of our incentive mechanism design. There are two factors influencing its reputation value: the quality of the local model and the completion of the training task. We have proposed the relationship between reputation value and model quality above. Next, we will analyze the impact of aggregation task completion on it:

– To encourage each participant to actively and honestly participate in the training task. They need to be rewarded for honest behavior and punished for malicious behavior. Make each participant try to complete the current round of training without quitting at will, and the participants who complete the task on time will be rewarded. Relatively speaking, if the participant fails to complete the work within the stipulated time, it will affect the generation of the sub-global model and should be punished. Therefore, it should be more rewarding to complete the work on time. We can use the existing incentive mechanism to promote participants to work actively. As shown in reference [36], assuming the reputation value reward is M and the penalty coefficient is c. For participant P_i in the group q, the influence of P_i's reputation value at this stage can be described as:

$$\begin{cases} rv_{P_i} = rv_{P_i} - c \times m, T(P_i) > T_q \\ rv_{P_i} = rv_{P_i} + c \times m, T(P_i) \leqslant T_q. \end{cases} \tag{1}$$

$T(P_i)$ is the time required by P_i to perform aggregation and decryption tasks in the current round. When P_i does not publish its decryption result within time T_q, it is considered to have timed out. rv_{P_i} is the reputation value of P_i. If the actual aggregation time $T(P_i)$ is larger than T_q, rv_{P_i} will be reduced. Otherwise, rv_{P_i} will get some improvement.

The reputation value of participants is an important factor for participants. Task publishers can decide whether to participate in the training task according to the reputation value of participants. Moreover, reputation value is also a key indicator to determine whether it is eligible for aggregation. However, the research work in this area is also very complex, which is not the focus of this paper and will not be described here in detail.

Algorithm 4. Privacy-preserving and Reliable Distributed Federated Learning.

Require: The sizes of datasets $s_1, ..., s_n$ cossesponding to $\mathcal{P} = \{\mathcal{P}_1, ..., \mathcal{P}_n\}$ respectively, Acc_{t-1}^i is the loacl model accuracy to the $t-1$-th round, t is the current training round.
Ensure: The global model M
1: FUNCTION MAIN:
2:     initialize ω_0
3:     For each round t=1,2,... do
4:         Local Training and Encryption (Algorithm 1)
5:         ω_t=Model Aggregation (t) (Algorithm 2)
6:         Model Evaluation (Algorithm 3)
7:         If $t == 1$:
8:             $s = \sum_{i=1}^{N} s_i$
9:             Global model $M \leftarrow \frac{\omega_t}{s}$
10:         Else:
11:             $Acc = \sum_{i=1}^{N} Acc_{t-1}^i$

12:         Global model $M \leftarrow \frac{\omega_t}{Acc}$
13:         Participants update local models based on the global model M.
14:     End for
15: END FUNCTION

As the training rounds change, participants will execute the Algorithm 4 which includes invokes to Algorithms 1-3. After obtaining the global update ω_t, participants need to perform a simple division operation on the global update locally to obtain the global update after weight division, and then use the global model update to update and optimize the local model.

5   Security Analysis

In this section, we show that our scheme achieves the design goals defined in Sect. 3.2, i.e., the privacy of the local data, high model quality, server freeness, and incentive fairness, and give security analysis accordingly.

5.1   Analysis of Local Data Privacy

Since there is no parameter server in our system, we only need to ensure that participants, task publishers, and external attackers do not infringe on the privacy of the local dataset.

The privacy of the participants' dataset depends on the privacy of the participants' local model. In our scheme, participants establish secure channels with TEE. Due to the security of TEE, it can ensure that the data on it is not leaked. The participant's model is encrypted under their public key using the MK-CKKS cryptosystem and uploaded to the blockchain. Although external attackers can obtain encrypted updates through eavesdropping and other methods, Lemma 1 can ensure that MK-CKKS is secure, so plaintext updates cannot be obtained, and the privacy of local data sets will not be violated. There will be some sub-global models in the plaintext state in the system. Since we restrict the number of participants in the generation of sub-global models to be greater than or equal to 3, other participants cannot infer any information about the local model dataset from the sub-global models.

Lemma 1. Based on the RLWE parameter assumption, the single encryption scheme is IND-CPA secure [6]. MK-CKKS scheme has single-key encryption algorithms. A ciphertext is generated by adding an encoded plaintext to a random encryption of zero. Hence, it suffices to show that random encryption of zero is computationally indistinguishable from a uniform random variable. We consider the random variable (a, b, c_0, c_1) over R_q^4 defined by $a \leftarrow U(R_q), b = -s \cdot a + e(mod\,q)$ for $s \leftarrow \chi$ and $e \leftarrow \psi$, and $(c_0, c_1) = v \ldots (b, a) + (e_0, e_1)(mod\,q)$ for $v \leftarrow \chi$ and $e_0, e_1 \leftarrow \psi$. First, we change the definition of b as $b \leftarrow U(R_q)$. It is computationally indistinguishable by the RLWE assumption with parameter (n, q, χ, ψ). Then (b, c_0) and (a, c_1) can be viewed as two RLWE samples of secret v which are computationally indistinguishable from the uniform random variable $U(R_q^2)$ under the same RLWE assumption. So we conclude that the MK-CKKS scheme is IND-CPA secure.

5.2   Analysis of High Model Quality

We use the idea of RONI to evaluate the quality of the participant's local model and use TEE to ensure that malicious participants do not falsify their models.

TEE creates an isolated execution environment on untrusted computers. The safety requirements of our model quality verification are based on one of the TEE implementations, namely Inter SGX. Intel SGX is an extended instruction set of Intel x86 processors. It supports the creation of an isolated memory area called an enclave. In this area, all programs and data can be decrypted, quickly processed, and transparently encrypted outside the CPU package by the memory encryption engine using the key that only the processor hardware can access. In this trusted space, the CPU prohibits any untrusted software from accessing to protect the confidentiality and integrity of programs and data in the enclave. SGX supports two types of authentication, namely local and remote authentication, which can verify the correct initial state and authenticity of the enclave's trusted environment. Participants can safely perform confidential calculations in enclaves. Each local participant uploads their model to the enclave, and the task publisher uploads the test dataset to TEE through a secure channel. It can ensure that the models of local participants are not leaked, and also ensure the security of the task publisher's test dataset.

Moreover, since TEE will sign the models uploaded by participants, it can ensure that the models participating in the aggregation are consistent with the uploaded models. Participants are unable to tamper with the local model at different stages. During the validation phase, the task publisher will upload the test dataset to TEE through a secure channel to protect the test dataset.

6   Evaluation

In this section, we systematically evaluate the performance of the scheme. The experiments were performed on a PC with Inter(R) Core(TM) i9-10850k CPU @ 3.60GHz with 32GB memory and NVIDIA GeForce RTX 3060 GPU. We evaluated the scheme properties, the overhead of the solution, the accuracy of the model, and the deployment of TEE. Specifically, we first evaluated the properties of our scheme and its advantages compared to other solutions. Next, we calculated the cost and model accuracy. Finally, to evaluate the effectiveness of the model quality evaluation, we configured the Intel SGX environment using the local PC terminal and implemented the corresponding ports. The experimental environment is based on Ubuntu 20.04 and Graphene-SGX. Graphene-SGX is a lightweight LibOS that adds a layer of the environment to SGX-driven applications. The advantage of this is that it can be executed in the SGX environment without modifying the code of the corresponding module. Graphene SGX enables users to run original applications directly in the SGX environment. It includes Tensorflow, Python, and other deep learning frameworks.

Firstly, we compare and analyze the properties of our scheme. As shown in Table 1, compared with other schemes, our scheme can prevent attacks from malicious participants in a serverless environment. And there is basically no loss

Table 1. Comparison of properties

Schemes	Scheme in [2]	Scheme in [4]	Scheme in [11]	Scheme in [30]	Our Scheme
Architecture	Single-server	Single-server	None-server	None-server	None-server
Participant type	Trusted	Semi-trusted	Semi-trusted	Semi-trusted	Malicious
High model quality	–	–	✓	–	✓
Privacy-preserving approach	HE-based	MPC-based	DP-based	DP-based	HE-based
Lossless accuracy	✓	✓	✗	✗	✓
Dynamic participant behavior	Join+Leave	Leave	Join+Leave	Join+Leave	Join+Leave

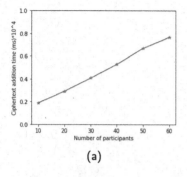 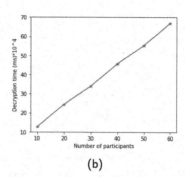

(a)   (b)

Fig. 3. The time cost of our scheme. (a) Ciphertext addition time as function of the number of participants (n); (b) Decryption time as function of the number of participants (n).

of model accuracy while supporting participants to join and leave. Compared with the existing solutions, our solution has more comprehensive functions, and the problems it addresses are more suitable for the actual situation.

Next, we evaluate the cost of the scheme. In multi-key homomorphic encryption schemes, the number of participants is one of the main factors affecting the cost. Therefore, we evaluated the relationship between ciphertext addition and decryption costs and the number of participants in the scheme. Because each experiment will have a certain error, we repeat the experiment 100 times to calculate the average value of the time required. We set the number of participants to 10, 20, 30, 40, 50, and 60. Data volume is set to 10^4. As shown in Fig. 3, the average time for ciphertext addition and decryption increases with the number of participants. The experimental results reflect the rationality of the grouping of participants. Different groups can perform aggregation and decryption work in parallel. Therefore, the actual time of each round of federated learning after grouping will be much less than the time overhead without grouping.

For model accuracy, we compare with traditional federated learning schemes. As shown in Table 2, we choose MNIST and CIFAR-100 as test datasets. We trained a three-layer neural network model on MNIST, where we set the learning rate to 0.02, the training batch size to 128, and the local epochs to 5 and 20, and used the SGD optimizer. In addition, we have trained a ResNet34 neural network

Table 2. Datasets used in the simulations

Dataset	Examples	Classes	Features	Models
MNIST	60000	10	784	3-layer neural network
CIFAR-100	60000	100	1024	ResNet34

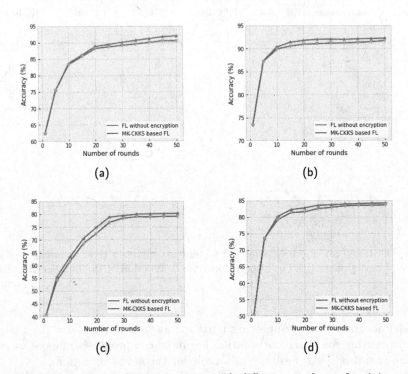

Fig. 4. Classification accuracy comparison with different numbers of training epochs locally (L) by each device in aggregation rounds. (a) Trained using MNIST (L = 5); (b) Trained using MNIST (L = 20); (c) Trained using CIFAR-100 (L = 5); (d) Trained using CIFAR-100 (L = 20).

model on the CIFAR-100 dataset, where the learning rate is set to 0.001, the training batch size is set to 128, and the number of local epochs is set to 5 and 20. Adam optimizer is used. The number of participants is set to 30, and the number of training rounds ranges from 1 to 50. We compared the accuracy of MK-CKKS based federated learning to traditional federated learning. As shown in Fig. 4, for each experiment, we repeated it twenty times and took the average value as the final result. Comparing the accuracy of our scheme with that of federated learning without encryption protection, our scheme has little accuracy loss and is very close to basic federated learning. The results prove the effectiveness of our scheme, which can keep the accuracy of the model while protecting the update of the model.

Finally, we simulated the process of model quality assessment. After each round of global model training is completed, the task publisher sends the global model and threshold to the participant's SGX through a secure channel. The SGX will execute the model evaluation algorithm and return the calculation result to the task publisher. The task publisher selects the participants who continue to participate in the training through the evaluation results.

7   Conclusion

We propose a privacy-preserving and reliable distributed federated learning. It fundamentally solves the single point of failure and the malicious problem of the parameter server. This scheme can protect the data privacy of participants and can detect the quality of model updates uploaded by participants. To ensure fairness, we have used an incentive mechanism based on reputation to encourage participants to undertake aggregation work and act honestly.

Acknowledgements. This work was supported by the NSF of China under Grants 61972159 and 62372177, Guangxi Key Laboratory of Cryptography and Information Security (no. GCIS202109), and the "Digital Silk Road" Shanghai International Joint Lab of Trustworthy Intelligent Software: Grant (no. 22510750100).

References

1. Abdelmoniem, A.M., Elzanaty, A., Alouini, M.S., Canini, M.: An efficient statistical-based gradient compression technique for distributed training systems. Proc. Mach. Learn. Syst. **3**, 297–322 (2021)
2. Aono, Y., Hayashi, T., Wang, L., Moriai, S.: Privacy-preserving deep learning via additively homomorphic encryption. IEEE Trans. Inf. Forens. Secur. **13**(5), 1333–1345 (2017)
3. Barreno, M., Nelson, B., Joseph, A.D., Tygar, J.D.: The security of machine learning. Mach. Learn. **81**, 121–148 (2010)
4. Bonawitz, K., Ivanov, V., Kreuter, B., et al.: Practical secure aggregation for privacy-preserving machine learning. In: Proceedings of the 2017 ACM SIGSAC Conference on Computer and Communications Security, pp. 1175–1191 (2017)
5. Chaudhuri, K., Monteleoni, C.: Privacy-preserving logistic regression. Adv. Neural Inf. Process. Syst. **21**, (2008)
6. Chen, H., Dai, W., Kim, M., Song, Y.: Efficient multi-key homomorphic encryption with packed ciphertexts with application to oblivious neural network inference. In: Proceedings of the 2019 ACM SIGSAC Conference on Computer and Communications Security, pp. 395–412 (2019)
7. Chen, L., Zhang, Z., Wang, X.: Batched multi-hop multi-key FHE from ring-LWE with compact ciphertext extension. In: Kalai, Y., Reyzin, L. (eds.) Theory of Cryptography. LNCS, vol. 10678, pp. 597–627. Springer, Cham (2017). https://doi.org/10.1007/978-3-319-70503-3_20
8. Fang, H., Qian, Q.: Privacy preserving machine learning with homomorphic encryption and federated learning. Future Internet. **13**(4), 94 (2021)

9. Fang, M., Cao, X., Jia, J., Gong, N.: Local model poisoning attacks to Byzantine-Robust federated learning. In: 29th USENIX Security Symposium (USENIX Security 20), pp. 1605–1622 (2020)
10. Fraboni, Y., Vidal, R., Lorenzi, M.: Free-rider attacks on model aggregation in federated learning. In: International Conference on Artificial Intelligence and Statistics, pp. 1846–1854. PMLR (2021)
11. Gao, Y., Zhang, L., Wang, L., Choo, K.K.R., Zhang, R.: Privacy-preserving and reliable decentralized federated learning. IEEE Trans. Serv. Comput. **16**(4), 2879–2891 (2023)
12. Gao, Y., Wang, L., Zhang, L.: Privacy-preserving verifiable asynchronous federated learning. In: Proceedings of the 2021 3rd International Conference on Software Engineering and Development, pp. 29–35 (2021)
13. Hazay, C., Mikkelsen, G.L., Rabin, T., Toft, T., Nicolosi, A.A.: Efficient RSA key generation and threshold paillier in the two-party setting. J. Cryptol. **32**, 265–323 (2019)
14. Hegedűs, I., Danner, G., Jelasity, M.: Gossip learning as a decentralized alternative to federated learning. In: Pereira, J., Ricci, L. (eds.) DAIS 2019. LNCS, vol. 11534, pp. 74–90. Springer, Cham (2019). https://doi.org/10.1007/978-3-030-22496-7_5
15. Hitaj, B., Ateniese, G., Perez-Cruz, F.: Deep models under the GAN: information leakage from collaborative deep learning. In: Proceedings of the 2017 ACM SIGSAC Conference on Computer and Communications Security, pp. 603–618 (2017)
16. Jayaraman, B., Evans, D.: Evaluating differentially private machine learning in practice. In: 28th USENIX Security Symposium (USENIX Security 19), pp. 1895–1912 (2019)
17. Kang, J., Xiong, Z., Niyato, D., Ye, D., Kim, D.I., Zhao, J.: Toward secure blockchain-enabled internet of vehicles: optimizing consensus management using reputation and contract theory. IEEE Trans. Veh. Technol. **68**(3), 2906–2920 (2019)
18. Konečný, J., McMahan, H.B., Yu, F.X., Richtárik, P., Suresh, A.T., Bacon, D.: Federated learning: strategies for improving communication efficiency. arXiv preprint arXiv:1610.05492 (2016)
19. Li, Y., Zhou, Y., Jolfaei, A., Yu, D., Xu, G., Zheng, X.: Privacy-preserving federated learning framework based on chained secure multiparty computing. IEEE Internet Things J. **8**(8), 6178–6186 (2020)
20. Liu, X., Li, H., Xu, G., Chen, Z., Huang, X., Lu, R.: Privacy-enhanced federated learning against poisoning adversaries. IEEE Trans. Inf. Forensics Secur. **16**, 4574–4588 (2021)
21. Lu, Y., Zhang, L., Wang, L., Gao, Y.: Privacy-preserving and reliable federated learning. In: Lai, Y., Wang, T., Jiang, M., Xu, G., Liang, W., Castiglione, A. (eds.) Algorithms and Architectures for Parallel Processing. LNCS, vol. 13157, pp. 346–361. Springer, Cham (2022). https://doi.org/10.1007/978-3-030-95391-1_22
22. Ma, Z., Ma, J., Miao, Y., Li, Y., Deng, R.H.: ShieldFL: mitigating model poisoning attacks in privacy-preserving federated learning. IEEE Trans. Inf. Forens. Secur. **17**, 1639–1654 (2022)
23. McMahan, B., Moore, E., Ramage, D., Hampson, S., Arcas, B.A.: Communication-efficient learning of deep networks from decentralized data. In: Artificial Intelligence and Statistics, pp. 1273–1282. PMLR (2017)
24. Nasr, M., Shokri, R., Houmansadr, A.: Comprehensive privacy analysis of deep learning: passive and active white-box inference attacks against centralized and federated learning. In: 2019 IEEE Symposium on Security and Privacy (SP), pp. 739–753. IEEE (2019)

25. Ormándi, R., Hegedűs, I., Jelasity, M.: Gossip learning with linear models on fully distributed data. Concurr. Comput. Pract. Exp. **25**(4), 556–571 (2013)
26. Prakash, A., Chitta, K., Geiger, A.: Multi-modal fusion transformer for end-to-end autonomous driving. In: Proceedings of the IEEE/CVF Conference on Computer Vision and Pattern Recognition, pp. 7077–7087 (2021)
27. Schirrmeister, R.T., Springenberg, J.T., Fiederer, L.D.J., et al.: Deep learning with convolutional neural networks for EEG decoding and visualization. Hum. Brain Mapp. **38**(11), 5391–5420 (2017)
28. Shokri, R., Shmatikov, V.: Privacy-preserving deep learning. In: Proceedings of the 22nd ACM SIGSAC Conference on Computer and Communications Security, pp. 1310–1321 (2015)
29. Vaidya, J., Kantarcıoğlu, M., Clifton, C.: Privacy-preserving naive bayes classification. VLDB J. **17**(4), 879–898 (2008)
30. Weng, J., Weng, J., Zhang, J., Li, M., Zhang, Y., Luo, W.: Deepchain: auditable and privacy-preserving deep learning with blockchain-based incentive. IEEE Trans. Depend. Secure Comput. **18**(5), 2438–2455 (2019)
31. Wu, M., Ye, D., Ding, J., Guo, Y., Yu, R., Pan, M.: Incentivizing differentially private federated learning: a multidimensional contract approach. IEEE Internet Things J. **8**(13), 10639–10651 (2021)
32. Xiong, W., Droppo, J., Huang, X., et al.: Achieving human parity in conversational speech recognition. arXiv preprint arXiv:1610.05256 (2016)
33. Xu, G., Li, H., Liu, S., Yang, K., Lin, X.: Verifynet: secure and verifiable federated learning. IEEE Trans. Inf. Forens. Secur. **15**, 911–926 (2019)
34. Yousuf, H., Lahzi, M., Salloum, S.A., Shaalan, K.: Systematic review on fully homomorphic encryption scheme and its application. In: Al-Emran, M., Shaalan, K., Hassanien, A.E. (eds.) Recent Advances in Intelligent Systems and Smart Applications. SSDC, vol. 295, pp. 537–551. Springer, Cham (2021). https://doi.org/10.1007/978-3-030-47411-9_29
35. Zhan, Y., Zhang, J., Hong, Z., Wu, L., Li, P., Guo, S.: A survey of incentive mechanism design for federated learning. IEEE Trans. Emerg. Top. Comput. **10**(2), 1035–1044 (2021)
36. Zhang, J., Wu, Y., Pan, R.: Incentive mechanism for horizontal federated learning based on reputation and reverse auction. In: Proceedings of the Web Conference 2021, pp. 947–956 (2021)
37. Zhong, L., Zhang, L., Xu, L., Wang, L.: MPC-based privacy-preserving serverless federated learning. In: 2022 3rd International Conference on Big Data, Artificial Intelligence and Internet of Things Engineering (ICBAIE), pp. 493–497. IEEE (2022)

Joint Optimization of Request Scheduling and Container Prewarming in Serverless Computing

Si Chen[1], Guanghui Li[1(✉)], Chenglong Dai[1], Wei Li[1], and Qinglin Zhao[2]

[1] School of Artificial Intelligence and Computer Science, Jiangnan University,
Wuxi 214122, Jiangsu, China
ghli@jiangnan.edu.cn
[2] School of Computer Science and Engineering, Macau University of Science
and Technology, Macau 999078, China

Abstract. Serverless computing has emerged as a compelling paradigm for deploying applications and services due to its elastic scalability in response to changing demand. However, it often suffers from cold-start problems due to the overhead of initializing code and data dependencies. Thus, it's necessary to keep a container warm after the container completes function processing and reuse it for subsequent requests. However, existing schedulers of Serverless platforms, which are usually load balancers, may not efficiently locate the reusable containers. Moreover, elastic scalers only scale passively based on predefined thresholds, making it challenging to handle burst requests. This paper introduces a scheduling algorithm called Consistent Hash-based Affinity Scheduling (CHAS), which aims to increase the chances of reusing warm containers by assigning functions to appropriate working nodes. We also propose a container prewarming strategy called LSTM-NB that uses a Long Short-Term Memory Network (LSTM) to predict the parameters of the negative binomial distribution (NB). This strategy performs joint learning of call time series of multiple functions and predicts future function calls to actively warm up or evict containers, thereby reducing cold start latency and excessive resource consumption. We build a serverless computing environment by using the SimPy discrete-event simulation framework to evaluate the proposed method. The results show that CHAS can reduce the cold start rate by an average of 10.5% compared to baseline approaches. Furthermore, an average reduction of 20.1% in the cold start rate can be achieved by prewarming with the LSTM-NB strategy.

Keywords: Serverless Computing · Request Scheduling · Proactive Prewarming · Elastic Scaling · Time Series Analysis

1 Introduction

The gap between the resources paid for by customers and the actual resource usage in cloud computing Infrastructure as a Service (IaaS) causes a waste of

Z. Tari et al. (Eds.): ICA3PP 2023, LNCS 14487, pp. 150–169, 2024.
https://doi.org/10.1007/978-981-97-0834-5_10

resources and increased costs for cloud customers. Function as a Service (FaaS), or serverless computing, was introduced to address this issue. It is based on container virtualization and describes a more fine-grained deployment model that divides traditional applications into fine-grained functions. Each function instantiates containers of that type on a working node to handle a particular request, which elastically scales the number of containers based on actual requests. However, due to the diversity of workloads, when a serverless platform starts a new container to handle a user request, it requires initialization steps such as setting up the runtime environment, loading the relevant function code and dependencies and so on, before it can serve the request. This process incurs a time overhead known as a cold start [17].

Many existing FaaS schedulers are uninspired load balancers that may ignore nodes with idle containers for load balancing, leading to cold starts. Additionally, existing elastic scalers, such as Kubernetes [3], are passive scaling systems where users predefine rules or thresholds and perform passive scaling operations by analyzing the current system state. However, this approach makes it difficult to cope with the bursts of requests. Moreover, cloud platforms such as AWS Lambda [1] and Azure function [2], as well as the open-source framework OpenWhisk [4], use a fixed "time-to-live" (TTL) policy that keeps resources in memory separately for a period after a function has been executed for reuse. Although this policy is simple and practical, it ignores the actual frequency and pattern of function calls, which can result in poor performance and waste of resources.

Recently, many studies have focused on optimizing cold start latency in serverless computing and generally can be divided into two categories: reusing idle containers to reduce the number of cold starts [9, 11, 19, 20, 22, 23] and pooling resources to reduce the time spent on starting up cold containers [6, 7, 16, 21, 24].

Reusing idle containers can be achieved in two ways. One is, reserving containers for a specific duration after function execution to prioritize the incoming requests to avoid cold starts. Another way is by predicting the expected number of function instances that may be called shortly, analyzing historical data and warming up the containers in advance. This approach enables the later requests to reuse the warmed-up containers. Shahrad et al. [19] proposed a combined prewarm and TTL (time-to-live) strategy that uses a histogram of arrival intervals to predict future function calls and eagerly remove the warm functions. Dang-Quang et al. [9] proposed an architecture for a Kubernetes-based system with an actively customized elastic scaler. Fuerst et al. [11] transformed the container retention mechanism into a caching tool and applied the concepts of cache reuse distance and hit curves to scale server resource provisioning automatically. Suo et al. [20] proposed HotC, a container-based runtime management framework that maintains a pool of active container runtimes and designs an adaptive hot container control algorithm by combining exponential smoothing and Markov chains. Vahidinia et al. [22] used reinforcement learning algorithms to discover function call patterns over time to determine the optimal time to keep alive. Future function calls were predicted using LSTM to determine the number of prewarm containers required. Wu et al. [23] proposed a container lifecycle-aware reuse policy by prioritizing the reuse of idle containers, avoiding the centralized placement leading to container eviction and separating the requests for new containers from container allocations.

Resource pooling means that instead of fully prewarming a container, the relevant resources needed to instantiate a container are prepared in advance, such as preloading the required packages. Abad and Aumala et al. [6,7] proposed an adaptive warm-up (AWU) strategy and an adaptive container pool scaling (ACPS) strategy to reduce function start-up time. Routing all function requests that require a specific packet to the same work node maximizes cache hit rates and reduces cold start times. Suresh et al. [21] proposed that the ENSURE system uses queueing theory techniques to handle CPU resource-holding activities and resource provisioning. Mampage et al. [16] suggested minimizing the service provider's cost while meeting the user's application requirements (i.e., deadlines)-dynamic management of resources to reduce function response times.

In summary, the main challenges in reusing container types lie in providing enough containers and scheduling requests to release containers once they are ready. This study proposes an approach to mitigate cold starts by combining request scheduling with active container warm-up. The main contributions of this work are summarized as follows:

(1) An improved Consistent Hash Affinity Scheduling Algorithm (CHAS) is proposed, which prioritizes the allocation of function instances to appropriate computing nodes to increase the probability of reusing containers. At the same time, it also avoids ignoring idle containers on other nodes, thus preventing overlooking work nodes with reusable containers.
(2) To ensure sufficient idle containers, a container prewarming strategy called LSTM-NB, which is based on long short-term memory networks (LSTM) to predict the parameters of the negative binomial distribution (NB), is presented. This strategy employs historical time-series analysis of request patterns to predict future function calls, enabling proactive prewarming or eviction of containers to reduce cold start latency and excessive resource consumption. To promote joint learning of time series data for multiple functions with varying calling rates, a scaling factor is introduced to reduce prediction errors.
(3) A simulation environment for serverless computing is built using the SimPy discrete-event simulation framework in Python [5]. Numerous simulation experiments shows that CHAS could reduce cold start rates by 4.6%, 17%, and 9.8%, respectively, on three workloads compared to baseline algorithms, while LSTM-NB could reduce cold start rates by 49.2%, 6.8%, and 4.3% respectively on the same three workloads compared to baseline strategies.

2 System Model

The serverless computing architecture proposed in this paper is shown in Fig. 1. The system consists of a Controller and a Cluster composed of several homogeneous computing nodes. Here, computing nodes represents the virtual machines or physical machines.

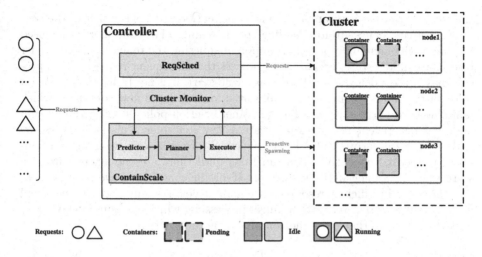

Fig. 1. System model.

At a high-level, incoming function requests are received by Controller. The three core components of the Controller that handle resource management across nodes are the monitoring component (Cluster Monitor), the scheduling component (ReqSched) and the elastic scaling component (ContainScale).

Cluster Monitor continuously tracks the available resources, processed requests, and metadata, such as container pools for each worker node in the cluster.

ReqSched uses the metadata collected by the Monitor component to select an appropriate node to which requests will be scheduled. At the chosen node, a function-specific container is launched if no free containers are available to execute the incoming request.

Additionally, to cope with the changes in request traffic, the Predictor in ContainerScale analyzes and predicts the next moment of requests based on the time series data of historical requests in Cluster Monitor. The Planner decides to warm up or evict containers based on the prediction and calculates the number of operations accordingly. The Executor receives the command to change the number of containers proactively. The corresponding metrics of Cluster Monitor will also change in response to traffic changes, avoiding under or over-provisioning containers.

Each node in the Cluster maintains a pool of containers with varying lifecycle states, as illustrated in Fig. 2. When a node receives a request, it undergoes two processing steps. Firstly, it searches for available containers that correspond to the required function. If a free container is found, i.e., a container with the status "Idle" exists for that function type, it is directly reused to process the request. The container's status is changed to "Running" during processing, and the process is warm-started. However, if a container with the status "Idle" is unavailable, a new container is created, and its status remains "Pending" until the creation process is completed. Once the creation process is finished, the container's status changes from "Pending" to "Running." The container is then used to process the arriving request, and upon completing the request, the container returns to an idle state, preparing to process the subsequent requests. This process may cause a delay and is known as a cold start. It is necessary to note that a cold start requires resources to be allocated for creating a new container, which limits node resource capacity. The container eviction policy is triggered when a node has insufficient resources to create a new container. This policy uses the LRU (Least Recently Used) algorithm, which prioritizes evicting the containers of the least frequently called functions until there are enough resources to start a new container after the eviction. If all idle containers have been evicted, and there are insufficient resources to create a new container, the request will need to wait for other requests to finish processing, which can cause delays. It is important to note that only idle containers can be evicted.

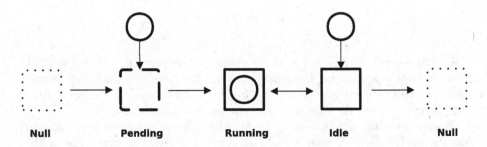

Fig. 2. Lifecycle of a container.

The research focuses explicitly on scheduling memory-constrained requests for single-function applications. In the subsequent sections, a working node is denoted by the symbol v, while the j-th node is represented as v_j, with each node having an available memory resource capacity of v_j^m. The available free memory resources in v_j at a given time t is denoted by $v_j^m(t)$. The ReqSched receives a sequence of function call requests, where r_i represents the i-th request. Each request carries two attributes: r_i^{type} indicates the type of function to be called and r_i^m indicates the memory resources required by the function instance (container) processing the request. To minimize cold starts by reusing as many containers in the cluster as possible, we have designed a request scheduling algorithm called CHAS in this paper. Moreover, in the cases where insufficient containers are available for reuse, the warm-up container policy, LSTM-NB, is required to prepare containers in advance.

3   Proposed Solution

3.1   Scheduling Requests Across Invokers via ReqSched

The proposed Consistent Hash Affinity Scheduling Algorithm (Algorithm 1) aims to reuse containers as many as possible and reduce cold starts. The main idea is to schedule the requests of same type to one or more nodes (one node for small request types and multiple nodes for large ones), facilitating the reuse of idle containers.

Algorithm 1: Consistent Hash Affinity Scheduling Algorithm (CHAS)

Input: Function requests r_i, $r_i = < r_i^{type}, r_i^m, t >$; List of nodes that can be selected for scheduling $V = \{v_1, v_2, v_3, ..., v_N\}$; Information for each node $v_j = < v_j^m, v_j^{am}(t) >$

Output: Selected computing nodes v

1 $j_{conhash} \leftarrow Get\_ConHash\_id(r_i^{type})$, $j \leftarrow j_{conhash}$, $A \leftarrow$ null, $B \leftarrow$ null, $m = -1$
2 **while** $(j + 1) \bmod N! = j_{conhash}$ **do**
3     $n \leftarrow Get\_Container\_Info(j, r_i)$
    // n_{idle} - number of idle containers of r_i^{type} on node j
    // $n_{running}$ - number of running containers of r_i^{type} on node j
4   **if** $n_{idle} > 0$ **then**
5 | **return** v_j
6   **end**
7   **if** $r_i^m \leq v_j^{am}(t)$ *and* $n_{running} > m$ **then**
8 | $A \leftarrow j$, $m \leftarrow n_{running}$
9   **else if** $Get\_Evictable\_resources(j) \geq r_i^m$ *and* $B = null$ **then**
10 | $B \leftarrow j$
11   **end**
12 **end**
13 **if** $A! = null$ **then**
14 | **return** v_A
15 **else if** $B! = null$ **then**
16 | **return** v_B
17 **else**
18 | **return** null
19 **end**

The goal of CHAS is to select a node and sequentially attempt to find an affinity node with enough resources, or a node that can accept the request by evicting containers. Line 1 employs the consistent hashing algorithm [14] to obtain the affinity node for the current request. Lines 4–6 identify the idle containers on the affinity node and select that node directly. Lines 7–8 select the node that has sufficient resources and the highest number of containers serving the same type of request. Lines 9–10 identify a node that can accept requests by container eviction. However, if the free resources released after evicting all idle containers are still insufficient to create a new container, the algorithm returns the empty node (Lines 17–18), indicating that scheduling has failed and the system must wait until enough requests have been processed and released. Once a node has been selected, the request is forwarded to the selected computing node for execution, where the free container is reused, or a new container for the requested function is created, and the request is subsequently serviced.

3.2 Elastically Scaling Serverless Capacity via ContainScale

The purpose of this component is to elastic scale the number of idle containers. The number of calls is predicted over time, and the number of prewarmed containers is determined. This is a time series problem, as the number of future concurrent calls depends on the pattern of previous calls.

There is a wide variety of functions in the serverless computing environment. We attempt to use a long short-term memory network (LSTM) based on a negative binomial distribution to perform joint time-series learning of multiple functions.

From the system architecture presented in Fig. 1, it is clear that the Executor layer of ContainScale component goes through a Prediction phase and a Planning phase before executing the number of scaling containers, which is described in detail as follows.

Prediction Phase - LSTM-NB Predicts the Number of Requests. In the prediction phase, the model uses the latest w-length time series data $d_{i,t-w+1},...,$ $d_{i,t-2}$, $d_{i,t-1}$, $d_{i,t}$ of HTTP requests collected by the Cluster Monitor to predict the next data point in the sequence, $d_{i,t+1}$. To achieve this, LSTM-NB neural network model is proposed. This model predicts the workload of future HTTP requests by predicting the negative binomial distribution parameters of future time series while also incorporating covariates $z_{i,t}$ and a scale factor v_i. Each function sequence is assigned a unique number as a covariate, and the embedding layer is added to the network. The scale factor addresses the problem of significant differences in magnitude between multiple time series, which commonly occurs in numerous time series forecasting.

A detailed illustration of the LSTM-NB neural network model is provided in Fig. 3. LSTM is a modified version of the Recurrent Neural Network (RNN) network. LSTM incorporates gates to regulate the input, output, and forgetting of memory units in order to address the issue of gradient vanishing observed in conventional RNNs. This enables the model to better handle long-term dependencies in sequential data. The gates include input gates, forgetting gates, and output gates.

In this study, we use LSTM to predict future tensors' probability distribution by introducing parameters. We chose the negative binomial distribution to be the most effective since the serverless computing scenario involves counting data. Although both Poisson and negative binomial distributions can characterize count-like data, the default mean of the Poisson distribution is equal to its

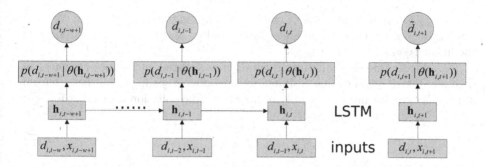

Fig. 3. LSTM-NB network diagram.

variance, which may not be the case in this scenario. Therefore, after experimenting with the negative binomial distribution, we found it more appropriate.

The negative binomial distribution describes the probability of success after a fixed number of failures, and its probability mass function is shown in Eq. (1):

$$P(k|r,p) = \binom{k+r-1}{r-1} p^r (1-p)^k \tag{1}$$

where r represents the required number of failures, p represents the probability of success, and $l-p$ represents the probability of failure. Equation (2) and Eq. (3) express r and p using the mean μ and variance σ^2, respectively.

$$p = \frac{\sigma^2 - \mu}{\sigma^2} \tag{2}$$

$$r = \frac{\mu^2}{\sigma^2 - \mu} \tag{3}$$

Therefore, μ and σ parameterize the probability mass function, as shown in Eq. (4).

$$P(k|\mu,\sigma) = \binom{k + \frac{\mu^2}{\sigma^2-\mu} - 1}{k} \left(\frac{\mu}{\sigma}\right)^{\frac{\mu^2}{\sigma^2-\mu}} \left(\frac{\mu^2}{\sigma^2 - \mu}\right)^k \tag{4}$$

In order to obtain the parameters μ and σ necessary for parameterizing the negative binomial distribution that governs the predicted values, the following approach is employed: Initially, a mean value μ and a shape parameter α are assigned to the output of the neural network. Then, σ is calculated using the values of μ and α. Since both μ and α are non-negative parameters, the output of the neural network is subjected to a linear layer and a softplus layer. The resulting outputs, represented as o_μ and o_α, are presented in Eq. (5) and Eq. (6):

$$o_\mu = log(1 + exp(\mathbf{W}_\mu^T \mathbf{h}_{i,t} + b_\mu)) \tag{5}$$

$$o_\alpha = log(1 + exp(\mathbf{W}_\alpha^T \mathbf{h}_{i,t} + b_\alpha)) \tag{6}$$

where $\mathbf{h}_{i,t}$ is the hidden value of the LSTM training output, $\mathbf{W}_\mu$ and $\mathbf{W}_\alpha$ are the parameter matrices of the linear layer, and b_μ and b_α are the bias terms of the linear layer.

To address the issue of varying time series lengths and severe skewness in their size distribution during training, we introduce a scaling factor v_i, as shown in Eq. (7), which cannot be simply normalized. The equation is given as follows:

$$v_i = 1 + \frac{1}{t_0} \sum_{t=1}^{t_0} d_{i,t} \tag{7}$$

where t_0 is the length of the current history window of non-missing time. This scaling factor compensates for the differences between time series. It allows non-uniform sampling during training set sampling, with more extensive sequences sampled more frequently than those with fewer calls. This helps to ensure that the trained model is generalizable.

It's worth noting that data needs to be preprocessed by a scaling factor before training, and its parameters are scaled and transformed. The inverse transform involves a transformation as shown in Eq. (8) and Eq. (9):

$$\mu = v_i log(1 + \exp(o_\mu)) \tag{8}$$

$$\alpha = log(1 + exp(o_\alpha)) \div \sqrt{v_i} \tag{9}$$

where o_μ and o_α the output parameters of the neural network as in Eq. (5) and Eq. (6), and v_i is the scaling factor calculated as in Eq. (7). Variance σ^2 is obtained by converting the scaled shape parameter o_α to the mean value μ, as shown in Eq. (10):

$$\sigma^2 = \mu + \mu^2 \alpha \tag{10}$$

The neural network is trained with a loss function, as shown in Eq. (11) to maximize the log-likelihood probability:

$$L = \sum_{i=1}^{N} \sum_{t=t_0}^{T} logP(d_{i,t}|\theta(\mathbf{h}_{i,t})) \tag{11}$$

where $d_{i,t}$ is known in the training set, $\mathbf{h}_{i,t}$ is the hidden value of the neural network output, and $\theta(\mathbf{h}_{i,t})$ indicates that $\mathbf{h}_{i,t}$ is calculated after the neural network parameters and Eqs. (5)–(10).

Finally, the upper bound of the 95% confidence interval of the negative binomial distribution is used to predict the value $\tilde{d}_{t+1}$ at the next moment. This prediction is obtained from μ and σ, as shown in Eq. (12).

$$\tilde{d}_{t+1} = \mu + 1.96\sigma \tag{12}$$

Planning Phase. This phase is to calculate the number of containers that need to be warmed up or evicted in terms of the workload of HTTP requests predicted in the previous step, as shown in Algorithm 2, i.e., the aim is to scale up or down the number of containers to meet future workloads.

Algorithm 2: Adaptively scaling containers

Input: Request load forecast for the next time period $\tilde{d}_{t+1}$; Number of containers of r_i^{type} in the cluster $num\_avail[r_i^{type}]$

Output: Scaling action

1 **if** $\tilde{d}_{t+1} > num\_avail[r_i^{type}]$ **then**
2 $\quad$ | $\quad$ Prewarm $\tilde{d}_{t+1} - num\_avail[r_i^{type}]$ function containers
3 **else**
4 $\quad$ | $\quad$ Evict $num\_avail[r_i^{type}] - \tilde{d}_{t+1}$ function containers
5 **end**

---

4 Experiments and Discussion

4.1 Setup of the Simulation Environment

To evaluate the proposed methods, To evaluate the proposed methods, we built a serverless simulation environment using Python's discrete-event simulation framework SimPy which was also used in the related work [6,7,15]. The simulator abstracts and models the various entities of a serverless computing environment, as shown in Table 1. SimPy processes were employed for the user request execution process, Broker, ReqSched, ContainScale, and Monitor. The Broker process submitted requests to the Cluster instance according to their submission time until all requests were submitted. ReqSched scheduled requests depend on time steps until the Simulation is marked as finished. Monitor observed and recorded the Simulation's state at regular time intervals until the Simulation was deemed completed. The ContainScale function performed pre-warming or eviction of containers and adjusted the container count using time steps. The remaining entities were standard classes responsible for managing relevant information. The simulation experiments were carried out on an AMD 3.20 GHz Core R7 system with 16 GB RAM.

4.2 Processing of the Dataset

A diverse sample of traces was drawn from the Azure Function Trace dataset [19]. This dataset encompasses over 50,000 functions and includes minute-level call details, execution times, application memory, and other attributes, covering 14 d from July 15 to 28, 2019. For the evaluation, the first day's data was used as the workload, focusing on the functions with more than two calls to facilitate ease of reuse. Before conducting simulations, the raw dataset underwent preprocessing

Table 1. Modelling and definition of simulation entities.

Entity	Definition
Request	The requests submitted by users
Broker	Submit requests to the cluster instead of the user
Container	The container that processes the request
LambdaData	The metadata of the request and container
Machine	The computing nodes
Cluster	Aggregate all nodes and all submitted requests
Algorithm	The interface to the proposed method
ReqSched	The request scheduler
Monitor	Detect and log the state of the simulation
Predictor	Calculate the number of requests for the next time
Planner	Decides whether to prewarm or evict containers and the number
Executor	Perform a pre-warming or eviction
ContainScale	One elastic scaling process
Simulator	One simulation process

to ensure its suitability for the simulation environment. First and foremost, an assumption was made regarding the arrival time of requests within a minute. If there was only one request in a minute, it was assumed to arrive at the beginning of that minute. This assumption was made because the function call attribute in the raw dataset represents the sum of calls at the minute granularity and does not provide the exact arrival time for each request. For cases involving multiple calls, it was assumed that they arrived at equal intervals within a minute. Furthermore, since the raw dataset only accounted for application-level memory consumption, and an application may consist of one or more functions, the memory was evenly distributed among all functions. Lastly, the simulation execution time for each function was set using the average execution time, and the cold start overhead was determined by subtracting the average runtime from the maximum runtime. After the dataset's preprocessing, smaller samples were taken for analysis on the simulator. The proposed method's performance was evaluated by sampling three different types of workloads (further details are provided in Table 2).

Table 2. Invocations and average inter-arrival time (IAT) for different workloads.

Trace	Num Invocations	Reqs per sec		Avg.IAT
		Avg	Max	
Rare	14,022	0.16 /s	39 /s	6157.46 ms
Representative	139,334	2.36 /s	45 /s	620 ms
Dense	2,033,870	23.54 /s	861 /s	42.48 ms

Rare: a random sample of the 100 least frequently called functions. These functions are requested at longer intervals on average, making it difficult to reuse containers.

Representative: a random sample of 100 representative functions, obtained by sampling 25 functions from each quartile of the dataset based on call frequency, resulting in higher functional diversity.

Dense: a sample of 100 functions, sampled from the fourth quartile of the dataset based on frequency.

4.3   Evaluation Metrics

Metrics for Cold Starts. We focus on three metrics related to cold starts: the number of cold starts, the average cold start rate, and the cumulative distribution function (CDF) of all requests' average delay or latency. The average cold start rate and average latency are calculated as the average of all function calls. These metrics will help us evaluate the performance of the proposed method in reducing cold start latency.

Metrics for Elastic Scaling. To evaluate the Elastic Scaling approach for the ContainerScale module, we use the performance metrics recommended by the Standard Performance Evaluation Corporation (SPEC) [8], as shown in Eqs. (13)–(17), which have also been used in references [12,13]. r_t and p_t represent the number of requests and the number of available idle containers at time t, respectively. Additionally, Δt denotes the time interval between scalings. The metrics Θ_u and Θ_o denote the proportion of under-provisioned and over-provisioned containers, respectively, while T_u and T_o indicate the duration when the system is under-provisioned and over-provisioned, respectively. Finally, ε_n serves as an indicator for assessing the relative performance of two methods. When $\varepsilon_n < 1$, it indicates that method n outperforms method a, and vice versa.

$$\Theta_u[\%] = \frac{100}{T} \sum_{t=1}^{T} \frac{max(r_t - p_t, 0)}{r_t} \Delta t \tag{13}$$

$$\Theta_o[\%] = \frac{100}{T} \sum_{t=1}^{T} \frac{max(p_t - r_t, 0)}{r_t} \Delta t \tag{14}$$

$$T_u[\%] = \frac{100}{T} \sum_{t=1}^{T} max(sgn(r_t - p_t), 0) \Delta t \tag{15}$$

$$T_o[\%] = \frac{100}{T} \sum_{t=1}^{T} max(sgn(p_t - r_t), 0) \Delta t \tag{16}$$

$$\varepsilon_n = \left(\frac{\Theta_{u,n}}{\Theta_{u,a}} \frac{\Theta_{o,n}}{\Theta_{o,a}} \frac{T_{u,n}}{T_{u,a}} \frac{T_{o,n}}{T_{o,a}} \right)^{\frac{1}{4}} \tag{17}$$

4.4　Experiment on Single-Node Container Warm-Up

To mitigate the effects of cold starts resulting from neglecting idle container nodes during request scheduling, we opted to assess the elastic scaling method (LSTM-NB) within a simulated serverless environment on a single node. The following baseline methods were compared:

Moving Average (MA): the default auto-scaling method for Kubernetes, which determines the scaling capacity by averaging the number of requests in the previous moving window. In the experiments, we set the container-concurrency-target parameter to 1, as we don't allow containers to process requests concurrently.

Simple Exponential Smoothing (SES): a weighted moving average method that assigns weights to recent observations during the observation period to enhance their influence on predicted values [10].

Double Exponential Smoothing (DES): a re-smoothing of primary exponential smoothing suitable for short-term request prediction problems [10,20].

Triple Exponential Smoothing (TES): a re-smoothing of the double exponential smoothing, better-capturing requests' fluctuation.

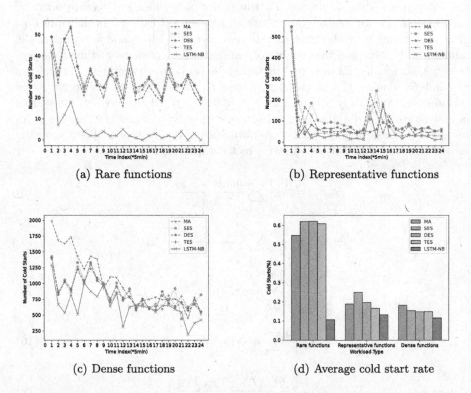

(a) Rare functions　　　　　　　　(b) Representative functions

(c) Dense functions　　　　　　　　(d) Average cold start rate

Fig. 4. Cold start with five warm-up strategies for three types of workloads.

To avoid cold start latency caused by container eviction due to insufficient resources, we ensured that the simulation experiments had sufficient resources by setting the VM's memory to 100G. Additionally, to eliminate the influence of scheduling on the results, we used a single node with sufficient resources to conduct experiments under three different workloads. The sub-dataset of the first 22 h was used as the training set, and the data of the succedent 2 h were used as the test set.

To obtain information on the number of cold starts, cold start rate, and request latency, we simulated requests for the last 2 h on the simulator. We set the monitor simulation time step to 5 min, meaning that we counted the data every 5 min for a total of 12 times. The change in the number of cold starts counted every 5 min under the three workloads listed in Table 2 is shown in Fig. 4.

For the Rare functions (Fig. 4(a)), the short-term prediction-based approach caused more cold starts due to the long request interval. LSTM-NB was significantly better than the short-term prediction based algorithms for reducing cold starts. The framework is designed to reuse containers to reduce cold starts effectively. However, the first visit of a specific type of request still requires a cold start, resulting in more cold starts during the 1st period of the experiment. After initializing the container, subsequent requests can be reused to reduce cold starts. Compared with the other methods whose cold starts increased over the 2nd to 4th period, LSTM-NB quickly analyzed the data and prewarmed the most effective containers. SES, DES, and TES depend on relatively recent time predictions, so the predictions are similar but not very effective for workloads with long call intervals.

For Representative functions (Fig. 4(b)), the typical workloads with high and low invocation frequencies, LSTM-NB was still effective in reducing cold starts. In addition, we observed that when the volume of requests fluctuated wildly from the 12th to the 16th time interval, LSTM-NB quickly captured the increase and decrease in requests and could warm up enough containers or evict excess containers soon, reducing cold starts and resource losses. Comparatively, the baselines showed delays in responding to load changes.

For Dense functions (Fig. 4(c)), there are small request intervals and large request fluctuations, LSTM-NB effectively captured request changes, prewarmed or evicted containers, and reduced cold starts with the best performance.

Figure 4(d) shows the average cold start rate for two hours of cold starts requested by the three modes, and it can be observed that LSTM-NB has the lowest cold start rate for all three types of loads. For the Rare mode, where the analysis of long-term historical data is essential due to the large request interval, LSTM-NB is significantly effective, followed by MA, while the short-term predictions of SES and DES are the least effective. However, for Dense patterns with shorter request intervals, SES and DES outperformed MA, except for LSTM-NB, which was more effective. In summary, LSTM-NB reduced cold starts with the best results in all three mobilization patterns. MA was more suitable for request patterns with more significant time intervals. SES, DES, and TES were ideal for smaller time intervals of Dense request patterns, but TES working better than SES and DES in all three workloads.

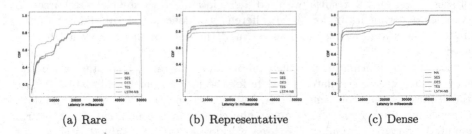

(a) Rare   (b) Representative   (c) Dense

Fig. 5. CDF plot of average latency in milliseconds for five policies under three workloads.

We conducted experiments to assess the impact of different warm-up methods on request latencies, as shown in Fig. 5. Figure 5 presents cumulative distribution functions (CDFs) of request latencies obtained using different warm-up methods for each workload type.

For Rare functions (Fig. 5(a)), on average, LSTM-NB reduces latency by 51%, 46%, and 45% compared to MA, SES, and DES, respectively. In addition, significant improvements can be observed in the percentile range [40–90] of CDFs, such as a reduction of 81% in request latency at the 60th percentile than MA and a decrease of 52% and 81% at the 95% tail latency than SES, DES, and TES, respectively.

For Representative functions (Fig. 5(b)), LSTM-NB demonstrates the most significant improvement in the percentile range [75–85] of CDFs, where request latency at the 75th, 80th, and 85th percentiles is 54%, 92%, and 98% less than SES, respectively. TES's latency is slightly less than LSTM-NB at the 90th percentile of CDFs, possibly due to the heterogeneity of the functions that make the request latency and cold start not strictly correlated.

For Dense functions (Fig. 5(c)), on average, LSTM-NB reduces latency by 28% compared to the least effective MA. At the same time, LSTM-NB reduces latency by 28%, 24% and 23% compared to MA, SES and DES, respectively. However, analyzing the CDFs, it can be noted that the most significant improvements are in the percentile range [80–92]. For example, request latency is reduced by 60% at the 80th percentile compared to MA, and LSTM-NB is reduced by approximately 38% at the 92-th percentile compared to SES, DES, and TES.

Table 3 presents a comparison between the elastic scaling method (LSTM-NB) and four baselines. LSTM-NB exhibits the lowest values for Θ_u and T_u under all three workloads, suggesting that it is less prone to under-provisioning, regardless of request interval. Furthermore, by analyzing the historical data of requests, LSTM-NB can accurately predict the next moment, enabling it to prewarm the container and reduce cold starts. Although LSTM-NB has higher

Table 3. Metric results for Elastic Scaling.

Workload	Policy	$\Theta_u[\%]$	$\Theta_o[\%]$	$T_u[\%]$	$T_o[\%]$	ε_n
Rare	LSTM-NB	**0.085**	3.183	**0.207**	99.793	1
	MA	6.188	**0.074**	6.733	21.483	0.539
	SES	6.228	0.186	6.310	**6.448**	0.586
	DES	5.380	0.102	6.681	36.440	0.452
	TES	5.303	0.100	6.664	36.491	0.456
Representative	LSTM-NB	**0.132**	10.792	**0.672**	99.328	1
	MA	7.517	**0.769**	11.000	18.681	0.532
	SES	6.270	1.182	7.069	**7.147**	0.710
	DES	5.636	1.143	10.741	28.560	0.468
	TES	5.570	1.260	10.776	28.595	0.458
Dense	LSTM-NB	**0.659**	22.397	**2.871**	97.129	1
	MA	8.269	6.379	18.078	21.474	0.670
	SES	6.778	6.069	11.276	**11.388**	0.940
	DES	6.057	**5.987**	17.483	34.216	0.660
	TES	6.051	6.022	17.474	34.310	0.659

values for Θ_o and T_o, indicating a greater risk of over-provisioning, the combination of the other four metrics, represented by ε_n, is less than 1 when compared to the other methods, demonstrating that LSTM-NB outperforms all baselines in elastic acceleration.

4.5   Evaluation of Multi-node Request Scheduling

In this section, we evaluate the performance of the proposed method, CHAS, in a more challenging and practical multi-caller scenario. To demonstrate the advantages of CHAS, we compare its performance against three baselines:

Least Loaded (LL), which assigns requests to the working node with the lowest number of active connections (similar to NGINX's Least Connected).

Random which distributes requests randomly among staffs based on user-defined staff weights.

PASch [7], which uses a combination of consistent hashing and the power of 2 choices [18] to schedule requests and selects the node that is currently processing the fewest requests for scheduling when the scheduled node is under-resourced.

We conducted experiments using diverse clusters consisting of 30, 40, 50, 60, 70, 80, and 90 VMs. Within each cluster, we performed experiments involving three distinct workload types: Rare, Representative, and Dense functions. Given the substantial differences in request volumes across these workloads, the memory allocation for VMs in each cluster was individually configured as 1GB, 3GB, and 5GB to cater to the requirements of the three load scenarios. Instead of adding a warm-up mechanism, we used a simple TTL policy to keep the containers warm. Here, we set the TTL time to 10 s to actively evict the containers.

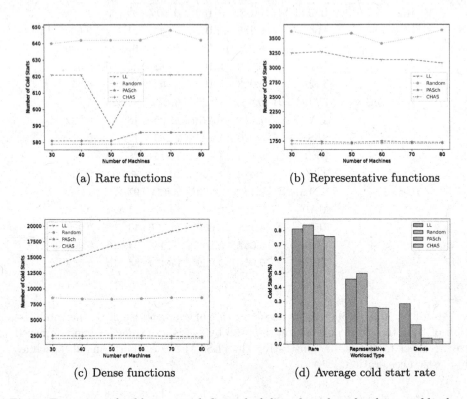

(a) Rare functions   (b) Representative functions

(c) Dense functions   (d) Average cold start rate

Fig. 6. Experimental cold starts with four scheduling algorithms for three workloads.

Figure 6(a)–6(c) illustrates the number of cold starts for various workloads, with the horizontal axis representing the number of working nodes in the cluster. CHAS significantly exceeds the load balancing algorithms LL and Random, and slightly outperforms the affinity-based scheduling algorithm PASch. While both PASch and CHAS are affinity-based scheduling techniques, PASch opts to process requests on the node with least loads when the affinity node is overloaded and cannot handle the request, which usually results in a cold start. By comparison, CHAS can locate free containers on non-affinity nodes via a hash step, reducing the cold start rate more effectively than PASch.

Figure 6(d) displays the cold start rate for each method under three different workloads using a cluster including 30 VMs. We observe that Rare functions have a larger time interval because they are the least frequently called function samples, and containers that have not been reused for a long time will be evicted, resulting in a higher cold start rate at the typical TTL compared to the other two loads. The Representative workloads also exhibit a higher cold start rate.

By comparison, Dense workloads have the lowest cold start rate due to their more intensive requests with a shorter average time interval, which increases the likelihood of reusing idle containers ahead of them.

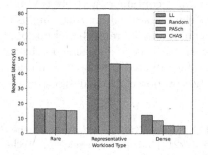

Fig. 7. Average completion time

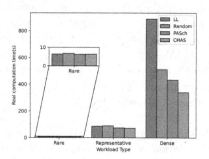

Fig. 8. Real computation time

Figure 7 displays the average latency of all requests for each method under three different workloads with a cluster including 30 VMs. Due to the heterogeneity of the functions, the request latency and cold start are not strictly correlated as each extracted function has a different cold start time for execution. The average latency of the Representative function appears to be higher than the other two kinds of workloads. However, by examining each type of load individually, we can find that the average latency of CHAS is lower than the baseline methods.

Finally, we analyze the time overhead of CHAS. CHAS's worst-case time complexity of the node selection operation is $O(n)$, which aligns with LL and PASch. Consequently, no additional CPU overhead is incurred. We utilize two integers to track the number of free containers and running containers on a node for a specific request. This results in minimal additional memory usage for CHAS. Our simulation experiments have recorded the real-time data, as presented in Fig. 8. The figure displays the actual computation time for a one-hour simulation of four scheduling strategies across three workloads with 30 VMs in the cluster. Notably, CHAS exhibits the shortest real computation time, highlighting the effectiveness of CHAS in improving the idle container hit rate for scheduling without introducing additional complexity. Furthermore, it effectively mitigates cold start occurrences caused by disregarding nodes with idle containers.

5   Conclusion

This work focuses on optimizing strategies for reducing cold-start latency in serverless computing. We propose a request scheduling algorithm and a container prewarming strategy, which effectively reuses idle containers and mitigates cold-start and long latency problems. The scheduling algorithm schedules requests to computing nodes with available idle containers, thus avoiding cold-star latency caused by neglecting nodes with idle containers. However, the algorithm can only reuse available idle containers through scheduling. In the event of sudden traffic bursts or container eviction due to long request intervals with no available idle containers for reuse, it is necessary to prewarm containers actively through prediction. The prewarming strategy analyzes the request calling rules, builds a time-series prediction model, and prewarms containers for reuse based on forecasts. We implemented a serverless environment simulator by using Python's SimPy discrete event simulation framework and conducted extensive simulation experiments. The results demonstrate that the proposed method significantly reduces cold-start latency.

Acknowledgements. This work was supported by the National Natural Science Foundation of China (No. 62072216 and No. 62372214) and the Science and Technology Development Fund, Macau SAR (File no. 0076/2022/A2).

References

1. Aws lambda. https://aws.amazon.com/lambda/, (Accessed 14 Apr 2023)
2. Azure functions. https://azure.microsoft.com/en-us/products/functions/, (Accessed 14 Apr 2023)
3. Kubernetes. https://kubernetes.io/docs/concepts/services-networking/service/, (Accessed 15 Apr 2023)
4. Openwhisk. https://azure.microsoft.com/en-us/products/functions/, (Accessed 14 Apr 2023)
5. Simpy. https://pythonhosted.org/SimPy/, (Accessed 14 Apr 2023)
6. Abad, C.L., Boza, E.F., Van Eyk, E.: Package-aware scheduling of faas functions. In: Companion of the 2018 ACM/SPEC International Conference on Performance Engineering, pp. 101–106 (2018)
7. Aumala, G., Boza, E., Ortiz-Avilés, L., Totoy, G., Abad, C.: Beyond load balancing: package-aware scheduling for serverless platforms. In: 2019 19th IEEE/ACM International Symposium on Cluster, Cloud and Grid Computing (CCGRID), pp. 282–291. IEEE (2019)
8. Bauer, A., Grohmann, J., Herbst, N., Kounev, S.: On the value of service demand estimation for auto-scaling. In: German, R., Hielscher, K.-S., Krieger, U.R. (eds.) MMB 2018. LNCS, vol. 10740, pp. 142–156. Springer, Cham (2018). https://doi.org/10.1007/978-3-319-74947-1_10
9. Dang-Quang, N.M., Yoo, M.: Deep learning-based autoscaling using bidirectional long short-term memory for kubernetes. Appl. Sci. **11**(9), 3835 (2021)
10. Fan, D., He, D.: Knative autoscaler optimize based on double exponential smoothing. In: 2020 IEEE 5th Information Technology and Mechatronics Engineering Conference (ITOEC), pp. 614–617. IEEE (2020)
11. Fuerst, A., Sharma, P.: Faascache: keeping serverless computing alive with greedy-dual caching. In: Proceedings of the 26th ACM International Conference on Architectural Support for Programming Languages and Operating Systems, pp. 386–400 (2021)
12. Herbst, N., et al.: Ready for rain? a view from spec research on the future of cloud metrics. arXiv preprint arXiv:1604.03470 (2016)

13. Imdoukh, M., Ahmad, I., Alfailakawi, M.G.: Machine learning-based auto-scaling for containerized applications. Neural Comput. Appl. **32**, 9745–9760 (2020)
14. Karger, D., et al.: Web caching with consistent hashing. Comput. Netw. **31**(11–16), 1203–1213 (1999)
15. Li, F., Hu, B.: Deepjs: job scheduling based on deep reinforcement learning in cloud data center. In: Proceedings of the 4th International Conference on Big Data and Computing, pp. 48–53 (2019)
16. Mampage, A., Karunasekera, S., Buyya, R.: Deadline-aware dynamic resource management in serverless computing environments. In: 2021 IEEE/ACM 21st International Symposium on Cluster, Cloud and Internet Computing (CCGrid), pp. 483–492. IEEE (2021)
17. Manner, J., Endreß, M., Heckel, T., Wirtz, G.: Cold start influencing factors in function as a service. In: 2018 IEEE/ACM International Conference on Utility and Cloud Computing Companion (UCC Companion), pp. 181–188. IEEE (2018)
18. Mitzenmacher, M.: The power of two choices in randomized load balancing. IEEE Trans. Parallel Distrib. Syst. **12**(10), 1094–1104 (2001)
19. Shahrad, M., et al.: Serverless in the wild: characterizing and optimizing the serverless workload at a large cloud provider. In: 2020 USENIX annual technical conference (USENIX ATC 20), pp. 205–218 (2020)
20. Suo, K., Son, J., Cheng, D., Chen, W., Baidya, S.: Tackling cold start of serverless applications by efficient and adaptive container runtime reusing. In: 2021 IEEE International Conference on Cluster Computing (CLUSTER), pp. 433–443. IEEE (2021)
21. Suresh, A., Somashekar, G., Varadarajan, A., Kakarla, V.R., Upadhyay, H., Gandhi, A.: Ensure: efficient scheduling and autonomous resource management in serverless environments. In: 2020 IEEE International Conference on Autonomic Computing and Self-Organizing Systems (ACSOS), pp. 1–10. IEEE (2020)
22. Vahidinia, P., Farahani, B., Aliee, F.S.: Mitigating cold start problem in serverless computing: a reinforcement learning approach. IEEE Internet Things J. **10**(5), 3917–3927 (2022)
23. Wu, S., et al.: Container lifecycle-aware scheduling for serverless computing. Software: Pract. Experience **52**(2), 337–352 (2022)
24. Xu, Z., Zhang, H., Geng, X., Wu, Q., Ma, H.: Adaptive function launching acceleration in serverless computing platforms. In: 2019 IEEE 25th International Conference on Parallel and Distributed Systems (ICPADS), pp. 9–16. IEEE (2019)

Multi-stage Optimization of Incentive Mechanisms for Mobile Crowd Sensing Based on Top-Trading Cycles

Jingjie Shang[1,2], Haifeng Jiang[1,2](✉) ⓘ, Chaogang Tang[1,2], Huaming Wu[3], Shuhao Wang[1,2], and Shoujun Zhang[1,2]

[1] Mine Digitization Engineering Research Center of Ministry of Education, China University of Mining and Technology, Xuzhou 221116, China
haifeng.jiang.cumt@foxmail.com
[2] School of Computer Science and Technology, China University of Mining and Technology, Xuzhou 221116, China
[3] The Center for Applied Mathematics, Tianjin University, Tianjin 300072, China

Abstract. For collaborative tasks requiring multiple users, in Mobile Crowd Sensing (MCS), low user interest in certain tasks usually results in insufficient user re-cruitment. However, the interest of the user directly affects the quality and effi-ciency of task completion. To address this issue, we propose a multi-stage incen-tive mechanism based on the Top-Trading Cycles (TTC) from economics, ena-bling users to participate in tasks that align with their interest through the optimi-zation of multiple stages. Firstly, we perform an initial screening of users using a reverse auction. Then, we adopt the Top-Trading Cycles algorithm to determine the optimal task-user pairs. For tasks with insufficient collaborators, an interest-based task recommendation algorithm is proposed, which calcu-lates user interest similarity in the social network, recommends tasks to other users, and evaluates rewards based on their contributions. The proposed incentive mechanism can theoretically guarantee computational effectiveness, truthfulness, and individual rationality in this paper. Sim-ulation experiments show that the proposed mecha-nism outperforms traditional incentives in terms of user participation rates, task coverage, and average user utility.

Keywords: Mobile Crowd Sensing · Incentive Mechanisms · Top-Trading Cycles · Multi-Stage Optimization

1 Introduction

With the rapid development of wireless networks, smartphones have become an in-dispensable necessity in people's lives. They integrate multiple sensors, such as mag-netic sensors, gyroscopes, GPS location sensors, and fingerprint sensors [1], which enable them to perceive and analyze data from the surrounding environment and complete Mobile Crowd Sensing tasks. Currently, MCS has a

Z. Tari et al. (Eds.): ICA3PP 2023, LNCS 14487, pp. 170–186, 2024.
https://doi.org/10.1007/978-981-97-0834-5_11

wide range of applica-tions in various fields, including environmental monitoring, disaster emergency search and rescue, and Smart Cities [2]. However, performing sensing tasks may incur signif-icant costs for users, including data traffic, CPU computation, time consumption, battery usage, and potential privacy issues [3, 4], which can lead to decreased user participation in sensing activities. Therefore, in order to encourage more users to par-ticipate in sensing activities and enhance their enthusiasm, it is necessary to design reasonable incentive mechanisms [5, 6].

The existing incentive mechanisms can be broadly classified into two catego-ries: monetary incentives and entertainment incentives. The former is the most direct and effective mean for motivation [7]. Reverse auction is the most widely adopted incentive method, although, it is an incomplete information mechanism [8]. For in-stance users can only submit bids based on their own costs, and the service platform make decisions on the optimal bid. Thus, users have only one chance to participate in a sensing task, which may reduce their willingness to participate. Furthermore, the user's sensing ability depends on various factors, such as the sensing device's capabil-ity and the user's interest in the task, and it is often impossible to recruit enough users to participate in collaborative tasks. Therefore, it is feasible to design incentive mech-anisms by considering the inter-est of the users.

To address the aforementioned issues, we consider the similarity of user inter-est in social network scenarios [9]. Merely considering the similarity of users' interest is insufficient to solve the aforementioned issues. We introduce the con-cept of Top-Trading Cycles from economics [10] and propose to use the You Request My House-I Get Your Turn (YRMH-IGYT) mechanism for secondary transactions based on users' preference sequences, enabling them to acquire tasks of higher interest levels. For tasks with insufficient collaboration, we put forward a task recommendation algorithm based on interest similarity. With this multi-stage strategy, users have more opportunities to participate in tasks, ultimately resulting in a higher task coverage rate.

The primary contributions of this paper are as follows:

- We proposed two trading models and developed user selection criteria for each. Additionally, in order to address tasks with a shortage of users, we proposed a task recommendation model. This model recommends tasks to other users based on the similarity of their interest.
- We provide evidence that the proposed mechanism meets the expected char-acteristics of computational efficiency, veracity, and individual rationality. Further-more, MO-TTC can output the optimal solution.

2   Related Works

Reverse auctions have been widely used in the incentive mechanisms of mobile crowdsensing, which are mainly designed to maximize social welfare with a user-centered incentive mechanism. Under the uncertain task execution, Zheng et al. [11] designed a reverse auction model and considered the completion of tasks, minimizing the social cost of user recruitment. Gu et al. [12] proposed an information quality incentive mechanism for multimedia crowdsensing, maximizing social welfare by designing a reverse auction model. Ji et al. [13] designed an incentive mechanism based on reverse auctions, which minimizes costs and retains users who are about to exit through the lottery mechanism. Furthermore, Luo et al. [14] designed two capaci-ty reputation systems to evaluate the online staff abilities and proposed an incentive mechanism based on reverse auctions and fine-grained capacity reputation. The reward is determined based on user bids and fine-grained capacity reputation. Jin et al. [15] introduced the Quality of Information indicator, designed an incentive mech-anism based on reverse combination auctions, accomplishing the optimal social wel-fare while meeting individual rationality and computational efficiency. The incentive mechanism based on reverse auctions has the advantages of budget controllability and maximizing social welfare, but ignores the characteristics of users, resulting in a decrease in data quality and user motivation.

Many scholars consider designing incentive mechanisms from the perspective of user interests. Li et al. [16] proposed an incentive mechanism for an inter-est tag-ging application program, which maximizes platform revenue through a three-stage decision process. Meanwhile, Xiong et al. [17] introduced a cosine similarity calcula-tion protocol with privacy protection, which computes the similarity between task and user vectors and conducts a second selection of users to ensure fairness of user perception of being selected. In social networks, Xu et al. [18] considered the collabo-ration compatibility of users for multiple collaborative tasks and designed two incen-tive mechanisms based on reverse auction. Additionally, they presented a user group-ing method based on neural network model and clustering algorithm to minimize social cost by introducing a neural network-based method for learning the similarity between users and clustering them accordingly. This approach helps avoid uneven task allocation and incomplete tasks that can result from varying user interest in tasks.

This paper proposes a Multi-Stage Optimization of Top-Trading Cycles (MO-TTC) incentive mechanism to address the issues of the existing incentive mechanism. We innovatively integrated the TTC into the incentive mechanism of MCS. TTC is a resource allocation algorithm that divides agents into non-overlapping sets or cycles, exchanging resources internally within each cycle, and ensuring that each agent ac-quires their preferred resources. The contribution based incentive mechanism deter-mines the reward based on the user's contribution in the perceived task, which can ensure the effectiveness of the platform, but does not consider the impact of user interest on perceived quality. The MO-TTC mechanism designed in this paper prelim-inarily screens users through a reverse auction model, and performs secondary selec-tion on users through TTC algorithm based on their submitted preference order. For tasks lacking collaboration, user inter-est similarities are calculated, and the tasks are assigned to users who may be intrigued by them. Finally, the reward is determined based on the degree of their contributions.

3   System Model

In this paper, all users in the MCS are considered to be part of a social network, and the platform serves as the auctioneer, and tasks are the auctioned items. Table 1 presents the primary symbols used in this paper.

The set of users is denoted as $U = \{u_1, u_2, ..., u_i, ..., u_n\}$. The corresponding user reputation values are expressed by $RP = \{rp_1, rp_2, ..., rp_i, ..., rp_n\}$. The perceptual task set is denoted as $T = \{t_1, t_2, ..., t_j, ..., t_m\}$. The task preference set of user u_i is F_{ui}. The number of users required for the task is referred to as the collaboration number and is expressed by the set $NUM = \{num_1, num_2, ..., num_j, ..., num_m\}$. n represents the number of users, while m represents the number of tasks. The contribution of users to tasks in this paper is integrated into the incentive mechanism, with the Quality of Information (QoI) they provide used as a measure. The users' contributions are categorized into five levels in this paper and represented by the set $Q = \{1, 2, 3, 4, 5\}$.

Table 1. Notations used.

Notations	Definitions
U	set of users
URW	set of winners in reverse auction
URL	set of users entering the second phase
$UTTC$	set of task-user in the second phase
UIQ	set of task owners with insufficient collaboration count
UIW	set of users finally participating in the task
T	set of tasks
TW	set of tasks with successfully bids
TL	set of tasks with failed bids
num_j	collaboration count required for task t_j
rp_i	reputation value of user u_i
$c_{i,j}$	true cost of user u_i for task t_j
$b_{i,j}$	bidding price made by user u_i for task t_j
$p_{i,j}$	reward for user u_i completing task t_j
$q_{i,j}$	contribution of user u_i to task t_j
r_j	maximum budget for task t_j in reverse auction

3.1   Design Objectives

For each task, denoted as $t_j \in T$, a limited budget is assigned to the incentive mechanism, which requires the selection of num_j users from the user set U to perform the task. If the number of collaborative users falls short, the task is recommended to other users in the social network based on their interest similarity, allowing them to decide whether to participate. To enhance the QoI of users, a credibility mechanism is implemented to update based on their contributions. The contribution values of users are linearly normalized, and defined in Eq. (1).

$$rp_i = rp_i + \frac{q_{i,j} - \min(q_{tj})}{\max(q_{tj}) - \min(q_{tj})} \tag{1}$$

where $\min(q_{tj})$ represents the minimum contribution value required by the platform for task t_j, $\max(q_{tj})$ denotes the maximum contribution value needed for the same task by the platform.

The utility of user u_i is the difference between the compensation provided by the service platform and the user's self-costs, given as below:

$$e_i = p_i - c_i, u_i \in UIW \tag{2}$$

where $p_i = \sum_{j=1}^{m} p_{i,j}$ represents the total reward that user u_i receives for completing all the tasks, and $c_i = \sum_{j=1}^{m} c_{i,j}$ represents the total costs incurred by the user for participating in the tasks.

This paper aims to identify the optimal set of tasks and users from task set T using two transaction models. Additionally, we utilize information submitted by users through the reverse auction process to recruit other users to collaborate with those in the $UTTC$. For tasks that lack sufficient collaboration, an interest-based task recommendation algorithm is applied to suggest tasks to users who may be interested. Finally, user compensation and reputation values are updated based on their contribution.

3.2   Reverse Auction Model

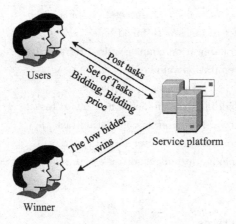

Fig. 1. Reverse auction.

During the first phase, the service platform releases tasks and each user, denoted as $u_i \in T$, sends a tuple (ζ_i, B_i) to the platform. ζ_i represents the task set that the user u_i is bidding on, and B_i represents the corresponding bid set. The platform selects the user with the lowest bid $b_{i,j}$, where $b_{i,j}$ is less than the maximum budget r_j, to be the task owner and added to the URW set. The tasks that have been successfully bid on are added to the TW set, as illustrated in Fig. 1.

At this stage, the winning participant u_i in URW is tasked with undertaking task t_j in the second phase of the project. The unsuccessful bidders' tasks will also move to the second phase and are collectively represented as TL.

3.3   Secondary Trading Model Based on TTC Algorithm

In the second phase, the platform sorts failed users in the reverse auction by their reputation values, denoted as rp_i. The platform then gathers a set of potential users, denoted as URL, by requesting their agreement to participate in the second phase based on their (ζ_i, B_i) values. Users in URL are not assigned any tasks, whereas tasks in TL are unclaimed. Users in URW and URL indicate their preferences for tasks to the platform, while tasks in TW and TL reveal their preferred users. The TTC algorithm is employed to construct the task-user set $UTTC$.

Before introducing the TTC algorithm, some concepts need to be introduced. In URW, users who are dissatisfied are labeled as unsatisfied users, while users who enter the third stage after their transactions are considered satisfied users. In TW, tasks are referred to as old tasks, while in URL, new users are added, and in TL, new tasks are added. This article utilizes graph theory to define the transactions between users and tasks in the YRMH-IGYT mechanism by constructing a TTC graph.

Definition 1. *(The construction rules of the graph): In the TTC graph, the set of vertices consists of users and task sets; directed edges represent task vertices pointing to their owner users. If there is no owner for a task, the user set without a task is sorted based on credibility, and all tasks without own-ers are assigned to the user with the highest credibility. User vertices point to their most interesting task.*

The following steps constitute the second transaction phase based on the TTC:

Step 1: Assign old tasks to their respective owners while assigning new tasks to the user with the highest reputation. Each user is linked to the task of their maximum interest.

Step 2: In case of no cycle, the new user moves to the next task in their prefer-ence list.

Step 3: Identify any cycle and eliminate it by allocating tasks linked to each user within the cycle.

Step 4: Assign old tasks to their respective owners while assigning new tasks to the remaining new users with the highest reputation, and each user is linked to the task of their maximum interest among the remaining tasks.

Step 5: Identify any cycle that may arise in this phase and eliminate it by allo-cating tasks linked to each user within the cycle.

Step 6: Repeat steps 2 to 5 until all users and tasks have been assigned. The as-signment process ends upon exhaustion.

3.4 Task Recommendation Model Based on Interest Level

In the third phase of the project, secondary trades will match all tasks with users. The service platform will select $num_j - 1$ users with the lowest $b_{i,j}$ that is below the maximum budget r_j, based on the collaboration number of each task and users' (ζ_i, B_i). These users will be added to the UIW set. If there are not enough users that meet the requirements, the task owner will be included in the UIQ set. As the social relationships between users are private, the interest level of users for a particular task will be calculated based on their (ζ_i, B_i), and those users who are potentially interested in a task will be recommended. The recommendation process, which is based on interest level, mainly comprises the following steps:

Step 1: involves constructing a user-task rating matrix (UT) based on the sets of users and tasks.

Step 2: the cosine similarity formula is used to calculate the similarity of interest among users in both UIQ and $U\backslash UIQ$, resulting in the interest similarity matrix W. The cosine similarity formula is shown in Eq. (3).

$$w_{uv} = \frac{|N(u) \bigcap N(v)|}{\sqrt{|N(u)||N(v)|}} \tag{3}$$

where $N(u)$ represents the set of tasks that user u is interested in within the set $U\backslash UIQ$, and $N(v)$ represents the set of tasks that user v is interested in within the UIQ set, and $w_u v$ is the similarity of interest between user u and user v.

Step 3: Sort users based on the similarity of their interest and recommend the tasks to them.

Step 4: When the required number of collaborators for a task is met, add the users to the UIW set.

Step 5: After recruiting enough users, tasks will be carried out and com-pensation will be determined based on their contributions. The compensation formula is defined by Eq. (4).

$$p_{i,j} = \max \left\{ \frac{q_{i,j} num_j r_j}{\sum_i^{num_j} q_{i,j}}, b_{i,j} \right\} \tag{4}$$

where $\frac{q_{i,j}}{\sum_i^{num_j} q_{i,j}}$ is the proportion of contribution from user u_i, and $num_j r_j$ is the total budget of task t_j.

3.5 Desirable Properties

This paper aims to design an incentive mechanism that achieves efficiency within a reasonable time frame. In a reverse auction, each user submits a tuple (ζ_i, B_i)

to the platform, which includes the set of tasks they are interested in bidding for ζ_i and their bidding price B_i. Users can submit a tuple (ζ_i', C_i) that deviates from the truthful value, where C_i represents the user's cost set.

It is critical for the incentive mechanism to ensure non-negative utility for each user to prevent negative utility from causing them to withdraw from the platform.

Based on the descriptions given above, this paper expects the incentive mecha-nism designed to satisfy the following features:

- **Computational Efficiency:** Algorithmic incentivization mechanisms can be completed in polynomial time.
- **Truthfulness:** The tuples (ζ_i, B_i) and preference order F_{ui} submitted by each user $u_i \in U$ are truthful.
- **Individual Rationality:** For every user $u_i \in UIW$, the variable $e_i \geq 0$.

4   Construction of MO-TTC

In this section, we present two algorithms: the TTC algorithm and an interest-based task recommendation algorithm. We use an example to demonstrate the process of the second phase using TTC. Finally, we demonstrate the character-istics possessed by the proposed incentive mechanism.

4.1   Secondary Trading Algorithm

The second-hand trading process based on the TTC is exemplified in Fig. 2. Users rank their preferred tasks, as indicated in Fig. 2(a), where u_1, u_2, and u_3 are users in the URW set who own tasks t_1, t_2, and t_3, respectively $(t_1, t_2, t_3 \in TW)$, u_4, u_5 are users in the URL set who don't own any tasks, and t_4, t_5 are tasks in TL that don't belong to any user. In Fig. 2(b), users point to their favorite tasks, which in turn point to their owners; t_4, t_5 do not have owners. Assuming that u_4 has a higher reputation score than u_5, they point to the new user with the highest reputation score, u_4. Hence, TTC leads to the cycle $u_1 \to t_5 \to u_4 \to t_1 \to u_1$. As a result, users u_1 acquire tasks t_5 and u_4 acquires task t_1, which are both subsequently removed from the cycle, completing the first round of trading. Figure 2(c) displays the updated preference orders for the users and tasks. In the second round, a cycle $u_5 \to t_4 \to u_5$ is discovered, and user u_5 gets task t_4, which is then removed from the system. The third round is illustrated in Fig. 2(d), where a cycle $u_2 \to t_3 \to u_3 \to t_2 \to u_2$ is found. As a result, user u_2 acquires task t_3 and user u_3 acquires task t_2, which are both subsequently removed from the cycle. The transaction comes to an end because there are no tasks or users left to trade.

After the completion of the transaction, all users involved in the subsequent exchange demonstrate a preference for the final allocated task that is equal to or greater than their initial task allocation, indicating a preference for higher interest level tasks during the exchange process.

This paper presents the TTC algorithm to implement the transaction process during the second phase. The specific procedures are outlined in Algorithm 1.

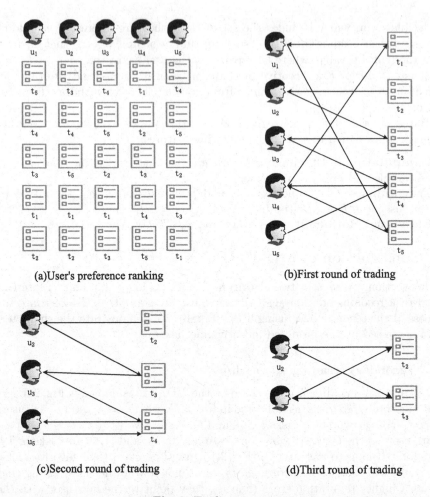

(a)User's preference ranking

(b)First round of trading

(c)Second round of trading

(d)Third round of trading

Fig. 2. Trading process.

4.2   Interest-Based Task Recommendation Algorithm

At this stage, the paper considers recommending tasks with insufficient user numbers based on similarity of user interest, to other users, to achieve sufficient user numbers.

In the first stage of the reverse auction, all users submitted bids to the platform, bidding on the task set ζ_i that they were interested in, which serves as their behavior towards tasks in the third stage. A user-task rating matrix, denoted as UT, is then constructed. For a user u_i, if a task $t_j \in \zeta_i$, the rating for this task by the user is considered 1; otherwise, it is considered 0. Similarly, user's similarity matrix W is calculated based on the UT matrix using Eq. (3). Finally, the recommendation list is generated by multiplying the UT matrix and the user similarity matrix W. Firstly, we identify the K users in the similarity matrix W

Algorithm 1. TTC algorithm

Input: U: set of user; T: set of task
Output: $UTTC$
1: Initialize users' preference sequence F_{ui}
2: Construct the TTC graph G based on Definition 1 and user preference order
3: **while** G Number of nodes > 0 **do**
4:     Find all *cycles* in G
5:     **for** *cycle* $\in$ *cycles* **do**
6:         UTTC $\leftarrow$ task-user
7:         Remove node and update graph G
8:         Update task owner
9:     **end for**
10: **end while**
11: **return** $UTTC$

who are most similar to user u_i. The set of K users, denoted as $S(u_i, K)$, are the owners of tasks in the UIQ set, where user recruitment has been insufficient. Subsequently, we extract all tasks from S that users are interested in and remove tasks which u_i has already expressed an interest in. The degree of interest that user u_i has towards task t_j is computed using Eq. (5).

$$p(u_i, t_j) = \sum_{v \in S(u_i, K) \cap N(t_j)} w_{uv} ut_{vt_j} \tag{5}$$

where ut_{vt_j} represents the degree of liking or rating that user v gives to task t_j, and $N(t_j)$ refers to users who have interacted with task t_j.

Algorithm 2. Interest-based Task Recommendation Algorithm

Input: U: set of user; UIQ: set of task owners with insufficient collaboration count
1: Build user-task rating matrix UT
2: Calculate similarity matrix W by Eq. (3)
3: **for** $u \in U$ and $v \notin U$ **do**
4:     **for** $v \in UIQ \cap N(t)$ **do**
5:         Find the K users in the UIQ set that are most similar to user u
6:         Calculate the level of interest of u in task t using Eq. (5)
7:     **end for**
8: **end for**
9: descending order sorting U base on $F(u)$
10: Select the top N users according to Eq. (5) to participate in the task and add them to the UIW collection
11: Update the reputation of the users in the UIW collection by using Eq. (1)
12: Determine the rewards for the users in the UIW collection by using Eq. (4)

4.3   Mechanism Analysis

The following section conducts theoretical analysis to demonstrate that the incentive mechanism proposed in this article can achieve the following characteristics:

Lemma 1. *MO-TTC is computationally efficient.*

Proof. Both Algorithm 1 and Algorithm 2 satisfy the computational effectiveness. In Algorithm 1, initializing the preference sequence of users and constructing the directed graph G both traverse the user and task sets, costing $O(n^2)$ time. The worst-case scenario for finding the set of winning users UIW (lines 3–10) is when only one task-user match occurs each time, costing $O(n^2)$ time in this case. Therefore, Algorithm 1 costs $O(n^2)$ time. In Algorithm 2, building the user-task scoring matrix and calculating the similarity matrix both traverse the user set, costing $O(n)$ time. Finding the top K similar users (lines 3–8) costs $O(n^2)$ time, sorting and selecting N users (lines 9–10) costs $O(n)$ time, and updating the reputation and reward (lines 11–12) costs $O(n)$ time. Therefore, Algorithm 2 costs $O(n^2)$ time.

Lemma 2. *MO-TTC is truthful.*

Proof. Each user submits a tuple (ζ_i, B_i) to the platform, which contains the set of tasks of interest ζ_i and the bid set B_i. Users have strategic considerations in bidding and may provide a tuple (ζ_i', C_i) that deviates from the true value. Since the filtering process in all three stages considers B_i, users' participation rates will decrease when submitting false bids. Additionally, if ζ_i and the user's preference sequence F_{ui} are false, the user cannot obtain the tasks that they truly desire because task trades will only occur based on the user's genuine preference sequence in the second stage, while the third stage constructs a user-scoring matrix based on ζ_i to search for users with high similarity.

Lemma 3. *MO-TTC is individually rational.*

Proof. It is evident, as shown by Eq. (2), that e_i is greater than or equal to zero.

5   Experimental Evaluation

5.1   Simulation Setup

In order to evaluate the effectiveness of the MO-TTC algorithm, this paper conducts simulation on three performance indicators: user participation rate, task coverage rate, and average user utility. The indicators are defined as follows:

(1) User participation rate: The ratio of the number of users participating in completing tasks to the total number of users, with a range of [0,1].
(2) Task coverage: The ratio of the number of tasks recruited with sufficient collaborations to the total number of tasks, with a range of [0,1].

Table 2. Simulation parameters.

Parameter	Range
Number of users n	[20,100]
Number of tasks m	[20,100]
Maximum budget for reverse auction r_j	[20,40]
Number of task collaborators num_j	[2,4]
User bid $b_{i,j}$	[10,50]
User contribution $q_{i,j}$	[1,5]
User reputation value rp_i	[0,1]
Whether the user accepts the task	0 or 1

(3) Average user utility: The ratio of the total user utility calculated by formula (2) to the number of tasks completed by the user.

In the simulation experiment of this article, the number of users n and the number of tasks m are within the range of [20,100], and their impact on the perfor-mance of the incentive mechanism is detected by changing the number of users n or the number of tasks m. In order to be closer to the actual scenario, whether the user accepts the task is represented by a random number of 0 or 1, where 0 represents rejection and 1 represents acceptance of the task. The simulation parameters are shown in Table 2. The simulation experiments run on Windows 11 system with hard-ware configuration of AMD R7-6800H CPU, RTX3060 GPU, and 16GB RAM.

The performance of MO-TTC is compared with three other incentive mechanisms, including the Incentive Mechanism Based on Reverse Auction (IMBRA), the Contribution Based Incentive Mechanism (CBIM), and the Reverse Vickrey Auction Incentive Mechanism (RVA-IM). In IMBRA, low bidders are selected to perform tasks, and if there are not enough users, the task is not executed, with user rewards being based on their bidding value. In CBIM, task perform-ers are randomly selected, with user acceptance being random as well. If there are not enough users available, the task is not executed, and rewards are deter-mined based on the user's contribution. In RVA-IM, compensation is determined using the Vickrey-Clark-Groves (VCG) mechanism. All four algorithms mentioned above were simulated 100 times and their average values were taken for display of simulation results.

5.2   User Participation Rate

Figure 3 show a comparison of user participation rates between MO-TTC, IMBRA, CBIM and RVA-IM. Figure 3(a) depicts the comparison of user participation rates at different task numbers, with 50 users. The figure shows that the proportion of partici-pating users increases with the number of tasks. This is because, when the number of tasks is smaller than the number of users, there are

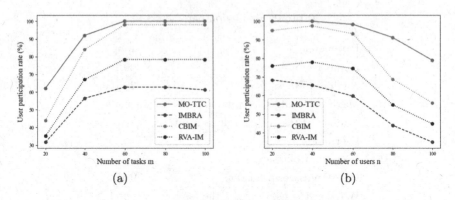

Fig. 3. Comparison of user participation rates.

relatively fewer users participating in sensing tasks, and as the number of tasks increases, so too does the number of users participating in sensing tasks. With 50 users, MO-TTC outperformed IMBRA, CBIM, and RVA-IM by an average of 36%, 6%, and 23%, respectively, in terms of user participation rate when the number of tasks varies. Figure 3(b) shows that user par-ticipation rates decrease with an increase in the number of users. When the number of tasks is constant, an increase in the number of users quickly recruits enough users to reach saturation in the tasks. With 50 tasks, MO-TTC outperformed IMBRA, CBIM and RVA-IM by an average of 39%, 12%, and 28%, respectively, in terms of user participation rate when there were varying numbers of users. For MO-TTC, a multi-stage optimization strategy was used to provide users with more opportunities to choose whether to participate in sensing tasks, thereby increasing the number of users executing sensing tasks. Therefore, user participation rates were always higher with MO-TTC than with IMBRA, CBIM and RVA-IM.

5.3 Task Coverage Rate

Figure 4 illustrate a comparison of task coverage rates between MO-TTC, IMBRA, CBIM, and RVA-IM. Figure 4(a) represents the comparison of task coverage rates for different task numbers when the user number is 50, while Fig. 4(b) shows the task coverage rate comparison for different user numbers when the task number is 50. The figures demonstrate that the task coverage rate decreases with an increase in the number of tasks due to the users' limited budget for task completion. Figure 4(a) reveals that MO-TTC outperforms IMBRA, CBIM, and RVA-IM by an average of roughly 17%, 9%, and 15%, respectively, regarding task coverage rate when the user number is 50 and the task number increases. Figure 4(b) demonstrates that the task coverage rate increases with an increase in the number of users, as recruiting more users means more task coverage opportunities. When the task number is 50, MO-TTC outperforms IMBRA, CBIM, and RVA-IM by an average of roughly 39%, 10%, and 21%, respectively, concerning the task coverage rate as the user number increases. Interest-based

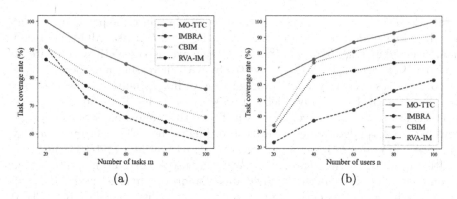

Fig. 4. Comparison of task coverage rates.

task recommendation algorithms can effectively recommend remain-ing tasks to users, thereby allowing MO-TTC to achieve higher task coverage rates than those of IMBRA, CBIM, and RVA-IM.

5.4 Average User Utility

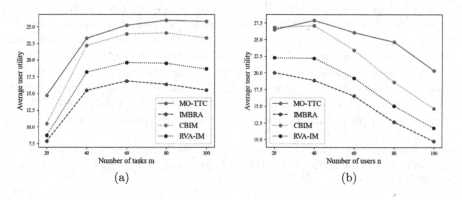

Fig. 5. Comparison of average user utility.

Figure 5 compare the average user utility of MO-TTC, IMBRA, CBIM, and RVA-IM. Figure 5(a) presents a comparison of the average user utility concerning different task numbers at constant user quantity of 50. Figure 5(b) shows a comparison of the average user utility with different user quantities but constant task number of 50. From Fig. 5(a), it is evident that the average user utility increases with task quantity due to the increased involvement of users when the number of tasks increases. When there are 50 users, MO-TTC shows an average increase of 63%, 14%, and 40% in average user utility compared to IMBRA, CBIM, and RVA-IM, respectively. In contrast, from Fig. 5(b), the average user utility decreases as the number of users increases be-cause the participation rate is reduced with an increase in tasks. When there are 50 tasks, MO-TTC results in an average increase of 67%, 17%, and 50% in average user utility compared to IMBRA, CBIM, and RVA-IM, respectively, with increasing num-bers of users. IMBRA shows relatively lower average user utility as only low bidding users are chosen. CBIM randomly selects users only once, leading to decreased participation rates and lower average user utility as the number of users increases. In contrast, RVA-IM determines rewards according to the VCG mechanism, achieving maximum welfare instead of high average user utility. MO-TTC utilizes a multi-stage optimization strategy, incentivizing users based on their interest and recommending tasks requiring inadequate collaboration to other users. This approach results in con-sistently higher average user utility than IMBRA, CBIM, and RVA-IM.

6   Conclusion

In social networks, perceptual tasks usually require multi-user collaboration. Consid-ering the low user interest in certain tasks, it is challenging for these tasks to recruit enough contributors. This paper proposes an incentive mechanism based on MO-TTC. The paper first uses a reverse auction model to select users and then employs the TTC algorithm to assign tasks to users with higher interest for secondary transac-tions. If tasks have insufficient collaborative efforts, we recommend them to other users with similar interest in the social network. Rewards are then allocated to users based on their contributions, and their reputation values are updated accordingly. We have proven that MO-TTC is computationally efficient, truthful, and individually rational. Furthermore, the simulation results show that the MO-TTC incentive mech-anism proposed in this paper outperforms other approaches such as IMBRA, CBIM, and RVA-IM in terms of user participation rate, task coverage, and average user utility.

Acknowledgements. This work is supported by the National Natural Science Foundation of China (No. 62071327) and Tianjin Science and Technology Planning Project (No. 22ZYYYJC00020).

References

1. Xu, J., Bao, W., Gu, H., Xu, L., Jiang, G.: Improving both quantity and quality: incentive mechanism for social mobile crowdsensing architecture. IEEE Access **6**, 44992–45003 (2018). https://doi.org/10.1109/ACCESS.2018.2860900
2. She, R.: Survey on incentive strategies for mobile crowdsensing system. In: 2020 IEEE 11th International Conference on Software Engineering and Service Science (ICSESS), Beijing, China, pp. 511–514 (2020). https://doi.org/10.1109/ICSESS49938.2020.9237745

3. Wang, Z., et al.: Towards privacy-driven truthful incentives for mobile crowdsensing under untrusted platform. IEEE Trans. Mobile Comput. **22**(2), 1198–1212 (2023). https://doi.org/10.1109/TMC.2021.3093552

4. Esmaeilyfard, R., Esmaili, R.: A privacy-preserving mechanism for social mobile crowdsens-ing using game theory. Trans. Emerging Telecommun. Technol. **33**(9), e4517 (2022)

5. Hao, L., Jia, B., Liu, J., Huang, B., Li, W.: VCG-QCP: a reverse pricing mechanism based on VCG and quality all-pay for collaborative crowdsourcing. In: IEEE Wireless Communications Networking Conference (WCNC), Seoul, Korea (South), pp. 1–99 (2020) https://doi.org/10.1109/WCNC45663.2020.9120841

6. Xu, J., Guan, C., Wu, H., Yang, D., Xu, L., Li, T.: Online incentive mechanism for mobile crowdsourcing based on two-tiered social crowdsourcing architecture. In: 2018 15th Annual IEEE International Conference on Sensing, Communication, and Networking (SECON), Hong Kong, China, 2018, pp. 1–9. https://doi.org/10.1109/SAHCN.2018.8397102

7. Ji, G., Zhang, B., Yao, Z., Li, C.: Multi-platform cooperation based incentive mech-anism in opportunistic mobile crowdsensing. In: GLOBECOM 2022–2022 IEEE Global Communications Conference, Rio de Janeiro, Brazil, pp. 3575–3580 (2022). https://doi.org/10.1109/GLOBECOM48099.2022.10001047

8. Guo, D., Feng, X., Zheng, H.: Incentive mechanism design for mobile crowdsens-ing considering social networks. In: 2020 IEEE 6th International Conference on Computer and Communications (ICCC), Chengdu, China, pp. 2345–2350 (2020). https://doi.org/10.1109/ICCC51575.2020.9345046

9. Gao, H., An, J., Zhou, C., Li, L.: Quality-aware incentive mechanism for social mobile crowd sensing. IEEE Commun. Lett. **27**(1), 263–267 (2023). https://doi.org/10.1109/LCOMM.2022.3204348

10. Alcalde-Unzu, J., Molis, E.: Exchange of indivisible goods and indifferences: the top trading absorbing sets mechanisms. Game. Econ. Behav. **73**(1), 1–16 (2011)

11. Zheng, Z., Yang, Z., Wu, F., Chen, G.: Mechanism design for mobile crowdsensing with execution uncertainty. In: 2017 IEEE 37th International Conference on Distributed Computing Systems (ICDCS), Atlanta, GA, USA, pp. 955–965 (2017). https://doi.org/10.1109/ICDCS.2017.230

12. Gu, Y., Shen, H., Bai, G., et al.: QoI-aware incentive for multimedia crowdsensing enabled learning system. Multimedia Syst. **26**, 3–16 (2020). https://doi.org/10.1007/s00530-019-00616-w

13. Ji, G., Zhang, B., Yao, Z., Li, C.: A reverse auction based incentive mechanism for mobile crowdsensing. In: ICC 2019–2019 IEEE International Conference on Communications (ICC), Shanghai, China, pp. 1–6 (2019). https://doi.org/10.1109/ICC.2019.8762030

14. Luo, Z., Xu, J., Zhao, P., et al.: Towards high quality mobile crowdsensing: incentive mechanism design based on fine-grained ability reputation. Comput. Commun. **180**, 197–209 (2021)

15. Jin, H., Su, L., Chen, D., et al.: Quality of information aware incentive mechanisms for mobile crowd sensing systems. In: Proceedings of the 16th ACM International Symposium on Mo-bile Ad Hoc Networking and Computing, pp. 167–176 (2015)

16. Li, Y., et al.: PTASIM: incentivizing crowdsensing with POI-tagging cooperation over edge clouds. IEEE Trans. Industr. Inf. **16**(7), 4823–4831 (2020). https://doi.org/10.1109/TII.2019.2954848

17. Xiong, J., Chen, X., Yang, Q., Chen, L., Yao, Z.: A task-oriented user selection incentive mechanism in edge-aided mobile crowdsensing. IEEE Trans. Netw. Sci. Eng. **7**(4), 2347–2360 (2020). https://doi.org/10.1109/TNSE.2019.2940958
18. Xu, J., Rao, Z., Xu, L., Yang, D., Li, T.: Incentive mechanism for multiple cooperative tasks with compatible users in mobile crowd sensing via online communities. IEEE Trans. Mobile Comput. **19**(7), 1618–1633 (2020). https://doi.org/10.1109/TMC.2019.2911512

A Chained Forwarding Mechanism for Large Messages

Jiaqi Lin[1], Tao Feng[1(✉)], Nanxin Zhou[1,2], Xianming Gao[1],
and Shanqing Jiang[1,3]

[1] Institude of System Engineering AMS PLA, Beijing, China
feng09@163.com
[2] School of Computer Science and Engineering, University of Electronic Science and
Technology of China, Chengdu, China
202122081015@std.uestc.edu.cn
[3] School of Cyber Science and Engineering, Southeast University, Nanjing, China
sqjiang@njnet.edu.cn

Abstract. The proliferation of wide-bandwidth network services, e.g.
high-definition video (HD video), has made the fragmentation mechanism necessary. However, conventional fragmentation mechanisms apply
the same treatment to fragmented and unfragmented packets, thus
impeding efficiency and impairing the quality of service. To address this
issue, we propose a novel chained forwarding mechanism that integrates
the typical features of large message transmission. The proposed mechanism is uniquely confined to the data plane, providing seamless chained
forwarding capabilities. Specifically, network devices exclusively process
the initial fragment, exempting subsequent fragments from further processing. This reduces the processing delay in network devices, attenuating
the forwarding latency caused by Access Control List (ACL) and Longest
Prefix Match (LPM). Evaluation results based on the P4 simulator show
that the proposed chained forwarding mechanism can reduce 10% of
the forwarding latency. Furthermore, the mechanism demonstrates substantial scalability and provides significant performance improvements
in large-scale networks.

Keywords: Large message · Fragments · QoS · P4 · Data plane

1 Introduction

In recent years, networking technologies have made remarkable progress, with
one of the most rapidly developing areas being high-definition (HD) video [1]. By
2023, video traffic is expected to constitute one-fifth of all Internet traffic [2]. This
growth can be attributed to the increasing popularity of new video use cases,
such as e-commerce and recreational gaming, leading to a surge in subscribers [3].
HD video, including 4K and 8K video, has extremely high resolutions and frame
rates with demanding throughput requirements, posing significant challenges
to network bandwidth, transmission, and processing speeds [4]. Therefore, it is

imperative to continuously enhance network technologies to accommodate these new data traffic demands.

Large video messages often exceed the maximum transmission unit (MTU) limit, which is the maximum packet size that a network can transmit and the maximum payload of the traditional network interface controller (NIC) in the data link layer. As a result, these messages must be fragmented before or during transmission and defragmented upon receipt, causing the receiver to wait for all fragments to be processed in the network. These characteristics introduce greater delay and jitter, ultimately impacting the quality of service (QoS) [5], especially in scenarios requiring high real-time performance.

As modern high-speed networks develop rapidly, routing and forwarding tasks have become quite heavy, and each fragments must be processed by network devices, leading to increased overhead. Optimization for large messages is usually associated with network fragmentation. Consequently, researchers have proposed various methods and techniques to enhance the efficiency of large message transmission, such as optimizing the fragmentation and reassembly process, reducing transmission overhead, and improving the congestion control algorithms.

Message fragmentation is an integral part of the Internet Protocol (IP) standard and efforts have been made to improve its forwarding efficiency. The Path Maximum Transmission Unit Discovery (PMTUD) protocol has been developed to discover link MTUs and disable fragmentation during transmission, to avoid the problem of secondary fragmenting that is common in Internet Protocol version 4 (IPv4). To further reduce the processing overhead of fragmentation and defragmentation, the concept of jumbo frames was proposed to expand the MTU of Ethernet, thereby better utilizing the performance of Gigabit Ethernet. Improving message delivery for large messages can be achieved by carefully choosing appropriate fragment sizes [6] and the amount of network coding added [7]. While these approaches succeed in diminishing the volume of fragments and partially curbing overhead, they fall short in tackling the persistent challenge of delays arising from operations throughout the transmission process. An optimal fragmentation strategy in multipath transmission has been proposed in [8] to prevent unnecessary retransmission. Additionally, a novel architecture for the future Internet called Cache-and-Forward (CNF) has been proposed in [9], which transports content as "packages" in a hop-by-hop manner toward the destination, instead of transporting a flow of fragmented packets along an established Transmission Control Protocol/Internet Protocol (TCP/IP) connection. Various research efforts have been made to improve real-time performance and reduce delays in Local Area Networks (LANs) or industrial networks, such as real-time scheduling, resource allocation, and data recovery. These efforts are described in [10–13]. However, these methods are not commonly used for large message transmission in Wide Area Networks (WANs). Some efforts have also been made to widen the standard to use the tunnel to transfer the fragments, as described in [14].

Unfortunately, the existing methods and mechanisms are still inadequate in several aspects.

- The processing efficiency of existing mechanisms is inadequate in several aspects. For example, network devices require a series of matching conditions to filter packets, but current methods still rely on Access Control Lists (ACLs). When a packet is received by a device's port, its fields are analyzed based on the ACL rules applied to the port. After identifying a specific packet, it allows or prohibits the packet to pass according to the predefined policy. Another issue is the Longest Prefix Match (LPM) algorithm used by routers. When a router receives an IP packet, it compares the destination IP address of the packet with all the routing tables in its local routing table, bit by bit, until it finds the entry with the longest match. However, operations such as LPM and ACL are set to operate on all packets, which creates inefficiencies during the transmission of large messages that have to pass through a considerable number of routers and devices. The processing of all packets by multiple devices contributes to this inefficiency.

- Delay is a critical issue in the transmission of large messages and is typically composed of four parts: sending, queuing, transmission, and processing. Unfortunately, current methods often poorly handle each of these parts, resulting in an accumulation of packet delay that can cause significant transmission delays for the entire message [8]. This delay accumulation can also lead to jitter, generated due to differences in packet delay values. When packets do not arrive uniformly at the receiving end, the receiver must compensate by buffering received packets and using other techniques. In some cases, packet loss may occur if the receiver cannot make proper corrections. Congestion can also occur when a network device starts dropping packets, and the terminals cannot receive them. The terminals may request the retransmission of lost packets, leading to congestion crashes. If the congestion becomes severe, it may even cause a network collapse. Additionally, these negative effects are amplified for packets with strong correlations that require defragmentation at the receiver. The defragmentation of such packets can further amplify these negative effects or even cause more severe problems. For instance, certain blocks of data may be lost at the receiver, and subsequent blocks may be unable to decode them properly.

To address these problems, researchers have conducted research and developed various solutions. For example, NPU (Network Processing Unit) cache machine instructions and values during program operation can help reduce delays, but they still cannot eliminate duplicate matches during message forwarding. Another solution is to use OpenvSwitch (OVS) to introduce a fast path, but it is protocol-related. In a recent study, the authors utilized caching in the data plane to identify the same stateless message flow and improve the forwarding rate and throughput of P4 devices [15].

Using a cache in the data plane is a common approach, but it requires additional hardware support, which can add cost and complexity. Alternatively, registers can achieve similar results without adding hardware cost or complexity. In network forwarding, using registers can avoid the performance impact of factors such as cache line size and conflicts, making it suitable for quickly accessing, storing, and updating metadata like forwarding tables and state machines. Therefore, using registers can result in higher forwarding performance and lower delay in certain scenarios. Additionally, compared to using caches, the use of registers provides better control over the flow table size and space, making it more adaptable to different network environments and requirements.

The proposed mechanism in this paper introduces a novel chained forwarding approach that is specifically designed for large message transmission, taking into consideration its unique characteristics. This mechanism utilizes registers to log the entire flow operation in the form of a quadruplet with identification, allowing for subsequent fragments to be processed in the same way as the first fragment that enters the devices. Unlike other approaches, this mechanism operates entirely in the data plane, providing chained forwarding capability, while also enabling the forwarding latency caused by operations such as LPM and ACL to be reduced or even eliminated in network devices. By separating forwarding and checking, the mechanism achieves fast forwarding and ensures QoS for large messages. Based on our analysis and simulation results, we successfully implemented the register method and demonstrated that our proposed chain forwarding mechanism outperforms the traditional mechanism in terms of forwarding latency and throughput.

The paper is organized as follows. Section 2 introduces the mechanism's idea and model. In Sect. 3, we explain the specific implementation of the mechanism. Section 4 introduces the simulations set up and presents the experimental results. The conclusion is provided in Sect. 5. Finally, in Sect. 6, future work is discussed.

2   System Model

The section describes the proposed mechanism that aims to enhance the forwarding efficiency of large messages by reducing unnecessary operations and accelerating forwarding. It mentions that various operations may be required in the process of network forwarding, such as matching of source address, a destination address, port number, traffic restriction, and filtering, among others. However, ACL and LPM are more common and, therefore, selected for optimization in this paper. The mechanism achieves this by performing checks and longest prefix matches only on the first fragment of each large message that enters the network devices while logging the operation and forwarding status. The subsequent fragments is processed without additional checks and longest prefix matches, and unfragmented packets are processed traditionally. The mechanism is further illustrated in Fig. 1.

Our chained forwarding strategy primarily focuses on large message transmissions, but it also takes into account the impact on prevailing forwarding algorithms and smaller message transmissions. The mechanism dynamically adapts to diverse network traffic and packet sizes, reverting to conventional forwarding techniques for smaller packets. This versatility maintains network performance efficiency while accommodating various packet dimensions. Although our primary objective revolves around enhancing forwarding performance, we acknowledge potential security risks and the importance of security in real-world applications. Our approach is suitable for authentic network traffic, which typically

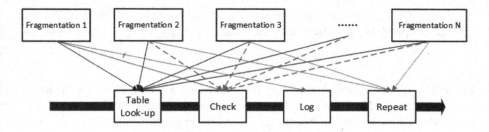

Fig. 1. The idea of chained forwarding mechanism

contains packets of diverse lengths, and provides scalability and adaptability for different network environments. In summary, our chained forwarding mechanism offers a versatile and adaptable solution for improving network performance, taking into consideration various packet sizes and network conditions.

We conducted an in-depth analysis of the mechanism and introduced a unique identifier, called a quadruplet, to resolve conflicts in identification within a certain period of time. When a packet enters the device, the device reads the frag information in the header and matches it with the information in the registers using the quadruplet as an index. The quadruplet is a four-tuple that uniquely marks the flow with an identification. In this paper, we use the term 'packet flow' to refer to a collection of packets that travel along a path in a network. A packet flow can include individual packets or components of packets, such as fragments. Specifically, we use this term to describe large packets that are fragmented into multiple pieces as they pass through a device, but are still considered part of the same flow. Packets entering the device are recorded as quadruplets based on the information, which is

$$p_k = \{sa_k, da_k, sp_k, dp_k\}. \tag{1}$$

In (1) the variable p_k represents the quadruplet for a packet. The quadruplet consists of four values: sa_k: the source address of the packet. da_k: the destination address of the packet. sp_k: the source port of the packet. dp_k: the destination port of the packet.

The identification information serves as a unique identifier for the packet flow and is used to match and reassemble the fragmented packets belonging to the same flow. When the quadruplet found in the register indexed by "identification" does not match the quadruplet of the current fragments, it indicates that although the identification of the current fragments is the same as the identification of the flow to which the quadruplet stored in the quadruplet register belongs, it is not the same flow. In this case, matching and checking should be initiated. The ACL defines the set of packets to which the rules of the statement apply, and we will refer to this set as the "ACL set". ACL: $L = \{r_1, r_2, r_3, \ldots, r_n\}$, where $r_i \in L$ represents the ith statement in L.

In an ACL, a statement defines the set of packets to which the rules of the statement apply. We denote this set of packets as "r_i", which is represented as

$$r_i = \{PO_i, SA_i, DA_i, SP_i, DP_i\}, \tag{2}$$

where PO_i is the protocol option, which may include information such as IP protocol version number, service type, datagram length, identifier, flags, fragment offset, time to live, protocol type, and header checksum. SA_i is the source address of the packet. DA_i is the destination address of the packet. SP_i is the source port number of the packet. DP_i is the destination port number of the packet. These elements are used to match and check the fields of packet headers in order to determine the forwarding path of the packet.

If the packet is within the range defined by the statement r_i then it is denoted as

$$p_k \in r_i \Leftrightarrow (sa_k \in SA_i) \wedge (da_k \in DA_i) \wedge (sp_k \in SP_i) \wedge (dp_k \in DP_i). \tag{3}$$

It indicates whether the device permits or denies the entire packet flow that the packet belongs to. Alternatively, the set of routing entries can also be represented as:

$$h_i = \{RE_i\}. \tag{4}$$

In (4), the variable "h_i" represents the set of routing entries that can be used to forward the packet. The set of routing entries is denoted as "RE_i". If the destination address matches the route entry, then it is noted as

$$p_k \in h_i \Leftrightarrow (da_k \in RE_i). \tag{5}$$

The register keeps track of which port should be used to send a packet and logs how fragment is processed. It stores a quadruplet as

$$p_{id} = \{sa_{id}, da_{id}, sp_{id}, dp_{id}\}, \tag{6}$$

where id represents the identification of the packet, which is a 16-bit value less than 65536. The register does not actively delete the quadruplet; instead, it waits for the fragments of the next large message with the same identification to enter the device and replace the existing quadruplet.

Assuming an ACL set $L = \{r_1, r_2, r_3, ..., r_n\}$, a routing entry set $H = \{h_1, h_2, h_3, ..., h_s\}$, a fragments set $P = \{p_1, p_2, p_3, ..., p_m\}$, and the time to match a packet with an ACL statement as T_{acl}, the elapsed time for a packet to ACL check is $T_{acl}f(k)$, where $f(k)$ is a function from the packet label to the statement number, and the value range of $f(k)$ is $(1, n)$. Therefore, the total time for all the fragments being ACL checked is $mT_{acl}f(k)$.

Additionally, the average time to match a packet with routing entries is denoted by T_{lpm}, which is related to the total number of route entries s and represents the size of the routing table in the host. To calculate the total time required to process a packet flow consisting of m fragments using the traditional mechanism, we need to consider the time required for each fragment to go through the network device. Assuming that each fragment goes through the same set of operations, we can calculate the total time as

$$T_{total} = \sum_{i=1}^{m} (T_{acl} f(k_i) + T_{lpm}),$$ (7)

where m is the number of fragments in the packet, $T_{acl} f(k_i)$ is the time required to match the ACL rule for the i-th fragment, and T_{lpm} is the time required for the LPM matching operation.

Since each fragment needs to go through the same set of operations, we can assume that $T_{acl} f(k_i)$ and T_{lpm} are the same for all fragments. Therefore, we can simplify the above equation as

$$T_{total} = m(T_{acl} f(k) + T_{lpm}),$$ (8)

where $T_{acl} f(k)$ and T_{lpm} are the time required to match the ACL rule and perform the LPM matching operation, respectively, for a single fragment.

The total time required to process a packet flow consisting of m fragments using the new mechanism is given by

$$T_{\text{total}} = T_{acl} f(k) + T_{lpm} + T_r + 2T_w + 2T_r(m - 1).$$ (9)

This formula takes into account various factors involved in packet processing. The term $T_{acl} f(k)$ represents the time required to match the ACL rule for the first fragment. The term T_{lpm} represents the time required for the LPM matching operation. The term T_r represents the time required for reading the registers, while $2T_w$ represents the time required for writing the registers since the first fragment needs to write twice. Finally, the term $2T_r(m - 1)$ represents the time required to read the registers for the subsequent $m - 1$ fragments. These fragments need to read the registers to obtain the ID and forward port information of the first fragment to write the registers.

In (8), the processing time increases linearly with the number of fragments, m. This is because each fragment must undergo the same set of operations, including matching the ACL rule and performing the LPM matching operation. In (9), the processing time has both a constant component and a component that increases linearly with the number of fragments, m. The constant component represents the time taken to process the first fragment, while the linear component represents the time taken to process the remaining $m - 1$ fragments. As m increases, the old mechanism's processing time increases linearly with m, while the new mechanism's processing time increases at a slower rate. This suggests that the new mechanism may have a performance advantage over the old mechanism for large m.

The actual performance of each mechanism depends on various factors, such as the specific network devices, packet characteristics, and load conditions. Thus, it's difficult to establish a general conclusion about their relative performance without considering these factors. In practical scenarios, the performance of each mechanism may be influenced by factors such as the speed of the network devices, the efficiency of the ACL rule matching and LPM matching operations, and the complexity of the network topology.

In summary, we can say that the new mechanism may have a performance advantage over the old mechanism for large m, but the actual performance of each mechanism depends on various factors. To draw a more definitive conclusion, it's necessary to consider these factors and perform experiments or simulations in specific network environments.

3   Implementation of the Mechanism

In the previous section, we described the mechanism flow of the chained forwarding mechanism. In this section, we present the concrete implementation of each part of the mechanism in the data plane of the Tofino architecture, using P4 language pseudo-code.

The Tofino architecture can be divided into two parts: programmable components, including Ingress and Egress pipeline, each of which contains a parser, Match-Action Pipeline, Deparser, Tofino-specific Intrinsic Metadata, and Tofino-Specific Externs; and fixed functional components, which are not programmable but configurable, such as Packet I/O (MAC/SerDes/DMA), Packet Replication Engine, and Traffic Manager.

Algorithm 1. Operation with quadruplets

Input: *Curquadruplet*, *Index*, *Reg* for the register
Output: *Prequadruplet*
1: *Prequadruplet=Register[Index]*
2: *Register[Index]=Curquadruplet*
3: Return(*Prequadruplet*)

We have implemented the reading and writing of quadruplets in registers, as well as the chained processing and dynamic storage and inheritance of actions in the ingress section of the Tofino architecture. For example, we use Algorithm 1 to show how we store the quadruplets in registers and read them by their index when the fragment enters the device. The stored operations are represented by numbers in the registers, and the same operation is applied to subsequent fragments in the same stream if the quadruplet check passes.

This mechanism is consistent with the idea of matches and checks. After reading the operation number, the system determines whether to execute the stored operation or write the operation after matching or checking. For example, we provide the pseudo-codes for LPM and ACL in Algorithm 2 and Algorithm 3, respectively.

In Algorithm 2, the write operation writes the forwarding port number, which is within the range of 0-65535 in the Match-Action table and can be adjusted

---

Algorithm 2. Longest Prefix Matches for large messages

Input: *CurQuat, Indentification, Reg*
1: PreQuat=LPMRegAction.Read(*CurQuat*)
2: **if** *PreQuat==CurQuat* **then**
3:     Meta.Port=PortReg.Read(*Indentification*)
4: **else**
5:     IpLPM.Apply()
6:     PortReg.Write(*Indentification*)
7: **end if**

Algorithm 3. Access Control Lists for large messages

Input: *CurQuat, Indentification, Reg*
1: **if** *CurQuat*.Hdr.Ip.IsValid() **then**
2:     PreQuat=ACLRegAction.Read(*Indentification*)
3:     **if** PreQuat==*CurQuat*.Hdr.Ip.Quat **then**
4:         Meta.Oper=OperReg.Read(*Indentification*)
5:         ACLCheck.Apply()
6:     **else**
7:         IpAcl.apply()
8:         OperReg.Write(*Indentification*)
9:     **end if**
10: **end if**

---

based on the device. In the simulated environment in this paper, the port numbers range from 1 to 64.

In Algorithm 3, the write operation writes either 0 or 1 in the Match-Action table, representing either discard or send, respectively. To speed up operations, we set exact lookups instead of ternary lookups.

Algorithm 4 describes a chain processing mechanism that uses LPM and ACL. The input parameters are the quadruplet of the current fragment (*CurQuat*), an identifier (*Indentification*), and a register (*Reg*). If the IP header in the current quadruplet is valid, the algorithm reads the previously stored quadruplet from the ACL register, compares it with the current quadruplet, and if they match, reads the corresponding values from the port register and the operation register, and then applies the flag check to the fragment. If the quadruplets do not match, the ACL rules are applied first, then the operation number is written into the operation register, the LPM rules are applied, and finally, the port number is written into the port register.

The algorithms we have developed are effortlessly integrated and deployed, functioning harmoniously with prevailing network devices and protocols. This characteristic renders our approach more viable in real-world implementations and can bolster network performance with minimal disruption to established network architectures. By implementing the chained forwarding mechanism in the data plane, we can reduce the overhead caused by control from the control plane. In addition, the mechanism reduces the number of subsequent fragments

---

Algorithm 4. LPM-ACL Chain

Input: *CurQuat, Indentification, Reg*

 1: **if** *CurQuat*.Hdr.Ip.IsValid() **then**
 2:    PreQuat=ACLRegAction.Read(*Indentification*)
 3:    **if** PreQuat==*CurQuat*.Hdr.Ip.Quat **then**
 4:       Meta.Port=PortReg.Read(*Indentification*)
 5:       Meta.Oper=OperReg.Read(*Indentification*)
 6:       FlagCheck.Apply()
 7:    **else**
 8:       IpACL.apply()
 9:       OperReg.Write(*Indentification*)
10:       **if** !Ipv4Host.Apply().Hit **then**
11:          IpLpm.Apply()
12:       **end if**
13:       PortReg.Write(*Indentification*)
14:    **end if**
15: **end if**

---

matching and checking, which is expected to improve the forwarding performance
of network devices, such as switches and routers.

4   Evaluation of Proposed Chained Forwarding Mechanism with P4 Simulator

In this section, we conduct simulations to verify the efficiency of the chained
forwarding mechanism and the proposed model. We use a P4 switch simulator
for this purpose. The simulation environment and scenarios are described in
detail in the following subsections.

4.1   Environment

The simulation was conducted using a P4 switch simulator with Tofino architecture, featuring multiple pipelines that can run different lookup logic (i.e.,
different P4 programs) and 12 Match-Action Units in each pipeline, which are
shared between incoming and outgoing pipelines. The network functions were
developed using data plane programming in the P4 language. In the context
of our paper, transmission efficiency refers to the effectiveness of the forwarding process in handling large messages, with a focus on reducing the overall
time taken for a message to traverse the network. This includes improvements
in various aspects of the forwarding process, such as processing, queuing, and
transmission latencies. The objective of the simulation was to compare the efficiency of the chained forwarding mechanism with the traditional large message
forwarding mechanism, focusing on the improvement in transmission efficiency,
primarily by reducing the forwarding latency for large messages while maintaining the effectiveness of handling smaller messages.

4.2   Simulation Setup

Our experiment seeks to evaluate the performance of traditional forwarding mechanisms and proposed chained forwarding mechanisms optimized for ACL and LPM when processing data packets of different fragment sizes. To achieve this, we set up various parameters and conditions during the experiment, including the flow table size for ACL and LPM, the number of fragments, and the number of routing entries.

To simulate traffic in actual networks, we selected 2000 routing entries and 2000 ACL rules, which reflect the typical number of routing tables and ACL rules in real-world networks. When selecting the flow table size for ACL, we considered that relatively few filtering rules in actual networks need to be processed, and the processing overhead of ACL is relatively high. Therefore, we set the flow table size for ACL to 4096. For the flow table size for LPM, we selected a larger value of 12288, as LPM matching operations are more common in modern networks, and modern hardware can increasingly support larger flow tables. Thus, we expect that increasing the size of the LPM flow table can improve the matching rate of LPM operations and maximize our optimization effect.

We designed three experimental groups and one control group, each with different mechanisms. Experimental Group 1 used the optimized ACL chained forwarding mechanism, Experimental Group 2 used the optimized LPM chained forwarding mechanism, and Experimental Group 3 used the fully optimized chained forwarding mechanism. The control group used the traditional forwarding mechanism, which is conducive to comparing the performance differences between different experimental groups. We separately examined the performance of User Datagram Protocol (UDP) and Transmission Control Protocol (TCP) under two protocols. Each group was tested under the same test environment. To compare the performance differences under different mechanisms, we measured the ratio of forwarding delay between the new mechanism and the traditional mechanism in each test group. For each test, we performed multiple runs to obtain more accurate results. We reported only the results of UDP to highlight the superiority of the chained forwarding mechanism because UDP does not have a complex processing process like TCP, thus emphasizing the benefits of optimization. However, comparing the results of TCP and UDP is still valuable for a comprehensive evaluation of the impact of IP layer optimization.

We selected different numbers of fragments as the horizontal coordinate and tested them in increments to control the input traffic size without changing other parameters, thus studying the performance of the new mechanism. At the same time, by comparing the forwarding delay under different conditions, we can prove the optimization effect of the new mechanism, which is not only for larger messages. Additionally, we set different sizes for ACL and LPM flow tables to compare the impact of different flow table sizes on performance in the experiment.

Through the above experimental design, we aim to prove that the chained forwarding mechanism can optimize network forwarding performance, especially in situations with many ACL rules and routing table entries. Our results will provide a quantitative evaluation and analysis of the optimization effect of the chained forwarding mechanism.

4.3   Experimental Results

Figures 2 and 3 provide a visual comparison of the forwarding delay ratios for UDP and TCP using both the traditional and chained forwarding mechanisms. Generally, higher delays are observed when forwarding more fragments. As seen in Fig. 2, the traditional mechanism with complete ACL and LPM displays the highest forwarding delay for UDP. The chained forwarding mechanism optimized solely for ACL shows an average delay ratio of 98.7% compared to the traditional mechanism. The chained forwarding mechanism optimized solely for LPM exhibits an average delay ratio of 95.4%, while the chained forwarding mechanism optimized for both has an average delay ratio of 89.9%. Owing to the inherent characteristics of TCP, the delay of all TCP segments in Fig. 3 is slightly higher than that of the UDP segments. The delay ratios of the three TCP segments to the traditional mechanism are 98.8%, 96.3%, and 90.2%, respectively.

Table 1. Forwarding delay

Number of fragmentation	1000	2000	3000	4000	5000	6000	7000	8000	9000	10000
Chain forwarding mechanism for ACL (UDP)	25.67233	51.09	76.72633	101.94233	128.293	153.55467	178.727	203.958	229.939	257.2275
Chain forwarding mechanism for LPM (UDP)	24.95333	49.42167	73.907	98.872	123.66067	148.135	172.772	197.491	221.959	246.64933
Chain forwarding mechanism for ACL and LPM (UDP)	23.42567	47.05467	69.43033	92.60733	116.91	139.669	161.50033	185.902	208.59133	236.067
Traditional forwarding mechanism (UDP)	26.26233	52.182	77.71967	103.26067	129.338	155.65233	179.86867	206.09033	232.13267	260.41133
Chain forwarding mechanism for ACL (TDP)	26.5	53.09133	78.15567	104.34	130.094	155.914	182.434	209.351	236.7535	259.7805
Chain forwarding mechanism for LPM (TDP)	26.02167	51.242	76.65433	101.852	126.69333	152.423	176.93467	204.41067	228.046	253.07667
Chain forwarding mechanism for ACL and LPM (TDP)	23.858	47.694	71.455	96.287	119.857	143.949	167.193	190.46367	214.45033	237.445
Traditional forwarding mechanism (TDP)	26.78233	53.751	79.25333	105.68567	131.527	158.49367	184.073	210.81933	238.92633	262.76433

In Table 1, we present actual delay data in milliseconds to more lucidly showcase the improvement in time offered by the proposed chained forwarding mechanism. Despite our experimental data revealing processing delays exceeding 100ms, we maintain that these results remain valuable in demonstrating the performance improvement trend of the chained forwarding mechanism. It is essential to recognize that although the experimental delay values may diverge from those in real-life network environments, these data underscore the potential advantages of the chained forwarding mechanism in minimizing processing delay. This pattern implies that when processing a significant volume of packets, employing a chained forwarding mechanism could lead to performance enhancements. For a more precise assessment of the chained forwarding mechanism's efficacy in real-world network environments, future research endeavors could aim to refine the experimental setup and simulator to better emulate real-world network conditions. This advancement would further substantiate the effectiveness of our chained forwarding mechanism in real-world applications.

Our study reveals that the impact of ACL optimization is not as substantial as that of LPM optimization. Our simulation results indicate that even with the capability of the chained forwarding mechanism to bypass subsequent checks,

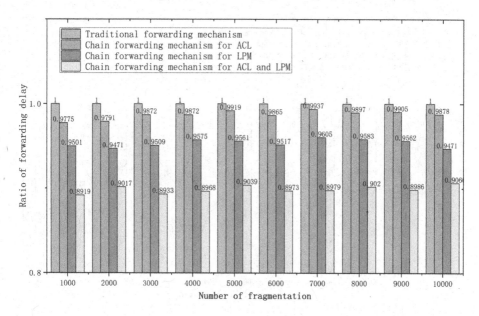

Fig. 2. Comparison of mechanisms for UDP

the forwarding delay is still affected by the number of filtering conditions and the size of the flow table. In cases where the number of ACL rules is constant and the flow table size of ACL is modified to 4096, 8192, and 12288, respectively, the forwarding delay increases to varying degrees. Given that the flow table size of LPM is larger, optimizing solely for LPM results in better outcomes. Our experimental findings indicate that employing chained forwarding for ACL and LPM results in processing delay savings of 2% and 5%, respectively. However, when we implement chained forwarding in both ACL and LPM concurrently, the processing delay savings escalate to 10%. This amplified performance enhancement can be ascribed to the following factors:

– Upon implementing the proposed chained forwarding in both ACL and LPM, each fragment stands to benefit from optimizations during the forwarding process in both domains. As a result, individual fragments can traverse ACL and LPM rules more swiftly, ultimately reducing the overall processing time.
– By optimizing both ACL and LPM processes, chained forwarding enables more efficient utilization of network resources. This fosters improved resource scheduling and load balancing, which in turn bolsters processing efficiency.
– Moreover, the simultaneous application of chained forwarding in both ACL and LPM could yield several potential synergistic optimizations. For instance, chained forwarding may enhance the alignment of packet fragments and forwarding rules during ACL and LPM processing, subsequently diminishing the computational overhead necessary for forwarding decisions.

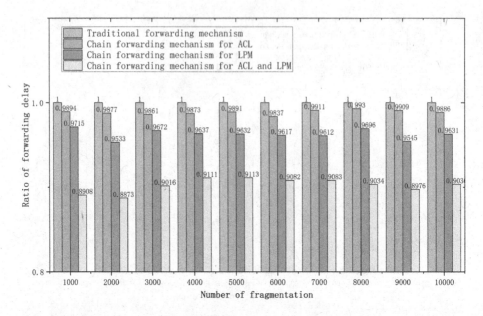

Fig. 3. Comparison of mechanisms for TCP

In larger network setups, it is common to have a higher number of route entries and ACL rules due to the increased network topology complexity. This complexity necessitates a greater number of routing information and ACL rules to maintain proper network functionality. We conducted experiments to assess the forwarding delay rate by varying the number of ACL rules and route entries while keeping other parameters constant. Figure 4 demonstrates that when the number of route entries is fixed at 2000, the forwarding delay ratio decreases as the number of ACL rules increases. Conversely, Fig. 5 compares the forwarding delay ratio with a constant number of ACL rules (2000) and varying route entries. The forwarding delay ratio tends to decrease with an increase in route entries, but not as much as when the number of ACL rules increases. This is because the LPM algorithm's time complexity is relatively low, indicating that the impact of increasing the number of route entries on forwarding delay may be relatively insignificant. Overall, these experimental findings demonstrate that the chained forwarding mechanism can reduce forwarding delay compared to the traditional mechanism, and the effect of the chained forwarding mechanism is more significant when the number of ACL rules increases. These findings provide valuable insights into the performance of the chained forwarding mechanism and can assist in network design and optimization.

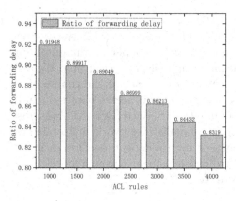

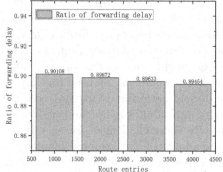

Fig. 4. The effect of ACL rules on forwarding delay

Fig. 5. The effect of route entries on forwarding delay

5 Conclusion

To alleviate potential performance bottlenecks associated with the processing of ACL rules and route entries, this study introduces a chained forwarding mechanism. This mechanism accelerates the forwarding of large messages by minimizing unnecessary checks and matches, such as ACL and LPM. Through an open discussion of the forwarding model and a practical implementation of the P4 switch in the data plane, we demonstrate the efficacy of our model via simulation. This highlights a significant improvement in the forwarding performance of large messages, with the chained forwarding mechanism achieving over a 10% improvement in forwarding delay compared to traditional mechanisms. Moreover, the mechanism performs exceptionally well in large-scale networks, satisfying the demands of large message services and ensuring Quality of Service (QoS) in a timely and efficient manner.

Our chained forwarding mechanism, by reducing processing latency for large message transmissions, equips the network to manage large messages more proficiently. This potentially enables the network to accommodate an increased volume of video traffic, catering to the escalating demand for multimedia services. Furthermore, this optimization allows for more efficient management of existing traffic within a shorter time frame, a crucial aspect for enhancing network throughput and responsiveness, particularly during peak hours. Reducing processing latency via the chained forwarding mechanism also reduces the likelihood of network congestion and packet loss. This is critical for applications requiring low latency and high reliability, such as online gaming and real-time communication. Moreover, efficient processing of large packets could potentially free up more network bandwidth, thereby supporting superior-quality video streaming and other multimedia services [16].

While our chained forwarding mechanism has proven its effectiveness in handling large message transmission, we acknowledge several key challenges that warrant exploration in future work. For instance, understanding the impact of our chained forwarding mechanism on CPU load and energy efficiency, and addressing congestion and collision issues more effectively in real network environments are necessary. Additionally, we need to delve deeper into maintaining data security and privacy protection while ensuring efficient forwarding. We also recognize that in order to further enhance the performance of the chained forwarding mechanism, we need to compare it with more advanced schemes, include additional performance metrics, conduct a Quality of Service (QoS) analysis, discuss compatibility issues, and evaluate it in more realistic network environments. Despite these challenges, we eagerly anticipate improving the performance and applicability of our proposed chained forwarding mechanism through more extensive experiments and analyses in our future work.

References

1. Yeo, H., Lim, H., Kim, J., Jung, Y., Ye, J., Han, D.: Neuroscaler: Neural video enhancement at scale. In: Proceedings of the ACM SIGCOMM 2022 Conference. New York, NY, USA (2022). https://doi.org/10.1145/3544216.3544218
2. Cisco: Cisco visual networking index report. Tech. rep. (2022). http://www.cisco.com/c/en/us/solutions/collateral/service-provider/visual-networking-index-vni/complete-white-paper-c11-481360.pdf
3. Li, J., et al.: Livenet: a low-latency video transport network for large-scale live streaming. In: Proceedings of the ACM SIGCOMM 2022 Conference, pp. 812–825 (2022)
4. Yang, J., Jiang, Y., Wang, S.: Enhancement or super-resolution: learning-based adaptive video streaming with client-side video processing. In: IEEE International Conference on Communications, vol. 2022-May, pp. 739–744 (2022). https://doi.org/10.1109/ICC45855.2022.9839262
5. Jin, X., Xia, C., Guan, N., Zeng, P.: Joint algorithm of message fragmentation and no-wait scheduling for time-sensitive networks. IEEE/CAA J. Autom. Sinica $8(2)$, 478–490 (2021). https://doi.org/10.1109/JAS.2021.1003844
6. Bialon, R., Tolkes, J., Graffi, K.: Improving message delivery in opportunistic networks with fragmentation and network coding. In: 2019 International Conference on Internet of Things (iThings) and IEEE Green Computing and Communications (GreenCom) and IEEE Cyber, Physical and Social Computing (CPSCom) and IEEE Smart Data (SmartData) (2019)
7. Tran Thai, T., Chaganti, V.G., Lochin, E., Lacan, J., Dubois, E., Gelard, P.: Enabling e2e reliable communications with adaptive re-encoding over delay tolerant networks. In: IEEE International Conference on Communications. vol. 2015-September, pp. 928–933 (2015). https://doi.org/10.1109/ICC.2015.7248441
8. Wang, J., Liao, J., Zhu, X.: On preventing unnecessary fast retransmission with optimal fragmentation strategy. In: IEEE International Conference on Communications, pp. 85–89 (2008)
9. Dong, L., Liu, H., Zhang, Y., Paul, S., Raychaudhuri, D.: On the cache-and-forward network architecture. In: IEEE International Conference on Communications (2009). https://doi.org/10.1109/ICC.2009.5199249
10. Fogli, M., Giannelli, C., Stefanelli, C.: Edge-powered in-network processing for content-based message management in software-defined industrial networks. In: IEEE International Conference on Communications. vol. 2022-May, pp. 1438–1443 (2022). https://doi.org/10.1109/ICC45855.2022.9838863

11. Xi, J., Kong, F., Kong, L., Wei, L., Peng, Z.: Reliability and temporality optimization for multiple coexisting wirelesshart networks in industrial environments. IEEE Trans. Indust. Electron. **PP**(8), 1–1 (2017)
12. Tang, J., Shim, B., Quek, T.Q.: Service multiplexing and revenue maximization in sliced c-ran incorporated with urllc and multicast embb. IEEE J. Sel. Areas Commun. **37**(4), 881–895 (2019)
13. Kong, L., et al.: Data loss and reconstruction in wireless sensor networks. IEEE Transactions on Parallel and Distributed Systems (2013)
14. Saldana, J., Wing, D., Fernandez-Navajas, J., Ruiz-Mas, J., Perumal, M.A.M., Camarillo, G.: Widening the scope of a standard: Real time flows tunneling, compressing and multiplexing. In: IEEE International Conference on Communications, pp. 6906–6910 (2012)
15. Ma, Z., Bi, J., Zhang, C., Zhou, Y., Dogar, A.B.: Cachep4: a behavior-level caching mechanism for p4. In: ACM Special Interest Group on Data Communication (2017)
16. Cisco: Cisco annual internet report (2018–2023) white paper. Tech. rep. (2021). https://www.cisco.com/c/en/us/solutions/collateral/executive-perspectives/annual-internet-report/white-paper-c11-741490.html

TDC: Pool-Level Object Cache Replacement Algorithm Based on Temperature Density

HuaCheng Lu[1], Yong Wang[1,2(⊠)], JunQi Chen[1,2], DaHuan Zhang[1], and ZhiKe Li[1]

[1] School of Computer and Information Security, Guilin University of Electronic Technology, Guilin 541004, China
ywang@guet.edu.cn
[2] Guangxi Educational Big Data and Cyberspace Collaborative Innovation Center, Guilin University of Electronic Technology, Guilin 541004, China

Abstract. For mixed HDD and SSD storage scenarios, Ceph Cache Tier provides a tiered caching feature that separates fast and slow storage pools to manage data objects more efficiently. However, due to the limited total capacity of the cache pool, only some data objects can be stored. Performance can be significantly improved when clients focus on accessing hot objects in the cache pool. If a client accesses the cache pool without hitting data, redundant IO operations occur, which increases client access latency and reduces throughput. To improve the hit rate of the Ceph Cache Tier cache pool, this paper proposes a temperature density-based cache replacement algorithm (TDC). The algorithm improves the hit rate of the cache pool by calculating the temperature density of the space consumed by each object and evicting objects with the lowest temperature density, thus evicting objects that contribute less to the hit rate. The algorithm mainly includes object temperature calculation, temperature density calculation and cache replacement policy. Subsequently, we evaluate the TDC algorithm on a real traces dataset using playback workload IO and demonstrate the efficiency of the algorithm in improving the cache hit rate. Finally, we applied the TDC algorithm to a Ceph distributed storage system and verified the performance of the Cache Tier based on the TDC algorithm.

Keywords: Distributed Storage · Hybrid Media · Caching Algorithm · Hierarchical Storage

1 Introduction

In the era of increasing popularity of cloud computing and big data technologies, the demand for efficient and reliable data storage and management services in data

This work was supported by the National Natural Science Foundation of China (61831013), the Guangxi Innovation-Driven Development Project (Science and Technology Major Special Project Gui Ke AA18118031) and Guangxi Graduate Education Innovation Project (YCSW2021179).

Z. Tari et al. (Eds.): ICA3PP 2023, LNCS 14487, pp. 204–223, 2024.
https://doi.org/10.1007/978-981-97-0834-5_13

centers is increasing [1]. To meet these demands, typical large-scale distributed storage systems such as GFS [2], Ceph [3], Azure storage [4], and Amazon S3 [5] have emerged. These storage systems spread data across multiple independent storage devices while providing a unified storage service interface. Ceph is an open source distributed storage system with high reliability, automatic load balancing, and fault recovery capabilities. Since its first release in 2012, Ceph has become the default storage backend for OpenStack [6]. Ceph uses the CRUSH algorithm to map data to storage nodes [7], thus replacing metadata servers in traditional distributed storage systems and allowing clients to locate data storage locations more quickly. Ceph also supports a variety of storage services such as object storage, block storage, and file storage, and thus is widely used.

With the continuous upgrading of hardware and software, large-scale distributed storage clusters may employ heterogeneous storage media such as mechanical hard disks, solid state drives, and NVMe SSDs. How to achieve high performance, high reliability and cost effective distributed storage among these media is an important issue [8]. Although full SSD storage systems have the highest performance, it is difficult to adopt all of them because of the high price of SSDs. Therefore, a hybrid architecture of SSDs and HDDs has become necessary. Ceph Cache Tier provides tiered caching between fast and slow storage pools to provide higher performance [9]. However, Ceph Cache Tier uses a cache replacement strategy based on frequency estimation, which underutilizes metadata information, resulting in data hotspot identification errors and insufficient cache hits to reach the theoretical limit. In addition, Ceph Cache Tier caches and writes back on an object-by-object basis, and if the cache hit rate is low, it will generate a large number of data pulls, occupy SSD bandwidth, make the actual IO path longer, and degrade performance.

In this paper, we design a temperature density cache replacement algorithm based on cache pool level objects based on the traditional Ceph Cache Tier tiered storage system. The algorithm improves the hit rate of the cache pool by calculating the temperature density of each object consuming space and evicting the object with the lowest temperature density. The algorithm mainly includes object temperature calculation, temperature density calculation, and cache replacement policy. Subsequently, we evaluate the performance of the TDC algorithm based on the traces dataset using playback of workload IO. Finally, we validate the effectiveness of the Cache Tier based on the TDC algorithm in a Ceph distributed storage system.

This paper is structured as follows: Sect. 1 provides an introduction; Sect. 2 highlights related work; Sect. 3 analyzes the limitations of the Ceph Cache Tier caching algorithm; Sect. 4 focuses on the architecture and implementation of the system; Sect. 5 provides an experimental evaluation of the TDC caching algorithm; and Sect. 6 concludes the paper.

2   Related Work

Today, many storage media with different characteristics and capacities are used in complex and heterogeneous storage systems. To address the challenges enterprises face in improving resource utilization and read/write rates of storage services, distributed storage systems provides caching and tiering mechanism to improve cluster performance in heterogeneous storage environments. The main difference between caching and tiering is that a copy of data in a caching system is kept in the cache, while in a tiered system, the original data is migrated between multiple tiers through both upgrade and downgrade operations.

Fares et al [10] evaluated three traditional cache replacement strategies, FIFO [11], LRU [12], and LFU [13], in large-scale data storage systems, and found that these traditional policies can be successfully used in large-scale data storage systems, while the LFU policy is the most tolerant to "cleanup" policies. On the other hand, the performance of cloud applications depends heavily on the hit rate of the data center cache. Data center caches typically use least recent use (LRU) as their cache retirement policy, but LRU hit rates are far from optimal under real-world workloads. Beckmann et al. [14] proposed a cache replacement policy called EVA, which is essentially a cost-benefit analysis based approach that considers whether the hit probability of a candidate data is worth the cache space it consumes.EVA treats each candidate data as an investment and tries to retain the candidate data with the highest expected profit (measured in terms of hit rate). And a cache replacement policy based on hit density [15] is proposed to measure the degree of contribution of an object to the cache hit rate. The policy predicts the expected hit rate or hit density per object per space consumed by dynamically predicting the hit rate and evicting the object with the lowest hit rate. Gu et al. [16] proposed a two-tier user space cache management mechanism to improve the read performance of distributed memory file systems. The first layer caches cached packet references to reduce frequent page fault outages; the second layer caches and manages small file data units to avoid redundant interprocess communication. Duong et al. [17] proposed a cache management policy based on protection distance that uses dynamic reuse distance to further improve the cache replacement policy. The policy protects cache lines and supports reuse by preventing the replacement of cache lines until a certain number of accesses are made to their cache set, but cannot exceed that length to avoid cache pollution. Yang et al. [18] proposed a storage selection policy based on a predictive model that analyzes objects at long time granularity, accurately determines the hotness of objects, and decides, based on the frequency of object accesses, whether objects request access to SSD pool or backend HDD pool to reduce unnecessary overhead due to cold data processing, thus improving the overall I/O performance of the cluster.

However, the metadata information carried by data objects is not fully utilized in the optimization study of hierarchical storage, resulting in data hotspot identification errors and insufficient cache hit ratio to reach the cache theoretical limit performance. To further improve the cache hit ratio, we consider the feature that data objects carry more information and design a new Cache algorithm to improve the cache hit ratio.

3   Background and Motivation

In this section, we describe how the Ceph Cache Tier works, the limitations of the caching algorithm, and initial ideas and approaches to address these issues.

3.1   Ceph Cache Tier Data Stored Process

Ceph storage clusters are usually built with inexpensive PCs and traditional mechanical hard disks, so the disk access speed is somewhat limited to achieve the desired IOPS performance level. To optimize the IO performance of the system, you can consider adding fast storage devices as cache to reduce the data access latency. In Ceph serverside caching, the Cache Tier tiered storage mechanism is a common solution to effectively improve the I/O performance of the backend storage tier. Cache Tier [19] requires the creation of a storage pool consisting of highspeed and expensive storage devices (such as SSDs) as the cache tier, and a backend storage pool consisting of relatively inexpensive devices as the economic storage tier. The cache tier uses a multi-copy model and the storage tier can use either multi-copy or corrective code model. The architecture is shown in Fig. 1.

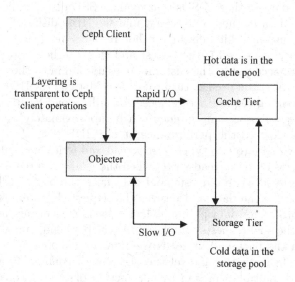

Fig. 1. Ceph Cache Tier cache pool processing process.

The theory behind Ceph's cache tiering is that there are hotspots of data that cause uneven access to data. Typically, 80% of applications access only 20% of the data, and this 20% is called hot data. To reduce response time, the hot data can be stored in a higher performance storage device (e.g., solid state drive), which is how Cache Tiering is implemented. In Cache Tiering, there is a tiered agent that is used to flush data kept in the cache tier to the storage tier and remove it from the cache tier when it becomes cold or no longer active. This operation is called a refresh or an eviction. When a client reads or writes data, Ceph's object processor is responsible for deciding where objects are stored, and the Cache Tier is responsible for deciding when objects in the cache tier are flushed back to the back-end storage tier. For write operations, the request arrives at the cache tier and answers the client directly after the write operation is completed, after which the proxy thread in the cache tier is responsible for writing the data to the storage tier. For read operations, if the cache layer is hit, it is read directly at the cache layer, otherwise it can be redirected to the storage layer for access. If the data has been accessed recently, it means it is hotter and can be elevated to the cache tier. For Ceph clients, the cache tier and the back-end storage tier are completely transparent. Therefore, all Ceph clients can use the cache tier, giving Cache Tier the potential to boost I/O performance for block devices, Ceph object stores, Ceph file systems, and native bindings.

3.2   Ceph Cache Tier Cache Hit Rate Issue

In Ceph, the tiered storage system is implemented by means of caches and storage pools, and Fig. 2 shows the Ceph Cache Tier cache pool processing. Hot resource pools can store data to those OSDs that manage SSD disks, while cold resource pools can store data to those OSDs that manage HDD disks. If the data being accessed by a customer hit falls in the hot resource pool, it can be accessed directly, when the IO speed is the fastest and close to the performance of SSD disks. If the data accessed by the customer does not fall in the hot resource pool, there is a cache miss, you need to turn to read the data on the HDD disk, and the HDD disk processing request access speed for millisecond level, so the network latency and request processing latency can be approximately ignored, that its access speed is close to the performance of the HDD disk. The processing at this point is divided into two types: proxy read and write and data pull. In the read-write process of proxy read-write, when the read-write request has a cache miss, the proxy read-write requests cold data from the backend, but the cache pool does not cache the data and returns the requested content directly to the client. In the process of data pulling, when a cache miss occurs in a read-write request, the cache pool requests cold data from the back end, and after requesting cold data from the back end, it reads the data into the cache pool, continues to process the client request and returns the request content. In addition, data that is accessed multiple times in a short period of time is considered as hot data and pulled into the hot pool, which consumes the read bandwidth of HDD disks and the write bandwidth of SSD disks. On the other hand, data that is in the hot pool needs to be written back to the cold pool periodically, at which point the write-back data will temporarily use some of the bandwidth of the SSD and HDD disks, a process called data write-back. Data that has not yet been written back to the cold pool is modified again in the hot pool. The more this happens, the more efficient the cache becomes, which is equivalent to the hot pool bandwidth being fully utilized and helping the cold pool block a large amount of write bandwidth. In short, 1% of the data is required to be flushed back dirty (i.e., 1% of the data is clean after flushing back, so subsequent hits will be non-dirty hits), and if all the data is not flushed back dirty and all the hits are accessed, then the dirty hit rate is 100%.

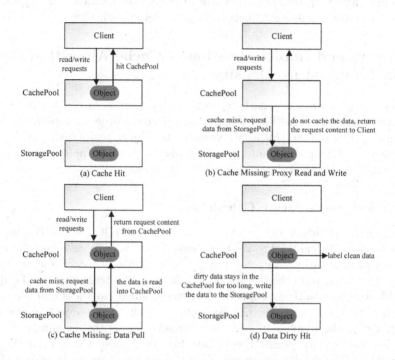

Fig. 2. Ceph Cache Tier cache pool processing process.

According to the above principle, it is easy to find that the performance of Ceph Cache Tier depends on the access hit ratio. The higher the access hit ratio, the closer the storage system is to the performance of SSD disks; conversely, the lower the access hit ratio, the closer it is to the performance of HDD disks. On the other hand, in Ceph, cache granularity is pulled and written back as objects. Therefore, in practice, if there are too many cache misses, a large amount of data will be pulled, thus taking up the bandwidth of SSD disks and making their access bandwidth worse than SATA disks. However, during actual production usage, the total data usage always increases gradually, and at the same time, the amount of hot data will also increase gradually. Therefore, as the amount of data increases throughout the usage cycle, it will inevitably go through the following process: First of all, when it is first used, the amount of data is still very small. At this time, all the data can be cached and the data hit rate is 100%, which is very effective. As the total data volume and the amount of hot data increases, the cache pool can no longer hold all the data and can only hold more hot data, and the cache hit ratio decreases gradually. As the data increases further, the cache hit ratio falls below the threshold we want, at which point keeping the same size cache pool no longer gives a good enough return on usage. Therefore, in Ceph Cache Tier, hit rate is a key metric to ensure the efficiency of the cache pool, and it needs to be monitored and alerted, and optimized and improved when appropriate. Optimizing the cache replacement algorithm of Ceph Cache Tier is one of the feasible solutions to improve its performance.

4   Design and Implementation of Cache Algorithm Based on Temperature Density

In this section, we propose a cache pool-level object temperature density caching algorithm for Ceph Cache Tier. We first introduce the object temperature based on Multiple Bloomfilter [20] and calculate the object temperature density by combining the size and age information of data objects. Finally, we propose a temperature density-based cache replacement algorithm.

4.1   Object Temperature Calculation

When a user uploads a file to the Ceph cluster, RADOS splits the file into multiple objects by calling the file_to_extents() function. To calculate the access frequency of each object, we use Multiple Bloomfilter to record the access frequency of each object in the storage pool in the face of a large number of objects, by which we can effectively capture more fine-grained closeness and frequency. Assuming the use of n versions of Bloomfilter, each hit_set object represents a Bloomfilter. if the number of historical hit_set exceeds the maximum number of versions set by the system, the earliest historical version is deleted. The following steps can be followed to calculate the temperature of an object: First, the system checks whether the current object exists in the current hit_set. If it exists, the temperature is set to a very high value (e.g., 1000000), otherwise the temperature is set to 0. Second, the system calculates the historical temperature by finding the n-1 most recent historical hit_set. If the read/write record of the object exists in the history hit_set, the current temperature is summed with the history temperature. The formula for calculating the periodic temperature based on Multiple Bloomfilter is as follows:

$$temperature_n = 1000000 * a^n \tag{1}$$

where temperature denotes the temperature value of the nth historical period. To reflect the change of hotspot data over time, the system uses a statistical period T to decay the temperature value of each object. Here, we define the elapsed time or record a certain number of requests as a statistical period T. For each statistical period, the object temperature value decays a times. Considering the stability of hotspot data, when calculating the object temperature, we consider the object temperature of the previous cycle and decay the historical statistics by a certain percentage of a. This approach provides a more comprehensive access to the statistics of object data and reduces the number of cache misses due to temperature accumulation. Figure 3 shows the variation of the object temperature with cycle and temperature decay coefficient.

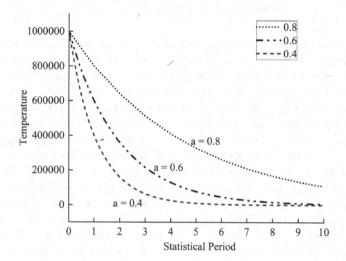

Fig. 3. Temperature changes with attenuation factor.

4.2   Object Temperature Density Calculation

In the cache pool of Ceph Cache Tier, different objects contribute differently to the cache hit rate. Traditional cache replacement policies (such as the LRU algorithm) only consider the access time of an object without taking into account factors such as the size of the object and the frequency of access. As a result, traditional cache replacement policies may keep some large, low-frequency objects in the cache, thus reducing the cache hit ratio. To solve this problem, a new cache replacement policy needs to be designed that can consider factors such as object size, access frequency and access time to determine the contribution of each object to the cache hit ratio. To improve the hit rate of the cache pool, it is necessary to consider both the historical statistics information in the cache pool and the data information carried by the cache objects, and to find a way to reconcile these factors in a single metric. Such a metric needs to satisfy three factors: the data information carried by cached objects prefers candidates that are easy to hit, and penalizes candidates that take longer to hit. To achieve these factors, the contribution of objects in the cache pool to the cache hit rate is transformed into a cost-benefit analysis by considering the time spent by object data in the cache and the cache pool occupied as cache cost. Temperature density is a metric that combines factors such as object size, access frequency, and access time, and can be used to evaluate each object's contribution to cache hit rate and perform cache replacement based on hit density. This can improve the cache hit rate and thus the performance of the application. For each candidate object, the temperature density is calculated as follows:

$$temperature\_density = \frac{temperature}{size * age} \tag{2}$$

where temperature denotes the temperature of the object, size denotes the size of the object, and age denotes the age of the object in the cache pool. The temperature density decreases as the object size increases and decreases as the age of the object increases. When selecting a swipe back object, this is achieved by comparing the temperature density of each candidate object and evicting the candidate object with the lowest temperature density. Since the temperature density is calculated considering both hit rate, size and age of the object, it can better balance various factors and ensure efficient use of the cache.

Figure 4 shows a time-dependent cache illustration depicting the lifecycle of a single object in the cache pool. The blocks in the graph represent different objects, and green and red colors indicate whether the life cycle of an object ends with a hit or an eviction. The size of each block reflects the amount of cache pool resources occupied by the corresponding object, while the temperature density is inversely proportional to the block size. The cache pool is vertically aligned and time increases from left to right, so each hit or eviction starts the lifecycle of a new block. Thus, each block represents the lifetime of a single object, i.e., the amount of free time the object spends in the cache between hits or evictions. In addition, the figure illustrates the challenge of the cache replacement strategy: with limited resources, the cache pool wants to maximize the hit rate, i.e., add as many green blocks as possible. The amount of resources each object takes up is proportional to its size and the time it spends in the cache. Therefore, cache replacement policies usually tend to keep small objects that hit quickly in order to maximize the hit rate within the space constraints of the cache pool. In this case, the cache replacement policy needs to consider the size of the objects and the frequency of accesses to determine which objects should be retained and to decide which objects should be replaced.

Temperature density can be used to measure the importance of objects in the cache and whether they will be accessed in the future, thus evaluating the efficiency and performance of the cache. Temperature density can be calculated by integrating uniformly over the entire graph. Here, hit contribution is used to define a measure of an object's contribution to the hit rate. For each green block, its hit contribution can be calculated by equation (3):

$$hit\_contribution = \frac{1hit}{size * age} \tag{3}$$

where size denotes the size of the object and age denotes the existence time of the object in the cache. For each red object, its hit contribution is 0. Therefore, the temperature density of the whole graph can be expressed as the sum of the contributions of all hit objects divided by the sum of all objects occupying cache resources, i.e.:

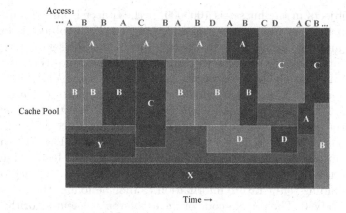

Fig. 4. The change of Cache Pool over time.

$$total\_temperature\_density = \frac{\sum_{time} \overbrace{1 + 1 + ... + 1}^{hits} + \overbrace{0 + 0 + ... + 0}^{evicts}}{\sum_{time} \underbrace{S_1 \times L_1 + ... + S_N \times L_N}_{hit\_resources} + \underbrace{S_1 \times L_1 + ... + S_M \times L_M}_{evict\_resources}} \quad (4)$$

In this equation, time denotes the life cycle of an object, hits denotes the number of hits, evicts denotes the number of evicted objects, S denotes the size of an object, and L denotes the time an object stays in the cache pool. Since the overall temperature density of the cache is proportional to its hit rate, maximizing the temperature density will also maximize the hit rate. Therefore, the goal of the cache replacement policy is to retain as many objects with higher temperature density as possible to improve the hit rate and performance of the cache.

4.3   TDC Cache Replacement Algorithm Design

When the cache pool runs out of space, it is necessary to select which objects to evict from the cache based on the cache replacement policy to make room for new objects. Choosing an inappropriate cache replacement policy may result in a lower cache hit rate, which in turn reduces the performance of the cache pool. Therefore, choosing the right cache replacement policy is critical to improve the performance of the cache pool. By controlling the caching of read and write request objects, the temperature density-based cache replacement policy achieves the goal of filtering out objects with higher temperature density to stay in the cache pool, reducing unnecessary cache replacement operations, and eliminating objects with lower temperature density to improve the cache pool hit rate. The cache replacement strategy of the TDC algorithm includes creating and initializing the cache pool data structure and hit_set, updating hit_set according to the temperature decay coefficient and statistical period, performing cache management, and executing read and write requests. When there is an IO request for a data object, it is inserted into the current hit_set, at which time the temperature of all the objects in the hit_set is 1000000. if the object is already in the cache pool, it can be read from it directly and returned; if the object is lost in the cache pool and the cache pool is full, some objects must be selected for deletion to make room. Select N objects starting from the next position of the chain table. Assume that the selected objects are $\{O_0, O_1, ..., O_n\}$, and the temperature density of N objects is calculated by the temperature density formula as $\{T_0, T_1, ..., T_n\}$. To free up cache pool for new objects, the temperature density of the cached objects can be compared and the objects with lower temperature density can be removed by a certain percentage. The TDC algorithm uses the temperature decay coefficient and statistical cycle history information to calculate the temperature density of the objects. For some cached objects, if the corresponding temperature information cannot be found, their temperature is assumed to be 0. The pseudo-code of the cache replacement algorithm is given in Algorithm 1.

5   Experimental Evaluation

In this section, we first evaluate the TDC algorithm performance in terms of cache hits, including algorithm sensitivity analysis and hit rate testing and analysis. Then we integrate the TDC algorithm in Ceph distributed storage system to evaluate the performance of TDC algorithm in Ceph Cache Tier.

5.1   Performance Test and Analysis of TDC Cache Algorithm

A. Experimental Setup and Workload. To verify the temperature density-based cache replacement algorithm designed in this paper, multiple sets of comparison experiments were conducted. The hardware environment used for the experiments is an Intel(R) Xeon(R) Silver 4110 @ 2.10 GHz processor, a server with 64GB memory, and simulated data reads and writes using simulation software written in C++. In the experiments, data reads and writes are simulated using playback of workload IO to evaluate the performance of the TDC algorithm.

We perform performance tests using real storage traces publicly available online [22], and Table 1 shows the details of both P5 and DS1 workload traces, including the number of requests, the percentage of write requests, and the average size of read and write requests. In particular, the P5 dataset is a read-dominated workload that captures disk read accesses from a large commercial search engine in response to various web search requests. This trace dataset was collected over a period of approximately one month and represents the disk read workload of a large search engine. the DS1 dataset is a writedominated workload collected from the database server on which a commercial ERP application is running. This trace dataset lasts for seven days and represents the disk read workload of the database server of a commercial ERP application.

Algorithm 1. handle_io_request and select_objects_to_be_deleted

---

- N: The number of pending objects in the cache pool.

```
 1: function HANDLE_IO_REQUEST(request)
 2:     hit_set.insert(request.object, 1000000)
 3:     if cache_pool.contains(request.object) then
 4:         return cache_pool.get(request.object)
 5:     end if
 6:     if cache_pool.is_full() then
 7:         objects ← cache_pool.get_next_objects(N)
 8:         temperatures ← []
 9:         for obj in objects do
10:             temperature ← hit_set.get_temperature(obj)
11:             temperatures.append(temperature)
12:         end for
13:         to_be_deleted ← select_objects_to_be_deleted(objects, temperatures)
14:         for obj in to_be_deleted do
15:             cache_pool.remove(obj)
16:         end for
17:     end if
18:     cache_pool.add(request.object)
19: end function
20: function SELECT_OBJECTS_TO_BE_DELETED(objects, temperatures)
21:     to_be_deleted ← []
22:     densitys ← calculate_density(temperatures, objects)
23:     max_density ← max(densitys)
24:     for i in len(objects) do
25:         if densitys[i] < (max_density * threshold) then
26:             to_be_deleted.append(objects[i])
27:         end if
28:     end for
29:     return to_be_deleted
30: end function
```

Table 1. Trace characteristics of workload.

Real load	Number of requests (times)	Write request ratio (%)	Average write request size	Average read request size
P5	1118323	12.15	29.63	19.24
DS1	2505947	40.6	16.00	18.42

B. Sensitivity Analysis of TDC Algorithm. This paper investigates the performance impact of two core parameters in the TDC caching algorithm - temperature decay coefficient and statistical period - in real-world application scenarios. Both determining the optimal range of the algorithm and further investigating its access characteristics by looking at the optimal values for different workloads are considered when evaluating the sensitivity of the algorithm.

Temperature decay coefficients are used to balance the weight of past and current requests, thus more accurately reflecting the temperature of the current request. The inclusion of temperature decay coefficients helps the caching system to maintain data temperature information more accurately and thus perform cache replacement and management more efficiently. The statistics period is a parameter that measures the frequency of hot data changes. An appropriate statistics period should be set according to the update frequency of hot data in order to prolong the residence time of hot data in the cache pool and improve the hit rate. Choosing the appropriate statistics period is very important, and factors such as system performance requirements and storage costs should be considered to balance system performance and resource consumption.

Figure 5 illustrates the effect of temperature decay coefficient and statistical period on the performance of the TDC algorithm. The temperature decay coefficient is tested in the range of 0.1 to 0.9. In the P5 workload test, the TDC algorithm is not sensitive to temperature, and temperature in this range has no significant effect on the algorithm performance. In the temperature attenuation coefficient range of 0.6 to 0.9, the hit rate gradually increases, indicating that the TDC algorithm is more sensitive to changes in temperature and requires more careful consideration of the selection of the temperature attenuation coefficient. In the DS1 workload test, the hit rate of the TDC algorithm is relatively stable, indicating that the temperature attenuation coefficient is not sensitive to the write-oriented workload.

The statistical cycle test ranges between 6,000 and 10,000 requests. In the P5 and DS1 workload tests, the algorithm hit rate gradually improves as the statistical period gets larger, but the improvement is relatively smooth. This indicates that the statistical period is sensitive to the hit rate of the TDC algorithm, but the sensitivity is not very large. In addition, the statistical period should not be too large in order to save memory resources and improve the efficiency of computing temperature. An excessively large statistical period will occupy more memory and increase the time to calculate the temperature. Therefore, it is necessary to balance the statistical cycle and the utilization of memory resources, the efficiency of computing temperature, and the algorithm hit rate to obtain the best performance and efficiency. In general, the temperature decay coefficient has a greater impact on the performance of the TDC algorithm than the statistical period. This is because the size of the temperature decay coefficient determines the temperature decay rate. A too small temperature decay coefficient cannot reflect the change of current hot data in time and cannot effectively identify the hot data, which leads to low hit rate. Therefore, in order to obtain a high hit rate, it is critical to select the appropriate temperature decay coefficient and statistical period according to different types of workloads.

C. Hit Rate Testing and Analysis. To verify the performance of the TDC algorithm in terms of hit rate, we selected the Ceph Cache Tier native caching algorithm CephTierCache algorithm, LRU and LFU typical caching algorithms

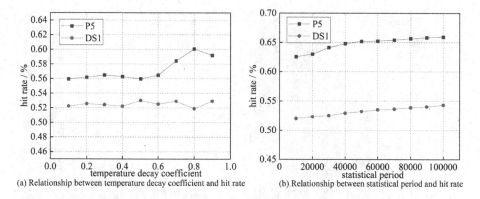

(a) Relationship between temperature decay coefficient and hit rate   (b) Relationship between statistical period and hit rate

Fig. 5. Influence of Temperature Decay Coefficient and Statistic Period on the Performance of TDC Algorithm.

as comparison objects and conducted comparison experiments in the same simulation environment. Among them, LRU algorithm is a recently used advanced caching algorithm that has been widely used in real systems, while CephTierCache algorithm is a variant of LRU algorithm. LFU algorithm is a caching algorithm based on access frequency. In the simulation test, we set the parameters of the TDC algorithm to a temperature decay factor of 0.8 and a statistical period of 10,000 requests. Figure 6 shows the hit rate of different algorithms under two workload environments, P5 and DS1. In the P5 workload, the hit ratio of the TDC algorithm improves by 9.74%, 25.64%, and 10.41%, respectively, compared to the LRU, LFU, and CephTierCache algorithms.The hit ratio of the TDC algorithm is close to that of the LRU algorithm and the CephTierCache algorithm. However, for workloads like DS1, which is write-oriented, the TDC algorithm outperforms the other algorithms when the cache is relatively small. However, as the cache increases, the hit rate of the TDC algorithm is lower than that of the LFU algorithm, but still better than the LRU and CephTierCache algorithms.

5.2   Ceph System Performance Test Based on TDC Cache Algorithm

A. Experimental Setup. In our experiments, we used four physical machines to build a complete Ceph cluster. Three of the machines are dedicated to providing storage services, while the other machine acts as a Ceph client as well as a Ceph source code compilation platform. Each storage node is equipped with 2 solid state drives and 4 mechanical drives, each managed by a corresponding OSD process, for a total of 18 OSDs. all 4 servers are located in the same server room and configured with two 10 Gigabit networks via SDN [21] switches: a public network to handle communication and data interaction between Ceph and clients, and a private network to handle Ceph cluster internal related traffic. The

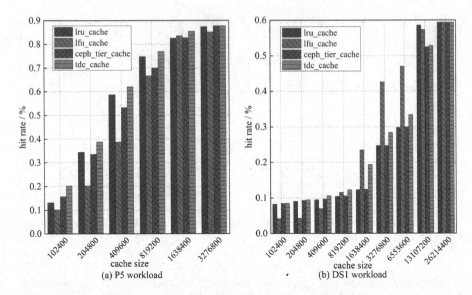

Fig. 6. Hit ratios of different cache algorithms under P5 and DS1 workloads.

topology of the whole cluster is shown in Fig. 7. The detailed server hardware and software configurations are shown in Table 2.

B. System Performance Verification Method. To verify the impact of the TDC algorithm on the performance of the Cache Tier, we will analyze the performance of different cache capacity sizes and compare it with the Cache Tier of the native caching algorithm. However, it is important to note that the read and write performance of large objects is mainly limited by network performance rather than disk performance. Therefore, it is not applicable to the evaluation of the TDC algorithm that aims to eliminate disk performance bottlenecks. On the other hand, TDC algorithms mainly target different sizes of data. Common storage testing tools, such as fio and Ceph rados bench, require specifying the size of test data when testing, which is not compatible with testing TDC algorithms. Therefore, we choose rados load-gen, a built-in benchmark test tool that comes with Ceph, to conduct our experiments. Rados load-gen can specify the size range of test data when testing cluster ballast, and the resulting data can meet the requirements of the TDC algorithm. However, its disadvantage is that it can only output throughput. We selected data in the size range of 4K~64K as small objects and data in the size range of 1M~4M as medium objects for the use case of testing Cache Tier performance. The test steps of the experiment are as follows:

1) Build the Ceph cluster and create the required crush rules: ssd_rule and hdd_rule for distributing the replicas on the ssd and hdd level OSDs. Create corresponding storage pools based on the different crush rules, including cache_pool and ssd_pool (based on ssd_rule) and storage_pool and hdd_pool

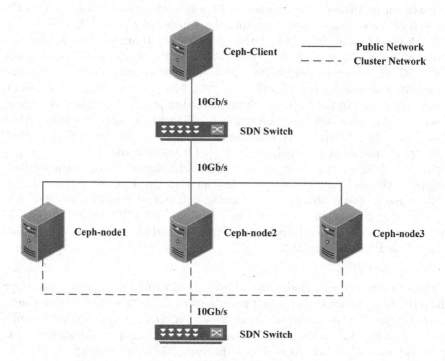

Fig. 7. Ceph cluster topology.

Table 2. The hardware and software configuration of the server.

Ceph node (*3)	
Processor	Intel Xeon Bronze 3204 @ 1.90 GHz
Memory	DDR4-2933 32 GB
Network	10 Gbps
OS	CentOS Linux
HDD(*4)	TOSHIBA MG07ACA12TE 12TB
SSD(*2)	WD Blue SA510 500GB
Ceph	14.2.22
Ceph client	
Processor	Intel(R) Xeon(R) Silver 4110 @ 2.10 GHz
Memory	DDR4-2933 64 GB
Network	10 Gbps
OS	CentOS Linux
Ceph	14.2.22

(based on hdd_rule). The number of PGs in each storage pool is set as the result of the formula calculation, and the number of copies is 2.

2) To clearly observe the effect of the caching algorithm, for the performance test of small objects, set the cache capacity of cache_pool to 1G, 2G, 3G, 4G, 5G, 6G. For the performance test of medium-sized objects, the cache capacity of cache_pool is 6G, 8G, 10G, 12G, 14G, 16G. use rados load-gen on the Cache Tier of the native caching algorithm is tested on the Ceph client server using rados load-gen. The test lasts 300 s and the data is recorded. After each experiment, the cache is flushed back and the storage_pool is cleared. The performance of ssd_pool and hdd_pool is tested using the same load-gen parameters as a comparison of the performance improvement of Cache Tier.

3) Replace the cache algorithm of Cache Tier with the TDC algorithm, replace the ceph-osd executable by rpm upgrade, and restart the OSD process.

4) Test the performance of Cache Tier based on the TDC caching algorithm using rados load-gen with the same steps as 2) and record the data. Compare this data with the data in 2).

C. Performance Evaluation and Analysis. In this paper, experiments are conducted to compare the system throughput of Cache Tier and native Cache Tier using the TDC cache replacement algorithm, and the experimental results are shown in Fig. 8. The test results show that the throughput of both the TDC Cache Tier and the native Cache Tier improves as the total capacity of the cache pool increases. In the small object test, TDC Cache Tier shows higher throughput than native Cache Tier, and when the cache pool capacity is 6 GB, TDC Cache Tier has the largest throughput improvement of 56.6%, but its performance can only reach 67.7% of ssd_pool. In the medium-sized object test, when the cache pool capacity is less than 10 GB, the throughput of TDC Cache Tier and native Cache Tier are similar. The reason is that when the cache is small, the data needs to be refreshed frequently, and the data written by the client needs to be cached at the same time, and the cache granularity is large, which leads to excessive load on the SSD OSD and affects the system performance. When the cache pool capacity is higher than 10 GB, the throughput improvement of TDC Cache Tier becomes larger and the performance is higher than that of the native Cache Tier. When the cache pool capacity is 16 GB, its performance is 10.9% higher than that of the native Cache Tier, but it still can only reach the performance level of 65.2% of ssd_pool.

In the small object test, Cache Tier based on the TDC algorithm was able to improve the performance of mechanical drives by 3.45 times. In the medium-sized object test, Cache Tier's performance was inadequate when the cache pool capacity was small, but when the cache pool capacity was large, Cache Tier based on the TDC algorithm was able to improve the performance of mechanical hard drives by 1.23 times. Considering the efficient configuration of storage systems, Cache Tier based on the TDC algorithm can make a trade-off between price and capacity to get a larger performance to capacity ratio while paying less for the price to capacity ratio. Although neither the native Cache Tier nor the TDC algorithm-based Cache Tier fully exploits the performance of SSDs, the price/performance ratio is still significant when compared to using 12 12T mechanical drives, which can be improved by using six 500G SSDs as cache.

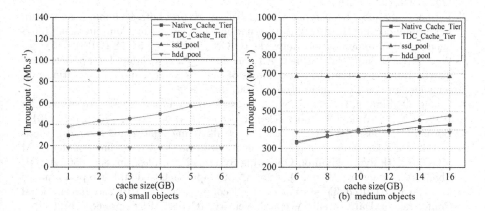

Fig. 8. Throughput comparison of objects of different sizes in Cache Tier based on TDC algorithm.

6   Conclusion

This study proposes a temperature density-based cache replacement algorithm (TDC) to address the limitations of the Ceph Cache Tier caching algorithm, and describes the algorithm design and implementation in detail. The TDC algorithm includes three aspects: object temperature calculation, object temperature density calculation, and cache replacement algorithm. In order to evaluate the performance of the TDC algorithm, this study uses the method of playing back workload IO and tests the IO records of two workloads, the read-based P5 and the write-based DS1, and performs sensitivity analysis on the temperature decay coefficient and statistical period of the TDC algorithm. Also, this study compares the performance of the TDC algorithm and other cache replacement algorithms under different cache size conditions, and the results show that the TDC algorithm outperforms other cache algorithms in terms of performance. Finally, this study compares the throughput of the native Cache Tier and the Cache Tier based on the TDC cache algorithm using the load-gen test tool in a four-node Ceph distributed storage system experimental environment, and finds that the Cache Tier based on the TDC cache algorithm has some performance improvement in storage reads and writes, further verifying the validity.

References

1. China Academy of Information and Communication Technology. Big Data White Paper [EB/OL].[2022-04]. http://www.caict.ac.cn/kxyj/qwfb/bps/202301/t20230104_413644.htm

2. Verma, C, Pandey, R.: comparative analysis of gfs and hdfs: technology and architectural landscape. In: 2018 10th International Conference on Computational Intelligence and Communication Networks (CICN). IEEE, pp. 54–58 (2018)

3. Weil, S.A., Brandt, S.A., Miller, E.L., et al.: Ceph: a scalable, high-performance distrib-uted file system. In: Proceedings of the 7th Symposium on Operating Systems Design and Implementation, pp. 307–320 (2006)

4. Huang, C., Simitci, H., Xu, Y., et al.: Erasure coding in windows azure storage. In: 2012 USENIX Annual Technical Conference (USENIX ATC 12), pp. 15–26 (2012)

5. Bornholt, J., Joshi, R., Astrauskas, V., et al.: Using lightweight formal methods to validate a key-value storage node in Amazon S3. In: Proceedings of the ACM SIGOPS 28th Symposium on Operating Systems Principles, pp. 836–850 (2021)

6. Zhang, X., Wang, Y., Wang, Q., et al.: A new approach to double i/o performance for ceph distributed file system in cloud computing. In: 2019 2nd International Conference on Data Intelligence and Security (ICDIS). IEEE, pp. 68–75 (2019)

7. Weil, S.A., Brandt, S.A., Miller, E.L., et al.: CRUSH: controlled, scalable, decentralized placement of replicated data. In: Proceedings of the 2006 ACM/IEEE Conference on Su-percomputing. 2006: 122-es (2006)

8. Hoseinzadeh, M.A.: survey on tiering and caching in high-performance storage sys-tems. arXiv preprint arXiv:1904.11560 (2019)

9. Meyer, S., Morrison, J.P.: Supporting heterogeneous pools in a single ceph storage clus-ter. In: 2015 17th International Symposium on Symbolic and Numeric Algorithms for Scientific Computing (SYNASC). IEEE, pp. 352–359 (2015)

10. Fares, R., Romoser, B., Zong, Z., et al.: Performance evaluation of traditional caching poli-cies on a large system with petabytes of data. In: 2012 IEEE Seventh International Conference on Networking, Architecture, and Storage. IEEE, pp. 227–234 (2012)

11. Guan, N., Yang, X., Lv, M., et al.: FIFO cache analysis for WCET estimation: a quantitative approach. In: 2013 Design, Automation & Test in Europe Conference & Exhibition (DATE). IEEE, pp. 296–301 (2013)

12. Fricker, C., Robert, P., Roberts, J.: A versatile and accurate approximation for LRU cache performance. In: 2012 24th international teletraffic congress (ITC 24). IEEE, pp. 1–8 (2012)

13. Hasslinger, G., Heikkinen, J., Ntougias, K., et al.: Optimum caching versus LRU and LFU: comparison and combined limited look-ahead strategies. In: 2018 16th International Symposium on Modeling and Optimization in Mobile, Ad Hoc, and Wireless Networks (WiOpt). IEEE, pp. 1–6 (2018)

14. Beckmann, N., Sanchez, D.: Maximizing cache performance under uncertainty. In: 2017 IEEE International Symposium on High Performance Computer Architecture (HPCA). IEEE, pp. 109–120 (2017)

15. Beckmann, N., Chen, H., Cidon, A.: LHD: Improving Cache Hit Rate by Maximizing Hit Density. In: NSDI, pp. 389–403 (2018)

16. Gu, R., Li, C., Dai, H., et al.: Improving in-memory file system reading performance by fi-ne-grained user-space cache mechanisms. J. Syst. Architect. **115**, 101994 (2021)

17. Duong, N., Zhao, D., Kim, T., et al.: Improving cache management policies using dynamic reuse distances. In: 2012 45Th Annual IEEE/ACM International Symposium on Microarchitecture. IEEE, pp. 389–400 (2012)
18. Yang, P., Wang, Z.: A distributed block storage optimization mechanism based on Ceph. In: 2019 4th International Conference on Automatic Control and Mechatronic Engineering(ACME 2019) (2019)
19. Cache Tiering Ceph Documentation. https://docs.ceph.com/en/reef/rados/operations/cache-tiering/. Accessed 15 Oct 2023
20. Park, D., Du, D.H.C.: Hot data identification for flash-based storage systems using multiple bloom filters. In: 2011 IEEE 27th Symposium on Mass Storage Systems and Tech-nologies (MSST). IEEE, pp.1–11 (2011)
21. Wu, D., Wang, Y., Feng, H., et al.: Optimization design and realization of ceph storage system based on software defined network. In: 2017 13th International Conference on Computational Intelligence and Security (CIS). IEEE, pp. 277–281 (2017)
22. Megiddo, N., Modha, D.S.: ARC: a self-tuning, low overhead replacement cache. In: Fast. 2003, 3(2003), pp. 115–130 (2003)

Smart DAG Task Scheduling Based on MCTS Method of Multi-strategy Learning

Lang Shu[1], Guanyan Pan[2], Bei Wang[3], Wenbing Peng[2], Minhui Fang[2], Yifei Chen[4], Fanding Huang[4], Songchen Li[4], and Yuxia Cheng[1(✉)]

[1] School of Computer Science, Hangzhou Dianzi University, Hangzhou, China
yxcheng@hdu.edu.cn
[2] Taizhou Urban and Rural Planning and Design Institute, Taizhou, China
[3] School of Computer Science, Zhejiang University, Hangzhou, China
[4] HDU-ITMO Joint Institute, Hangzhou Dianzi University, Hangzhou, China

Abstract. In distributed heterogeneous computing systems, efficient algorithms are crucial for improving system performance. We know that the task scheduling problem based on Directed Acyclic Graph (DAG) has been proven to be an NP-complete problem, making it difficult to find the optimal solution. Although previous research proposed many greedy strategies to solve this problem, these algorithms often have very limited search spaces. To overcome these limitations, we propose a smart DAG task scheduling algorithm based on the Monte Carlo Tree Search (MCTS) method of multi-strategy learning. We design effective state, action, and reward functions to train agents and allow them to adaptively adjust their search strategies. Experimental results prove the effectiveness of our algorithm. Specifically, our algorithm is superior to PSLS, PEFT, and HEFT algorithms in scheduling the maximum completion time under a large number of randomly generated and real-world applications.

Keywords: DAG scheduling · MCTS · multi-strategy learning

1 Introduction

With the development of high-speed networks, computers that are logically integrated are actually distributed in different geographical locations and connected to each other by high-speed networks. This new computing platform is called a distributed heterogeneous computing system [16]. The appearance of this system has reduced the difficulty for people to handle complex parallel and distributed applications [28], as idle and scattered computing resources can be used to tackle these problems. However, an urgent problem is how to effectively schedule the tasks of the application to the available computing resources, which is also one of the key technologies for the system to achieve high performance.

Usually, the task scheduling problem can be divided into two sub-problems [24]: how to prioritize tasks in the application, and how to assign tasks to appropriate processors for execution. The directed acyclic graph (DAG) model is used

Z. Tari et al. (Eds.): ICA3PP 2023, LNCS 14487, pp. 224–242, 2024.
https://doi.org/10.1007/978-981-97-0834-5_14

to represent applications, where vertices in the graph represent tasks, and edges represent data dependencies between tasks. The key challenge is to schedule all tasks in an application to appropriate processors and minimize the overall execution time of tasks. However, it has been proven that the scheduling problem is an NP-complete problem [7,21], making it extremely difficult to find optimal solutions in most cases.

Due to its importance, numerous researchers have dedicated their efforts to studying the DAG scheduling problem and have proposed a large number of heuristic algorithms [4,24]. Most of these algorithms adopt greedy scheduling strategies and have limited application scenarios. Reinforcement learning [22] has been introduced into task scheduling in order to obtain better scheduling results by learning from past experiences and improving current actions. However, recently proposed task scheduling algorithms based on reinforcement learning typically require large amounts of computing resources for training or significantly simplify the scheduling model.

Based on the above research background, we proposed a smart DAG task scheduling algorithm based on the MCTS method of multi-strategy learning. During the scheduling process, the search strategy of the algorithm can adaptively adjust the balance the weight of exploration and exploitation. The experimental results demonstrated the effectiveness of the proposed algorithm.

The main contributions of this paper are as follows:

- We propose a smart DAG task scheduling algorithm based on the MCTS method of multi-strategy learning. During the algorithm's operation, the weight of exploration and exploitation for each node is adaptively adjusted. This allows the algorithm to effectively search for as many nodes as possible with short task completion times, ultimately reducing the overall program completion time.
- Valid state, action, and reward functions are defined to avoid the problem of time-consuming or infeasible convergence of the algorithm caused by a large state space.
- We conducted numerous simulated and practical experiments to compare our algorithm with Base-MCTS (BMCTS) [8], Breadth-First Search (BFS) [1], PEGA [2], PSLS [29], PEFT [3], HEFT [24], and Greedy algorithms using DAGs with various types and heterogeneous configurations.

The remainder of this paper is organized as follows: Sect. 2 presents the related work on DAG task scheduling. Section 3 introduces the fundamental concepts and models of task scheduling. Section 4 describes the task scheduling process of the MCTS algorithm. Section 5 provides experimental results and comparative analyses. Finally, Sect. 6 summarizes the contributions and conclusions of this paper.

2 Related Work

There are roughly two categories of scheduling algorithms based on whether all the information about DAG tasks has been determined before the scheduling starts: static scheduling algorithm and dynamic scheduling algorithm. In the case of the dynamic scheduling algorithm, the information related to tasks is not predetermined, so the processor used for executing a task can only be specified during the task execution process.

Static task scheduling algorithm are divided mainly into four categories: list scheduling algorithm [20,24], clustering heuristic task scheduling algorithm [5,17], task replication-based scheduling algorithm [11,27], and random search algorithm based on genes and evolution [10,18].

The list scheduling algorithm is mainly divided into two stages. In the first stage, the priority of each task is calculated, and a priority queue is generated based on these priorities. In the second stage, tasks are selected according to their priorities and allocated to appropriate processors using a processor allocation strategy. Recently, an excellent task scheduling algorithm named the Pre-Scheduling-based List Scheduling algorithm (PSLS) has been proposed by Yi Zhao [29]. This algorithm introduces a new concept called Downward Length Table (DLT), which prioritizes tasks based on DLT. Compared with classic list scheduling algorithms, PSLS requires a pre-scheduling stage that uses a list scheduling algorithm before the task sorting stage, and DLT is generated based on the result obtained during the pre-scheduling stage. However, the PSLS algorithm relies heavily on the pre-scheduling algorithm used and lacks generality for different application scenarios.

The core idea of a clustering scheduling algorithm is to map task clusters to as many processors as possible in order to optimize the scheduling time. The task replication-based scheduling algorithm distributes tasks in the DAG to different processors for processing, which reduces the communication overhead between nodes with dependencies and minimizes the number of messages, thus shortening the overall scheduling time. A random search algorithm [2] based on genes and evolution is a method that uses a combined problem space scheduling solution to simulate biological evolution mechanisms for generating better scheduling strategies.

With the development of heterogeneous systems and complexity of applications, the traditional DAG task scheduling algorithm can no longer meet performance requirements [25,26]. In order to make the task scheduling algorithm intelligent, researchers have proposed a machine learning-based algorithm.

Zheng [30] proposed the use of Monte Carlo methods for random DAG scheduling due to uncertainty in task execution time for static DAG task scheduling. This improved the static DAG scheduling by continuously adjusting the scheduling through simulation. Orhean [19] also proposed a reinforcement learning-based scheduling method for heterogeneous distributed systems.

3   Task Scheduling Problem

An application is typically represented by a directed acyclic graph, $G = (V, E)$, where the set of m tasks in the application is denoted by $V = \{n_1, n_2, ..., n_m\}$. The set of s directed edges is denoted by $E = \{e_1, e_2, ..., e_s\}$. The priority constraint between task n_i and task n_j is represented by $edge\,(i, j) \in E$, which means task n_j cannot be executed until task n_i is completed. A task without a predecessor node is called an entry task n_{entry}, while a task without a successor node is called an exit task n_{exit}. The computation overhead matrix W represents a $v \times q$ matrix, where each $w_{i,j}$ represents the computational cost of executing task n_i on processor p_j. The average computational cost of task n_i is defined by

$$\overline{w_i} = \frac{\left(\sum_{j=1}^{q} w_{i,j}\right)}{q}. \tag{1}$$

B is a $q \times q$ matrix that represents the transfer rate between processors. L is a q-dimensional vector that represents the communication initiation cost of each processor. $Data$ is a $v \times v$ matrix of communication data, where $data_{i,j}$ represents the quantity of data sent from n_i to n_j. The communication overhead of $edge\,(i,j)$ is defined by

$$c_{i,j} = L_m + \frac{data_{i,j}}{B_{m,n}}. \tag{2}$$

Where L_m represents the communication startup cost of processor p_m. When task n_i and n_j are processed on the same processor, $c_{i,j} = 0$. The average communication overhead of $edge\,(i,j)$ is defined by

$$\overline{c_{i,j}} = \overline{L} + \frac{data_{i,j}}{\overline{B}}. \tag{3}$$

Where $\overline{B}$ represents the average transfer rate between processors, and $\overline{L}$ represents the average communication startup overhead.

Before introducing the final scheduling function, let's first understand the properties of the earliest start time (EST) and earliest finish time (EFT). $EST(n_i, p_j)$ indicates the earliest time at which task n_i can begin on processor p_j.

$$EST(n_i, p_j) = max \left\{ avail[j], \max_{n_m \in pred(n_i)} (AFT(n_m) + c_{m,i}) \right\}. \tag{4}$$

Where $pred(n_i)$ represents the set of all direct precursor tasks of task n_i, $AFT(n_m)$ represents the actual processing end time of the precursor task n_m, and $avail[j]$ represents the earliest time that processor p_j can start executing tasks. The predecessor value of the entry task is 0, and the successor value of the exit task is also 0. For the entry task, $EST(n_{entry}, p_j) = 0$. The $EFT(n_i, p_j)$, which indicates the latest possible time at which task n_i can be executed on processor p_j, is defined by taking into account the actual processing end time $AFT(n_m)$ of each predecessor task n_m in the set of direct predecessor tasks $pred(n_i)$, adding the processing time $t_{i,j}$ of task n_i to the earliest possible processor start time $avail[j]$.

$$EFT(n_i, p_j) = w_{i,j} + EST(n_i, p_j). \tag{5}$$

The makespan refers to the completion time after all tasks in a DAG are scheduled to be executed on proper processor.

$$makespan = max\left\{AFT(n_{exit})\right\}. \tag{6}$$

The ultimate objective of the DAG task scheduling problem is to allocate all tasks in a given application to suitable processors, with a minimum scheduling length.

4 MCTS Algorithm Design

In this section, we first introduce the reinforcement learning process and propose the design of three key elements of reinforcement learning: state function, action function, and reward function. Then, we introduce the running process of the smart DAG task scheduling algorithm based on the MCTS method of multi-strategy learning.

4.1 Reinforcement Learning

After defining the scheduling problem, we propose utilizing the reinforcement learning [13,14] method to solve it. Figure 1 illustrates the schematic diagram of the scheduling model based on reinforcement learning. At time t, the scheduler observes the environment and obtains the observed state value O_t; then, based on O_t, the scheduler determines the scheduling action A_t. After executing A_t, the scheduler receives a reward R_t. This process continues until the end of the scheduling. The observed value O_t can also be expressed as the state S_t.

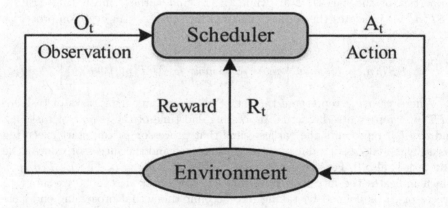

Fig. 1. Reinforcement Learning Based Scheduling Model.

In reinforcement learning, the design of state function, action function and reward function is crucial to improving the performance of the algorithm [9]. We define these three functions as follows:

- **State function:** The state space of scheduling problems can be very large. We devised a simple and effective representation of the state space by setting the state flag $S(n_i)$, and visit counts $C(n_i)$, where $S(n_i)$ and $C(n_i)$ are initialized to 0. Each time the state is updated, we set $S(n_i)=1$ and $C(n_i) = C(n_i) + 1$. When $S(n_i)$ is set to 1, certain condition must be met, which is defined by

$$\forall S(n_m) = 1, n_m \in pred(n_i) \tag{7}$$

where $pred(n_i)$ represents the set of all direct precursor tasks of task n_i, the state of task n_i is allowed to be set to 1 only when all the direct precursor tasks n_m are in the state of 1. The logic of changing the state flag $S(n_i)$ and visit counts $C(n_i)$ is simple and effective, contributing to high execution efficiency.

- **Reward function:** The design of the reward function is crucial in determining the scheduling strategy, and it has significant implications for strategy training. The reward obtained for completing the current task should guide subsequent scheduling actions, while the cumulative reward should reflect the ultimate scheduling objective. According to the above description, we set the initial value of the cumulative reward $R(n_i)$ to 0, and define an updating method for $R(n_i)$ is defined by

$$R(n_i) = R(n_i) + EST(n_i). \tag{8}$$

where $EST(n_i)$ represents the earliest start time of task n_i. Cumulative reward $R(n_i)$ can ensure that tasks with higher value are prioritized for execution, resulting in a shorter makespan.

- **Action function:** The action function is used to select the next scheduled task. Before introducing the action function, it is necessary to understand the concept of $rank_u(n_i)$ in the HETF algorithm. $rank_u(n_i)$ refers to the upward rank of task n_i and is defined by

$$rank_u(n_i) = \overline{w_i} + \max_{n_j \in succ(n_i)} (\overline{c_{i,j}} + rank_u(n_j)). \tag{9}$$

where $succ(n_i)$ represents the set of immediate successor tasks for task n_i. Specifically, for the exit task, $rank_u(n_{exit}) = \overline{w_{exit}}$. Essentially, $rank_u(n_i)$ is the length of the critical path from task n_i to the exit task, including the computation cost of task n_i.

We define the action function $A(n_i)$, where the task with the largest $A(n_i)$ value is scheduled next. The definition of $A(n_i)$ is

$$A(n_i) = rank_u(n_i) + \frac{R(n_i)}{m} + \beta \frac{\sqrt{C(n_m)}}{1 + C(n_i)}. \tag{10}$$

where $A(n_i)$ is divided into two parts. The left part consists of $rank_u(n_i) + \frac{R(n_i)}{m}$ which represents the average upward reward value under the upward rank of task n_i, where m represents the number of DAG tasks. The larger this value, the more valuable the current node is. The right variable represents the sum $C(n_m)$ of direct precursor visit counts of the current node divided by the visit counts $C(n_i)$ of the current node. If the current node has few visits, this value will increase, indicating that the current node is worth exploring further. β is an essential hyperparameter used to balance the weight between exploration and exploitation.

The action function is effective because it achieves a balance between exploration and exploitation. Increasing the number of visits to a current node may lead to higher reward values. If a current node has never been visited, it must be prioritized for the next visit to ensure that any newly expanded nodes are visited at least once. This approach gives lower-reward nodes a chance to be explored and prevent the algorithm from getting stuck in local optimization, ultimately improving performance and expanding the search across the entire state space.

4.2 MCTS

Monte Carlo Tree Search (MCTS) [6,15] is a search algorithm that employs a tree structure. It mainly consists of four stages, as presented in Fig. 2.

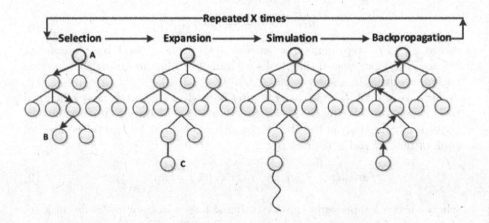

Fig. 2. The fundamental MCTS process.

The MCTS algorithm comprises four main steps. Firstly, selection recursively selects the best child node from the root node A until it reaches the leaf node B. Secondly, expansion creates one or more new child nodes when the selected leaf node B is not a terminal node, and then selects one of the newly created child nodes C. Thirdly, simulation artificially plays out game scenarios starting from the selected child node C until the end of the game, producing a simulation result. Finally, backpropagation updates node information for all nodes along the path to the simulated node with the simulation result. This summarizes the basic idea of MCTS.

The DAG scheduling problem can be modeled as an MCTS process, where the root node of the MCTS tree represents the initial state to start scheduling the DAG tasks. The subsequent child nodes describe the various states that MCTS might reach after selecting the following action, and the edges imply the available scheduling actions. This study introduces a novel multi-strategy learning MCTS algorithm treating the DAG task graph as a tree structure and scheduling tasks directly on it. The scheduling process is divided into four stages, as depicted in Fig. 3.

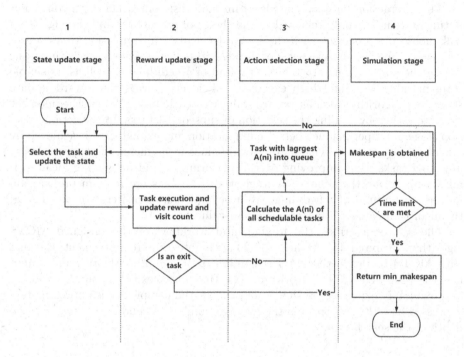

Fig. 3. The scheduling task process in MCTS algorithm.

- **State update stage:** The initialization process begins by adding only the entry task n_{entry} to the ready queue. Next, a task n_i is chosen from the queue to be scheduled, and the state $S(n_i)$ and visit count $C(n_i)$ of the selected task are updated.
- **Reward update stage:** Once a task n_i is selected from the queue, it will be executed. After the successful execution of n_i, its cumulative reward $R(n_i)$ and visit count $C(n_i)$ are updated accordingly. These values are then propagated back to the entry task n_{entry}. When the scheduled task is n_{exit}, it indicates that this round of scheduling is over and the simulation phase is entered.

- **Action selection stage:** Starting from the scheduled task, the values of $A(n_i)$ for all schedulable tasks are calculated, and then the task n_i with the largest $A(n_i)$ value is selected and added to the ready queue.
- **Simulation stage:** Repeat the first three stages until the exit task n_{exit} is scheduled, and calculate the makespan value. Then reset the state flag $S(n_i)$ to 0 for the next round of scheduling.

Finally, repeat the above phases until either the maximum iteration number or time limit is reached. Ultimately, return the min_makespan value.

The interaction between the algorithm and distributed heterogeneous computing system is mainly reflected in the choice of processors, and the algorithm will choose the processor with the shortest calculation completion time to execute every time. The effectiveness of the algorithm lies in avoiding revisiting tasks during the state update stage after a round of scheduling, as well as its simple state marking logic and high execution efficiency. During the reward update stage, the algorithm assigns reward values to each task that is scheduled for execution, thereby guiding the selection of subsequent tasks. In the action selection stage, a proper balance between exploration and exploitation is achieved by giving priority to tasks with high reward values while increasing the weights of unvisited tasks, thus improving overall performance and preventing local optimization. During the simulation stage, the state flags are reset in preparation for the next round of scheduling. After several rounds of iterative scheduling, the algorithm can get a better makespan value.

The above describes the fundamental workflow of the designed MCTS algorithm. Inspired by AlphaZero [23], we integrate reinforcement learning with MCTS to tackle the DAG scheduling problem [12]. Algorithm 1 formalizes the proposed MCTS algorithm. The time complexity of the algorithm is approximately expressed as $O(n^3)$, and the spatial complexity is approximately expressed as $O(b^d)$, where b is the average number of child nodes and d is the depth of the search tree.

5   Experiment

In this section, we compare the efficiency of the MCTS task scheduling algorithm with several classical task scheduling algorithms, including BMCTS [8], BFS [1], PEGA [2], PSLS [29], PEFT [3], HEFT [24], and Greedy, using a large number of randomly generated and real-world DAGs. The BMCTS algorithm refers to the basic MCTS algorithm, which utilizes the general MCTS process and is used in ablation experiments. By comparison, the effectiveness of our designed algorithm is clearly reflected.

5.1   Random Graph Generator

We initially developed a random DAG graph generator, which can generate a corresponding DAG based on user-defined parameters [3,24]. To evaluate

Algorithm 1. The MCTS Algorithm
___
Input: DAG tasks
Output: A near-optimal scheduling strategy
 1: Create empty *queue* and put n_{entry} as initial task
 2: **while** simulation times is NOT reached the upper limit **do**
 3:     Schedule the task n_i in the queue to execute
 4:     **if** n_i is NOT n_{exit} **then**
 5:         Put task with the largest $A(n_i)$ in the queue
 6:     **else**
 7:         Get makespan of this round of scheduling
 8:     **end if**
 9:     **if** $min\_makespan > makespan$ **then**
10:         $min\_makespan \leftarrow makespan$
11:     **end if**
12: **end while**
13: **return** $min\_makespan$
___

performance, we utilized identical DAGs for all task scheduling algorithms tested in this research. β was fixed at 2. Specific parameters are presented below:

- (v) The number of nodes in the DAG.
- CCR parameter quantifies the average communication versus computation demands of nodes present in the DAG. A high CCR value suggests a communication-intensive DAG, while a low CCR value implies a computationally intensive DAG.
- max_out parameter denotes the highest number of child nodes that a single node is allowed to have.
- fat parameter value has an effect on the overall width and height of a DAG. When the parameters are high, the resulting DAG tends to be tall and narrow, while low parameter values result in wider DAGs.
- $regularity$ parameter value describes the consistency of the number of task nodes present within each layer of a DAG. Larger values indicate a more uniform distribution of nodes across the layers of the DAG, whereas lower values suggest less distribution uniformity.
- P parameter value represents the number of processors.

In this study, using different combinations of the following parameters, we can obtain 10260 different types of DAGs. 10 DAGs of each type have been randomly generated. Finally, the total number of DAGs generated is approximately 102.6K. In order to ensure fairness in the analysis, we provided MCTS, BMCTS, BFS, and PEGA algorithms with nearly identical running times.

- $SET_v = \{10, 20, 30, 40, 50, 60, 70, 80, 90, 100, 200, 300, 400, 500, 600, 700, 800, 900, 1000\}$
- $SET_{CCR} = \{0.1, 0.5, 1.0, 5.0, 10.0\}$
- $SET_{max\_out} = \{1, 3, 5\}$

- $SET_{fat} = \{0.5, 1.0, 2.0\}$
- $SET_{regularity} = \{0.1, 0.5, 1.0\}$
- $P = \{2, 4, 8, 16\}$

5.2 Performance Metrics Analysis

To evaluate the performance of scheduling algorithms, we employed five metrics [3,24].

- **Schedule Length Ratio(SLR).** Due to variations in DAGs, scheduling length for different tasks can vary significantly, which hinders effective performance comparisons of scheduling algorithms. To address this challenge, we introduced the concept of SLR. SLR is defined as the ratio of the actual scheduling time to the ideal scheduling time obtained from a given scheduling algorithm.

$$SLR = \frac{makespan}{\sum_{n_i \in CP_{MIN}} min_{p_j \in Q} \{w_{i,j}\}}. \tag{11}$$

The denominator of the SLR metric represents the cumulative value of the minimum computing overhead across processors for all critical path tasks within the DAG. This value serves as the ideal minimal scheduling time limit. Consequently, SLR is always greater than 1 as scheduling times are generally longer than the ideal limit. A smaller SLR indicates better algorithmic performance compared to larger SLRs.

- **Speedup.** The speedup ratio is a measure of parallel performance and is defined as the ratio of the total time overhead resulting from serial execution of all tasks to the corresponding overhead value derived from parallel execution methods.

$$Speedup = \frac{min_{p_j \in Q} \{\sum_{n_i \in V} w_{i,j}\}}{makespan}. \tag{12}$$

Speedup is calculated by dividing the minimum scheduling time for processing all tasks sequentially on a single processor by the scheduling time achieved through the use of a scheduling algorithm. Higher speedup values indicate better parallel performance of the scheduling algorithm.

- **Efficiency.** The efficiency is defined as the ratio of the speedup ratio to the actual number of processors used (denoted by p^*). It reflects the utilization of the processor in the task scheduling process, and a higher efficiency indicates greater processor utilization.

$$Efficiency = Speedup/p^* \tag{13}$$

- **Running Time of the Algorithms.** The running time refers to the duration required by a scheduling algorithm to generate its corresponding scheduling policy from a given DAG. A smaller running time implies that the algorithm is more efficient.

– **Quantity of Better Solutions.** In this performance evaluation approach, the number of scheduling outcomes rated as "better," "worse," or "equal" are measured for each algorithm relative to other competing algorithms when applied to a specific DAG.

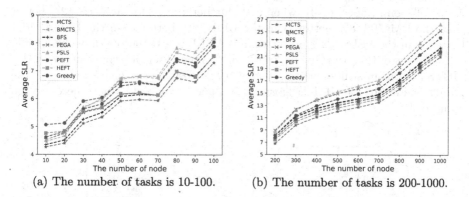

(a) The number of tasks is 10-100.　　　(b) The number of tasks is 200-1000.

Fig. 4. The correlation between the average SLR and the task number.

Figure 4(a) displays the correlation between various scheduling algorithms used in the study and the average scheduling length ratio (SLR) when the number of tasks is 10–100. Results demonstrate that the MCTS algorithm performed the best amongst all tested algorithms. When the number of node is 10, BMCTS, BFS, and PSLS algorithms have average SLR values of 4.47, 4.35, and 4.51, respectively. In contrast, the MCTS algorithm shows an 17% decrease in average SLR value with 4.27, the SLR of our designed MCTS algorithm is 20% lower than that of the BMCTS algorithm, the effectiveness of the algorithm has been demonstrated. In Fig. 4(b), when the number of tasks is 1000, the MCTS algorithm still maintains a good performance advantage, which shows that the MCTS algorithm has good scalability and robustness and can adapt to the scale of the problem with a huge number of tasks.

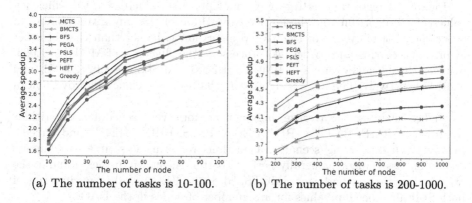

(a) The number of tasks is 10-100.　　　(b) The number of tasks is 200-1000.

Fig. 5. The correlation between the average speedup and the task number.

Figure 5(a) illustrates the relationship between average speedup ratio and corresponding task numbers when the number of tasks is 10–100. A higher speedup ratio indicates better efficiency at scheduling tasks for parallel computing purposes. As per our experimental observations, MCTS scheduling algorithm outperformed all other tested algorithms by producing the highest parallel execution efficiency. Specifically, when task number is 10, The MCTS is 12% better than the BMCTS, 8% better than the BFS, 15% better than the PEGA, 14% better than the PSLS, 19% better than the PEFT, 23% better than the HEFT, and 33% better than the Greedy. In Fig. 5(b), When there are thousands of tasks, MCTS algorithm still shows good speedup performance, which shows that the parallelism ability of MCTS algorithm has good scalability when there are too many tasks.

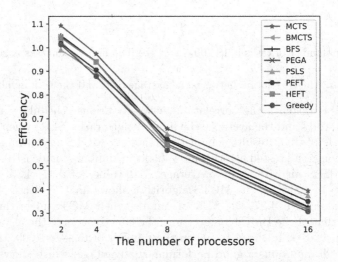

Fig. 6. The correlation between the efficiency and the processor number.

Figure 6 depicts that as the number of processors increases, the efficiency values of all algorithms gradually decrease. However, for any given number of processors, the efficiency of the MCTS algorithm is higher than that of other algorithms. For instance, with two processors, the efficiency of MCTS is 1.093, whereas the efficiency values of BMCTS, BFS, and PSLS are 1.027, 1.051, and 0.991, respectively. The MCTS algorithm achieves an efficiency improvement of 7%.

Figure 7 exhibits the average running times for several scheduling algorithms comprising MCTS, BMCTS, BFS, PEGA, PSLS, PEFT, HEFT, and Greedy. In any given scheduling scenario, algorithms requiring less time to complete scheduling assignments generally perform better. As shown in Figs. 4 and 5, the MCTS algorithm outperforms BMCTS, BFS, and PEGA algorithms in terms of both SLR and speedup values for any number of nodes in the DAG.

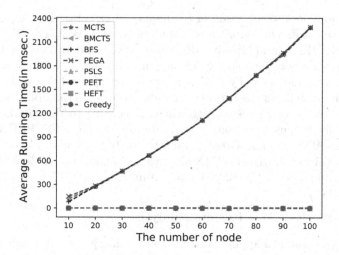

Fig. 7. The correlation between the average running time and the task number.

Table 1. Pairwise Schedule Length Comparison of the Scheduling Algorithm.

		MCTS	BMCTS	BFS	PEGA	PSLS	PEFT	HEFT	Greedy	Combined
MCTS	better		91076	69142	82318	88452	88334	86260	101348	80%
	worse	*	9688	30402	15530	17712	17626	10638	4752	14%
	equal		7236	8456	10152	1836	2040	11102	1900	6%
BMCTS	better	9688		13500	28772	54810	41656	32692	53222	31%
	worse	91076	*	88582	76000	51138	64142	71582	51732	65%
	equal	7236		5918	3228	2052	2202	3726	3046	4%
BFS	better	30402	88582		63460	83690	81172	67132	97060	67%
	worse	69142	13500	*	40986	22378	24776	37152	8380	29%
	equal	8456	5918		3554	1932	2052	3716	2560	4%
PEGA	better	15530	76000	40986		74844	71000	25012	87816	52%
	worse	82318	28772	63460	*	31364	34862	1772	18802	34%
	equal	10152	3228	3554		1792	2138	81216	1382	14%
PSLS	better	17712	51138	22378	31364		37746	34874	53234	33%
	worse	88452	54810	83690	74844	*	65060	77144	53752	65%
	equal	1836	2052	1932	1972		5194	1382	1014	2%
PEFT	better	17626	64142	24776	34862	65060		38556	64292	41%
	worse	88334	41656	81172	71000	37746	*	67510	42002	57%
	equal	2040	2202	2052	2138	5194		1934	1706	2%
HEFT	better	10638	71582	37152	1772	77144	67510		83840	46%
	worse	86260	32692	67132	25012	34874	38556	*	21558	40%
	equal	11102	3726	3716	81216	1382	1934		2602	14%
Greedy	better	4752	51732	8380	18802	53752	42002	21558		27%
	worse	101348	53222	97060	87816	53234	64292	83840	*	71%
	equal	1900	3046	2560	1382	1014	1706	2602		2%

Table 1 presents a comparison of the quality distribution of scheduling decisions for seven different algorithms including MCTS, BMCTS, BFS, PEGA, PSLS, PEFT, HEFT, and Greedy. Algorithms are evaluated based on their ability to consistently produce successful outcomes during scheduling, with superior performance indicated by a higher proportion of good decision outcomes. The last column reflects the ratio of each algorithm's decision outcome success rate relative to the other tested algorithms. According to the results, the top-performing algorithms based on success rate are MCTS, BFS, PEGA, HEFT, PEFT, PSLS, BMCTS, and Greedy, respectively. In particular, the performance improvement of our designed MCTS algorithm is significant, with a 49% higher better rate compared to the BMCTS algorithm.

5.3   Real-World DAGs Results Analysis

To validate the practical application.of the algorithm proposed in this paper, we conducted experiments using a dataset of 51,894 real DAGs obtained from the Zenodo dataset (https://doi.org/10.5281/zenodo.4667690).

The dataset consists of jobs containing multiple tasks, represented as a large number of DAGs with the corresponding task nodes. The extracted DAGs from the Zenodo dataset are classified according to their size: small-scale DAGs (1–20 task nodes), medium-scale DAGs (21–50 task nodes), and large-scale DAGs (more than 50 task nodes). The performance of the MCTS scheduling algorithm in comparison with BMCTS, PSLS, PEFT, and HEFT is evaluated for these three types of DAGs. The experimental results are analyzed as follows.

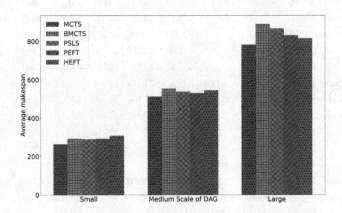

Fig. 8. The average task makespan comparison.

Figure 8 displays the average makespan achieved by the five task scheduling algorithms for three DAGs of varying sizes. The MCTS algorithm consistently outperforms the other algorithms across all scales. In particular, for small-scale DAGs, the MCTS algorithm achieves an average makespan value that is 9.8% lower than the second best performing PSLS algorithm, 16.7% lower than the worst performing HEFT algorithm, and 11% lower than the BMCTS algorithm.

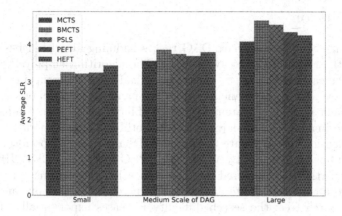

Fig. 9. The average SLR comparison.

Figure 9 illustrates the average SLR values obtained by the five task scheduling algorithms for three DAGs of varying sizes. The average SLR value is a normalized measure of the average makespan, with smaller values indicating better algorithm performance. For small-scale DAGs, the MCTS algorithm achieves an average SLR value that is 20.9% lower than that of the BMCTS algorithm, and 17% lower than that of the PSLS algorithm.

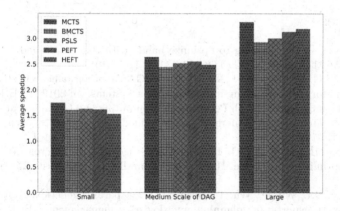

Fig. 10. The average speedup ratio comparison.

Figure 10 presents the average speedup values obtained by the five task scheduling algorithms for three DAGs of varying sizes. For small-scale DAGs, the MCTS algorithm achieves an average speedup ratio that is 14.4%, 12.2%, 13.6%, and 22.2% higher than that of BMCTS, PSLS, PEFT, and HEFT, respectively. These results demonstrate that the MCTS algorithm has a strong parallelization effect.

6   Conclusion

This section introduces a smart DAG task scheduling algorithm based on the MCTS method of multi-strategy learning. The algorithm employs four MCTS stages specifically designed to tackle the challenges inherent in DAG task scheduling by adaptively balancing exploration and exploitation of search space. This approach aims to mitigate the problem of large search spaces failing to converge. Experimental results carried out on both randomly generated and real-world applications demonstrate that our MCTS algorithm outperforms classical task scheduling algorithms BMCTS, BFS, PEGA, PSLS, PEFT, HEFT, and Greedy. Nevertheless, compared to classical heuristic algorithms, our algorithm requires more time to execute and with the increase of the depth and breadth of MCTS search tree, the search difficulty increases exponentially. Therefore, future works for applications of this method may include: (1) exploring techniques for parallelizing the search process to enhance the overall efficiency of our task scheduling algorithm, (2) pruning the search tree to avoid unnecessary expenses caused by too much space in the search tree. (3) find the optimization method of simulation times to greatly reduce the simulation time overhead of the algorithm.

Acknowledgements. This research was funded by the Basic Public Welfare Research Project of Zhejiang Province grant number LY20F020014 and the National Science Foundation for Young Scientists of China grant number 61802096.

References

1. Adams, A., et al.: Learning to optimize halide with tree search and random programs. ACM Trans. Graph. (TOG) **38**(4), 1–12 (2019)
2. Ahmad, S.G., Munir, E.U., Nisar, W.: PEGA: a performance effective genetic algorithm for task scheduling in heterogeneous systems. In: 2012 IEEE 14th International Conference on High Performance Computing and Communication & 2012 IEEE 9th International Conference on Embedded Software and Systems, pp. 1082–1087. IEEE (2012)
3. Arabnejad, H., Barbosa, J.G.: List scheduling algorithm for heterogeneous systems by an optimistic cost table. IEEE Trans. Parallel Distrib. Syst. **25**(3), 682–694 (2013)
4. Arkhipov, D.I., Wu, D., Wu, T., Regan, A.C.: A parallel genetic algorithm framework for transportation planning and logistics management. IEEE Access **8**, 106506–106515 (2020)
5. Arunarani, A., Manjula, D., Sugumaran, V.: Task scheduling techniques in cloud computing: a literature survey. Future Gener. Comput. Syst. **91**, 407–415 (2019)
6. Best, G., Cliff, O.M., Patten, T., Mettu, R.R., Fitch, R.: DEC-MCTS: Decentralized planning for multi-robot active perception. Int. J. Rob. Res. **38**(2–3), 316–337 (2019)
7. Bittel, L., Kliesch, M.: Training variational quantum algorithms is np-hard. Phys. Rev. Lett. **127**(12), 120502 (2021)
8. Browne, C.B., et al.: A survey of monte carlo tree search methods. IEEE Trans. Comput. Intell. AI Games **4**(1), 1–43 (2012)

9. Cheng, Y., Wu, Z., Liu, K., Wu, Q., Wang, Y.: Smart dag tasks scheduling between trusted and untrusted entities using the mcts method. Sustainability **11**(7), 1826 (2019)
10. Defersha, F.M., Rooyani, D.: An efficient two-stage genetic algorithm for a flexible job-shop scheduling problem with sequence dependent attached/detached setup, machine release date and lag-time. Comput. Ind. Eng. **147**, 106605 (2020)
11. Djigal, H., Feng, J., Lu, J., Ge, J.: IPPTS: an efficient algorithm for scientific workflow scheduling in heterogeneous computing systems. IEEE Trans. Parallel Distrib. Syst. **32**(5), 1057–1071 (2020)
12. Hu, Z., Tu, J., Li, B.: Spear: optimized dependency-aware task scheduling with deep reinforcement learning. In: 2019 IEEE 39th International Conference on Distributed Computing Systems (ICDCS), pp. 2037–2046. IEEE (2019)
13. Kiran, B.R., et al.: Deep reinforcement learning for autonomous driving: a survey. IEEE Trans. Intell. Transp. Syst. **23**(6), 4909–4926 (2021)
14. Lei, K., et al.: A multi-action deep reinforcement learning framework for flexible job-shop scheduling problem. Expert Syst. Appl. **205**, 117796 (2022)
15. Li, K., Deng, Q., Zhang, L., Fan, Q., Gong, G., Ding, S.: An effective mcts-based algorithm for minimizing makespan in dynamic flexible job shop scheduling problem. Comput. Ind. Eng. **155**, 107211 (2021)
16. Li, T., Sahu, A.K., Zaheer, M., Sanjabi, M., Talwalkar, A., Smith, V.: Federated optimization in heterogeneous networks. Proc. Mach. Learn. Syst. **2**, 429–450 (2020)
17. Mao, H., Schwarzkopf, M., Venkatakrishnan, S.B., Meng, Z., Alizadeh, M.: Learning scheduling algorithms for data processing clusters. In: Proceedings of the ACM Special Interest Group on Data Communication, pp. 270–288 (2019)
18. Mirjalili, S., Mirjalili, S.: Genetic algorithm. In: Evolutionary Algorithms and Neural Networks: Theory and Applications, pp. 43–55 (2019)
19. Orhean, A.I., Pop, F., Raicu, I.: New scheduling approach using reinforcement learning for heterogeneous distributed systems. J. Parallel Distrib. Comput. **117**, 292–302 (2018)
20. Panda, S.K., Jana, P.K.: An energy-efficient task scheduling algorithm for heterogeneous cloud computing systems. Clust. Comput. **22**(2), 509–527 (2019)
21. Prates, M., Avelar, P.H., Lemos, H., Lamb, L.C., Vardi, M.Y.: Learning to solve np-complete problems: a graph neural network for decision tsp. In: Proceedings of the AAAI Conference on Artificial Intelligence, vol. 33, pp. 4731–4738 (2019)
22. Schrittwieser, J., Hubert, T., Mandhane, A., Barekatain, M., Antonoglou, I., Silver, D.: Online and offline reinforcement learning by planning with a learned model. Adv. Neural. Inf. Process. Syst. **34**, 27580–27591 (2021)
23. Silver, D., et al.: Mastering the game of go without human knowledge. Nature **550**(7676), 354–359 (2017)
24. Topcuoglu, H., Hariri, S., Wu, M.Y.: Performance-effective and low-complexity task scheduling for heterogeneous computing. IEEE Trans. Parallel Distrib. Syst. **13**(3), 260–274 (2002)
25. Wu, J., Chen, X.Y., Zhang, H., Xiong, L.D., Lei, H., Deng, S.H.: Hyperparameter optimization for machine learning models based on bayesian optimization. J. Electron. Sci. Technol. **17**(1), 26–40 (2019)
26. Yang, L., Shami, A.: On hyperparameter optimization of machine learning algorithms: theory and practice. Neurocomputing **415**, 295–316 (2020)
27. Yao, F., Pu, C., Zhang, Z.: Task duplication-based scheduling algorithm for budget-constrained workflows in cloud computing. IEEE Access **9**, 37262–37272 (2021)

28. Zhang, C., Song, W., Cao, Z., Zhang, J., Tan, P.S., Chi, X.: Learning to dispatch for job shop scheduling via deep reinforcement learning. Adv. Neural. Inf. Process. Syst. **33**, 1621–1632 (2020)
29. Zhao, Y., Cao, S., Yan, L.: List scheduling algorithm based on pre-scheduling for heterogeneous computing. In: 2019 IEEE International Conference on Parallel & Distributed Processing with Applications, Big Data & Cloud Computing, Sustainable Computing & Communications, Social Computing & Networking (ISPA/BDCloud/SocialCom/SustainCom), pp. 588–595. IEEE (2019)
30. Zheng, W., Sakellariou, R.: Stochastic dag scheduling using a monte carlo approach. J. Parallel Distrib. Comput. **73**(12), 1673–1689 (2013)

CT-Mixer: Exploiting Multiscale Design for Local-Global Representations Learning

Yin Tang, Xili Wan$^{(\boxtimes)}$, Xinjie Guan, and Aichun Zhu

College of Computer and Information Engineering, Nanjing Tech University, Nanjing, China
{ty996911,xiliwan,xjguan,aichun.zhu}@njtech.edu.cn

Abstract. Convolutional neural networks (ConvNets) have been widely used for feature extraction in various computer vision tasks, such as image classification, object detection, and instance segmentation. Recently, Vision Transformers (ViTs) have demonstrated exceptional performance on upstream vision tasks, such as image classification, due to their effectiveness in modeling long-range dependencies. However, ViTs' performance is limited by weak inductive biases in modeling two-dimensional data and under-explored multi-scale representation learning, both of which are fundamental strengths of ConvNets. In this paper, we introduce CT-Mixer, a novel architecture that combines convolution-based and transformer-based modules to exploit the advantages of both and compensate for their respective weaknesses. The CT-Mixer architecture cross-stacks convolution-based and transformer-based modules, where each module is assigned an order to better learn local information and global context. Additionally, we incorporate a dynamic mechanism into the convolution-based modules to model adaptive dependence. We also improve the multi-scale representation learning strategy by adopting the multi-branch structure of MPViT. Experimental results demonstrate that CT-Mixer achieves competitive performance compared to existing methods.

Keywords: Multi-scale Processing · Self-Attention · Convolutional Neural Networks

1 Introduction

Convolutional Neural Networks (ConvNets) have been the standard paradigm for most computer vision tasks, owing to their design philosophy for image data such as translation invariance. Nevertheless, over the past two years, attention has shifted to Vision Transformers (ViTs) [10], and many ViTs variations, *e.g.,* [4, 24, 34, 40] have been proposed for handling vision tasks. ViTs, with the self-attention mechanism, are developed for building long-range dependencies, in contrast to ConvNets, which have the advantage of learning local context. However, because

Z. Tari et al. (Eds.): ICA3PP 2023, LNCS 14487, pp. 243–262, 2024.
https://doi.org/10.1007/978-981-97-0834-5_15

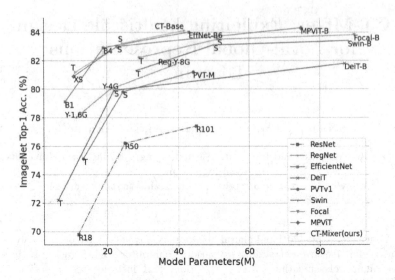

Fig. 1. Comparisons of model performance using ImageNet-1K classification. The proposed CT-Mixer surpasses most state-of-the-art models while requiring fewer parameters.

the computation consumption of ViTs is quadratic to the image size, it also imposes a significant computational burden.

To apply ViTs on high-resolution tasks like detection and segmentation, modern variants *e.g.,* [11,24,34,40] follow the design principles of ConvNets to build a hierarchical architecture by adapting some inductive biases of ConvNets to ViTs. An insightful and successful attempt is to leverage the local attention [24] to mitigate the computational burden problem, similar to the sliding window mechanism of ConvNets. However, the local attention mechanism is usually applied within *non-overlapping* windows, which still requires extra effort to expand the receptive field.

ConvNets have inherent multi-scale capabilities due to the design of their convolution kernels [5,12] and feature maps [21], but achieving multi-scale learning in ViTs remains a challenge. While some approaches, *e.g.,* [11,34], learned hierarchies and used pooling attention to downsample features at different scales, these operations can lead to information loss and hinder modeling of fine-grained representations. Other approaches [2,19] adopted the multi-branch structure to learn features of different scales and perform the features fusion across all branches. Nevertheless, these approaches under-utilize the property of the multi-branch structure and leave much room for improvement.

To address these issues, we propose the CT-Mixer, a general vision backbone comprising convolution-based and self-attention-based modules. Specifically, we introduce an improved multi-scale interaction scheme for the self-attention module named **h**ybrid-**s**cale **s**elf-**a**ttention (HSSA), which is based on the multi-branch structure [19]. Inspired by the approach in [4], the CT-Mixer is cross-stacked with alternating local and global modules. The local part uses two depthwise convolutions with a dynamic mechanism to obtain a larger receptive field than previous window-based approaches. The global part adopts an efficient self-attention mechanism with linear complexity to input resolution and a hybrid-scale projection scheme to improve the capacity of modeling multi-scale representations. The Re-embedding layer is applied at the initial inputs, and the query, key, and value tensors are projected from the embedded results with different scales. Our experiments on the ImageNet-1K classification show that the proposed CT-Mixer outperforms state-of-the-art models while using only half the model size (as demonstrated in Fig. 1). The contribution of this work can be summarized as three-fold:

1. We introduce Local Conv-Attention (LCA) which adopts the depthwise convolution with a dynamic mechanism to efficiently learn local relations, mimicking the standard window-based local self-attention. ·
2. We develop Hybrid-Scale Self-Attention (HSSA) to improve multi-scale interaction based on the multi-branch structure. HSSA enables more flexible and efficient multi-scale learning, resulting in significant performance improvements.
3. With LCA and HSSA, we build the CT-Mixer, a model that can effectively capture both local and global information, and learn fine-grained and coarse-grained texture features.

The rest of this is organized as follows. Section 2 reviews related works on vision transformers and self-attention mechanisms. Section 3 presents the details of the proposed model CT-Mix and its main modules. Section 4 illustrates the experimental dataset and evaluation metrics, then analyzes the experimental results. Finally, we conclude this work and present future work in Sect. 5.

2   Related Work

In this section, we review related work on vision transformers and self-attention mechanisms. We also discuss studies that aim to augment ViTs with convolutional modules to incorporate inductive biases and improve their modeling capabilities. Moreover, we explore approaches that leverage multi-scale representations to enhance the effectiveness of vision transformers in various computer vision tasks.

Attention Mechanism in CNNs. ConvNets have been widely used for various vision tasks due to their inductive biases such as translation invariance and locality, which make them efficient for processing 2D data. Unlike the self-attention mechanism in transformer-based models, convolution is a position-dependent rather than data-dependent operation and can be considered as a static network. Previous work has proposed various approaches to develop attention mechanisms for ConvNets. For instance, SENet [16] introduced an adaptive weight allocation strategy among channels to bring dynamics to ConvNets, while Non-local blocks [36] were designed to aggregate information from other locations to enhance the characteristics of the current location. These approaches have shown improved performance on different vision tasks compared to their static counterparts.

Vision Transformers. ViT was proposed as a pioneering approach for image classification [10], inspired by the success of transformers in NLP. It keeps the original structure as consistent as possible and shows competitive performance compared to convolution-based models. However, it is challenging to directly apply the vanilla, no-frills ViT to high-resolution tasks like detection and segmentation. Therefore, many methods have emerged, *e.g.*, [4, 24, 34, 40] to further explore the potential of ViTs and apply the classification-only model to various visual tasks. However, most studies focused on reducing the computing burden, while, to some extent, neglecting further exploration in terms of modeling capabilities.

Efficient Vision Transformers with Multi-scale Design. As computational complexity is being addressed, recent work has shifted focus to explore the potentials of multi-scale representations learning of vision transformers. For instance, CrossViT [2] introduced multi-scale learning by partitioning an image using two different patch sizes and separately adopting two branches for cross-scale interaction. Shunted self-attention [30] allowed interaction of query, key/-value with different scales in a single attention layer. Recently, MPViTs [19] leveraged the multi-path structure for the patch merging layer and fed these multi-scale embedded features into the corresponding transformer encoder. However, the main difference in each path is only in the scale of the input features, which is relatively inefficient and inflexible. Instead, we encode the hybrid-scale interaction among query, key, and value at a branch level, enabling multiple combinations of receptive fields.

3   Method

In this section, we provide an overview of CT-Mixer and describe its two main components, namely Local Context Aggregation (LCA) and Hierarchical Scale Spatial Attention (HSSA). We also present the architecture specifications and variants with different FLOPs and parameters.

3.1   Overview Architecture

CT-Mixer follows a hierarchical architecture that consists of overlapping patch merging layers and either transformer or convolution encoders in each stage. The spatial resolution is progressively reduced by a factor of 32 as the output size of the last stage, and the channel dimension increases with depth. A feature fusion layer connects each branch, and an extra branch is added in LCA stages to expand the receptive field. The overview of CT-Mixer is shown in Fig. 2.

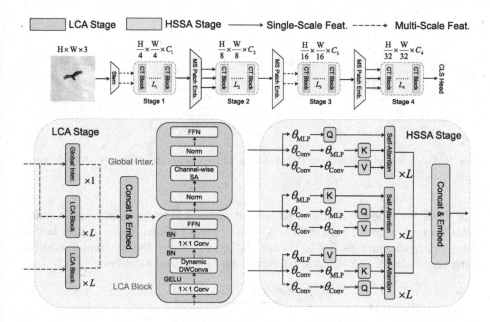

Fig. 2. Overall Architecture of CT-Mixer. It comprises three main parts: the Convolution Stem module, Stages for modeling relations, and the Multi-Branch Patch Merging module. The Convolution Stem module contains two consecutive convolutions with a stride of 2, increasing the number of channels from 3 to C_1 and reducing the resolution by a factor of 4. The Stages of CT-Mixer are composed of two different units: the Local Context Aggregation (LCA) unit and the Hybrid Scale-Shift Attention (HSSA) unit. The Multi-Branch Patch Merging module is responsible for downsampling the features by a factor of 2, and the scale (receptive field) of its output varies depending on the type of unit used. Feature maps of different scales are indicated with different colors.

3.2 Patch Merging Layer

We adopt a similar design to [19] by stacking multiple successive convolutions with fixed kernel sizes to obtain different receptive fields. However, we modify the original design to accommodate different stages. In stages that model local relations (*i.e.*, stages with LCA encoder), the output scales are different. In contrast, in stages that model global context (*i.e.*, stages with HSSA encoder), the embedded results initially have the same scale, which is then used to generate features of different scales before being fed into the corresponding transformer encoder.

For Stages with LCA Encoder. In stages with the LCA encoder, the CT-Mixer uses a patch merging layer to merge learned features $\mathbf{X}_l \in \mathbb{R}^{H_l \times W_l \times C_l}$ from the previous stage l by $2\times$ downsampling ratio. Here, $H_l \times W_l$ indicates the number of tokens while C_l refers to the channel. The merging operation uses multiple successive convolutions with fixed kernel sizes to obtain different recep-

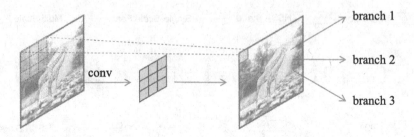

Fig. 3. Patch Merging layers for HSSA stages. It first generates features with the same scale (SS).

tive fields for different scales of embedded features. Fine-grained levels require small receptive fields, such as 3×3, while coarse-grained levels require larger receptive fields, such as 5×5. We use $\mathcal{F}_{s,k}(\cdot)$ to denote the merging operation with kernel size k and stride s, which can be formulated as follows:

$$\mathbf{X}_{l+1}^{(i)} = \underbrace{\mathcal{F}_{k,s1}(\cdots \mathcal{F}_{k,s2}(\mathbf{X}_l))}_{i}, \tag{1}$$

where $i \in [1, P]$ denotes the index of embedded features for each branch. It is worth noting that the stride corresponding to the innermost layer is set as 2 to downsample the features, while the stride of the subsequent layers is set as 1 to maintain the same resolution.

Therefore, in stages with the LCA encoder, we take a learned feature $\mathbf{X}_l \in \mathbb{R}^{H_l \times W_l \times C_l}$ from the previous stage l and apply a series of patch merging operations to obtain a set of embedded results $[\mathbf{X}_{l+1}^1, \cdots, \mathbf{X}_{l+1}^P] \in \mathbb{R}^{\frac{H_l}{2} \times \frac{W_l}{2} \times C_l}$, each with the same resolution and channels but different receptive fields. Specifically, we calculate the receptive fields as follows:

$$RF_1 = k_1, \underset{1 < i \leq P}{RF_i} = RF_{i-1} + s_i(k_i - 1), \tag{2}$$

where RF_i refers to the receptive field of the i-th feature and k_i, s_i are the kernel size and stride of the i-th convolution operation. These features are then fed into each branch separately.

For Stages with HSSA Encoder. In Fig. 3, the patch merging layers located at the front of the transformer-based stages generate features with a single scale (such as a 3×3 receptive field) which are fed separately into the transformer encoder. These features are then mapped into different scales for query $\mathbf{Q}$, key $\mathbf{K}$, and value $\mathbf{V}$, allowing for multiscale interaction within the multi-branch structure.

3.3 Preliminaries

It is widely accepted that convolution and self-attention are two different operators for representation learning. Convolution is designed to capture local relations, while self-attention mechanisms aim to model long-range dependencies. The global nature of the transformer has been shown to be one of the primary reasons for its superior performance compared to ConvNets. However, the computational burden of self-attention is enormous. Recent research [20,27] has revealed that there are many redundant computations in the self-attention process. This suggests that locality still plays a significant role in vision tasks, and this *de facto* preference for locality has fueled numerous efforts to address this issue.

Window-Based Local Self-attention. Window-based vision transformers [9,24,40] are representative approaches that address the issue of locality. In a standard self-attention module with N heads, let $\mathbf{X}_{i,j} \in \mathbb{R}^{C_{in}}$ be any patch within a local window centered at pixel (i, j) under the 2D coordinate system, and let $\mathbf{Z}_\Omega \in \mathbb{R}^{C_{out}}$ be the resulting feature corresponding to the given patch X_Ω. The window-based self-attention can be formulated as:

$$\mathbf{Z}_\Omega = \overset{N}{\underset{h=1}{\|}} \left(\sum_{i,j \in \aleph_w} \mathcal{A}(\mathbf{W}_h^Q \mathbf{X}_\Omega, \mathbf{W}_h^K \mathbf{X}_{i,j}) \mathbf{W}_h^V \mathbf{X}_{i,j} \right) \mathbf{W}^O, \tag{3}$$

where $\|$ denotes the operation of concatenation, and $\mathbf{W}_h^Q, \mathbf{W}_h^K, \mathbf{W}_h^V$ are the projection matrices for the query, key, and value in the h-th head. $\aleph_w$ represents a local region with a window size of w, and $\mathcal{A}\left(\mathbf{W}_h^Q \mathbf{X}_\Omega, \mathbf{W}_h^K \mathbf{X}_{i,j}\right)$ is the corresponding attention weight for the patch $\mathbf{X}_{i,j}$ within $\aleph_w$. The attention weight is calculated as:

$$\mathcal{A}\left(\mathbf{W}_h^Q \mathbf{X}_\Omega, \mathbf{W}_h^K \mathbf{X}_{i,j}\right) = \underset{ij}{\Phi} \left(\frac{(\mathbf{W}_h^Q \mathbf{X}_\Omega)^T (\mathbf{W}_h^K \mathbf{X}_{i,j})}{\tau} \right), \tag{4}$$

where $\underset{ij}{\Phi}(\cdot)$ denotes the attention weight between $\mathbf{X}_\Omega$ and $\mathbf{X}_{i,j}$ after softmax processing. τ is the scaling factor, usually set to $\sqrt{d}$, where d is the feature dimension. Window-based approaches alleviate the computational burden and make transformers behave more like CNNs by leveraging locality.

Qualitative Comparison. Convolution and self-attention are both operators used for feature weighting. Self-attention is a dynamic operator, where the weight is data-dependent, making the transformer a naturally dynamic network. On the other hand, convolution is a static operator. However, recent research [16,36,37] has introduced dynamic mechanisms to ConvNets with significant effect, which can be seen as the "attention mechanism" in ConvNets.

In terms of feature weighting, a vanilla depthwise convolution with a kernel size of $k \times k$ can be seen as a position-aware and overlapped window-based attention module with a window size of k and N heads, where N is equal to the dimension of the channel. Therefore, local self-attention can be further simply streamlined as:

$$\mathbf{Z}_{a,b}^{\text{attn}} = \mathop{\Big\|}_{h=1}^{N} \left(\sum_{i=1,j=1}^{w} w_{i,j}^h \mathbf{X}_{i,j} \right) \mathbf{W}^O, \tag{5}$$

$$\text{s.t.} \quad i, j, a, b \in \aleph_w$$

where $w_{i,j}^h$ is the corresponding *data-dependent* attention score of head h to patch $\mathbf{X}_{i,j}$. In contrast, depthwise convolution uses *position-dependent* weight scalars for each single-dimension patch vector as follows:

$$\mathbf{Z}_{a,b}^{dwc} = \mathop{\Big\|}_{h=1}^{C} \left(\sum_{i=1,j=1}^{k} w_{i,j}^h \mathbf{X}_{i,j} \right) \begin{bmatrix} 1 & & \\ & \ddots & \\ & & 1 \end{bmatrix}_{C \times C}, \tag{6}$$

where C is the channel dimension and $w_{i,j}^h$ is the corresponding *position-dependent* weight scalar to the single-dimension patch vector $\mathbf{X}_{i,j}$.

To leverage the efficiency of depthwise convolution and local self-attention, we try to mimic local self-attention with convolution to obtain more efficient local information learning ability. Furthermore, to better reflect the data-dependent property of the self-attention mechanism, we introduce extra adaptivity into convolution.

3.4   Local Conv-Attention Stages (LCA)

To efficiently capture local relations, we build the LCA encoders a the multi-branch structure. As in Fig. 2, the LCA encoder receives several inputs of various scales and feeds them into their respective Conv-Attention branch. An additional branch is included to further expand the receptive field, which contains a global channel-level transformer layer. The features learned from each branch are then aggregated in the channel dimension to a size of $C_i \times (P + 1)$ and transformed to the dimension C_{i+1} in the next stage via a linear projection layer.

Depthwise Convolution with Dynamic Injection. Depthwise convolution is widely used in lightweight networks due to its fewer parameters and FLOPs compared to standard convolution. However, the main difference between depthwise convolution and local self-attention may be their data adaptability (see Appendix for a qualitative comparison). Inspired by the adaptive dynamic mechanism introduced in SENet [16] (Fig. 4(b)), we introduce the dynamic mechanism into our D-DWConv module (Fig. 4(a)), but our approach differs in two ways: 1) The SE module captures the attention weight at the spatial level, where each channel has the same attention weight generated by all locations. In contrast, our D-DWConv module allows learning an independent adjustment factor at all locations, resulting in each location obtaining a distinct weight. 2) The SE branch works after a transformation function (*e.g.*, residual, inception), while our dynamic branch starts at the original input features.

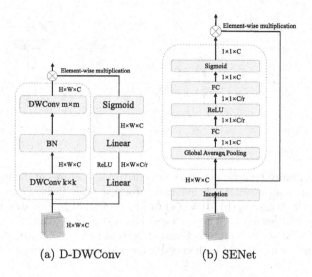

(a) D-DWConv   (b) SENet

Fig. 4. Comparison of SENet and our Dynamic Depthwise Convolution Module (D-DWConv). Our method includes a conv-attention branch with two depthwise convolutions, followed by batch normalization, and a dynamic branch that adapts channel weights across all locations. The output of both branches is multiplied element-wise to produce the final feature maps.

Conv-Attention Block. Our Conv-Attention Block, illustrated in Fig. 2, follows the structure of the transformer, comprising layers of projection, normalization, relation modeling, feedforward network (FFN), and skip connections. For the projection layer, we use 1×1 pointwise convolution to translate the original input and features from depthwise convolution. For the relations modeling layers, we use dynamic depthwise convolution as a local attention mechanism, which contains two successive depthwise convolutions with different kernel sizes and dilation, and a batch normalization layer. For the FFN layer, we also use pointwise convolution for learning channel correlation at each location. We use batch normalization instead of layer normalization, which is commonly used in convolution-based modules. It can be formulated as:

$$\hat{\mathbf{X}}^l = \text{D-DWConv}\left(\text{BN}(\mathbf{X}^{l-1})\right) + \mathbf{X}^{l-1},$$
$$\mathbf{X}^l = \text{C-MLP}\left(\text{BN}(\hat{\mathbf{X}}^l)\right) + \hat{\mathbf{X}}^l, \tag{7}$$

where $\mathbf{X}^l$ is the output feature of block l.

Channel-Level Attention for Global Interaction. We improve the capability of ConvNets to model long-range dependencies by adding a branch based on grouped channel-level self-attention [8] as follows:

$$\text{Attn}_g = \left(\mathop{\Big\|}_{g=1}^{G} \text{softmax}\left(\frac{\mathbf{Q}_g \mathbf{K}_g^T}{\sqrt{\mathbf{d}_g}}\right) \mathbf{V}_g \right)^T \mathbf{W}^O \tag{8}$$

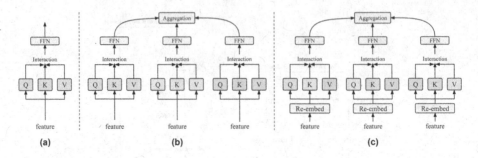

Fig. 5. Comparison of transformer encoder structures. Features of different scales are marked with different colors. (a) Previous approaches (*e.g.,* Swin, Focal, Twins) only have a single-scale structure. (b) MPViTs introduced a multi-branch structure to aggregate multiscale features but each encoder still has only a single scale. (c) HSSA enables features of different scales to interact within each self-attention module, resulting in *Hybrid-scale* output features of each encoder for multi-branch aggregation.

$$\mathbf{Q}_g, \mathbf{K}_g, \mathbf{V}_g \in \mathbb{R}^{G \times D_g \times L}, \mathbf{W}^O \in \mathbb{R}^{D \times D},$$

where $\|$ is the operation of orderly gathering along the channel dimension, and the projection matrix $\mathbf{W}^O$ "shuffles" the gathered non-overlapped channels. To obtain a global view of each token, we set $heads = 1$. The complexity of self-attention applied to the whole channel dimension is $\mathcal{O}(6LD^2)$, while the grouped alternative has reduced complexity of $\mathcal{O}(4LD^2 + 2GLD_g^2)$.

The LCA stages are capable of modeling both multiscale local relations and global context, as expressed by the following equation:

$$\mathbf{X}_{lca} = \mathrm{LN} \left(\overset{P}{\underset{p=1}{\|}} \left(\mathbf{E}_{conv}^{(1)}, \cdots, \mathbf{E}_{conv}^{(P)}, \mathbf{E}_{sa} \right) \mathbf{W}^O \right), \tag{9}$$

where $\mathbf{E}_{conv}^{(i)}$ is the output of the last conv-attention block corresponding to the i-th branch, $\mathbf{E}_{sa}$ is the output feature of the self-attention branch, and $\mathbf{W}^O$ is the projection matrix.

Table 1. Model configuration. For Stage 1 and Stage 3, the parameter cfg determines the kernel size and dilation used in the LCA unit. B_i, L_i, and C_i represent the number of branches, blocks, and channels at Stage i, respectively. The pooling layer and classifier head after Stage 4 are not shown for simplicity.

Stage	Output Size	Layer	CT-Tiny: 11.0M/2.9G	CT-Small: 24.4M/5.3G	CT-Base: 41.71M/10.0G
stage 1	56×56	LCA	cfg $= \begin{cases} k=5, d=1 \\ k=7, d=3 \end{cases}$	cfg $= \begin{cases} k=5, d=1 \\ k=7, d=3 \end{cases}$	cfg $= \begin{cases} k=5, d=1 \\ k=7, d=3 \end{cases}$
			$B_1=2, L_1=1, C_1=64$	$B_1=2, L_1=2, C_1=96$	$B_1=2, L_1=1, C1=96$
stage 2	28×28	HSSA	$B_2=3, L_2=2, C_2=96$	$B_2=3, L_2=3, C_2=128$	$B_2=3, L_2=3, C_2=192$
stage 3	14×14	LCA	cfg $= \begin{cases} k=5, d=1 \\ k=3, d=3 \end{cases}$	cfg $= \begin{cases} k=5, d=1 \\ k=3, d=3 \end{cases}$	cfg $= \begin{cases} k=5, d=1 \\ k=3, d=3 \end{cases}$
			$B_3=2, L_3=5, C_3=176$	$B_3=2, L_3=6, C_3=224$	$B_3=2, L_3=7, C_3=288$
stage 4	7×7	HSSA	$B_4=3, L_4=2, C_4=232$	$B_4=3, L_4=3, C_4=320$	$B_4=3, L_4=3, C_4=360$

3.5   Hybrid-Scale Self-attention Stages (HSSA)

As depicted in Fig. 5(c), HSSA stages take the outputs of the patch merging layer with the same scale (default: 3×3) as input. Each branch of HSSA applies re-embedding layers to the input features to obtain features with different scales. A linear layer is then applied to project the features into query, key, and value tensors. For a feature map $\mathbf{F} \in \mathbb{R}^{H \times W \times C}$, the resulting query, key, and value tensors in the first branch can be defined as:

$$\mathbf{F}_Q = \mathbf{F}, \mathbf{F}_K = \mathrm{PE}(\mathbf{F}), \mathbf{F}_V = \mathrm{PE}(\mathbf{F}_K), \tag{10}$$
$$\mathbf{Q} = \mathrm{RS}(\mathbf{F}_Q)\mathbf{W}^Q, \mathbf{K} = \mathrm{RS}(\mathbf{F}_K)\mathbf{W}^K, \mathbf{V} = \mathrm{RS}(\mathbf{F}_V)\mathbf{W}^V.$$

Here, PE refers to the patch embedding layer, and RS represents the reshape operation. In the implementation, we use convolution layers with a kernel size of 3, a stride of 1, and a padding of 1.

Given the computation complexity of the adopted multi-branch structure, we follow the efficient factorized self-attention mechanism proposed in CoaT [38], which also be utilized in MPViTs. This mechanism is defined as follows:

$$\mathrm{FSA}(Q, K, V) = \mathbf{Q}\mathbf{W}^P,$$
$$\mathbf{W}^P = \mathrm{softmax}_L(\mathbf{K})^T \mathbf{V}/\tau. \tag{11}$$

Here $\mathbf{W}^P \in \mathbb{R}^{D \times D}$ can be regarded as a projection matrix generated by $\mathbf{K}$ and $\mathbf{V}$, and the subscript of softmax function denotes that the function operates on the dimension L (*i.e.,* sequence length). We also perform ablation studies to compare the performance of different self-attention mechanisms. It is worth mentioning that the LCA stages mainly consist of convolution, and have excellent multi-scale flexibility in terms of kernel size, which means that we do not need to introduce too much multi-scale information with additional approaches. For the LCA stages, we use only two branches to achieve a moderate performance-computation trade-off. We present three kinds of model configurations for CT-Mixer, named CT-Tiny, CT-Small, and CT-Based. Table 1 shows the output size and layer name in each stage. For the LCA stages, we list the selection of the kernel size and dilation for the dynamic depthwise convolution. Parameters and GFLOPs of each model are given in the first row of the table.

4   Experiments

We evaluate the performance of the proposed CT-Mixer network on three mainstream vision tasks: image classification on ImageNet-1k [7], object detection, and instance segmentation on COCO2017 [23]. Furthermore, we conduct extensive ablation experiments to study the impact of the CT-Mixer's components on the overall performance.

Table 2. Image classification on ImageNet-1K [7]. The parameters and FLOPs of all models are calculated with the resolution of 224 × 224.

Model	#Params (M)	FLOPs (G)	Top-1 Acc (%)
ResNet-18 [14]	11.7	1.8	69.8
RegNet Y-1.6G [29]	11.2	1.6	78.0
EfficientNet-B1 [32]	8	0.7	79.1
DeiT-T [33]	6	1.3	72.2
PVT-T [34]	13.2	1.9	75.1
XCiT-T24/16 [1]	12.0	2.3	79.4
CrossViT-Ti [2]	6.9	1.6	73.4
MPViT-XS [19]	10.5	2.9	80.9
CT-Tiny	11.0	2.9	**81.2**
ResNet-50 [14]	25.0	4.1	76.2
RegNet Y-4G [29]	21.0	4.0	80.0
EfficientNet-B4 [32]	19	4.2	82.9
ConvNeXt-T [25]	29	4.5	82.1
DeiT-S [33]	22	4.6	79.9
CrossViT-S [2]	26.7	5.6	81.0
PVT-S [34]	24.5	3.8	79.8
Swin-T [24]	28.3	4.5	81.3
CoaT-S [39]	22.0	12.6	82.1
TwinsP-S [4]	24	3.8	81.2
CvT-13 [38]	20	4.5	81.6
CoAtNet-0 [6]	25	4.2	81.6
Focal-T [40]	29	4.9	82.2
MPViT-S [19]	22.9	4.8	83.0
CT-Small	24.4	5.4	**83.2**
ResNet-101 [14]	45.0	7.9	77.4
RegNet Y-8G [29]	39.0	8.0	81.7
EfficientNet-B6 [32]	43.0	19.0	84.0
ConvNeXt-S [25]	50	8.7	83.1
DeiT-B [33]	87.0	17.6	81.8
Swin-S [24]	50	8.7	83.0
PVT-M [34]	44.2	6.7	81.2
CrossVit-18 [2]	43.3	9.0	82.5
Focal-S [40]	51.1	9.1	83.5
CT-Base	41.7	10.0	**84.1**

4.1   Image Classification on ImageNet-1k

We evaluate the performance of our proposed model on the ImageNet-1k [7] dataset, which consists of approximately 1.3M training samples and 50K validation samples. To ensure a fair comparison, we follow the training protocols used in Swin Transformer [24] and DeiT [33]. Specifically, we train our model for 300 epochs using the AdamW [26] optimizer with a weight decay of 0.05. The batch size and base learning rate are set to 1024 and 0.001, respectively, and we adopt the cosine learning rate scheduler with 5 linear warm-up epochs. Our data augmentation strategy follows the methods used in [19,33], which includes random cropping, random flipping (with horizontal flip only by default), MixUp [42], repeated augmentation [15], CutMix [41], and random erasing [43]. In addition, we apply regularization techniques during training, including Stochastic Depth [17] with drop rates of 0.1, 0.2, and 0.4 for CT-Tiny, CT-Small, and CT-Base, respectively, as well as Label Smoothing [31] with a smoothing factor of 0.1. We do not use Exponential Moving Average (EMA) [28] because we find that it does not improve the performance of our model. The detailed hyper-parameter settings are provided in Table 3. Our model is implemented using the `Pytorch` and `timm` codebase.

Table 2 shows a comparison of our models with state-of-the-art convolution- and transformer-based architectures on the ImageNet-1k dataset. CT-Mixer outperforms previous state-of-the-art methods by a significant margin. Specifically, CT-Tiny achieves similar performance to Swin-T and TwinsP-S with less than half the number of parameters (11.0 *vs.* 28/24) and requires 55% ~ 65% of the FLOPs. CT-Small surpasses the Swin Transformer and the Focal Transformer by 1.9% and 1.0% while requiring fewer parameters, and it also outperforms recent modern pure-convolution (ConvNeXt) and hybrid (CoAtNet) models. Similarly, CT-Base surpasses modern pure attention and convolution models, even hybrid ones. Note that we follow the design of the multi-branch structure proposed in MPViT, but our model delivers competitive results against it.

4.2   Object Detection and Instance Segmentation

We evaluate the performance of CT-Mixer on object detection and instance segmentation tasks using the COCO dataset [23]. We use all configurations of CT-Mixer as backbones for the RetinaNet [22] and Mask R-CNN [13] frameworks, pre-trained on ImageNet-1k. We follow the common training recipe used in [24] that includes two training schedules: 1× schedules and 3× schedules. For the 1× schedule, we resize the shorter side of the input image to 800 and keep the longer side no more than 1333, with a total of 12 training epochs. For the 3× schedule, we use a multi-scale training strategy that resizes the shorter side of input images from 480 to 800, and train for 36 epochs. We use the AdamW optimizer with a weight decay of 0.05 for all configurations. The total batch size and the initial learning rate are set to 16 and 0.0001, respectively. Similar to the classification experiments, the stochastic depth drop rate is set to 0.1, 0.2, and 0.3 for CT-Tiny, CT-Small, and CT-Base. We implement the object detection

Table 3. Hyper-parameter settings adopted for the CT-Mixer family on ImageNet-1K.

Hyper-parameter	CT-Tiny	CT-Small	CT-Base
Training Epochs	300		
Batch size	1024		
Optimizer	AdamW		
Learning rate	1e–3		
Lr decay scheme	Cosine		
Warmup epochs	5		
Weight decay	5e–2		
Repeat Augment	✔		
MixUp	0.8		
CutMix	1.0		
Erasing prob	0.25		
Stochastic depth rate	0.1	0.2	0.4
Label smoothing	0.1		
EMA decay rate	None		

and instance segmentation experiments using the `MMDetection` toolbox [3]. The detailed hyper-parameters and experimental settings are provided in Table 6.

Table 4 shows the results of RetinaNet with different methods that have similar parameters. Compared to previous state-of-the-art approaches, our CT-Mixer achieves outstanding results across all object scales (small, medium, and large), demonstrating the effectiveness of our model in capturing multi-scale features.

In Table 5, we report the box mAP (AP^b) and the mask mAP (AP^m). For the 1× schedule, CT-Tiny achieves 10.6% higher AP^b than ResNet-18 and 7.9% higher AP^b than PVTv1-T, while also achieving 2.0% higher AP^m than PVTv2.

Table 4. Object detection on COCO dataset using RetinaNet framework. The presented models are trained using 1x and 3x schedules. All backbones are pre-trained on ImageNet-1K.

Model	#Params(M)	RetinaNet 1× schedule						RetinaNet 3× schedule + MS					
		AP^b	AP^b_{50}	AP^b_{75}	AP_S	AP_M	AP_L	AP^b	AP^b_{50}	AP^b_{75}	AP_S	AP_M	AP_L
ResNet-50 [14]	38	36.3	55.3	38.6	19.3	40.0	48.8	39.0	58.4	41.8	22.4	42.8	51.6
PVT-S [34]	34.2	40.4	61.3	43.0	25.0	42.9	55.7	42.2	62.7	45.0	26.2	45.2	57.2
PVTv2-B2 [35]	35	44.6	65.6	47.6	27.4	48.8	58.6	–	–	–	–	–	–
Swin-T [24]	38.5	42.0	63.6	44.7	26.6	45.8	55.7	45.0	65.9	48.4	29.7	48.9	58.1
Focal-T [40]	39.4	43.7	65.2	46.7	28.6	47.4	56.9	45.5	66.3	48.8	31.2	49.2	58.7
MPViT-S [19]	32.3	45.7	67.3	48.8	28.7	49.7	59.2	47.6	68.7	51.3	**32.1**	51.9	61.2
CT-Small (ours)	34.7	**45.9**	**67.6**	**49.1**	**28.7**	**49.9**	**59.6**	**47.8**	**68.8**	**51.6**	31.9	51.9	61.4

Table 5. Object detection and instance segmentation on COCO dataset using Mask R-CNN framework. The backbones are pre-trained on ImageNet-1K. The symbol "*" indicates that some results marked with "–" are not reported in the original paper.

Model	#Params(M)	Mask R-CNN 1× schedule						Mask R-CNN 3× schedule + MS					
		AP^b	AP^b_{50}	AP^b_{75}	AP^m	AP^m_{50}	AP^m_{75}	AP^b	AP^b_{50}	AP^b_{75}	AP^m	AP^m_{50}	AP^m_{75}
ResNet-18 [14]	31	34.0	54.0	36.7	31.2	51.0	32.7	36.9	57.1	40.0	33.6	53.9	35.7
PVT-T [34]	33	36.7	59.2	39.3	35.1	56.7	37.3	39.8	62.2	43.0	37.4	59.3	39.9
PVTv2-B1 [35]	33.7	41.8	64.3	45.9	38.8	61.2	41.6	–	–	–	–	–	–
XCiT-T12/8* [1]	25.8	–	–	–	–	–	–	44.5	66.4	48.8	40.3	63.5	43.2
CoaT-Mini* [39]	30.2	**45.1**	–	–	40.6	–	–	46.5	–	–	41.8	–	–
CT-Tiny (ours)	30.8	44.6	**66.9**	**48.9**	40.8	63.2	43.7	46.8	68.5	51.4	42.6	66.0	46.1
ResNet-50 [14]	44	38.0	58.6	41.4	34.4	55.1	36.7	41.0	61.7	44.9	37.1	58.4	40.1
PVT-S [34]	44	40.4	62.9	43.8	37.8	60.1	40.3	43.0	65.3	46.9	39.9	62.5	42.8
Swin-T [24]	48	42.2	64.6	46.2	39.1	61.6	42.0	46.0	68.2	50.2	41.6	65.1	44.8
Twins-S [4]	44	43.4	66.0	47.3	40.3	63.2	43.4	46.8	69.2	51.2	42.6	66.3	45.8
PVTv2-B2 [35]	45	45.3	67.1	49.6	41.2	64.2	44.4	–	–	–	–	–	–
Focal-T [40]	49	44.8	67.7	49.2	41.0	64.7	44.2	47.2	69.4	51.9	42.7	66.5	45.9
XCiT-S12/16*[1]	44	–	–	–	–	–	–	45.3	67.0	49.5	40.8	64.0	43.8
Shuffle-T* [18]	48	–	–	–	–	–	–	46.8	68.9	51.5	42.3	66.0	45.6
MPViT-S [19]	43	46.4	**68.6**	51.2	42.4	65.6	45.7	48.4	70.5	52.6	**43.9**	67.6	47.5
CT-Small (ours)	45	**46.6**	**68.6**	**51.6**	**42.6**	65.7	46.0	**48.8**	**70.8**	**53.1**	43.8	67.4	**47.5**

CT-Small outperforms the Swin Transformer by 4.4% and 2.8% AP^b with the 1× and 3× schedule, respectively. Our model also achieves 3.5% and 2.2% higher AP^m than Swin-T in the instance segmentation task, further demonstrating the superiority and effectiveness of our model in downstream tasks.

4.3 Ablation Studies

To better understand the CT-Mixer, we conduct experiments to examine the impact of some key components on performance. We evaluate different components of the CT-Small model size on the ImageNet-1K classification task.

Hybrid Scale Interaction Scheme. We proposed the hybrid scale interaction scheme for the transformer encoder, which allows the query **Q**, key **K**, and value **V** to interact at different scales. Since we follow the multi-branch structure, we use different scale combinations according to different branches. Table 7 shows that our proposed hybrid scale self-attention improves the original structure by about 0.3%.

Self-attention Mechanism. As the multi-branch structure leads to a high computational burden when using the standard token-level self-attention mechanism ($\frac{QK^T}{\sqrt{d_k}}V$), recent works have proposed more efficient solutions that scale linearly with image size. Table 8 presents a comparison of different attention mechanisms.

Table 6. Hyper-parameters of object detection and instance segmentation on COCO dataset.

Setting & hp-params	CT-Tiny	CT-Small
Training Scheme	1×/3×(MS)	
Batch size	16	
Optimizer	AdamW	
Learning rate	1e-4	
Lr decay scheme	Linear	
Warmup epochs	[8,11]/[27,33]	
Weight decay	5e–2	
Stochastic depth rate	0.1	0.2

Table 7. Comparison of different schemes for transformer encoder.

Interaction Scheme	#Params	FLOPs	Top-1 Acc
multi-branch + single-scale	23.8M	5.2G	82.9%
multi-branch + hybrid-scale (Ours)	24.4M	5.4G	**83.2%(+0.3)**

Dynamic Mechanism for LCA Stages. We investigate the proposed dynamic mechanism used in LCA stages. Previous studies [16,37] have shown that introducing an adaptive mechanism could enhance the representational power of ConvNets. Here, we evaluate the impact of this design on model performance by comparing three different configurations: without dynamic mechanism, channel-wise interdependencies, and our approach. Table 9 presents the results of CT-Small with different dynamic mechanisms. Our proposed approach achieves significant improvement with a negligible increase in the number of parameters and FLOPs.

Table 8. Comparison of different attention mechanisms. The "CRPE" notation in the table refers to the convolution relative position encoding used in the attention module proposed in [39]. We omit the feature dimensions for non-self-attention stages.

Self-Attention Mechanism	#Params	FLOPs	Activations	Top-1 Acc
Dimension: [–, 128, –, 320]				
Standard Attention	22.2M	6.5G	54.2M	83.3%
Cross-Covariance Attention	22.3M	5.2G	33.1M	82.9%
Channel-level Attention	22.2M	5.1G	32.0M	83.0%
Factorized Attention (with CRPE)	24.4M	5.4G	37.3M	83.2%

Table 9. Comparison of configurations with different dynamic mechanisms.

Dynamic Mechanism	#Params	FLOPs	Top-1 Acc
No Dynamic	24.1M	5.3G	82.8%
Channel-Wise Interdependencies	24.4M	5.3G	83.0%(+0.2)
Ours	24.4M	5.4G	**83.2%(+0.4)**

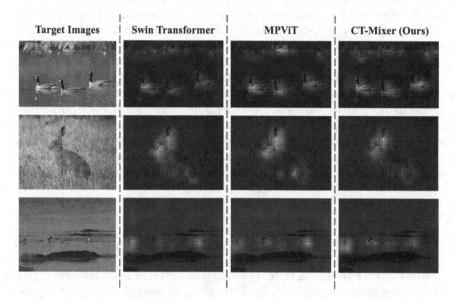

Fig. 6. Visualization of attention heatmaps for Swin Transformer [24], MPViT [19], and our proposed CT-Mixer. The first column displays the input image, and the following three columns show the attention heatmaps corresponding to each model.

4.4 Visualization Analysis

Figure 6 depicts the attention heatmaps generated by different networks, including the single-scale embedded Swin Transformer [24], multi-scale MPViT [19], and our hybrid-scale CT-Mixer. Dark regions indicate low attention scores, while bright regions indicate high attention scores.

Comparing our model with the single-scale embedded Swin Transformer, our CT-Mixer can focus on objects with different scales, such as the drake in the first and third given images. Our hybrid-scale embedded strategy is also more flexible and robust than MPViT. For example, in the first target image, our CT-Mixer captures more instances located at the margin of the image. The same phenomenon is observed in the third sample where our CT-Mixer renders more attention confidence for the desired objects.

5   Conclusion

In this work, we have proposed a new hybrid convolutional and transformer-based architecture, named CT-Mixer, for various computer vision tasks. Our approach leverages the strengths of convolutional networks and self-attention mechanisms to complement each other, resulting in improved performance over previous models. CT-Mixer introduces a novel hybrid-scale interaction scheme to enhance the flexibility of multi-scale learning, and an effective dynamic mechanism to bridge the gap of data adaptivity in the convolutional stage. Our extensive experiments on several benchmark datasets show that CT-Mixer achieves better performance on various vision tasks compared to state-of-the-art models.

However, the solid performance of CT-Mixer comes with an increase in computational cost and memory consumption, which may affect its practicality for deployment on certain hardware. Furthermore, while our model outperforms existing models on various metrics, there is still room for improvement in some specific metrics. In our future work, we plan to explore more efficient and effective ways to incorporate convolution and self-attention and investigate other techniques to further enhance the performance of CT-Mixer.

References

1. Ali, A., et al.: XCIT: cross-covariance image transformers. Adv. Neural. Inf. Process. Syst. **34**, 20014–20027 (2021)
2. Chen, C.F.R., Fan, Q., Panda, R.: Crossvit: cross-attention multi-scale vision transformer for image classification. In: Proceedings of the IEEE/CVF International Conference on Computer Vision, pp. 357–366 (2021)
3. Chen, K., et al.: Mmdetection: open mmlab detection toolbox and benchmark. arXiv preprint arXiv:1906.07155 (2019)
4. Chu, X., et al.: Twins: revisiting the design of spatial attention in vision transformers. Adv. Neural. Inf. Process. Syst. **34**, 9355–9366 (2021)
5. Dai, J., et al.: Deformable convolutional networks. In: Proceedings of the IEEE International Conference on Computer Vision, pp. 764–773 (2017)
6. Dai, Z., Liu, H., Le, Q.V., Tan, M.: Coatnet: marrying convolution and attention for all data sizes. Adv. Neural. Inf. Process. Syst. **34**, 3965–3977 (2021)
7. Deng, J., Dong, W., Socher, R., Li, L.J., Li, K., Fei-Fei, L.: Imagenet: a large-scale hierarchical image database. In: 2009 IEEE Conference on Computer Vision and Pattern Recognition, pp. 248–255. IEEE (2009)
8. Ding, M., Xiao, B., Codella, N., Luo, P., Wang, J., Yuan, L.: Davit: dual attention vision transformers. arXiv preprint arXiv:2204.03645 (2022)
9. Dong, X., et al.: Cswin transformer: a general vision transformer backbone with cross-shaped windows. In: Proceedings of the IEEE/CVF Conference on Computer Vision and Pattern Recognition, pp. 12124–12134 (2022)
10. Dosovitskiy, A., et al.: An image is worth 16×16 words: transformers for image recognition at scale. arXiv preprint arXiv:2010.11929 (2020)
11. Fan, H., et al.: Multiscale vision transformers. In: Proceedings of the IEEE/CVF International Conference on Computer Vision, pp. 6824–6835 (2021)

12. Gao, S.H., Cheng, M.M., Zhao, K., Zhang, X.Y., Yang, M.H., Torr, P.: Res2net: a new multi-scale backbone architecture. IEEE Trans. Pattern Anal. Mach. Intell. **43**(2), 652–662 (2019)
13. He, K., Gkioxari, G., Dollár, P., Girshick, R.: Mask r-cnn. In: Proceedings of the IEEE International Conference on Computer Vision, pp. 2961–2969 (2017)
14. He, K., Zhang, X., Ren, S., Sun, J.: Deep residual learning for image recognition. In: Proceedings of the IEEE Conference on Computer Vision and Pattern Recognition, pp. 770–778 (2016)
15. Hoffer, E., Ben-Nun, T., Hubara, I., Giladi, N., Hoefler, T., Soudry, D.: Augment your batch: improving generalization through instance repetition. In: Proceedings of the IEEE/CVF Conference on Computer Vision and Pattern Recognition, pp. 8129–8138 (2020)
16. Hu, J., Shen, L., Sun, G.: Squeeze-and-excitation networks. In: Proceedings of the IEEE Conference on Computer Vision and Pattern Recognition, pp. 7132–7141 (2018)
17. Huang, G., Sun, Yu., Liu, Z., Sedra, D., Weinberger, K.Q.: Deep networks with stochastic depth. In: Leibe, B., Matas, J., Sebe, N., Welling, M. (eds.) ECCV 2016. LNCS, vol. 9908, pp. 646–661. Springer, Cham (2016). https://doi.org/10.1007/978-3-319-46493-0_39
18. Huang, Z., Ben, Y., Luo, G., Cheng, P., Yu, G., Fu, B.: Shuffle transformer: rethinking spatial shuffle for vision transformer. arXiv preprint arXiv:2106.03650 (2021)
19. Lee, Y., Kim, J., Willette, J., Hwang, S.J.: Mpvit: multi-path vision transformer for dense prediction. In: Proceedings of the IEEE/CVF Conference on Computer Vision and Pattern Recognition, pp. 7287–7296 (2022)
20. Li, K., et al.: Uniformer: unified transformer for efficient spatiotemporal representation learning. arXiv preprint arXiv:2201.04676 (2022)
21. Lin, T.Y., Dollár, P., Girshick, R., He, K., Hariharan, B., Belongie, S.: Feature pyramid networks for object detection. In: Proceedings of the IEEE Conference on Computer Vision and Pattern Recognition, pp. 2117–2125 (2017)
22. Lin, T.Y., Goyal, P., Girshick, R., He, K., Dollár, P.: Focal loss for dense object detection. In: Proceedings of the IEEE International Conference on Computer Vision, pp. 2980–2988 (2017)
23. Lin, T.-Y., et al.: Microsoft COCO: common objects in context. In: Fleet, D., Pajdla, T., Schiele, B., Tuytelaars, T. (eds.) ECCV 2014. LNCS, vol. 8693, pp. 740–755. Springer, Cham (2014). https://doi.org/10.1007/978-3-319-10602-1_48
24. Liu, Z., et al.: Swin transformer: hierarchical vision transformer using shifted windows. In: Proceedings of the IEEE/CVF International Conference on Computer Vision, pp. 10012–10022 (2021)
25. Liu, Z., Mao, H., Wu, C.Y., Feichtenhofer, C., Darrell, T., Xie, S.: A convnet for the 2020s. In: Proceedings of the IEEE/CVF Conference on Computer Vision and Pattern Recognition, pp. 11976–11986 (2022)
26. Loshchilov, I., Hutter, F.: Decoupled weight decay regularization. arXiv preprint arXiv:1711.05101 (2017)
27. Meng, L., et al.: Adavit: adaptive vision transformers for efficient image recognition. In: Proceedings of the IEEE/CVF Conference on Computer Vision and Pattern Recognition, pp. 12309–12318 (2022)
28. Polyak, B.T., Juditsky, A.B.: Acceleration of stochastic approximation by averaging. SIAM J. Control. Optim. **30**(4), 838–855 (1992)
29. Radosavovic, I., Kosaraju, R.P., Girshick, R., He, K., Dollár, P.: Designing network design spaces. In: Proceedings of the IEEE/CVF Conference on Computer Vision and Pattern Recognition, pp. 10428–10436 (2020)

30. Ren, S., Zhou, D., He, S., Feng, J., Wang, X.: Shunted self-attention via multi-scale token aggregation (2021)

31. Szegedy, C., Vanhoucke, V., Ioffe, S., Shlens, J., Wojna, Z.: Rethinking the inception architecture for computer vision. In: Proceedings of the IEEE Conference on Computer Vision and Pattern Recognition, pp. 2818–2826 (2016)

32. Tan, M., Le, Q.: Efficientnet: rethinking model scaling for convolutional neural networks. In: International Conference on Machine Learning, pp. 6105–6114. PMLR (2019)

33. Touvron, H., Cord, M., Douze, M., Massa, F., Sablayrolles, A., Jégou, H.: Training data-efficient image transformers & distillation through attention. In: International Conference on Machine Learning, pp. 10347–10357. PMLR (2021)

34. Wang, W., et al.: Pyramid vision transformer: a versatile backbone for dense prediction without convolutions. In: Proceedings of the IEEE/CVF International Conference on Computer Vision, pp. 568–578 (2021)

35. Wang, W., et al.: Pvt v2: improved baselines with pyramid vision transformer. Comput. Visual Media $\mathbf{8}$(3), 415–424 (2022)

36. Wang, X., Girshick, R., Gupta, A., He, K.: Non-local neural networks. In: Proceedings of the IEEE Conference on Computer Vision and Pattern Recognition, pp. 7794–7803 (2018)

37. Woo, S., Park, J., Lee, J.Y., Kweon, I.S.: CBAM: convolutional block attention module. In: Proceedings of the European Conference on Computer Vision (ECCV), pp. 3–19 (2018)

38. Wu, H., et al.: CVT: introducing convolutions to vision transformers. In: Proceedings of the IEEE/CVF International Conference on Computer Vision, pp. 22–31 (2021)

39. Xu, W., Xu, Y., Chang, T., Tu, Z.: Co-scale conv-attentional image transformers. In: Proceedings of the IEEE/CVF International Conference on Computer Vision, pp. 9981–9990 (2021)

40. Yang, J., et al.: Focal self-attention for local-global interactions in vision transformers. arXiv preprint arXiv:2107.00641 (2021)

41. Yun, S., Han, D., Oh, S.J., Chun, S., Choe, J., Yoo, Y.: Cutmix: regularization strategy to train strong classifiers with localizable features. In: Proceedings of the IEEE/CVF International Conference on Computer Vision, pp. 6023–6032 (2019)

42. Zhang, H., Cisse, M., Dauphin, Y.N., Lopez-Paz, D.: mixup: beyond empirical risk minimization. arXiv preprint arXiv:1710.09412 (2017)

43. Zhong, Z., Zheng, L., Kang, G., Li, S., Yang, Y.: Random erasing data augmentation. In: Proceedings of the AAAI Conference on Artificial Intelligence, vol. 34, pp. 13001–13008 (2020)

FedQL: Q-Learning Guided Aggregation for Federated Learning

Mei Cao, Mengying Zhao, Tingting Zhang, Nanxiang Yu, and Jianbo Lu[✉]

School of Computer Science and Technology, Shandong University, Qingdao, China
{meicao,zhang_tingting,yu_nanxiang}@mail.sdu.edu.cn,
zhaomengying@sdu.edu.cn, ziduke@163.com

Abstract. Federated learning is a distributed machine learning paradigm, which is able to achieve model training without sharing clients' private data. In each round, after receiving a global model, selected clients train the model with local private data and report updated parameters to server. Then the server performs aggregation to generate a new global model. Currently, aggregations are generally conducted in a heuristic manner, and show great challenges with non-Independent and Identically Distributed (non-IID) data. In this paper, we propose to employ Q-learning to solve the aggregation problem under non-IID data. Specifically, we define state, action as well as reward in the target aggregation scenario, and fit it into Q-learning framework. With the learning procedure, effective actions indicating weights assignment for aggregation can be figured out according to certain system states. Evaluation shows that the proposed FedQL strategy can improve the convergence speed obviously under non-IID data, when compared with existing schemes FedAvg, FedProx and FedAdp.

Keywords: Federated Learning · Aggregation · non-IID Data · Q-learning

1 Introduction

In recent years, our daily life has been greatly changed due to the enormous success and advancement of deep learning, which enables smart decision making in scenarios such as self-driving, health monitoring and product recommendation. This relies on well-trained models driven by a large amount of data.

With a rapidly increasing number of Internet-of-Thing devices, data tend to be in a distributed manner. One way to utilize these distributed data for model training is sending them to the server for centralized training. However, this has drawbacks of high communication costs and privacy risks. To deal with these issues, Google proposes a framework called federated learning [1].

Figure 1 shows a representative framework of federated learning, with a centralized server and distributed clients. In the beginning, the server initializes a global model and sends it to selected clients. Then the clients use local private

© The Author(s), under exclusive license to Springer Nature Singapore Pte Ltd. 2024
Z. Tari et al. (Eds.): ICA3PP 2023, LNCS 14487, pp. 263–282, 2024.
https://doi.org/10.1007/978-981-97-0834-5_16

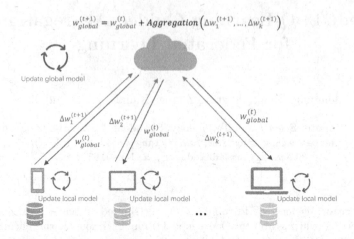

Fig. 1. A representative framework of federated learning [12].

data to train the model and report the updated parameters to the server. The parameters are aggregated in the server to generate a new global model. This procedure is iterated until a satisfying global model is achieved or a fixed number of rounds are accomplished [2–4]. Federated learning has been applied in scenarios such as clinical decision assistance [5,6], language model training [7], virtual keyboard searching [8], keyword spotting [9] and so on.

Aggregation is a vital step conducted by server during federated learning. It can be achieved by assigning weights to uploaded local parameters. A widely used aggregation mechanism is federated averaging (FedAvg), which defines the weight as the ratio of data the client contains. It indicates that a larger number of data makes higher contribution to the global model. Studies show that FedAvg performs well with balanced data distribution but results in obvious accuracy drop when data are non-Independent and Identically Distributed (non-IID). What is worse, the model may even be difficult to converge [1,10,11]. With non-IID data, the type, volume and distribution of data in clients are greatly different, which leads to great challenge for federated learning since non-IID data are very common in reality. For example, in computer-assisted diagnostics, classification models need to be trained for medical image analysis. Clinical samples, such as whole slide images of cancers, from different hospitals tend to exhibit great variance.

This paper focuses on effective aggregation in federated learning with non-IID data. We follow the classic framework that assigns weights to local models to aggregate, where the key issue is how to determine these weights. Instead of using heuristic strategies such as FedAvg, we propose to employ Q-learning, one kind of reinforcement learning, to explore a better combination of local models, so that the convergence speed can be improved while maintaining accuracy of the global model. Specifically, we map the weight decision problem to Q-learning framework. With definition of corresponding state, action and reward, the agent can learn to set weight according to certain system state. The main contributions of this paper are as follows.

- We propose to employ Q-learning to solve the aggregation problem in federated learning, with the objective of fast convergence of the global model.
- We design corresponding state space, action space, reward function, and backtracking update method to fit the target problem into the Q-learning framework.
- We develop a simulation platform to evaluate the efficacy of the proposed scheme and compare it with existing strategies.

The remaining sections of this paper are structured as follows. Section 2 provides a summary of preliminaries and related work in federated learning. Section 3 presents details of the proposed scheme. Evaluation results and discussions are described in Sect. 4. Finally, Sect. 5 concludes this paper.

2 Preliminaries and Related Work

Federated learning enables model training with distributed private data protected. The main steps of one round in federated learning can be summarized as the following steps[1], as illustrated in Fig. 1.

1) **Client Selection.** In the tth round of communication, the server selects some clients as participating clients and distributes the global model $\omega_{global}^{(t)}$ to them.
2) **Local Training.** The selected clients use their private data to train the model locally for a certain number of epochs.
3) **Parameter Uploading.** Each client uploads the model update $\Delta\omega_k^{(t+1)}$ to the server, where k represents the index of the client.
4) **Aggregation.** After receiving the model updates of all clients participating in the current round of training, the server aggregates them to generate a new global model $\omega_{global}^{(t+1)}$.

This procedure iterates until satisfactory accuracy is achieved or number of communication rounds reaches a predefined threshold [13]. In addition to the above centralized federated learning, there are also researches focusing on the decentralized framework, which has quite different working procedures [14–18]. In this paper, we study the aggregation issue in the context of centralized federated learning.

---

[1] Note that the starting point of one round is that the server has a global model, either an initialized one or aggregated one from the last round.

2.1   Client Selection

Generally, there are a lot of clients taking part in federated learning. A straightforward way is to select a subset of active clients randomly to participate in the training of each round [19–22]. In addition, some researches tending to choose clients intentionally rather than randomly. Considering the heterogeneity of clients in communication, statistics and computation, Nishio et al. [23] propose an approach named FedCS to choose clients based on their resources. Lin et al. [24] propose a contribution-based selection algorithm that dynamically adjusts the selection weights according to the influence of the client's data. Fang et al. [25] propose a client confidence reweighting approach that gives higher weight to clients with clean data sets and effective models. Furthermore, reinforcement learning is applied to federated learning client selection in several researches recently. For example, Wang et al. [26] propose a mechanism based on DQN that learns to select a subset of devices in each communication round to maximize a reward that encourages the increase of accuracy. Zhang et al. [27] propose a federated learning framework relying on multiple reinforcement learning agents to perform efficient client selection under non-IID data. These works verify the effectiveness of reinforcement learning used in federated learning. In this paper, we also employ reinforcement learning for optimizing federated learning. Instead of client selection, we focus on parameter aggregation, which is totally a different task and thus needs redesign of state, action as well as reward function.

2.2   Local Training

Aiming at speeding up federated learning, there are researches focusing on the acceleration of local training in clients. Li et al. [28] make use of the information of the global model in last round to reduce the local performance drift due to data heterogeneity. Xu et al. [29] propose an algorithm FedReg, which regularizes locally trained parameters with the loss on created pseudo data to accelerate federated learning. Divyansh et al. [30] present an adaptive quantization approach designed to obtain a low error floor by varying the number of quantization levels during training. Liu et al. [31] propose an MFL scheme in which momentum gradient descent (MGD) is used to accelerate model convergence. Ullah et al. [32] deploy two different approaches for local models. One is filtering out batch normalization parameters and removing them from weight decay, and the other is initializing the final batch normalization in each residual branch (in case of Resnet). After the clients complete the training of the local model, they will report the updated parameters to the server.

2.3   Parameter Uploading

Transmission cost is also a key issue in federated learning, especially for clients with limited bandwidth. Xu et al. [33] present a protocol named T-FedAvg which decreases the upstream and downstream transmission of federaated learning system. Cui et al. [34] present an adaptation framework to adjust the compression rate for model updates strategically in each iteration. Reisizadeh et al. [35] propose FedPAQ which reduces client-to-server communication by regularly averaging and quantifying. Caldas et al. [36] introduce two new methods for improving communication efficiency. One is federated dropout, which lowers the cost of client-to-server communication, and the other is lossy compression of the global model transmitted from the server to the client. Giovanni et al. [37] propose a new algorithm aiming at lowering the amount of shared parameters throughout the federated learning process in order to reduce communication costs. Abasi et al. [38] present a method named FedGWO to decrease data communication cost by transferring scoring principles rather than the weights of all clients' models.

2.4   Aggregation

FedAvg [1] is a representative aggregation algorithm, which aggregates local models using weighted average, where the weights are determined by the number of data in each client, as shown in Eq. (1).

$$\omega_{global}^{t+1} = \omega_{global}^{t} + \sum_{i \in S^t} \frac{n_i}{n} \Delta\omega_i^{t+1} \qquad (1)$$

$$\Delta\omega_i^{t+1} = \omega_i^{t+1} - \omega_{global}^{t} \qquad (2)$$

Here, ω_{global}^{t} and ω_{global}^{t+1} are the global models at the tth round and the $(t+1)$th round. n_i and n are the amount of data owned by the ith client, and the total amount of data of all clients participating in the training in this round, respectively. S^t is a set that contains clients selected in the current round. $\Delta\omega_i^{t+1}$ is the uploaded model updates from the ith client at $(t+1)$th round, as shown in Eq. (2). It represents the difference between the local training model and the global model of the previous round.

There are also other aggregation strategies. Li et al. [39] introduce a framework, FedProx, which determines aggregation weights considering both data heterogeneity and system heterogeneity. Nguyen et al. [40] use absolute average aggregation algorithm in the wireless Internet of Things network. Ma et al. [11] adjust the weights of local models according to contributions made by clients to each class. Song et al. [41] think clients with a good reputation should be given greater weight when aggregating since they contribute more than low-performance clients do. Huang et al. [42] integrate the information amount of training accuracy and training frequency to determine the weights. Tan et al. [43] adjust each client's weight in the aggregation according to their involvement history.

Mohri et al. [44] achieve a conclusion that weight assignment in federated learning is agnostic. Instead of using heuristic algorithms like all above researches, in this paper, we study the agnostic aggregation problem from a different point of view. We propose to employ reinforcement learning to help with weight assignments during aggregation. A learning agent is developed to learn to do the task of weight assignments, with the objective of fast convergence of the global model. The proposed scheme is orthogonal and can be combined with optimization techniques for client selection, local training and parameter uploading summarized in Sect. 2.1–2.3.

3   FedQL: Q-Learning Guided Aggregation

In this section, we give the detailed design of the proposed Q-learning guided aggregation, named FedQL, for federated learning.

3.1   Framework of FedQL

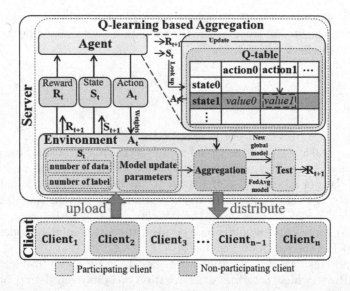

Fig. 2. The framework of FedQL.

Figure 2 shows the framework of FedQL. It involves a Q-learning framework to generate weights used for aggregation. In a standard Q-learning system, after the agent receives current state S_t from the environment, it selects an action A_t according to Q values from the Q-table. Then the environment returns a reward or punishment to the agent, indicating whether it is a good action or not for that certain state. And the agent will update values in the Q-table accordingly.

In order to employ Q-learning to determine the weights for aggregation, we need to appropriately define state, action as well as reward to fit the target problem into the Q-learning framework. In this way, the agent can construct an effective Q-table as a reference to guide which combination of weights to use in aggregation to achieve an efficient training model. Next, the design of each component will be discussed.

3.2   Agent

The agent acts as the system's central component. In a certain state, the agent chooses one action based on the learning strategy. After executing the action, the environment provides the agent with feedback of reward or punishment. The objective is to get the maximum total rewards after completing a series of actions. To achieve this, the agent should explore, try and learn as much as possible.

The Q value is updated as Eq. (3) [45], where $Q_{t+1}(S_t, a_t)$ is the quality after executing a_t under state S_t, and $Q_t(S_t, a_t)$ is the quality of previous round. α is the learning rate which measures how much the agent learns from the current training results. γ is the discount factor, which expresses the future impacts, e.g., a factor of 0 would cause the agent to solely evaluate current rewards.

$$Q_{t+1}\left(S_t, a_t\right) = Q_t\left(S_t, a_t\right) +$$
$$\alpha \cdot \left(R_{t+1} + \gamma \cdot \max_{(\forall a \in A)} \left[Q_t\left(S_{t+1}, a_{t+1}\right)\right] - Q_t\left(S_t, a_t\right) \right) \tag{3}$$

Given a state, the agent chooses an action according to Eq. (4). In general, the agent is able to choose the action with the largest Q value. For exploration consideration, we employed ϵ-greedy method [45] to escape from local optimum.

$$a_{t+1} = \begin{cases} \text{random action from } A_{(S_{t+1})}, & \text{if } \rho < \epsilon \\ \text{argmax}\left(Q_{(S_{t+1},\ a)}\right), & \text{otherwise} \end{cases} \tag{4}$$

Here ρ is a random number ranging between 0 and 1, and ϵ is a predetermined random rate. The agent will randomly select an action in this state if ρ is less than ϵ. Otherwise, it will execute the action with the highest Q value.

3.3   State

State should be designed to accurately describe the condition of system at certain moment. In federated learning with non-IID data, clients have great variance in the number and category of data. Inspired by this, we propose to use two features to describe client i, i.e., $\langle n_i, c_i \rangle$, where n_i represents the number of data, and c_i is the number of labels in this client. In one certain round of federated learning, some clients are selected to participate, so that the state can be defined as the sets of $\langle n, c \rangle$ of all selected clients, i.e., $\{\langle n_i, c_i \rangle \mid$ client i is selected in this round$\}$.

In order to build the state, clients should report their n_i and c_i to server. Note that while sharing these two features, data are still kept private in each client. Instead of using an exact number of data in a certain client, we can use the portion of data it has to define state, so that n_i can fall in range of (0,1]. Then the state is changed to $\{\langle f_i, c_i \rangle \mid$ client i is selected in this round$\}$, where f_i is the ratio of data client i has, which can be calculated by the server after receiving n_i of selected clients. In order to eliminate redundant state, $\langle f_i, c_i \rangle$ pairs in the set can be sorted according to f_i, i.e., in either ascending or descending order.

The total number of state in the state space is mainly determined by two factors. The first one is the number of selected clients, and the other one is the precision of f_i. A larger number of selected clients in each round, and higher precision of f_i would lead to a larger number of state.

3.4   Action

With state well defined, the agent will learn to take action according to the current state. In the context of aggregation, we have two ways to define action. One is to fine-tune weights on the basis of some initialized weights. Then the action can be defined as increasing/decreasing the weight of one client by δ. Under this definition, assuming n clients are selected in this round, there would be $2n$ actions in total. In this way, the action set can cover all possibilities of tuning. However, it may need a long procedure to achieve a good weight assignment since it tunes one single weight by one step in one round.

The other way to define action is that, instead of fine tuning, we directly define possible weight combinations and make the agent choose one from them, i.e., $a_i = \{p_1, p_2, ...p_n\}$, where p_n is the weight to be assigned for the nth client participating in a round of federated learning training and a_i is ith action in action set. For example, assuming that three clients are selected for each round of federated learning, action set can be defined as a permutation combination of three numbers whose sum is 1 and agent chooses one from them, i.e., an action can be $\{0.2, 0.3, 0.5\}$, where 0.2, 0.3 and 0.5 are assigned weights for these three clients respectively. Similar to f_i in state definition, precision of p_n affects the total number of actions.

In this work, we take the latter one for action definition.

3.5   Reward

After the agent chooses an action according to certain state, the environment gives feedback, i.e., reward, to the agent, based on which the agent would update corresponding Q value in the Q-table.

In the target problem, we regard it as a good action if it can deliver high-quality aggregation, which can be evaluated by accuracy of the global model after aggregation. For example, a reference of accuracy can be used for comparison. If the action delivers higher accuracy than the reference, a positive reward can be given. Otherwise, a negative punishment can be issued. In addition, considering the inherent fluctuation in model accuracy, average value of continuous rounds can be used to derive reward. In this work, we use FedAvg as the reference to give reward and punishment.

3.6   Put Them Together

With all the components well defined, a Q-learning guided aggregation scheme can be constructed. We will first summarize the learning procedure of Q-table and then show the flow of aggregation with the proposed FedQL.

Learning Procedure of Q-Table. Algorithm 1 shows the procedure to build a stable Q-table. The global model is initialized with parameters $\omega^{(0)}$ (Lines 1–2) by the server. Moreover, Q-table and the action space A are also initialized (Line 3). The server then begins the iterative communications with clients (Lines 4–25) and performs the following operations with each communication round. Firstly, it chooses m available clients (Lines 5–7). Following that, it delivers the global model to a group of selected clients for local model training (Lines 8–12).

In selected clients, a mini-batch SGD technique is used to update the model with learning rate η, where the local data are divided into batches with the preset size of bs (Lines 28–31). After ep epochs, each client sends the parameter updates, the number of data n_k, and the number of classes c_k to the server (Line 33). Usually, to prevent infinite waiting, the server typically sets a deadline for clients to submit parameter updates.

With information from clients, the server can do aggregation to derive a new global model, and at the same time, update the Q-table (Lines 13–24). First, current state S_t can be derived according to reported n_k and c_k from clients (Line 11). If S_t already exists in Q-table, its value corresponding to actions will be updated. Otherwise, S_t will be inserted into Q-table as a new row, with all Q values initialized. Then the agent selects an action for the current state S_t according to the ϵ-greedy strategy as shown in Eq. 4 (Lines 13–17). The action gives weights for aggregation, leading to a new global model $\omega_{ql}^{(t+1)}$ (Lines 18–19). In order to give rewards, FedAvg is used to derive another global model $\omega_{fed}^{(t+1)}$ (Lines 20–21). These two global models are compared in terms of model accuracy on a testing set. If $\omega_{ql}^{(t+1)}$ outperforms $\omega_{fed}^{(t+1)}$, a positive reward will be given. Otherwise, if the accuracy of $\omega_{ql}^{(t+1)}$ is lower than $\omega_{fed}^{(t+1)}$, a negative reward, i.e., punishment, will be given. Then the reward is used to update corresponding Q value (Lines 22–24). This procedure is iterated until the Q-table is convergent.

In this way, the agent can build an efficient Q-table, based on which, a satisfactory action can be pinned out according to current state, leading to high-quality aggregation and thus fast model training in federated learning. Note that the Q-table can be further updated when necessary during usage.

Federated Learning with FedQL. The flow of federated learning with the proposed FedQL is shown in Algorithm 2. The framework is the same as traditional federated learning on the basis of FedAvg, but with the following differences. First, selected clients need to report their data volume, as well as the number of classes they have, to the server. Based on this, the server can derive current state (Line 8). Second, the server uses Q-table to do aggregation by selecting action according to system state (Lines 10–12).

Algorithm 1. *Learning procedure of Q-table. ep* and *bs* represent local epoch number and local mini-batch size, respectively. m is the number of clients that participate in one training round. n_k is the number of data in kth client. c_k is the number of class in kth client and η devotes the learning rate.

Server Executes:
1: Initialize global model with parameters $\omega_{fed}^{(0)}$;
2: $\omega_{ql}^{(0)} = \omega_{fed}^{(0)}$;
3: Initialize Q-table and action space A;
4: **for** communication round $t=0,1,...$ **do**
5:    $Av_{client}^{(t)} \leftarrow$ ask and get available clients set;
6:    $num\_client^{(t)} \leftarrow \min(m, |Av_{client}^{(t)}|)$;
7:    $S^{(t)} \leftarrow$ select $num\_client^{(t)}$ clients from $Av_{client}^{(t)}$;
8:    **for** each client $k \in S^{(t)}$ in parallel **do**
9:       $\Delta\omega_{fed}^k \leftarrow ClientUpdate_k(\omega_{fed}^{(t)})$;
10:      $\Delta\omega_{ql}^k \leftarrow ClientUpdate_k(\omega_{ql}^{(t)})$;
11:      Derive state S_t according to n_k and c_k from clients;
12:    **end for**
13:    **if** $\rho < \epsilon$ **then**
14:      $a_t \Leftarrow random\ action\ from\ A_{(S_t)}$;
15:    **else**
16:      $a_t \Leftarrow argmax\left(Q_{(S_t,\ a)}\right)$;
17:    **end if**
18:    $\Delta W_{ql} \leftarrow \sum\limits_{k \in S^{(t)}} a_t\,[k]\,\Delta\omega_{ql}^k$;
19:    $\omega_{ql}^{(t+1)} \leftarrow \omega_{ql}^{(t)} + \Delta W_{ql}$;
20:    $\Delta W_{fed} \leftarrow \sum\limits_{k \in S^{(t)}} \frac{n_k}{\sum n_k} \Delta\omega_{fed}^k$;
21:    $\omega_{fed}^{(t+1)} \leftarrow \omega_{fed}^{(t)} + \Delta W_{fed}$;
22:    $r \Leftarrow Reward(\omega_{fed}^{(t+1)}, \omega_{ql}^{(t+1)})$;
23:    $q' \Leftarrow r + \gamma \cdot max(Q\_table[S_{t+1}, a_{t+1}])$;
24:    $Q\_table[S_t, a_t] + = \alpha \cdot (q' - Q\_table[S_t, a_t])$;
25: **end for**

Client k Executes $ClientUpdate_k(\omega_{global})$:
26: Copy ω_{global} to ω_k;
27: **for** epoch 1 to ep **do**
28:    $\Xi \leftarrow$ Split local data with size bs;
29:    **for** mini batch $b \in \Xi$ **do**
30:      $\omega_k \leftarrow \omega_k - \eta\nabla L(\omega_k; b)$;
31:    **end for**
32: **end for**
33: Send $(\omega_{global} - \omega_k)$, n_k and c_k to server;

Algorithm 2. *Federated learning with FedQL.* Hyperparameter notation is the same as Algorithm 1.

Server Executes:
1: Initialize global model with parameters $\omega^{(0)}$;
2: **for** communication round $t=0,1,...$ **do**
3:     $Av_{client}^{(t)} \leftarrow$ ask and get available clients set;
4:     $num\_client^{(t)} \leftarrow \min(m, |Av_{client}^{(t)}|)$;
5:     $S^{(t)} \leftarrow$ select $num\_client^{(t)}$ clients from $Av_{client}^{(t)}$;
6:     **for** each client $k \in S^{(t)}$ in parallel **do**
7:         $\Delta\omega_k \leftarrow ClientUpdate_k(\omega^{(t)})$;
8:         Derive state S_t with n_k and c_k;
9:     **end for**
10:     $a_t \Leftarrow argmax\left(Q_{(S_t,\ a)}\right)$;
11:     $\Delta W \leftarrow \sum_{k \in S^{(t)}} a_t[k]\Delta\omega_k$;
12:     $\omega^{(t+1)} \leftarrow \omega^{(t)} + \Delta W$;
13: **end for**

---

4   Experiments

In this section, we first report the evaluation results of the proposed design, and then give overhead analysis and discussions.

4.1   Experimental Setup

We simulate 100 clients for federated learning during the evaluation. Considering the conclusion that large fraction has a weak effect for model convergence in non-IID settings [17], we set fraction C as 0.03 in evaluation. For simplicity, we choose clients randomly in each round, and it can be easily extended to other client selection schemes. The learning rate in local training is set as 0.01. The proposed scheme is compared with three existing aggregation strategies pertaining to FedAvg [1], FedProx [39] and FedAdp [46].

Datasets. MNIST [47] is a well-known dataset, which is composed of 60,000 training samples and 10,000 test samples. Each sample is a 28×28 pixels gray handwritten digit picture and represents a number from 0 to 9. CIFAR-10 [48] is made up of 60,000 color images, including of 50,000 training samples and 10,000 test samples. Each image is a 32×32 pixels in size and covers 10 categories, with a total of 6,000 samples in each category. Fashion-MNIST [49] contains a training set and a test set, in which the training set has a total of 60,000 images, and the test set has a total of 10,000 images. Each image is a single-channel black-and-white image with a size of 28×28 pixels and belongs to 10 categories.

Model Structure. For MNIST, we adopt a model named CNNMnist, which has two 5×5 convolution layers with channels of 10 and 20, a fully connected layer with 50 units, and a softmax output layer. For CIFAR-10, we use the CNNCifar model for evaluation which consists two 5×5 convolution layers (both contain 64 channels), each followed by a max pool layer with kernel size of 3 and stride of 2, two fully connected layers with 384 and 192 neurons, and a final softmax output layer. For Fashion-MNIST, we use the LeNet model with two 5×5 convolution layers, each followed by a max pool layer with kernel size of 2 and stride of 2, and three fully-connected layers with 160, 84, and 10 neurons.

4.2   Experimental Results

Two metrics are used to compare the proposed FedQL with three baselines. One is the accuracy of the global model on test datasets. The other is communication rounds needed to achieve a specified level of accuracy. Because the accuracy fluctuates in federated learning, we keep track of it across a number of continuous rounds, e.g., 30 in our evaluation, and regard it as steadily reaching a certain precision if all the continuous rounds have accuracy higher than this value. For further comparison, we report the number of rounds to achieve the target accuracy for the first time, too.

We evaluate aggregation schemes on basis of various grades of non-IID data, noted by non-IID(x), where x represents how many classes of data each client contains. Thus non-IID(1) indicates significant unbalance in terms of data distribution while non-IID(2) is comparatively more balanced since clients may have overlaps in data class.

Tables 1, 2 and 3 present the results of MNIST, CIFAR-10 and Fashion-MNIST under non-IID(1), respectively. At the fourth and fifth column, data are reported in the form of a/b, where a is the number of rounds to reach steady target accuracy, and b is the number of rounds to achieve the accuracy for the first time. The target accuracies of the three datasets are set as 87%, 71%, and 82%, which are steady accuracies that most models can achieve under different settings.

Test Accuracy. FedAvg, FedProx, FedAdp and FedQL have comparable accuracy of final global models. As shown in the third column of Tables 1, 2 and 3, the proposed FedQL has slight advantage in final accuracy. The same observation can be derived from Fig. 3. It indicates that FedAvg, FedProx, FedAdp and FedQL have alike capability in the ultimate classification.

Convergence Speed. It can be observed that for most cases, FedQL achieves the target accuracy with a faster speed of convergence than FedAvg, FedProx and FedAdp. Taking MNIST as an example, as shown in Table 1, FedQL saves 9.64% communication rounds when compared with FedAvg while FedProx uses 16.36% more communication rounds to deliver steady accuracy of 87%, the "–" in the fourth column here means that FedAdp could not reach the target accuracy stably within 3000 rounds. Figure 3(a) gives more details about accuracy of global models. It can be seen that, even with fluctuations, FedQL is able to maintain accuracy higher than 87% for continuous 30 rounds while FedAvg, FedProx and FedAdp drop below 87% from time to time.

Table 1. Results on MNIST under non-IID(1)

Setting	Algorithm	Accuracy of model	Rounds	Differences in Rounds
non-IID(1)	FedAvg	91.47%	2396/383	–
	FedProx	91.15%	2788/385	+16.36%/+0.52%
	FedAdp	89.47%	–/371	–/–3.13%
	FedQL	**91.95%**	**2165/383**	**–9.64%/0.00%**

Table 2. Results on CIFAR-10 under non-IID(1)

Setting	Algorithm	Accuracy of model	Rounds	Differences in Rounds
non-IID(1)	FedAvg	71.42%	45701/35555	–
	FedProx	70.72%	–/48382	–/+36.08%
	FedAdp	71.61%	53749/42598	+17.61%/+19.81%
	FedQL	**71.74%**	**41556/31026**	**–9.07%/–13.85%**

Table 3. Results on Fashion-MNIST under non-IID(1)

Setting	Algorithm	Accuracy of model	Rounds	Differences in Rounds
non-IID(1)	FedAvg	84.01%	14920/1558	–
	FedProx	82.81%	–/2360	–/+51.48%
	FedAdp	83.60%	–/3017	–/+93.65%
	FedQL	**84.04%**	**11652/1725**	**-21.90%/+10.72%**

Results Under Non-IID(1). In this evaluation, we randomly split each class of data into 10 shards. Since MNIST, CIFAR-10, and Fashion-MNIST all have 10 classes, so we have 100 shards of data and randomly assign them to 100 clients. The numbers of rounds of federated learning are set to be 3,000, 60,000, and 15,000 for MNIST, CIFAR-10, and Fashion-MNIST, respectively.

For CIFAR-10, the proposed FedQL achieves 9.07% reduction in communication rounds when compared with FedAvg, as shown in Table 2. FedProx can not accomplish a steady accuracy of 71% and FedAdp increases 17.61% communication rounds for CIFAR-10 under non-IID(1) setting. The advantage of FedQL is obvious as shown in Fig. 3(b).

For Fashion-MNIST, FedQL delivers 21.90% reduction in the number of communication rounds. FedProx and FedAdp cannot achieve steady accuracy of 82% within 15,000 rounds. As illustrated in Fig. 3(c), FedProx and FedAdp have more significant fluctuation and cannot maintain high accuracy for continuous 30 rounds.

Comparing the number of rounds to achieve target accuracy for the first time, FedQL has the same performance as FedAvg under MNIST, 13.85% reduction under CIFAR-10 and 10.72% increment under Fashion-MNIST. FedQL outper-

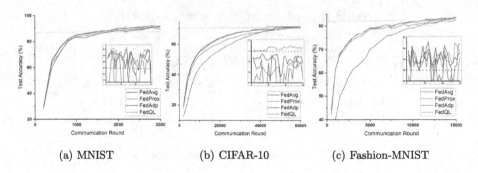

(a) MNIST   (b) CIFAR-10   (c) Fashion-MNIST

Fig. 3. Results under non-IID(1) setting.

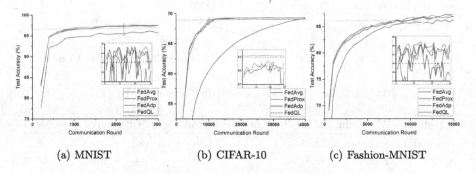

(a) MNIST   (b) CIFAR-10   (c) Fashion-MNIST

Fig. 4. Results under non-IID(2) setting.

forms FedProx for all tested cases. And under CIFAR-10 and Fashion-MNIST, FedAdp is inferior to FedQL. These results are quite different from those for steady accuracy because all models have obvious fluctuations, and results with steady accuracy are more meaningful in terms of model usage.

Results Under Non-IID(2). To generate data distributions for evaluation on non-IID(2), we divide each dataset into 200 shards. Then, each client randomly selects two shards as its local data, i.e., each client has two labels at most. Tables 4, 5 and 6 report the results, with their respective target accuracies are 97%, 69%, and 86%. The number of rounds of federated learning are set to be 3,000, 40,000, and 15,000 for MNIST, CIFAR-10, and Fashion-MNIST, respectively.

For MNIST, FedQL delivers steady accuracy of 97% with 2,213 rounds, while FedAvg and FedProx cannot reach this steady accuracy within 3,000 rounds. For CIFAR-10 and Fashion-MNIST, FedQL has 9.27% and 8.97% reductions in rounds to reach steady accuracies, when compared with FedAvg. FedProx has the best performance in Fashion-MNIST but does not perform well in CIFAR-10 in terms of convergence speed. FedAdp cannot achieve stable target accuracy within the setting rounds under three datasets. This is because the contributions

Table 4. Results on MNIST under non-IID(2)

Setting	Algorithm	Accuracy of model	Rounds	Differences in Rounds
non-IID(2)	FedAvg	97.50%	–/194	–
	FedProx	97.57%	–/168	–/–13.40%
	FedAdp	95.72%	–/133	–/–31.44%
	FedQL	**97.55%**	**2213/174**	**–/–10.31%**

Table 5. Results on CIFAR-10 under non-IID(2)

Setting	Algorithm	Accuracy of model	Rounds	Differences in Rounds
non-IID(2)	FedAvg	69.37%	9329/8551	–
	FedProx	69.14%	10770/8303	+15.45%/–2.90%
	FedAdp	69.25%	-/23042	–/+169.47%
	FedQL	**69.40%**	**8464/8305**	**–9.27%/–2.88%**

Table 6. Results on Fashion-MNIST under non-IID(2)

Setting	Algorithm	Accuracy of model	Rounds	Differences in Rounds
non-IID(2)	FedAvg	86.91%	12981/1343	–
	FedProx	86.88%	10229/1343	–21.20%/0.00%
	FedAdp	86.30%	–/1752	–/+30.45%
	FedQL	**87.38%**	**11816/1146**	**–8.97%/-14.67%**

of nodes cannot be well measured with high data heterogeneity, which confirms the conclusion addressed in [46].

For the first time to reach target accuracies, FedQL has reductions in number of rounds when compared with FedAvg under three datasets. FedProx uses fewer communication rounds under MNIST and CIFAR-10, but same communication rounds under Fashion-MNIST than FedAvg. FedAdp has obvious increase under CIFAR-10 and Fashion-MNIST. Figure 4 shows the accuracy of global models for all four strategies under non-IID(2).

Table 7. Communication cost per round&client

model	FedAvg	FedProx	FedAdp	**FedQL**
CNNMnist	21,840+1	21,840+1	21,840+1	**21,840+2**
CNNCifar	574,848+1	574,848+1	574,848+1	**574,848+2**
LeNet	71,014+1	71,014+1	71,014+1	**71,014+2**

4.3   Overhead Analysis

Table 7 summarizes the communication costs of four strategies in each round. CNNMnist, CNNCifar and LeNet have 21,840, 574,848 and 71,014 parameters, respectively, whose updates should be transmitted from clients to server in each communication round. In addition, a client running FedAvg, FedProx or FedAdp needs to send the number of data it contains to server. With Q-table successfully built, FedQL needs to send the number of data and the number of classes to server. The size of these additional information is negligible compared with CNNMnist/CNNCifar/LeNet, and the communication cost is dominated by the model itself.

During the learning procedure of Q-table, the communication cost is doubled since we need to simultaneously build two global models for FedQL and FedAvg. However, once the Q-table is established, it can be used for pretty long time with no or light updates.

4.4   Discussions

Table 8. Results on MNIST non-IID(1) with LeNet

Setting	Algorithm	Accuracy of model	Rounds	Differences in Rounds
non-IID(1)	FedAvg	97.36%	2678/812	–
	FedProx	95.13%	2778/1106	+3.73%/+36.21%
	FedAdp	98.04%	2638/925	−1.49%/+13.92%
	FedQL	**98.32%**	**1902/730**	**−28.98%/−10.10%**

Universality of Q-Table. A Q-table is essentially constructed to select good actions with certain state, and ideally can be used with different AI models. So we set a group of evaluations to test the universality of Q-table. We construct a Q-table using CNNMnist model on MNIST under non-IID(1), and test the performance on LeNet with this Q-table with all other settings unchanged. Table 8 shows the results. It can be seen that FedQL still has significant advantages in terms of convergence speed while maintaining accuracy of the global model. Figure 5 confirms more steady model accuracy of FedQL.

Privacy Issue. The proposed FedQL requires clients to report their number of data as well as number of classes to server. Even with this additional information provided, the data are still private with no share. From this point of view, privacy is preserved.

Actually, in order to improve performance of federated learning, there are related works adopting similar ideas, such as reporting data distribution information [50–52], or even averaged local data in [53]. This helps the server make smarter decisions towards higher model accuracy or faster convergence speed.

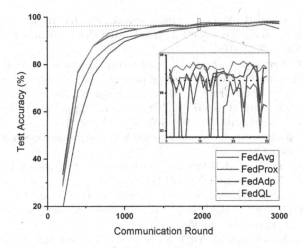

Fig. 5. Results under MNIST non-IID(1) with LeNet

5   Conclusion

In this paper, we propose an aggregation algorithm, called FedQL, which leverages Q-learning to guide weight assignments for clients. In FedQL, the number of data and the number of data classes of the client are derived as state of Q-learning, and the weights of clients are defined as action. The system can figure out satisfactory action in specific state with the guidance of Q-table. Evaluation results show that the proposed FedQL can reduce the number of communication rounds compared with FedAvg, FedProx and FedAdp, indicating faster convergence speed of the global model.

Acknowledgements. This paper is supported by National Natural Science Foundation of China (Grant No. 61973214), Shandong Provincial Natural Science Foundation (Grant No. ZR2020MF069), and Shandong Provincial Postdoctoral Innovation Project (Grant No. 202003005).

References

1. McMahan, B., Moore, E., Ramage, D., Hampson, S., Arcas, B.A.: Communication-efficient learning of deep networks from decentralized data. In: Artificial Intelligence and Statistics, pp. 1273–1282 (2017)
2. Xia, Q., Ye, W., Tao, Z., Wu, J., Li, Q.: A survey of federated learning for edge computing: research problems and solutions. High-Confidence Comput. **1**(1), 100008 (2021)
3. Yang, Q., Liu, Y., Cheng, Y., Kang, Y., Chen, T., Yu, H.: Federated learning. Synth. Lect. Artif. Intell. Mach. Learn. **13**(3), 1–207 (2019)
4. Xie, Z., Huang, Y., Yu, D., Parizi, R.M., Zheng, Y., Pang, J.: Fedee: a federated graph learning solution for extended enterprise collaboration. IEEE Trans. Ind. Inf. **19**(7), 8061–8071 (2023)

5. Brisimi, T.S., Chen, R., Mela, T., Olshevsky, A., Paschalidis, I.C., Shi, W.: Federated learning of predictive models from federated electronic health records. Int. J. Med. Inf. **112**, 59–67 (2018)

6. Sheller, M.J., Reina, G.A., Edwards, B., Martin, J., Bakas, S.: Multi-institutional deep learning modeling without sharing patient data: a feasibility study on brain tumor segmentation. In: Crimi, A., Bakas, S., Kuijf, H., Keyvan, F., Reyes, M., van Walsum, T. (eds.) BrainLes 2018. LNCS, vol. 11383, pp. 92–104. Springer, Cham (2019). https://doi.org/10.1007/978-3-030-11723-8_9

7. Hard, A., et al.: Federated learning for mobile keyboard prediction. arXiv preprint arXiv:1811.03604 (2018)

8. Yang, T., et al.: Applied federated learning: improving google keyboard query suggestions. arXiv preprint arXiv:1812.02903 (2018)

9. Leroy, D., Coucke, A., Lavril, T., Gisselbrecht, T., Dureau, J.: Federated learning for keyword spotting. In: IEEE International Conference on Acoustics, Speech and Signal Processing, pp. 6341–6345 (2019)

10. Konečný, J., McMahan, H.B., Yu, F.X., Richtárik, P., Suresh, A.T., Bacon, D.: Federated learning: strategies for improving communication efficiency. arXiv preprint arXiv:1610.05492 (2016)

11. Chen, S., Wang, Y., Yu, D., Ren, J., Xu, C., Zheng, Y.: Privacy-enhanced decentralized federated learning at dynamic edge. IEEE Trans. Computers **72**(8), 2165–2180 (2023)

12. Ma, Z., Zhao, M., Cai, X., Jia, Z.: Fast-convergent federated learning with class-weighted aggregation. J. Syst. Architect. **117**, 102125 (2021)

13. Wang, Y., et al.: Theoretical convergence guaranteed resource-adaptive federated learning with mixed heterogeneity. In: KDD, pp. 2444–2455 (2023)

14. Arachchige, P.C.M., Bertok, P., Khalil, I., Liu, D., Camtepe, S., Atiquzzaman, M.: A trustworthy privacy preserving framework for machine learning in industrial iot systems. IEEE Trans. Ind. Inf. **16**(9), 6092–6102 (2020)

15. Kim, H., Park, J., Bennis, M., Kim, S.-L.: Blockchained on-device federated learning. IEEE Commun. Lett. **24**(6), 1279–1283 (2019)

16. Li, Y., Chen, C., Liu, N., Huang, H., Zheng, Z., Yan, Q.: A blockchain-based decentralized federated learning framework with committee consensus. IEEE Netw. **35**(1), 234–241 (2020)

17. Lu, Y., Huang, X., Zhang, K., Maharjan, S., Zhang, Y.: Blockchain empowered asynchronous federated learning for secure data sharing in internet of vehicles. IEEE Trans. Veh. Technol. **69**(4), 4298–4311 (2020)

18. Majeed, U., Hong, C.S.: Flchain: federated learning via mec-enabled blockchain network. In: 20th Asia-Pacific Network Operations and Management Symposium, pp. 1–4 (2019)

19. Hu, F., Zhou, W., Liao, K., Li, H.: Contribution-and participation-based federated learning on non-iid data. IEEE Intell. Syst. **37**(4), 35–43 (2022)

20. Xu, J., Chen, Z., Quek, T.Q., Chong, K.F.E.: Fedcorr: multi-stage federated learning for label noise correction. In: Conference on Computer Vision and Pattern Recognition, pp. 10 184–10 193 (2022)

21. Zhang, L., Shen, L., Ding, L., Tao, D., Duan, L.-Y.: Fine-tuning global model via data-free knowledge distillation for non-iid federated learning. In: Conference on Computer Vision and Pattern Recognition, pp. 10 174–10 183 (2022)

22. Zheng, Y., Lai, S., Liu, Y., Yuan, X., Yi, X., Wang, C.: Aggregation service for federated learning: an efficient, secure, and more resilient realization. IEEE Trans. Depend. Secure Comput. **20**(2), 988–1001 (2022)

23. Nishio, T., Yonetani, R.: Client selection for federated learning with heterogeneous resources in mobile edge. In: IEEE International Conference on Communications, pp. 1–7 (2019)
24. Lin, W., Xu, Y., Liu, B., Li, D., Huang, T., Shi, F.: Contribution-based federated learning client selection. Int. J. Intell. Syst. **37**(10), 7235–7260 (2022)
25. Fang, X., Ye, M.: Robust federated learning with noisy and heterogeneous clients. In: Conference on Computer Vision and Pattern Recognition, pp. 10 072–10 081 (2022)
26. Wang, H., Kaplan, Z., Niu, D., Li, B.: Optimizing federated learning on non-iid data with reinforcement learning. In: IEEE Conference on Computer Communications, pp. 1698–1707 (2020)
27. Zhang, S.Q., Lin, J., Zhang, Q.: A multi-agent reinforcement learning approach for efficient client selection in federated learning. In: Proceedings of the AAAI Conference on Artificial Intelligence, vol. 36, no. 8, pp. 9091–9099 (2022)
28. Li, Z., Zhou, Y., Wu, D., Wang, R.: Local model update for blockchain enabled federated learning: approach and analysis. In: International Conference on Blockchain, pp. 113–121 (2021)
29. Xu, C., Hong, Z., Huang, M., Jiang, T.: Acceleration of federated learning with alleviated forgetting in local training. In: Conference on Learning Representations, ICLR (2022)
30. Jhunjhunwala, D., Gadhikar, A., Joshi, G., Eldar, Y.C.: Adaptive quantization of model updates for communication-efficient federated learning. In: IEEE International Conference on Acoustics, Speech and Signal Processing, pp. 3110–3114 (2021)
31. Liu, W., Chen, L., Chen, Y., Zhang, W.: Accelerating federated learning via momentum gradient descent. IEEE Trans. Parallel Distrib. Syst. **31**(8), 1754–1766 (2020)
32. Ullah, S., Kim, D.: Federated learning convergence on IID features via optimized local model parameters. In: International Conference on Big Data and Smart Computing, pp. 92–95 (2022)
33. Xu, J., Du, W., Jin, Y., He, W., Cheng, R.: Ternary compression for communication-efficient federated learning. IEEE Trans. Neural Netw. Learn. Syst. **33**(3), 1162–1176 (2022)
34. Cui, L., Su, X., Zhou, Y., Liu, J.: Optimal rate adaption in federated learning with compressed communications. In: Conference on Computer Communications, pp. 1459–1468 (2022)
35. Reisizadeh, A., Mokhtari, A., Hassani, H., Jadbabaie, A., Pedarsani, R.: Fedpaq: a communication-efficient federated learning method with periodic averaging and quantization. In: International Conference on Artificial Intelligence and Statistics, pp. 2021–2031 (2020)
36. Caldas, S., Konečny, J., McMahan, H.B., Talwalkar, A.: Expanding the reach of federated learning by reducing client resource requirements. arXiv preprint arXiv:1812.07210 (2018)
37. Paragliola, G.: Evaluation of the trade-off between performance and communication costs in federated learning scenario. Future Gener. Comput. Syst. **136**, 282–293 (2022)
38. Abasi, A.K., Aloqaily, M., Guizani, M.: Grey wolf optimizer for reducing communication cost of federated learning. In: IEEE Global Communications Conference 2022, pp. 1049–1154 (2022)

39. Li, T., Sahu, A.K., Zaheer, M., Sanjabi, M., Talwalkar, A., Smith, V.: Federated optimization in heterogeneous networks. Proc. Mach. Learn. Syst. **2**, 429–450 (2020)

40. Nguyen, V.-D., Sharma, S.K., Vu, T.X., Chatzinotas, S., Ottersten, B.: Efficient federated learning algorithm for resource allocation in wireless IoT networks. IEEE Internet Things J. **8**(5), 3394–3409 (2020)

41. Song, Q., Lei, S., Sun, W., Zhang, Y.: Adaptive federated learning for digital twin driven industrial internet of things. In: IEEE Wireless Communications and Networking Conference 2021, pp. 1–6 (2021)

42. Huang, W., Li, T., Wang, D., Du, S., Zhang, J.: Fairness and accuracy in federated learning. arXiv preprint arXiv:2012.10069 (2020)

43. Tan, L., et al.: Adafed: optimizing participation-aware federated learning with adaptive aggregation weights. IEEE Trans. Netw. Sci. Eng. **9**, 2708–2720 (2022)

44. Mohri, M., Sivek, G., Suresh, A.T.: Agnostic federated learning. In: International Conference on Machine Learning, pp. 4615–4625 (2019)

45. Prauzek, M., Mourcet, N.R., Hlavica, J., Musilek, P.: Q-learning algorithm for energy management in solar powered embedded monitoring systems. In: IEEE Congress on Evolutionary Computation 2018, pp. 1–7 (2018)

46. Wu, H., Wang, P.: Fast-convergent federated learning with adaptive weighting. IEEE Trans. Cogn. Commun. Network. **7**(4), 1078–1088 (2021)

47. LeCun, Y., Bottou, L., Bengio, Y.: Gradient-based learning applied to document recognition. Proc. IEEE **86**(11), 2278–2324 (1998)

48. Krizhevsky, A.: One weird trick for parallelizing convolutional neural networks. arXiv preprint arXiv:1404.5997 (2014)

49. Xiao, H., Rasul, K., Vollgraf, R.: Fashion-mnist: a novel image dataset for benchmarking machine learning algorithms. arXiv preprint arXiv:1708.07747 (2017)

50. Duan, M., et al.: Astraea: self-balancing federated learning for improving classification accuracy of mobile deep learning applications. In: International Conference on Computer Design, pp. 246–254 (2019)

51. Jiao, Y., Wang, P., Niyato, D., Lin, B., Kim, D.I.: Toward an automated auction framework for wireless federated learning services market. IEEE Trans. Mob. Comput. **20**(10), 3034–3048 (2020)

52. Yonetani, R., Takahashi, T., Hashimoto, A., Ushiku, Y.: Decentralized learning of generative adversarial networks from non-iid data. arXiv preprint arXiv:1905.09684 (2019)

53. Yoon, T., Shin, S., Hwang, S.J., Yang, E.: Fedmix: approximation of mixup under mean augmented federated learning. arXiv preprint arXiv:2107.00233 (2021)

An Adaptive Instruction Set Encoding Automatic Generation Method for VLIW

Xin Xiao and Zhong Liu[(✉)]

College of Computer, National University of Defense Technology, Changsha, China
{xiaoxin21,zhongliu}@nudt.edu.cn

Abstract. The tight integration of hardware and software enables very long instruction word (VLIW) architectures to vastly outperform superscalar architectures in performance. The performance of VLIW architectures largely depends on the careful design of their instruction sets. The design of VLIW instruction sets is an iterative process. In the instruction format design phase, issues such as encoding inefficiency, format adjustment, and decoding complexity may arise. In the evaluation and verification phase of the instruction set, the challenge of rapid modeling of iterative instruction set architectures (ISA) may emerge. To address the aforementioned issues, we propose an architecture description language. By automatically generating instruction formats, it solves problems such as encoding inefficiency, format adjustment, and decoding complexity. By generating encoding schemes, it provides an interface to solve the problem of rapid modeling of iterative ISA. This paper delineates this technology and evaluates it on the MT-3000 ISA. The experimental results demonstrate that it can automatically and expeditiously generate encoding schemes and optimize the issues in instruction format design. This can effectively accelerate the progress of the instruction set.

Keywords: VLIW instruction sets · architecture description language · Instruction Formats Design

1 Introduction

The very long instruction word (VLIW) architecture [4], an advanced parallel instruction set architecture, was proposed in 1980 s. It intelligently determines the parallelism of instructions at compile time using a parallel compiler and tightly packs multiple instructions into a single very long instruction. This highly integrated design of hardware and software enables the VLIW architecture to vastly outperform the scalar architecture. Therefore, the VLIW mechanism is more often used in mainstream high-performance processors, such as the vector processor Matrix developed by the National University of Defense Technology [2,15], TI's TMS320C64x [12], and ADI's TigerSHARC series [6].

The performance of the VLIW architecture heavily relies on the careful design of its instruction set, which needs to consider how to efficiently utilize the abundant hardware resources for highly parallel instruction scheduling to take full

© The Author(s), under exclusive license to Springer Nature Singapore Pte Ltd. 2024
Z. Tari et al. (Eds.): ICA3PP 2023, LNCS 14487, pp. 283–300, 2024.
https://doi.org/10.1007/978-981-97-0834-5_17

advantage of the computational power of the VLIW architecture. [13] There-fore, the design of the VLIW instruction set is arguably the most critical and challenging part of the VLIW architecture design process.

VLIW instruction set design is a cyclic and iterative process that requires continuous adjustment and optimization to meet the requirements of architec-ture and compilation efficiency. This process brings new challenges. Instruction format design has been a core component of VLIW instruction set research. The instruction format design of the VLIW instruction set has involved a trade-off between code density and hardware decoding ease. An overly complex instruc-tion format can express richer parallelism but is more difficult to decode. In contrast, a simple instruction format is easy to implement in hardware but can-not achieve higher code density, which limits the potential for instruction-level parallelism. Striking a balance between these two factors is key to the VLIW instruction set's instruction format design.

Issues in the instruction format design phase. Encoding inefficiency: Instruc-tions usually use longer opcode bit-width to represent more opcode types to accommodate more opcodes. However, actual programs usually only use some of these opcodes, resulting in wasted opcode bit-width. Format adjustment: Once an instruction format is designed, the maximum opcode parallelism it can express is fixed. If more opcode types or numbers are added later, it is challenging to continue expanding in the same format, and the entire instruction format needs redefining. Decoding complexity: Due to multiple iterations or for compatibility, a particular instruction type requires multiple bit-width with different coding fields at different locations to determine the instruction format, increasing the complexity of the subsequent software toolchain design and hardware decoder.

During the instruction set evaluation and verification phase, the instruction set needs to be considered in terms of compilation efficiency, throughput, code density, instruction space utilization, power and area, hardware implementation complexity, etc. Modeling the instruction set architecture (ISA) with the archi-tecture description language (ADL) can quickly generate a toolchain to evaluate the current version of the instruction set. The architecture designer uses the ADL to build an ISA model of the ISA description. Then the ISA model can be automatically generated by the language's compiler as either hardware descrip-tion language source code or high-level language source code for the software toolchain. Currently popular ADL include nML [3,16], ISDL [7,8], EXPRES-SION [9], LISA [11,18,20], and ArchC [1,17], etc. However, these ADL use a hard-coded way to describe the instruction format and encoding of the ISA. Once there are instruction additions, deletions, and instruction format adjust-ments, the ADL description file needs extensive re-modification, which seriously slows the instruction set iteration progress.

To address the aforementioned issues, we propose an architecture description language that can describe the instruction set encoding scheme requirements. It can automatically determine the shortest bit-width required for encoding in the instruction format design phase and adjust the instruction format according to requirement changes, effectively solving the encoding inefficiency and format adjustment problems. We unify the functions of encoding fields so that instructions in the instruction format analysis, instruction dispatch, and instruction semantic recognition phases all only rely on this. This reduces the hardware decoding logic and effectively solves the decoding complexity problem. In the evaluation and verification stage of the instruction set, we provide an interface on the generated encoding scheme. The ADL description model obtains the instruction format and encoding through signature retrieval, so the ADL description model does not need to be modified even if the instruction set iterates several times to adjust the format and encoding. This can effectively accelerate the evaluation and verification progress of the instruction set.

In summary, our main contributions are the followings:

- **Architecture description language for VLIW.** The current popular ADLs have a number of flaws and limitations. nML does not support VLIW; EXPRESSION focuses on architectural design space exploration for SoC and automatic generation of compiler or simulator toolkits; LISA has no public release compiler or toolkit; ISDL cannot model variable-length multi-cycle instructions; ArchC is designed for SystemC users; Also their descriptions of instruction sets are hard-coded and cannot be applied to fast iterations of instruction sets. We propose a VLIW-oriented ADL that takes a signature indexing approach to instruction encoding and can quickly adapt to the fast iteration of instruction sets.
- **Automatically generating instruction format.** We propose an encoding scheme requirement model that describes the format and number of instructions in the target instruction set and automatically generates either a fixed-length instruction format scheme or a variable-length instruction format scheme.
- **Instruction format classification algorithm and variable length encoding.** We propose 2 instruction format classification algorithms, which aim to group similar instruction formats into one category, reduce the number of instruction formats, and facilitate the generation of regular instruction format schemes. We propose a variable-length encoding detection algorithm for quickly detecting whether there is a variable-length encoding that meets the requirements. We propose 2 variable-length encoding generation algorithms, one for generating feasible variable-length encodings and one for generating optimal variable-length encodings that meet user requirements.

2   Related Work

Fisher et al. [5] first articulated three goals for VLIW instruction design: high instruction-level parallelism, simple hardware architecture, and compiler-friendly instruction sets. This laid the theoretical foundation for the development of the VLIW instruction format. Hennessy et al. [10] proposed four principles for VLIW instruction design: parallelism, orthogonality, simplicity, and code density. These principles provide the basic framework for VLIW instruction format design. The subsequent emergence of VLIW instruction sets, such as LILY [19] and MT-3000

[14], follow these design goals and principles, but we found no relevant research oriented towards the design of VLIW instruction set formats. The modeling of instruction sets by architecture description languages allows the rapid generation of corresponding tool chains for instruction set evaluation. However, there are many shortcomings and limitations in the current popular ADLs. nML [3] does not support VLIW; EXPRESSION [9] focuses on architectural design space exploration for soc and automatic generation of compiler or simulator toolkits; LISA [18] does not have a public release compiler or toolkit; ISDL [8] cannot model variable-length multi-cycle instructions; ArchC [1] is designed for SystemC users; Also their descriptions of instruction sets are hard-coded and cannot be applied to rapid instruction set iterations. Our proposed architecture description language models coding requirements and automatically generates instruction formats to avoid the problems of Encoding inefficiency, Format adjustment and Decoding complexity that arise during instruction set iteration.

3   Encoding Scheme Requirements Model

Automatic generation of an encoding scheme for VLIW requires knowledge of functional unit information, information about the instructions to which each functional unit belongs, and the required operation fields for each instruction. An encoding scheme requirements model is introduced to represent the requirements of the architecture designer. The encoding scheme requirements model contains **a declaration description** and **an encoding requirements description**.

The declaration description contains the instruction format scheme description, the function unit description and the encoding field description. In the instruction format scheme description, the keyword *Format* defines the instruction format, which has two values FIXED fixed-length instruction format and VARIABLE variable-length instruction format. The keyword *Length* defines the instruction format length, when *Length=auto* means automatically generating an instruction format length, when *Length=C* means generating a fixed-length instruction format scheme with instruction format length C or generating a variable-length instruction format scheme with instruction length C/2C. The keyword *InstructionFormat* defines the order of each encoding field of the instruction format. In the function unit description, the keyword *FunctionUnit* defines the declared function unit. In the encoding field description, the keyword *Field* declares local encoding fields, the keyword *GlobalField* declares global encoding fields, and the keyword *FiledSet* declares the set of defined encoding fields.

The Coding Requirement Description. The keyword *InstructionRequirementModel* defines the instruction requirement, which comprises multiple instruction requirement tuples. An instruction requirement tuple consists of four parts: the set of functional units, the set of encoding fields, the number of instructions, and the default instructions. The optimize function specifies N *FieldSet* to maximize the length of the opcode (OP) encoding fields.

A coding requirement scheme is shown below, it means: generate a 40/80 variable length instruction scheme with functional units ful, fu2, fu3 and fu4. contains 4 local coding fields R, IMM6, OR and AR0_7. Contains 3 global encoding fields REG, Z and P. Contains fs1, fs2 and fs3 a total of 3 encoding field sets. The instruction format field order is given by the keyword InstructionFormat. It is desired to maximize the OP bit-width of fs1. ful and fu2 have a total of N instructions each. fu3 and fu4 contain M instructions each.

```
Length= 40
FunctionUnit= ful, fu2, fu3, fu4
Filed R= 6, IMM6= 6, OR= 4, AR0_7= 3
GlobalField REG= 3, Z= 1, P= 1
FieldSet fs1= {R, R, R}
FieldSet fs2= {R, R, IMM6}
FieldSet fs3= {OR, AR0_7, R}
InstructionFormat= {REG, Z, LocalFieldSet, P}
Optimize( fs1 )
InstructionRequirementModel{
        {
                FunctionUnit= {ful, fu2},
                FiledSet= fs1,
                Number= N,
                Default= {instN1, instN2, instN3}
        },
        {

                FunctionUnit= {fu3, fu4},
                FiledSet= fs2,
                Number= M,
                Default= {instM1, instM2, instM3}
        }
}
```

4   Fixed-Length Instruction Format Automatic Generation Algorithm

Reads the encoding scheme requirement model and automatically generates the algorithm using the fixed-length instruction format when *Format=FIXED*. When *InstructionFormat* is not specified specifically, the default instruction format of VLIW has the following distribution of encoding fields: { Global Field Set (GFS), Local Field Set (LFS), Encoding Field OP, Encoding Field U, Encoding Field T }. The encoding field T corresponds to the local encoding field set one by one, the encoding field U corresponds to the functional unit one by one, and the encoding field OP corresponds to the instruction semantics one by one. The format encoding T is used to locate the LFS, the encoding field U is used to locate the functional unit, and the encoding field OP is used to locate the instruction.

Algorithm 1: Fixed-length instruction format generation algorithm

---

Input : Encoding scheme requirement model
Output: Fixed-length instruction format Encoding scheme

1 Count the number N of *LFS*
2 T=*BinLen(N)*
3 Calculate the value of *LFS.InstLen*
4 **if** *LFS.InstLen > Length* **then**
5 | return FAILURE
6 **else**
7 | **if** $BinLen(\|LFS_i.OP\|) <$
   | $BinLen(\|LFS_i.U\|) + MAX(BinLen(\|LFS_i.U_j.OP\|))$ **then**
8 | | *Use the field MIX to locate instructions and functional units*
9 | **else**
10 | | *Use the field OP and field U to locate instructions and functional units*
11 | **end**
12 | *Analyze field OP and field U to determine the positioning method*
13 | *Extend the length of $LFS_i.OP$ so that Instruction Length Equals*
   | *LFS.InstLen*
14 | *return SUCCESS*
15 **end**

---

where *LFS.InstLen* is the length of the current fixed-length instruction format, calculated by the Eqs. 1, 2.

$$\|LFS_i.LOU\| = \|LFS_i\| + \|LFS_i.OP\| + \|LFS_i.U\| \tag{1}$$

$$LFS.InstLen = \left\lceil \frac{\|GFS\| + MAX\left(\|LFS_i.LOU\|\right) + T}{8.0} \right\rceil * 8 \tag{2}$$

GFS is the global field set, *LFSi* is the i-th local field set, *LFSi.OP* is the set of instructions using the local field set, $\|LFSi.OP\|$ is the number of instructions using the local field set, *LFSi.U* is the set of functional units using the local field set, $\|LFSi.U\|$ is the number of functional units using the local field set, *BinLen* is the formula used to calculate the field bit-width, and the calculation formula is shown in Eq. 3.

$$BinLen(n) = \begin{cases} \lceil log_2 n \rceil & n > 1 \\ 0 & n = 0, 1 \end{cases} \tag{3}$$

The way to locate functional units and instructions is analyzed in step 7. Where $LFS_i.U_j.OP$ is the set of instructions that use LFS_i in functional unit $U_j.MIX$ Field is used to locate both functional units and instructions with one field.

5   Variable-Length Instruction Format Automatic Generation Algorithm

In the fixed-length instruction format automatic generation algorithm, we use the expansion of $LFS_i.OP$ to align the instruction length, but this also generates more free bits that are not used, thus increasing the Code Size of the program. a small number of long instructions in the instruction set makes the fixed-length instruction format very long. Some very long instructions are necessary due to the system architecture. For example, a long immediate assignment instruction with a register bit length of 64 must require a 64-bit immediate assignment instruction. The solution to this problem is to use a variable-length instruction format, where short instructions are aligned to N and long instructions are aligned to 2N length. To reduce the size of the program code, we want the variable-length instruction format length N to be as small as possible. In each field of the instruction format, GFS and LFS shortening would result in a lack of instruction functionality, so we can only explore Field OP, Field U and Field T. Field OP and Field U tightening can be achieved by step 7 of the fixed-length instruction format automatic generation algorithm, and Field T tightening can be achieved by a variable-length encoding algorithm.

5.1   Variable Length Coding Detection Algorithm

We propose an algorithm to quickly detect the existence of a set of variable-length codes 2. The most important requirement of variable length encoding is that a certain code word cannot be a prefix of other code words, and a code of length n can generate 2 codes of length n+1. Assume that a container holds the length of the variable-length code. If there are 2 equal elements t1, t2 in the container, remove these 2 elements and add t1-1 until there are no 2 equal elements in the container. If the container is equal to {0}, or if the elements of the container are all unequal positive integers, then a variable-length encoding exists that satisfies the current requirement.

5.2   Tree-Based Variable Length Coding Generation Algorithm

The variable length encoding detection algorithm is used to verify the existence of a set of variable length encodings, and then we propose a tree-based variable length encoding generation algorithm. Firstly, several core concepts need to be explained, encoding binary tree, encoding vector, encoding binary tree solution space CT_H^N and encoding binary tree solution space solving.

Encoding Binary Tree. An binary tree has 2 subtrees for all non-leaf nodes, then the tree is an encoding binary tree. **Encoding vector**: The path from the root node to all leaf nodes, ordered by path length, is called the vector corresponding to the encoding vector of the encoding binary tree.**The encoding binary tree solution space CT_H^N**: refers to the set consisting of all encoding

---

Algorithm 2: Variable-length encoding detection algorithm

Input : A container S contains the length of each variable-length encoding
Output: Can variable-length encoding be generated
1 while *S contains two equal elements t_1 and t_2* **do**
2 | $S - \{t_1, t_2\}$;
3 | $S \cup \{t_1 - 1\}$;
4 end
5 if $S == \{0\}$ **then**
6 | return TRUE
7 end
8 if *All elements in S are positive integers* **then**
9 | return TRUE
10 else
11 | return FALSE
12 end

---

vectors with the number of N components of the encoding binary tree with layer height H.

The relationship between γ_m and CT_H^N Given a $\gamma_m = [r_1, r_2, ..., r_m], 0 \le i < j \le m, r_i \le r_j$ and CT_H^N is encoding binary tree solution space.

- If $m \ne n$, then γ_m does not satisfy CT_H^N.
- If $m = n$, $\exists \xi_i = [a_{i1}, a_{i2}, ..., a_{im}]^T \in CT_H^N$ $s.t.$ $r_k \ge \|a_{ik}\|, 1 \le k \le m$, then γ_m satisfy CT_H^N , else γ_m does not satisfy CT_H^N.

Encoding Binary Tree Pruning Operations. The encoding binary tree can achieve a 1-bit space saving by reducing the ability of one encoding mapping. Prunable node: the parent of a leaf node and all the children of that node are leaf nodes. Pruning operation: remove all children nodes of the prunable node. CT_H^N solving algorithm

- If $n > 2^H$,then $CT_H^N = \emptyset$
- If $n = 2^H$,then $CT_H^N = \{[0...0, ..., 1...1]^T\}$
- If $n < 2^H$,then $CT_H^N \overset{Pruning}{\to} CT_H^{N-1}$
- If Prunable nodes in CT_H^N are empty , then $CT_H^{N-1} = \emptyset$

Variable Length Encoding Solving. Generate an m vector by sequentially generating the length of each variable length encoding to be solved, where the number of elements of m is n, and the maximum value of the elements is H.If $CT_H^N = \emptyset$, then there is no solution; if m does not satisfy $CT_H^N = \emptyset$, then there is no solution; if m satisfies $CT_H^N = \emptyset$, the variable length encoding is the encoding vector corresponding to m.

5.3   Field Sets Fully Compatible Algorithm

The length of the format encoding field T in the Fixed-length instruction format generation algorithm is determined by the number of FieldSet. If similar FieldSet are treated as the same kind and the length of the format encoding field T is determined by the kind of FieldSet instead, then the value that determines the length of the format encoding field T will be significantly reduced, and thus the length of the format encoding field T will also be shortened. The core concept of this algorithm is how to distinguish two FieldSets as compatible and find m FieldSets, which are compatible with all the remaining (N-m) FieldSets. the smaller m is, the shorter the format encoding field T is likely to be.

The Core Concepts of Compatible. Given FieldSetA=$\{A_1, A_2, ..., An\}$ and FieldSetB=$\{B_1, B_2, ..., Bn\}$.

- If A_i and B_j are the same, then A_i and B_j are compatible.
- If A_i and B_j have the same length, then A_i and B_j are compatible.
- If the length of field A_i is greater than the length of field B_j, then A_i is not fully compatible with B_j, and B_j is incompatible with A_i.
- If $\forall B_j \epsilon$ FieldSetB , $\exists A_{aj} \epsilon$ FieldSetA $s.t.$ A_{aj} is fully compatible with B_j, and FieldSetA=$\{A_{a1}, ..., A_{am}\}$, then FieldSetA fully compatible with FieldSetB.
- If $\forall B_j \epsilon$ FieldSetB , $\exists A_{aj} \epsilon$ FieldSetA $s.t. A_{aj}$ is fully compatible with B_j or A_{aj} is not fully compatible with B_j , and FieldSetA=$\{A_{a1}, ..., A_{am}\}$, then FieldSetA is not fully compatible with FieldSetB.
- If $\exists B_j \epsilon$ FieldSetB , $\exists A_{aj} \epsilon$ FieldSetA $s.t. A_{aj}$ is not compatible with B_j, then FieldSetA is not compatible with FieldSetB.

The input of the fully compatible algorithm is the FieldSet to be classified, and the output is the classified FieldSets. a compatibility table is generated inside the algorithm, and the different positions of the compatibility table represent the relationship of different FieldSets, if Compatible[i][j]==1, it means FieldSeti is fully compatible with FieldSetj; if Compatible[i][j]==2, FieldSeti is not fully compatible with FieldSetj; if Compatible[i][j]==0, FieldSeti is not compatible with FieldSetj.

5.4   Field Sets Not Fully Compatible Algorithm

When the number of mapping states required for format encoding field T is reduced, then the maximum bit-width of format encoding field T may be reduced, and it is possible to give the spare bits to the semantic encoding field OP, and then more instructions can be supported. We propose a Field Set Not Fully Compatible (NFC) algorithm, which aims to further reduce the variety of encoding field sets and the number of mapping states required for a format encoding field T.

One encoding field set can be not fully compatible by multiple encoding field sets, so a very wide variety of classification results can be produced. In order to produce better classification results, we propose the principle of not fully

---

Algorithm 3: Feild Set Fully Compatible

    Input : FieldSets to be classfied
    Output: classfied FieldSets

```
1  for i ← 0 to N do
2  │   for j ← i + 1 to N do
3  │   │   if FullyCompatible(Field_i,Field_j) then
4  │   │   │   Compatable[i][j] ← 1;
5  │   │   else
6  │   │   │   if NotFullyCompatible(Field_i,Field_j) then
7  │   │   │   │   Compatable[i][j] ← 2;
8  │   │   │   else
9  │   │   │   │   Compatable[i][j] ← 0;
10 │   │   │   end
11 │   │   end
12 │   end
13 end
```

---

compatibility classification: the encoding field set being classified is generally selected as the encoding field set with a small number of instructions and unique functional units; the functional units of the encoding field set being classified are a subset of the functional units of the encoding field set being classified to; If the encoding field set A is not fully compatible classified into encoding field set B, the instruction length of encoding field set B is better not be affected.

5.5   Priority Queue-Based Variable-Length Code Generation

In the process of designing the instruction set, some instructions rarely change, such as access instructions, and some instructions change a lot, such as double-operand instructions and triple-operand instructions. Therefore, there is a need to reserve the maximum instruction space for one or more instruction formats. The tree-based variable length coding generation algorithm can only generate a set of variable length codes to meet the demand, and it cannot handle this demand.

If $LFSi.OP$ is required to be maximum, i.e., let $LFSi.T$ be minimum, reduce $LFSi.T$ in order to find a minimum $LFSi.T$ that makes the variable length coding detection algorithm to be true.

5.6   Variable-Length Instruction Format Automatic Generation Algorithm Implementation

The input of the algorithm is the encoding scheme requirement model and the output is the variable-length instruction format encoding scheme. First, the requirement model is parsed and the number of FieldSet is reduced by FC and NFC algorithms. Then calculate the length of the format field T of each Field-Set, if a variable-length code satisfying the variable-length instruction format

---

Algorithm 4: Variable-length encoding generation based on priority queue

---

1 Input:The length of each variable-length encoding is stored in the request container;
2 Output:The result of variable-length encoding is stored in the result container;

3 Sort (result)
4 pq = priority_queue({"0", "1"})
5 **for** $i \leftarrow 0$ **to** $request.size()\text{-}1$ **do**
6     $prefix = pq.front()$ **if** $\|prefix\| == request[i]$ **then**
7        $result.push(prefix)$
8   **else**
9        $n = request[i] - \|prefix\|$
10   Postfixs=**GenerateSequence**(n)
11   code=**Concatenate**($prefix, Postfixs$)
12       $result.push(code[0])$
13       $code = code - code[0]$
14       $pq = pq \cup code$
15   **end**
16 **end**

---

is found, then continue, otherwise adjust a short instruction to a long instruction and recalculate the length of the format field T. If all instructions are long instructions, the generation of the variable-length instruction format encoding scheme fails. If the Optimize keyword is present in the encoding scheme requirement model, the priority queue-based variable length encoding generation algorithm is used, otherwise the tree-based variable length encoding generation algorithm is used.

6 Experiment

6.1 Evaluation Indicators

Field_T: The minimum number of fields required to identify the instruction format. Field_FU: The minimum number of fields required to encode the instruction function unit. Field_OP: The minimum number of fields required to encode the corresponding operation of the instruction. Bit_T: The minimum number of bits required to identify the instruction format. Bit_FU: The minimum number of bits required to encode the instruction function unit. Bit_OP: The minimum number of bits required to identify the operation of the instruction. Length: The length of the instruction. Extended_OP: The maximum width to which the operation (OP) field of the current instruction type can be increased while the length of the format field of other instruction types remains unchanged.

6.2 Current MT-3000 Encoding Scheme

As shown in Table 1 [14], because single and double operand instructions and scalar vector non-long immediate offset LDST instructions share a common for-

---

Algorithm 5: Implementation of variable-length instruction format automatic generation

---

Input   : Encoding scheme requirement model
Output: variable-length instruction format

1 Parsing model Classify the FieldSet using the FC algorithm Classify the FieldSet using the NFC algorithm

2 **while** *TRUE* **do**

3   *foreach* FS_i *generate* t_i **if** *detection algorithm(T)* **then**

4   break

5   **else**

6   **if** *All* $FS_i.t{==}2N$ **then**

7   return FALSE

8   **else**

9   find t_i which $FS_i.t{==}N$ set $FS_i.t{=}2N$

10   **end**

11   **end**

12 **end**

13 **if** *model contains the keyword Optimize* **then**

14   For the FieldSet that needs to be optimized, continuously reduce the length of their corresponding t, and generate the encoding field t using the priority queue-based variable-length encoding generation algorithm, until all the corresponding t cannot be reduced.

15 **else**

16   Generate the encoding field T using the tree-based variable length encoding generation algorithm.

17 **end**

---

mat encoding field T, 3 encoding field T, U and V are required to determine the format of single and double operand instructions. The scalar vector non-long immediate offset LDST instruction also contains 2 instruction formats, which need MODE encoding field to distinguish, so at least 4 encoding fields T, U, V, and MODE are required to determine the instruction format. All instruction formats require U and V to determine the function unit, except for the long immediate branch instruction which requires only V to determine the function unit. The long immediate MOV instruction 40 bits and the long immediate MOV instruction 80 bits require 2 coding fields, U and V, to determine exactly what operation to perform. The scalar vector non-long immediate offset LDST instruction requires OP and MODE 2 encoding fields to determine exactly what operation. The original MT-3000 instruction set instruction format, single and double operand instructions and scalar vector non-long immediate offset LDST instructions share a common format encoding field increases the logical complexity of identifying instructions; in the scalar vector non-long immediate offset LDST instruction MODE field is used to identify both the instruction format and the type of operation, increasing the logical complexity of identifying operations.

Table 1. Current MT-3000 encoding scheme

Instruction Type	Field_T	Field_FU	Field_OP	Bit_T	Bit_FU	Bit_OP	Len
Long imm MOV (40 bits)	1	2	2	2	3	0	40
Long imm MOV (80 bits)	1	2	2	2	3	0	80.
Long imm branch	1	1	1	3	1	2	40
Single and double operand	3	2	1	7	4	10	40
Scalar or vector non-long imm LS	4	2	2	11	4	8	40
Three operand	1	2	1	3	3	5	40
Scalar or vector long imm LS	1	2	1	3	2	4	40

6.3   Fixed-Length Instruction Format Experiment

Table 2. Fixed-length instruction format - no classification algorithm used

Instruction type	Field_T	Field_FU	Field_OP	Bit_T	Bit_FU	Bit_OP	Len
FieldSet1	1	1	1	5	*	*	88
...	...	...	...	...	...	...	...
FieldSet23	1	1	1	5	*	*	88

The Table 2 shows the encoding scheme of the fixed-length instruction formats generated according to the encoding scheme requirement model. It can be seen that the automatically generated instruction format is very regular, and the Field_T, Field_FU and Field_OP indicators of all instruction formats are **only one**. Due to the existence of long immediate assignment instructions, the length of GFS+LFS+OP+U+T has exceeded 80 bits, so it can only be aligned to 88 bits according to bytes. Although the instruction format is simple and the decoding logic is simple, most of the instructions have wasted coding space

6.4 Classification Algorithm Experiment

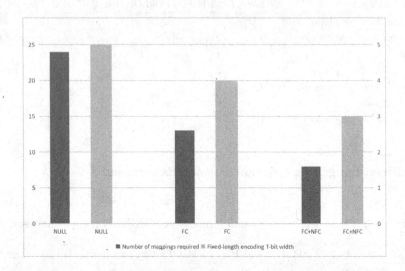

Fig. 1. Classification algorithm experiment

As shown in Fig. 1, parsing the encoding scheme requirement model, when the FieldSet classification algorithm is not used, there are 23 different FieldSet, and the fixed-length format encoding field T bit-width needs at least 5 bits. NULL means that no classification algorithm has been used. After using the FC algorithm, the FieldSet is classified into 13 FieldSets in total, and the fixed-length format encoding field T bit-width requires at least 4 bits. After continuing to use NFC to categorize FieldSet, it is finally divided into 8 types of FieldSet, which makes the fixed-length format encoding field T bit-width at least 3 bits. After the classification algorithm, the length of T is **drastically reduced.**

6.5 Different Instruction Formats Combined with Experiments on Classification Algorithms

Table 3. Different instruction length and different classification algorithms

Instruction length	classification algorithms	Instruction format length
fixed	null	88
fixed	FC	88
fixed	FC+NFC	88
Variable	null	40/80
Variable	FC	40/80
Variable	FC+NFC	40/80

As shown in Table 3, the encoding scheme for generating fixed-length instruction formats, even with FC or FC+NFC algorithms, only aligns the instruction format length to 88 bits. However, when the format field T is encoded with variable length, the instruction format length can be aligned to 40/80 even without the FieldSet classification algorithm, effectively reducing the instruction length and thus the Code Size of the program.

6.6 Variable-Length Instruction Formats Experiment

Table 4. variable-length instruction formats experiment

Instruction Type	Field_T	Field_FU	Field_OP	Bit_T	Bit_FU	Bit_OP	Len	Extended_OP
SBR long imm	1	1	1	4	2	2	40	2
Scalar vector OR LS	1	1	1	12	2	7	40	10
Scalar vector UCST10 LS	1	1	1	5	2	7	40	3
Scalar vector UCST18 LS	1	1	1	3	2	4	40	2
Single and double operand	1	1	1	7	3	7	40	5
Three-operand	1	1	1	4	3	4	40	2
MOV64	1	1	1	2	3	3	80	1
MOV24	1	1	1	4	1	1	40	2

As shown in the Table 4, it can be seen that the automatically generated instruction formats are very well organized, with Field_T, Field_FU and Field_OP indicators of all instruction formats being 1. Extended OP is the maximum width that the OP field of the current instruction type can continue to increase while the length of the format field of other instruction types remains the same. The extended OP shows that the OP encoding field of each instruction type can continue to be lengthened, increasing the number of instructions that the current instruction type can hold.

6.7   Optimize Keyword Tags Multiple FieldSets

Table 5. Optimize keyword tags multiple FieldSets

Instruction Type	Bit_OP1	Bit_OP2	Bit_OP3	Bit_OP4	Bit_OP5	Bit_OP6
SBR long imm	2	2	2	2	2	2
Scalar vector OR LS	7	7	7	7	7	7
Scalar vector UCST10 LS	7	7	7	7	7	7
Scalar vector UCST18 LS	4	4	4	4	4	4
Single and double operands	7	**12**	7	**12**	**11**	**10**
Three-operand	4	4	**6**	5	6	5
MOV64	3	3	3	3	3	3
MOV24	1	1	1	1	1	1

As shown in Table 5, from a practical development point of view, we know that some instruction types hardly add or delete, and the instruction types that add or delete more instructions are the few instruction types. For example, single and double operand instructions or triple operand instructions. If we mark these three types of instructions in the encoding requirement scheme with the Optimize keyword, then multiple instruction schemes will be generated for us to choose. If we expect the most single- and double-operand increment-delete instructions, then we can choose the Bit_OP2 scheme. If we expect to have the most triple-operand increment and decrement instructions, then we can choose the Bit_OP3 scheme. If we want to be more balanced, we can choose from Bit_OP4, Bit_OP5 and Bit_OP6.

7   Conclusion

In this paper, we propose a fixed-length instruction format automatic generation algorithm and a variable-length instruction format automatic generation algorithm for VLIW. Regarding variable-length encoding, we propose a variable-length encoding detection algorithm, a tree-based variable-length encoding generation algorithm, and a priority queue-based variable-length encoding generation algorithm. To reduce the number of mappings required for the format encoding field T, we classify the instruction formats into categories and propose two algorithms, FC and NFC, to classify the FiledSet. The experimental results show that the fixed-length instruction format automatic generation algorithm can generate regular instruction formats, and the FC and NFC algorithms can effectively classify the FieldSet and greatly reduce the variety of instruction formats. The variable-length instruction format automatic generation algorithm can generate regular instruction formats and effectively reduce the logic of decoding instructions. 2 variable-length coding generation algorithms can generate instruction formats that meet the needs of designers and increase the number of instructions that can be accommodated by a certain instruction. It solves the problems of coding waste, format adjustment and decoding complexity that occur in the instruction format design stage. Since the current architecture description language does not support the acquisition of instruction formats and encodings by indexing, in future work, we hope to automate the generation of subsequent required tool chains based on our current architecture description language.

Acknowledgment. This work is supported by NUDT Research Project (No. 23-ZZCX-JDZ-11), PDL Research Project (No. 2021-KJWPDL-11), Key Laboratory of Advanced Microprocessor Chips and Systems.

References

1. Azevedo, R., Rigo, S., Bartholomeu, M., Araujo, G., Araujo, C., Barros, E.: The archc architecture description language and tools. Int. J. Parallel Prog. **33**, 453–484 (2005)
2. Chen, S., et al.: Ft-matrix: a coordination-aware architecture for signal processing. IEEE Micro **34**(6), 64–73 (2013)
3. Fauth, A., Van Praet, J., Freericks, M.: Describing instruction set processors using nML. In: Proceedings the European Design and Test Conference. ED&TC 1995, pp. 503–507. IEEE (1995)
4. Fisher, J.A.: Very long instruction word architectures and the eli-512, pp. 140–150 (1983)
5. Fisher, J.A.: Very long instruction word architectures and the ELI-512. In: Proceedings of the 10th Annual International Symposium on Computer Architecture, pp. 140–150 (1983)
6. Fridman, J., Greenfield, Z.: The tigersharc dsp architecture. IEEE Micro **20**(1), 66–76 (2000)
7. Hadjiyiannis, G., Hanono, S., Devadas, S.: ISDL: an instruction set description language for retargetability. In: Proceedings of the 34th Annual Design Automation Conference, pp. 299–302 (1997)
8. Hadjiyiannis, G., Russo, P., Devadas, S.: A methodology for accurate performance evaluation in architecture exploration. In: Proceedings of the 36th Annual ACM/IEEE Design Automation Conference, pp. 927–932 (1999)
9. Halambi, A., Grun, P., Ganesh, V., Khare, A., Dutt, N., Nicolau, A.: Expression: a language for architecture exploration through compiler/simulator retargetability. In: Proceedings of the Conference on Design, Automation and Test in Europe, pp. 100-es (1999)
10. Hennessy, J.L., Patterson, D.A.: Computer Architecture: A Quantitative Approach. Elsevier, Boston (2011)
11. Hoffmann, A., et al.: A novel methodology for the design of application-specific instruction-set processors (ASIPS) using a machine description language. IEEE Trans. Comput. Aided Des. Integr. Circuits Syst. **20**(11), 1338–1354 (2001)
12. Inc., T.I.: Tms320c64x/c64x+ DSP CPU and instruction set reference guide (2010)
13. Jordans, R.: Instruction-set architecture synthesis for vliw processors. Elect. Eng., Embedded Syst. Group, Eindhoven Univ. Technol., Eindhoven, The Netherlands (2015)

14. Lu, K., et al.: Mt-3000: a heterogeneous multi-zone processor for HPC. CCF Trans. High Perform. Comput. **4**(2), 150–164 (2022)
15. Institute of Microelectronics: School of Computer Science. Marix DSP architecture manual, N.U.o.D.T. (2013)
16. Rajesh, V., Moona, R.: Processor modeling for hardware software codesign. In: Proceedings Twelfth International Conference on VLSI Design. (Cat. No. PR00013), pp. 132–137. IEEE (1999)
17. Rigo, S., Araujo, G., Bartholomeu, M., Azevedo, R.: Archc: a systemc-based architecture description language. In: 16th Symposium on Computer Architecture and High Performance Computing, pp. 66–73. IEEE (2004)
18. Schliebusch, O., Hoffmann, A., Nohl, A., Braun, G., Meyr, H.: Architecture implementation using the machine description language lisa. In: Proceedings of ASP-DAC/VLSI Design 2002. 7th Asia and South Pacific Design Automation Conference and 15h International Conference on VLSI Design, pp. 239–244. IEEE (2002)
19. Shen, Z., He, H., Yang, X., Jia, D., Sun, Y.: Architecture design of a variable length instruction set vliw DSP. Tsinghua Sci. Technol. **14**(5), 561–569 (2009)
20. Yang, S., Qian, Y., Tie-Jun, Z., Rui, S., Chao-Huan, H.: A new hw/SW co-design methodology to generate a system level platform based on lisa. In: 2005 6th International Conference on ASIC, vol. 1, pp. 163–167. IEEE (2005)

FastDet: Detecting Encrypted Malicious Traffic Faster via Early Exit

Jiakun Sun, Jintian Lu, Yabo Wang, and Shuyuan Jin[✉]

Sun Yat-Sen University, Guangzhou, China
{sunjk3,lujd6,wangyb58}@mail2.sysu.edu.cn,
jinshuyuan@mail.sysu.edu.cn

Abstract. Encrypted malicious traffic detection, which aims to identify encrypted malicious traffic from vast amounts of network traffic, is critical to network security. Existing detection techniques are difficult to improve detection speed to meet the needs of practical applications while ensuring high detection rates. This paper proposes a fast detection approach - FastDet, to detect encrypted malicious traffic. Based on the observation that the most of the network traffic is benign, FastDet designs an early exit mechanism in the detection, resulting in a significant increase in average detection time. The experimental results on four datasets show that FastDet achieves significant efficiency while maintaining comparable detection accuracy. The paper further illustrates the effectiveness of FastDet by discussing the characteristics of encrypted benign and malicious traffic samples.

Keywords: Encrypted malicious traffic detection · Early exit · Fastformer

1 Introduction

Encrypted malicious traffic detection, which aims to identify different types of encrypted malicious traffic from vast amounts of network traffic, is a critical technique to network security. Most existing traffic detection methods have difficulties in detecting encrypted malicious network traffic. The latest deep learning based detection methods have achieved notable detection performance in experimental environments, but in practical applications, these methods cannot avoid facing the problem of increasing computational costs caused by processing large amounts of network traffic. How to design an encrypted malicious traffic detection with a high detection accuracy and low computational overhead is a challenge.

Many works in the field of encrypted malicious traffic detection have been proposed in the literature. Early works [1], used fingerprint features extracted from the remaining plaintext in encrypted traffic to detect malicious traffic. These methods have gradually become ineffective with the evolution of encryption techniques. Some studies [2,3] employ machine learning algorithms to detect encrypted malicious network traffic using manually extracted features based on expert knowledge. The detection performance of these methods is greatly influenced by the features used. To eliminate the effects of feature selection on detection performance, deep learning-based detection methods [4,5] have utilized DNNs (Deep Neural Networks) to extract deep features in encrypted traffic, resulting in their excellent performance. Nevertheless, However, the deep learning based detection methods require a large amount of computational resources, resulting in their detection speed being much slower than non deep learning based detection methods.

In recent years, the early exit mechanism has made some progress in accelerating learning models. The early exit mechanism allows "easy" samples to terminate inference early so that the average inference time for a whole dataset can be reduced. Specifically, for a DNN, an "easy" sample refers to a sample where the features learned in the shadow layers of the DNN are enough to make an inference. As a result, the easy samples can exit the inference network in advance. A "hard" sample refers to a sample that needs to go through the entire DNN before making an inference.

This paper proposes a novel detection approach based on the early exit mechanism, called FastDet. The main idea of FastDet is to enable benign samples to exit the network as quickly as possible. Note that most encrypted network traffic is benign rather than malicious, therefore FastDet greatly improves detection speed and saves computation resources. FastDet comprises four layers: preprocessing layer, coarse-grained learning, early exit layer, and Fastformer-based multi-classification. Firstly, we propose a traffic embedding method in the preprocessing layer, which converts a traffic sample into a vector representation to serve as the input for model training. Secondly, the coarse-grained learning layer learns the shallow feature representation by repeatedly downsampling labeled encrypted traffic. Thirdly, based on the output of the coarse-grained learning layer, the cascade early exit layer determines whether the sample is an easy sample or not. If the sample is easy, the cascade early exit layer will give a conclusion that the sample is benign or malicious. Otherwise, the sample will enter the Fastformer-based multi-classification layer. Finally, the multi-classification layer utilizes Fastformer to classify the samples that are not classified in the previous layer (i.e. hard samples).

This paper makes the following main contributions:

1) FastDet designs an early exit mechanism, which overall shortens the detection time of malicious encrypted traffic and meets the needs of practical applications.
2) FastDet proposes a budget-constrained training strategy that reduces the computational cost of encrypted malicious traffic detection under budget constraints.
3) The experimental results show that FastDet significantly improves detection speed while maintaining a high detection rate. We further illustrate the effectiveness of FastDet by discussing the characteristics of encrypted benign and malicious traffic samples.

The remainder of this paper is as follows: In Sect. 2, we discuss related work. In Sect. 3, we introduce FastDet in details. In Sect. 4, we present and discuss the experimental results and comparative analysis. Finally, we summarize the paper, and future work is provided.

2   Related Work

2.1   HTTPS Background

HTTPS (Hyper Text Transfer Protocol over Secure socket layer) guarantees communication security via data encryption. The HTTPS process consists of three phases: the handshake phase, keys exporting phase, and data transportation phase.

During the handshake phase, Client C sends a list of supported encryption methods along with its unique, non-repetitive number. Server S selects a symmetric algorithm, a public key algorithm, and a MAC algorithm from the received list. Additionally, S sends back the selected algorithms along with a certificate and its own unique, non-repetitive number. C verifies the certificate, extracts the public key from the message to generate a pre-master secret (PMS), encrypts the PMS with the public key, and sends it back to S. From the PMS and a decryption method, C and S generate the master secret (MS), respectively, which is then split to create two ciphers and two MAC keys. Finally, C and S send MAC addresses from all handshake messages. The handshake phase is illustrated in Fig. 1.

In an HTTPS session, the majority of the plaintext information is transmitted during the handshake phase, while the remainder of the session is highly secure and encrypted, making it difficult to be deciphered. As such, this component does not require consideration within our study and will not be reiterated.

2.2   Encrypted Malicious Traffic Detection

Encrypted malicious traffic detection aims at detecting encrypted malicious traffic from massive network traffic, different from encrypted traffic classification, researchers are more interested in malicious sample types (e.g., malware traffic, attacks via HTTPS) than benign ones (e.g., app traffic, web service).

Machine Learning-Based Methods. Most machine learning-based studies on detecting encrypted malicious traffic can be divided into two steps. The first step is manually extracting features from raw traffic using expert knowledge, such as certificates and encryption algorithms, which are plaintext during the handshake phase [1,2]. The second step involves training a machine learning model based on the extracted features, which can be a random forest, Markov [6,7], Long Short-Term Memory (LSTM) [1], or other methods. Some studies also extract statistical features from the raw traffic, such as packet length, packet arrival time, and other contextual information, which have yielded promising

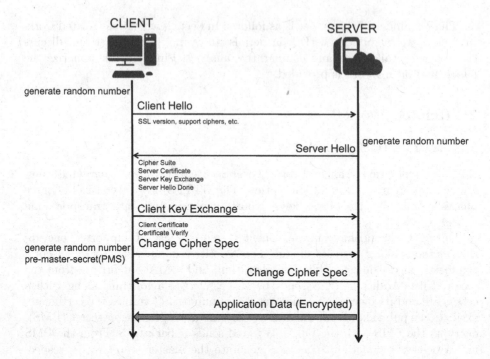

Fig. 1. Handshake Phase of HTTPS

results. However, these studies heavily rely on the manual design of features, and different expert knowledge used in training data can result in different feature selections that significantly impact the model's performance. Poor feature selection can seriously affect detection accuracy or lead to other issues, such as over-fitting.

Deep Learning-Based Methods. To address the issues mentioned above, deep learning-based methods have been proposed to extract discriminative features automatically using DNNs. For instance, Lotfollahi et al. [8] employed a 1D-CNN and Stacked AutoEncoder (SAE) to detect encrypted malicious traffic by using raw network flow as input, achieving a recall rate of 98%. In another study, Prasse et al. [4] used the raw traffic in the HTTP handshake phase to detect malware traffic using the LSTM model. Furthermore, Zeng et al. [5] proposed a Deep-Full-Range model to detect encrypted malicious traffic using CNN, LSTM, and SAE. Tong et al. [3] proposed a new bidirectional flow sequence network called BFSN based on LSTM. It only uses the long sequence of packets in the direction of encrypted traffic as input and outperforms CNN networks as verified through experimental verification. Dong et al. [9] combined packet payload and payload statistics in encrypted traffic to classify encrypted traffic. They introduced a comprehensive traffic analysis framework, CETAnalytics, for analyzing traffic information. Despite their high accuracy, deep learning models contain a large number of parameters and require significant computation resources, resulting in a trade-off between accuracy and detection speed, making it challenging to apply them in practice. In addition, the training process of deep learning methods is less interpretable, and the complexity of model convergence can lead to instability in detection performance.

2.3 Early Exit Mechanism

While adding more hidden layers to Deep Neural Networks (DNNs) can enhance model performance in classification tasks, it is not always the most efficient solution. Increasing the depth of the network requires more computation resources, which in turn lengthens the inference time. Additionally, for some cases where the classification task is easy and the input data has simple features, the use of deep layers results in excessive computation that is wasteful. Instead, an early exit mechanism has been developed to alleviate these concerns. This mechanism generates decisions based on shallow features to determine whether the output of the DNN model should terminate early and bypass subsequent, deeper off-ramps. By doing so, the early exit mechanism reduces redundant computation and improves the efficiency of the DNN model's inference time.

Cascade architecture is a simple technique to deepen networks and produce multiple decisions for determining whether a sample should be processed further. This approach adaptively directs "hard" samples beyond the exit threshold of shallow networks towards deeper ones in the cascade [10, 11]. To avoid computing waste caused by redundant features in shallow networks, intermediate classifiers can be added to the backbone network. This approach is used in cascade networks, which are mutually independent, and in branch networks, where features can be propagated to deeper layers [12–14].

Furthermore, chain-structured networks lacking global information in deep layers, and early classifiers would generate task-specialized features, bringing out performance degradation. The multi-scale dense network (MSDNet) [15] addresses this problem by utilizing multi-scale architecture and dense connections, resulting in improved accuracy across all classifiers in the network. To accelerate the model and construct an adaptive dynamic deep neural network, an effective approach is the use of early exit mechanisms, which are commonly applied in NLP, CV, and other fields.

3 FastDet

In this paper, we aim to detect encrypted malicious traffic fleetly in "real world" setting, which differs significantly from the experiment scenario. According to encrypted traffic data statistics, a large proportion of benign samples and a few malicious samples with various types consist of an unbalanced dataset. To effectively leverage the early exit mechanism for model acceleration, we mainly propose four main layers in FastDet as shown in Fig. 2.

1) Preprocessing layer transforms encrypted traffic into vectors as input.

2) Coarse-grained learning layer is proposed to generate shallow features with cascade downsampling operations.

3) Early exit layer, consisting of multiple early-exit units, is proposed to make an exit decision for samples that get the intermediate confidence exceed the preset threshold. Each unit is densely connected to all preceding layers to reuse shallow features.

4) Multi-classification layer consists of Fastformer blocks. Each block is composed of multi-head self-attention layers, which capture the implicit relationships between the encoded traffic units.

Each of the above layers is detailed in the following subsections. Additionally, we propose a budgeted-constrained training strategy, that yields less computing cost for "easy" samples, to obtain the minimized average inference time.

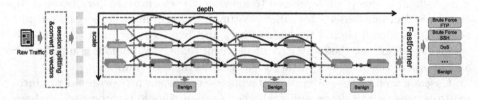

Fig. 2. Overview of FastDet.

3.1   Preprocessing Layer

In the preprocessing layer, raw encrypted traffic is embedded into normalized vectors to speed up inference time and reduce the impact of feature selection. In real-world scenarios where TLS sessions involve bidirectional flows between clients and servers that last for a while, waiting for the entire session to end before taking it as input would negatively affect detection efficiency. Additionally, taking the entire session as input in TLS sessions without decryption would result in model convergence difficulty by the disordered sequences of encrypted data, further reducing performance. Previous research [16] has shown that packets in the handshake phase contain more valuable information than subsequent packets. Therefore, we select packets from the head of traffic for input.

To begin with, we extract sessions from raw encrypted traffic using the 5-tuple in network traffic, comprising source IP, destination IP, source port, destination port, and transport layer protocol. As this study is focused on encrypted network traffic, we specifically analyze traffic utilizing the TLS/SSL protocol. A successful TLS/SSL session involves bidirectional flows between clients and servers, spanning from connection build to release, and would be taken as a sample for valid input. Other sessions that terminate unexpectedly will be disregarded as invalid data, and no further analysis will be conducted.

Secondly, while sessions can be directly converted to vectors by adopting packets as input, an extra operation is required for variable-length packets to maintain the inference process. For each session, we select the first B bytes of sequence data where B is a hyperparameter defined prior to model training and set to 1024 in our experiments. The effects of different values of B will be compared in Sect. 4.5. To construct fixed-length inputs, we pad with zeros if the packet length is less than B and truncate if it exceeds B. We also replace the Server Name Indication (SNI) fields in packets with zeros to prevent their effects. Then every valid traffic can be transformed into a 1024-byte length data and encoded such that each byte corresponds to an integer value between 0 and 255.

Thirdly, to improve the convergence speed and accuracy of the model, we divide 255 along each of the items in vectors to normalize to $[0, 1]$ from $[0, 255]$, where X is the normalized data, denoted by $X = [b_1, b_2, \ldots, b_{1024}], b_n \in [0, 1]$.

Moreover, compute the mean of all training samples and then deduct it from each sample. This approach offers the benefit of normalizing each dimension based on the attributes of the complete training set. The calculation can be performed using the following equation.

$$X' = X - \overline{X} \qquad (1)$$

where $\overline{X}$ is the mean vector of all X in train data. Then we get the input format X' for training.

3.2 Coarse-Grained Learning Layer

The coarse-grained learning layer applies 1D-CNN down-sampling to extract more generalized features from raw network traffic. This down-sampling technique combines information from adjacent time steps while retaining the crucial characteristics. Moreover, it includes vertical connections between feature maps of different scales to create representations across all scales for the inputs. The layer comprises a vertical structure of 1D-CNNs with S-scale, where S is set to 3 in our study. The output feature maps are denoted as x_l^S, representing the original input traffic data at layer l and scale S, which is labeled as x_0^1. Furthermore, this layer exhibits dense connectivity with subsequent layers, enabling the reuse of shallow features in the following network.

3.3 Early Exit Layer

The early exit layer consists of a couple of early exit units. Each early exit unit contains an analogous 1D-CNN down-sampling layer and a binary classifier composed of two 1D-CNN layers, an average pooling layer, and one linear layer. The output feature maps x_l^S from the l-th early exit unit are connected to the previous feature maps of scale S and $S - 1$, the number of early exit units L set to 3 in our study. The kth classifier, represented as $f_k(\cdot)$, utilizes all the shallow features $x_1^S, \ldots, x_l^S$ to determine a route choice. Samples go through the early exit units and decide whether to exit from the classifier $f_k(\cdot)$ based on the prediction confidence P_k and a preset threshold θ_k. The prediction confidence P_k for classifier $f_k(\cdot)$ is defined as the softmax function applied to the product of a learnable weight matrix w_k and the k_{th} feature map. Early exits happen when a sample surpasses θ_k before fully traversing the network. Additionally, we can classify samples of different scales by employing multiple early exit units in a cascade. In our study, we utilized three early exit units and achieved a well-balanced performance. For samples that do not meet the θ_k threshold, the inference process continues until the network exits.

3.4   Fastformer-Based Multi-classification

Once samples continue inference after early exit layer, they could exit from the final multi-classification layer, which contains Fastformer architecture. Fastformer [17] is an efficient extension of Transformer based on additive attention. The details of the Fastformer layer are shown in Fig 3.

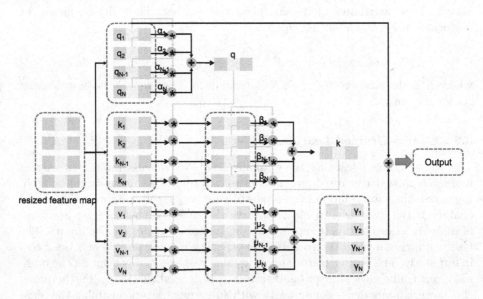

Fig. 3. Overview of Fastformer.

In Fastformer, each attention head transforms the features generated by the final early exit unit into attention query, key, and value matrices (Q, K, and V). These matrices are of size $\mathbb{R}^{N \times d}$ and are denoted as $Q = [q_1, q_2, \ldots, q_N]$, $K = [k_1, k_2, \ldots, k_N]$, and $V = [v_1, v_2, \ldots, v_N]$, respectively.

First, we apply additive attention to condense the global contextual information in the attention query into a global query vector $\mathbf{q} \in \mathbb{R}$. The attention weight α_i for the i-th query vector is calculated using the following equation:

$$\mathbf{q} = \sum_{i=1}^{N} \alpha_i \mathbf{q}_i = \sum_{i=1}^{N} \frac{\exp(\mathbf{w}_q^T \mathbf{q}_i / \sqrt{d})}{\sum_{j=1}^{N} \exp(\mathbf{w}_q^T \mathbf{q_j} / \sqrt{d})} \mathbf{q}_i, \tag{2}$$

where $\mathbf{w_q}$ represents a parameter vector that can be learned. The i-th vector in a global context-aware key matrix is represented as $\mathbf{p}_i$. This vector is obtained by taking the element-wise product of $\mathbf{q}$ and $\mathbf{k}_i$, denoted as $\mathbf{p}_i = \mathbf{q} * \mathbf{k}_i$, where the symbol $*$ indicates element-wise multiplication. The calculation of the additive attention weight for the i-th vector can be explained as follows:

$$\beta_i = \frac{\exp(\mathbf{w}_k^T \mathbf{p}_i)/\sqrt{d}}{\sum_{j=1}^{N} \exp(\mathbf{w}_k^T \mathbf{p}_j/\sqrt{d})}. \tag{3}$$

The global key vector $\mathbf{k}$ is similar to $\mathbf{q}$. The utilization of multiple layers of Fastformer allows for comprehensive modeling of contextual information. Furthermore, the sharing of value and query transformation parameters helps in minimizing memory usage. Ultimately, the output produced will serve as the final prediction for the given input sample.

3.5 Budget-Constrained Strategy

We propose a budget-constrained strategy to train our model with the least global computing cost.

Firstly, we compute the computational cost C_k for $f_k(\cdot)$. The exit probability m_k in a classifier, which remains constant across all layers, is denoted as follows:

$$m_k = z(1 - m)^{k-1}m, 0 \leq m \leq 1, \tag{4}$$

where z is a normalizing constant that ensures that $\sum_1^k m_k = 1$.

In our expectation, the entire sample inference should be less than $Budget$ in the test procedure. Formally, the following constraint needs to be met:

$$|D_{test}| \sum_1^k m_k C_k \leq Budget. \tag{5}$$

We can determine a threshold θ_k on the validate set to satisfy the constraint to ensure that $|D_{test}|m_k$ samples exit at the kth early exit unit. In the training procedure, we use softmax loss functions $L(f_k)$ for all classifiers, including Fastformer layer, and minimize a weighted cumulative loss.

$$L_{cumulation} = \frac{1}{|D|} \sum_{(x,y) \in D} \sum_k w_k L(f_k). \tag{6}$$

In this context, the training set is denoted as D, and the weight matrix of the k-th early exit unit is represented by w_k.

In the training process, we can incorporate prior knowledge about $Budget$ by assigning weight w_k based on the real-world setting. Through empirical observation, we have determined that using the same weight for all loss functions yields the best performance in our scenario.

4   Experiments

In this section, we compare the performance of FastDet with three methods and conduct four experiments using different data distributions to solve the problem in various scenarios. Additionally, we analyze the interpretability of the significant performance achieved by FastDet.

The experiments were conducted on a PC running Ubuntu. Table 1 provides details of the environment configuration.

Table 1. Experimental environment configuration

item	configuration
operating system	Ubuntu 20.04
hardware configuration	Intel(R) Core(TM) i7-7820X CPU @ 3.60 GHz, NVIDIA GeForce RTX 2080 SUPER
Python version	Python 3.5.2
PyTorch version	Pytorch 1.9.1

Our implementation of FastDet is adapted from the MSDNet [15] and Fast-former [17].

During training, the batch size is set to 64, and the cost function is softmax. We use the Adam optimizer with an initial learning rate of $1e - 3$ and a decay rate of 0.96 in every epoch. The training procedure runs for 50 epochs.

4.1   Datasets

To evaluate the performance of FastDet in general encrypted malicious traffic detection, unknown malicious traffic detection, TLS 1.3 traffic detection, and "real-world" malicious traffic detection, respectively, we construct four experiments to simulate different scenarios with CICIDS2017 [18], USTC-TFC [19], Open HTTPS Dataset [20], and CSTNET-TLS 1.3 [21].

Dataset1: CICIDS2017. In our study, we focus exclusively on HTTPS traffic from this dataset, which effectively imitates real-world data by encompassing various malicious activities such as DDoS attacks, infiltration attempts, and botnet attacks. This dataset comprises captured traffic from the main switch, memory dump, and system call records of all victim machines involved in an attack.

Dataset2: USTC-TFC. This dataset contains 10 malicious traffic types, such as Credex, Geodo, Htbot, and includes 10 types of benign traffic from common applications.

Dataset3: Open HTTPS Dataset. This dataset focuses on HTTPS traffic in a highly restricted environment and includes complete network traffic data.

It comprises 250,185 HTTPS flows, with 237,127 captured from Chrome and Firefox, and covers 779 of the most commonly visited websites.

Dataset4: CSTNET-TLS 1.3. This dataset was collected by [21] from March to July 2021 and includes 120 applications under CSTNET. It is the first TLS 1.3 dataset available, gathered from Alexa Top-5000 deployed with TLS 1.3.

4.2 Comparisons to Baselines

We compare FastDet with 3 baselines, including (1) machine learning-based method: Random Forest. (2) deep learning-based method: BGRUA [16]. (3) pre-training method: ET-BERT [21].

1) Random Forest is a cutting-edge machine learning model used for detecting encrypted malicious traffic. It employs a classifier consisting of multiple decision trees that generate an overall prediction by aggregating votes from each individual tree. Like other machine learning methods, it requires manual feature selection from raw network traffic prior to training. In this study, we utilize the feature selection approach developed by [1] to train this model.

2) BGRUA [16] is an end-to-end classification model that learns representative features from the raw traffic flows to classify encrypted traffic. The model incorporates GRU to extract sequence features between consecutive packets and uses an attention mechanism to enhance its ability to remember important features while ignoring irrelevant ones. The model demonstrates impressive results on a dataset of 18 different apps, making it suitable for real-world applications.

3) ET-BERT [21] pre-trains deep contextualized datagram-level representation from large-scale unlabeled data. This pre-trained model can then be fine-tuned using only a small number of task-specific labeled data, and has demonstrated state-of-the-art performance on five encrypted traffic classification tasks.

4.3 Evaluation Metrics

We use *Accuracy* (AC), *Precision* (PR), *Recall* (RC), and $F1 - score$ (F1) to evaluate the performance of FastDet. *Accuracy* refers to the proportion of correctly classified samples. *Precision* means the number of samples actually in class A divided by the total number of samples classified as class A. *Recall* is defined as a metric for a specific class. $F1 - score$ is a weighted average precision and recall. All the above indicators are typical metrics, we will not repeat them.

Additionally, we use the average **FL**oating point **OP**erations ($FLOPs$), the average inference time on test set, and the number of parameters to evaluate the efficiency of the model. The $FLOPs$ means the complexity of the model, inference time on test set reflects the performance in practice. Furthermore, samples terminate inferencing from different exits in our study, so we take the average of the two indicators as metrics. The number of parameters indicates the computing resource intuitively.

For a convolutional neural network, we can get $FLOPs$ as follows:

$$FLOPs = 2 \times H \times W \times (C_{in} \times K^2 - 1) \times C_{out}, \tag{7}$$

where H, W, and C_{in} are the height, width, and the number of channels of the input feature map, K is the kernel width, and C_{out} is the number of output channels. $FLOPs$ for fully connected layers could be computed as:

$$FLOPs = (2 \times I - 1) \times O, \tag{8}$$

where I is the input neuron numbers, and O is the output neuron numbers. In addition, machine learning-based models cannot compute $FLOPs$ using the above equations. $Time_{average}$ computed as:

$$Time_{average} = \frac{1}{n} \sum_{i=1}^{n} Time_i, \tag{9}$$

where $Time_i$ means the inference time of the i_{th} sample on test set.

4.4   Model Comparison and Analysis

Exp1: General Encrypted Malicious Traffic Detection. In this experiment, we construct a new dataset based on CICIDS2017 and USTC-TFC, which contains encrypted malicious attacks and encrypted malware traffic to evaluate the performance of general encrypted malicious traffic detection.

Figure 4 contains 3 subgraphs showing the AC, PR, and RC values for 9 classes in Exp1. FastDet achieves performance comparable to that of BGRUA and ET-BERT, indicating that methods based on deep learning can learn features from raw traffic to get better performance than random forest. This proves the validity of early exit units and Fastformer for encrypted malicious traffic detection. In Fig 4 (a), the AC values for all 9 classes exceed 90%, and FastDet achieves an average accuracy higher than that of the random forest model by up to 11%. As depicted in Fig 4 (b), FastDet achieves a 95% precision rate in detecting Botnet encrypted malicious traffic, demonstrating a 3% improvement compared to ET-BERT. In Fig 4 (c), our model outperforms other models in terms of recall values for most classes. Furthermore, 6 classes show similar recall rates. However, all models perform poorly in the DDos class and the Botnet class. These experimental results demonstrate that our model maintains consistent performance compared to other methods.

Exp2: Unknown Malicious Traffic Detection. In a "real world" scenario, encrypted malicious traffic continues to bring forth the new through the old. It is challenging to detect unknown types of malicious traffic with limited training data. Therefore, we conducted comparison experiments with various data distributions on a new dataset sampled from Dataset1 and Dataset2 to analyze the robustness of our model. The dataset is divided into two groups, the first

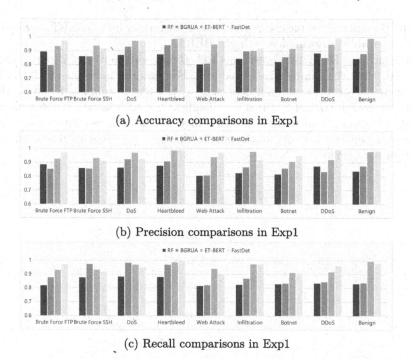

(a) Accuracy comparisons in Exp1

(b) Precision comparisons in Exp1

(c) Recall comparisons in Exp1

Fig. 4. Results comparisons in Exp1

includes 15 malicious types as the training set, whereas the second contains 5 distinct malicious types as a test set. During the testing phase, we classified the samples from the 5 malicious types into one unknown malicious type. We carried out 4 experiments by randomly sampling the training set, accounting for 100%, 40%, 20%, and 10%.

In Fig. 5, the comparison results illustrate that ET-BERT is least affected by the data size reduction. When the training set distribution is relatively small, ET-BERT achieves higher accuracy than the model trained from scratch. This is due to the lack of sufficient features in the training data to represent the actual data distribution. Leveraging relevant prior knowledge to learn new information can compensate for this deficiency. Regarding BGRUA, in cases with limited training samples, the weak generalization ability of the model trained from scratch results in lower accuracy. When the proportion of training samples is relatively large, the performance gap between ET-BERT and FastDet decreases since the learning ability of the model trained from scratch improves.

The F1-score of FastDet with 100%, 40%, 20%, and 10% data proportions are respectively 91%, 83%, 72%, and 73%, which achieves close performance to ET-BERT. The Fastformer architecture in FastDet is a Transformer variant based on additive attention that can handle long sequences data like traffic efficiently with linear complexity. With the help of Fastformer, common features between encrypted malicious traffic are learned like Transformer in ET-BERT, so FastDet gets comparable performance with ET-BERT.

In contrast, Random Forest method shows substantial F1 performance degraded when the sample size is reduced. For random forest method, the F1-score value is much lower than others. This is because random forest method utilizes session statistical features for classification, such as packet intervals and session durations. The numerical instability of these time-related features is apparent when the network environment changes, resulting in degraded classification performance.

Fig. 5. Comparison Results on Unknown Malicious Traffic Detection

Exp3: Encrypted Traffic Classification on TLS 1.3. In this experiment, our objective is to classify encrypted traffic using the new encryption protocol TLS 1.3. Our previous experiments demonstrated that FastDet achieved impressive performance in detecting encrypted malicious traffic. However, the datasets used in those experiments were obtained from prior work and are now somewhat outdated. Up to now, TLS has been updated to version 1.3, which includes features like TLS False Start and Zero Round Trip Time, designed to speed up connection establishment and reduce information exchange requirements, therefore making the task of encrypted malicious traffic detection more challenging. However, deep learning-based methods, which take raw traffic as input, can still extract relevant features from implicit information, reducing the impact of TLS version updates and maintaining robustness.

As shown in Fig. 6, FastDet is improved by 12.2% from 85.41% over the existing methods as the results on CSTNET-TLS 1.3, and pushes F1 to 97.6% by Fastformer. This indicates that the encrypted traffic over TLS 1.3 still has implicit patterns, which are better leveraged by FastDet for classification.

Exp4: Real-World Malicious Traffic Detection. According to the 2022 Malicious Bot Traffic report released by application security vendor Imperva

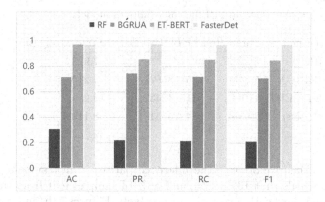

Fig. 6. Comparison Results in Exp3

[22], bot traffic accounted for 42.3% of all Internet activity in 2021. Among them, the proportion of malicious machine traffic is about 27.7%, which is about twice that of regular machine traffic. According to the above statics, we construct a new dataset based on Dataset1, Dataset2, and Dataset3 to simulate the "real world" setting, which contains 25,000 malicious sessions and 75,000 benign sessions sampled randomly. The new dataset is divided into the training set, the validation set, and the test set according to the ratio of 8:1:1. We train FastDet based on the data distribution in this experiment.

As shown in Table 2, amongst all the methods, random forest demonstrates the minimum average testing time, which includes the feature extraction procedure. However, it exhibits the worst performance among all models. Regarding BGRUA, while it has fewer parameters than ET-BERT and FastDet, it incurs more $FLOPs_{average}$ and $Time_{average}$ than FastDet. Because FastDet can terminate the classification early for "easy" samples, reducing the overall testing time. Being a complex deep learning model, ET-BERT has the most significant number of parameters, $FLOPs_{average}$, and $Time_{average}$ but achieves unparalleled performance, making it challenging to apply in resource-constrained environments.

Under our experiment setting, results indicate that FastDet achieves an average inference time of 8.119 milliseconds, marking a 13.49% and 50.92% improvement over BGRUA and ET-BERT, respectively. While slightly slower than random forest, its dense layer connections allow for the reuse of shallow layer features, and the Fastformer layer can learn contextual traffic features, maintaining high detection performance. FastDet also reduces testing time for benign samples by learning common features in early exit layers while also benefiting from benign sample rates being typically higher than malicious ones. The time consumption is positively correlated with the $FLOPs$ of the model. However, the $FLOPs$ of BGRUA and ET-BERT are fixed for each input sample but variances for FastDet. The average $FLOPs$ of all samples resulted in an almost 300% improvement, which indicates optimal efficiency.

Table 2. Efficient results of models

method	Parameters	$FLOPs_{average}$	$Time_{average}$ (ms)
RF	-	-	7.923
BGRUA	5.14M	5148582	9.215
ET-BERT	110M	15142432	12.254
FastDet	20.3M	4215351	8.119

To achieve this, we observe the inference paths for all test benign samples. Figure 7 (a) reveals that 13%, 25%, and 36% of samples exit from the first, the second, and the third early exit unit, respectively. Figures 7 (b) indicate the type proportions in each early exit unit, with audio and video traffic mostly exiting in the first two layers, suggesting that these types of traffic are easily detected with common features. Our experimental results demonstrate that time consumption in our FastDet model is times faster than other models when processing the whole dataset. Additionally, average $FLOPs$ time and inference time get reduced substantially. These findings suggest that the early exit mechanism can significantly enhance detection speed, enabling better detection performance for real-world applications.

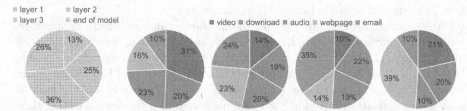

(a) Sample numbers exit from each layer   (b) Sample categories exit from layer 1, layer 2, layer 3, and end of model

Fig. 7. Statics of samples exit from each layer

4.5 Effects of Hyperparameters

In Fig. 8, we explored how various values of hyperparameters B and L influence the balance between speed and accuracy. The results of our experiments suggest that different hyperparameter values have a significant impact on the performance of FastDet. As illustrated in Fig 8 (a), our model's performance is highly sensitive to the value of B. Between 500 to 1024, higher values mean more information for feature extraction and yield an accuracy of up to 98% when L is set to 3. However, beyond 1024, there is a visible decrease in speed while accuracy increases by only 1%. This finding indicates that using 1024 bytes for general detection tasks achieves the best balance. In Fig 8 (b), we observe a significant increase in accuracy as the number of early exit units rises from 1 to 3. Other metrics exhibit similar trends, indicating that at least three layers should be utilized in practice. However, as the number of layers exceeds 3, the indicator growth rate slows down, and the normalized $FLOPs$ increases.

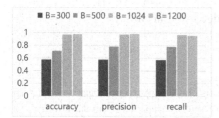

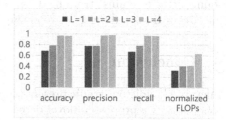

(a) Effects of different B values on experimental results

(b) Effects of different L values on experimental results

Fig. 8. Effects of different hyperparameters values on experimental results

4.6   Interpretability

Our experiment results indicate that benign samples possess distinct features from malicious samples, making them easily discernible. Benign samples commonly exhibit similar traits, such as the latest TLS version, continuous connections in both client and server, and sequential features. We transformed each feature map into a fixed size and used the average feature map of each exit to generate a single-channel image for visualization, as shown in Fig. 9. Our analysis revealed considerable differences between benign and malicious samples. Samples that exited from off-ramps were lighter in color than those at the end of the model, consistent with expectations that benign samples have more common features than malicious ones.

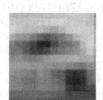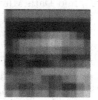

Fig. 9. Average feature map exit from layer 1, layer 2, layer 3, and end of model, respectively.

5   Discussion

This section reflects on some limitations of our work and possible implications for future research in this field. Although FastDet can detect encrypted malicious traffic quickly, it still has some limitations. Firstly, FastDet relies on labeled training samples. For samples with new label that do not appear in the training dataset, FastDet cannot detect them. Secondly, FastDet is unable to resist poisoning attacks against training data. If the attacker poisons the training dataset, FastDet will have difficulties to ensure the detection accuracy.

6   Conclusion

Encrypted malicious traffic presents continuous challenges to network security. In this paper, we propose a novel detection approach called FastDet that employs an early exit mechanism to detect encrypted malicious traffic faster. FastDet enables "easy" samples to terminate inference early via early exit mechanism, thereby improving the overall detection speed. The experimental results show that FastDet is considerably more efficient than the existing encrypted malicious detection techniques while maintaining comparable performance. We experimentally evaluate the sensitivity of FastDet to different hyperparameters, and further explain the effectiveness of FastDet by visualizing the differences between encrypted benign and malicious samples.

Acknowlegements. This paper is supported by GuangDong Basic and Applied Basic Research Foundation 2022B1515120072.

References

1. Torroledo, I., Camacho, L.D., Bahnsen, A.C.: Hunting malicious tls certificates with deep neural networks. In: Proceedings of the 11th ACM Workshop on Artificial Intelligence and Security, pp. 64–73 (2018)
2. Anderson, B., McGrew, D., Acm: Machine learning for encrypted malware traffic classification: Accounting for noisy labels and non-stationarity. In: Kdd'17: Proceedings of the 23rd Acm Sigkdd International Conference on Knowledge Discovery and Data Mining, pp. 1723–1732 (2017)
3. Tong, X., Tan, X., Chen, L., Yang, J., Zheng, Q.: Bfsn: a novel method of encrypted traffic classification based on bidirectional flow sequence network. In: 2020 3rd International Conference on Hot Information-Centric Networking (HotICN), pp. 160–165. IEEE (2020)
4. Prasse, P., Knaebel, R., Machlica, L., Pevny, T., Scheffer, T.: Joint detection of malicious domains and infected clients. Mach. Learn. **108**(8–9), 1353–1368 (2019)
5. Zeng, Y., Gu, H., Wei, W., Guo, Y.: $deep - full - range$: a deep learning based network encrypted traffic classification and intrusion detection framework. IEEE Access **7**, 45182–45190 (2019)
6. Korczynski, M., Duda, A., Ieee: Markov chain fingerprinting to classify encrypted traffic. In: 2014 Proceedings IEEE Infocom, pp. 781–789 (2014)

7. Shen, M., Wei, M., Zhu, L., Wang, M., Li, F.: IEEE: Certificate-aware encrypted traffic classification using second-order markov chain. 2016 IEEE/ACM 24th International Symposium on Quality of Service (Iwqos) (2016)

8. Lotfollahi, M., Siavoshani, M.J., Zade, R.S.H., Saberian, M.: Deep packet: a novel approach for encrypted traffic classification using deep learning. Soft. Comput. **24**(3), 1999–2012 (2020)

9. Dong, C., Zhang, C., Lu, Z., Liu, B., Jiang, B.: Cetanalytics: comprehensive effective traffic information analytics for encrypted traffic classification. Comput. Netw. **176**, 107258 (2020)

10. Park, E., et al.: Big/little deep neural network for ultra low power inference. In: 2015 International Conference on Hardware/software Codesign and System Synthesis (codes+ isss), pp. 124–132. IEEE (2015)

11. Bolukbasi, T., Wang, J., Dekel, O., Saligrama, V.: Adaptive neural networks for efficient inference. In: International Conference on Machine Learning, pp. 527–536. PMLR (2017)

12. Teerapittayanon, S., McDanel, B., Kung, H.T.: Branchynet: Fast inference via early exiting from deep neural networks. In: 2016 23rd International Conference on Pattern Recognition (ICPR), pp. 2464–2469. IEEE (2016)

13. Leroux, S., et al.: The cascading neural network: building the internet of smart things. Knowl. Inf. Syst. **52**(3), 791–814 (2017)

14. Wang, X., Luo, Y., Crankshaw, D., Tumanov, A., Yu, F., Gonzalez, J.E.: Idk cascades: Fast deep learning by learning not to overthink. arXiv preprint arXiv:1706.00885 (2017)

15. Huang, G., Chen, D., Li, T., Wu, F., Van Der Maaten, L., Weinberger, K.Q.: Multi-scale dense networks for resource efficient image classification. arXiv preprint arXiv:1703.09844 (2017)

16. Liu, X., et al.: Attention-based bidirectional gru networks for efficient https traffic classification. Inf. Sci. **541**, 297–315 (2020)

17. Wu, C., Wu, F., Qi, T., Huang, Y., Xie, X.: Fastformer: additive attention can be all you need. arXiv preprint arXiv:2108.09084 (2021)

18. Panigrahi, R., Borah, S.: A detailed analysis of cicids2017 dataset for designing intrusion detection systems. Int. J. Eng. Technol. **7**(3.24), 479–482 (2018)

19. Wang, W., Zhu, M., Zeng, X., Ye, X., Sheng, Y.: Malware traffic classification using convolutional neural network for representation learning. In: 2017 International Conference on Information Networking, ICOIN 2017, Da Nang, Vietnam, January 11–13, 2017, pp. 712–717. IEEE (2017)

20. Wazen, S., Thibault, C., Jerome, F., Isabelle, C.: Https websites dataset. 4 http:// betternet.lhs.loria.fr/datasets/https/ (2016)

21. Lin, X., Xiong, G., Gou, G., Li, Z., Shi, J., Yu, J.: Et-bert: A contextualized datagram representation with pre-training transformers for encrypted traffic classification. In: Proceedings of the ACM Web Conference 2022, pp. 633–642 (2022)

22. 2022 Bad Bot Report | Evasive Bots Drive Online Fraud | Imperva

Real-EVE: Real-Time Edge-Assist Video Enhancement for Joint Denoising and Super-Resolution

Liming Ge[✉], Wei Bao, Dong Yuan, and Bing Bing Zhou

Faculty of Engineering, The University of Sydney, Sydney, Australia
{liming.ge,wei.bao,dong.yuan,bing.zhou}@sydney.edu.au

Abstract. Real-time video applications have received much attention in recent years. However, the perceived quality of real-time videos in many situations is far from ideal due to two major obstacles: noise in the video frames caused by limited camera hardware and low resolution caused by bandwidth-limited networks. A straightforward solution is a direct investment in photography and networking hardware, but it is obviously cost-ineffective and unscalable. We are motivated to develop an alternative solution by leveraging edge AI. We propose a new Real-time Edge-assist Video Enhancement (Real-EVE) framework. It includes two key designs: The video-enhancement deep neural network (VE-DNN), which jointly eliminates noise and super-resolves videos in real time with a small inference delay; and the video-enhancement-aware adaptive bitrate streaming (VEA-ABR), which adapts sending rate in response to changing network conditions to optimize the video quality posterior to video enhancement. We develop a real-world prototype of the proposed Real-EVE, demonstrating Real-EVE outperforms all benchmarks, and both the VE-DNN and VEA-ABR bring drastic performance gain.

Keywords: Real-time video · Edge computing · Video super-resolution · Video denoising · Video bitrate adaptation

1 Introduction and Motivation

Video cameras have been widely deployed to support everywhere and every-time applications. The footage is streamed in real time to provide vital guidance under a wide range of circumstances (e.g., searching for missing persons, rescue operations amid natural disasters, etc.). There is an urged demand for high-quality real-time video playout with small delays, as faulty, unclear, or outdated information may be valueless or even misleading. However, there are two major obstacles impeding the fulfillment of high-quality real-time video: Noise and low resolution. First, many cameras are with inexpensive hardware for cost-effective deployment on large scales. The quality of the captured video is far from satisfactory, with noisy frames often observed especially during the night [3]. Second, due to massive and geo-distributed deployment, the available network

© The Author(s), under exclusive license to Springer Nature Singapore Pte Ltd. 2024
Z. Tari et al. (Eds.): ICA3PP 2023, LNCS 14487, pp. 320–339, 2024.
https://doi.org/10.1007/978-981-97-0834-5_19

bandwidth for each camera is also limited, especially in rural regions where the cellular network is not stable. The captured video has to be further down-sampled in a lower resolution for transmission in real time. The aforementioned factors drastically deteriorate the quality of the captured video and significantly affect the quality of experience (QoE).

One straightforward approach to address the aforementioned issue is to directly invest more in photography and networking hardware. In other words, more expensive high-performance cameras (wide aperture lens and large CMOS) are deployed and higher network bandwidth (e.g., base stations and cables) is installed. However, such an approach is cost-ineffective and unscalable. In many situations, it is also not feasible to upgrade the bandwidth or cameras. For exam-ple, cameras may be located in rural areas; a large number of devices are sharing the wireless channel in peak hours; the cameras themselves are embedded in other equipment (e.g., a drone has a limited takeoff weight). To this end, it is critical to develop alternative solutions to achieve high-quality real-time video delivery without investing in photography and networking hardware.

In this article, we propose to leverage the recent advances in edge AI to provide high-quality real-time video delivery, without investment on hardware. First, recent years have witnessed the rapid advancement of deep neural net-work (DNN) based video enhancement methods. Among them, DNN-based video denoising [26,27] and video super-resolution [2,12] methods are two promising solutions to enhance real-time video without the need to upgrade hardware. Such an approach gives us an unprecedented opportunity to deliver low-quality video through existing systems but to enhance it at a different place. Second, edge com-puting is an ideal place to provide such video enhancement services. Edge servers are to be placed close to video receivers so that the enhanced video, with a high data rate, is only transmitted within the local high-speed network. Only low-quality videos with small bandwidth requirements are transmitted through the unstable wireless hop or through the congested Internet. Please note that even though edge computing servers are not as powerful as cloud computing servers, a household-grade workstation is already sufficient to process video denoising and video super-resolution inference tasks in real time.

We propose Real-time Edge-assist Video Enhancement (Real-EVE), includ-ing two key designs: (1) video enhancement DNN (VE-DNN), our newly devel-oped DNN specifically for joint real-time video denoising and super-resolution in an edge-computing environment, and (2) video enhancement aware adaptive bitrate streaming (VEA-ABR), a new video adaptation framework in the exis-tence of DNN enhancement service.

VE-DNN outputs noise-free video frames in high resolution from noisy low-resolution frames with a small delay so that the video enhancement service provides real-time performance. Our design is essential since directly combin-ing existing state-of-the-art (SOTA) video denoising and video super-resolution algorithms (in tandem) fails in real-time scenarios. These DNNs are not designed for real-time video and their delay performance is too large. Most of them [2,26] cannot finish processing a frame within 200 ms. Also, the direct combination will cause amplified negative effects between them. For example, an inaccurate pixel after super-resolution will confuse the denoising DNN, causing serious artifacts in the final output. Our designed VE-DNN can well address this issue to achieve much more improved output video quality.

We also develop VEA-ABR to jointly work with the VE-DNN to optimize the overall video quality. Different from existing ABR algorithms, we now target to maximize the video quality *posterior* to the video enhancement, so that the joint effects on network bandwidth limitation and the quality of experience (QoE) enhancement by VE-DNN are jointly considered.

To evaluate the performance of our proposed Real-EVE, we establish a prototype implementing the VE-DNN and VEA-ABR at the edge. The prototype is leveraged to conduct comprehensive experiments. We compare our method with existing SOTA benchmarks. In terms of video quality performance, our proposed Real-EVE achieves drastic performance gain. In terms of quality-delay trade-off, our VE-DNN and VEA-ABR substantially improve the retrieved video quality and adapt to the fluctuating network bandwidth.

2   Background and Related Work

2.1   Video Enhancement DNNs

With their rapid development, DNNs have been applied to many areas of video enhancement, including video denoising and video super-resolution.

Video denoising aims to recover clean frames from noisy ones. Due to errors and inaccuracies, noise is inevitably observed in the video shooting process, especially at night. There are three sources of noise, namely the shot noise [3], the dark current noise [8], and the readout noise [29]. They are introduced due to the uncertain number of photons to arrive in the photosensor, the randomly generated erroneous electrons in the photosensor, and the circuit readout error, respectively. Traditionally, video denoising DNNs yield promising results by exploiting similar pixels or patches within the same frame and using the spatial similarity within each frame to identify and reduce noise. Recent DNNs [26,27] also exploit the temporal similarity among frames for noise reduction. However, existing denoising DNNs are mostly for offline video or images, where inference delay required for real time is not considered. Also, they are not jointly optimized with video super-resolution.

Video super-resolution aims to produce high-resolution video frames using low-resolution ones. Video super-resolution DNNs [2,12] exploit the details embedded in multiple low-resolution observations of the same scene and produce high-resolution frames. They apply temporal alignment prior to the aggregation to resolve the discrepancy in multiple low-resolution observations. Followed by aggregation and upsampling processes, they produce frames with sharper details and higher visual quality. Nevertheless, they are not jointly optimized with video denoising, and many of them [19] did not target to reduce the inference delay.

Combining existing denoising and super-resolution DNN models does not produce satisfactory results, and suffers from long processing times. Video super-resolution is sensitive to the artifacts brought by the denoising DNN, while video denoising is unable to eliminate the noise mistakenly amplified by the super-resolution DNN. The sequential combinations yield poor results, while bringing long processing times due to the two DNNs repetitively extract features and construct frames, making it unsuitable for real-time applications. In this work, we propose a novel combined DNN to address the aforementioned issues.

2.2 Adaptive Bitrate Algorithms and Limitations

Without video enhancement, a common approach to optimize the performance of real-time video streaming given the fluctuation of network bandwidth is adaptive bitrate streaming (ABR) [1,9]. It chooses an appropriate video quality (bitrate level) in response to changing network conditions. Different algorithms have been designed, such as rule-based algorithm [5], model predictive control based [16,32], and reinforcement learning based [20,31,34]. However, these algorithms are not designed with video enhancement. They only consider the network fluctuation and optimize the post-transmission video quality, while disregarding the video enhancement. These algorithms under-utilize the performance gain brought by video enhancement if we directly employ them in our system.

2.3 Edge Computing for Real-Time Video

Edge computing is a promising paradigm to perform computational tasks close to end devices at the network edge. Edge servers can perform as a video proxy or content delivery network (CDN) server [28], so that it reduces the delay and network workload. However, such an approach does not fully exploit the computing capability of edge computing.

Video enhancement DNNs consume additional amount of computing power for video denoising and video super-resolution [18]. These computational tasks should not be done on resource-constrained end devices, but can be fulfilled by edge computing servers [17]. Due to privacy, delay, and pricing/availability concerns, cloud computing is less appropriate to serve this purpose. Therefore, for video enhancement, we can utilize the advantages of edge computing from both the computing and the networking perspectives.

3   Design Outline

We introduce our system design to realize edge-assist real-time video enhancement. We first give an overall picture of the system design, and we then present our two key designs in video enhancement DNN (VE-DNN) and video enhancement aware adaptive bitrate streaming (VEA-ABR).

Due to noise and low resolution, the video quality in real-time video streaming systems is low at night. We aim to design a system to improve the quality of video transmission in real-time one-way communication scenarios. The system to realize the AI-enhanced real-time video delivery consists of two parts: the sender and the receiver as shown in Fig. 1. The sender is a simple device with

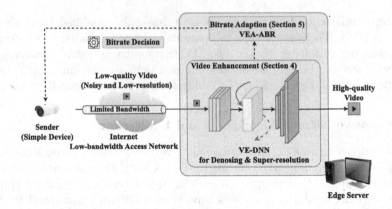

Fig. 1. System module design.

minimal computational capacities, and the receiver is an edge-computing server with sufficient computational power. Please note that one or multiple viewers can be connected to the edge server through high-speed cables or networks to play out the real-time videos. This part is straightforward and is irrelevant to our design and will be omitted in the rest of the paper. The bandwidth between the sender and the edge server is limited, either because the wireless access link of the sender is not stable, or because the path from the sender to the edge server is through a congested network. The sender sends the low-quality (low-resolution) real-time video to the receiver so that the data rate is not necessarily large. Then, the receiver will enhance the video in real time and the high-quality video is restored.

Our design is mainly implemented in the receiver (with more powerful computational capacity), and the modifications in the sender are minimized. The sender side captures real-time video, trans-codes it into the quality level with the required bitrate (as instructed by the receiver), and sends it to the receiver. There are no further complicated designs in the sender to allow fast implementation and propagation in practice. We focus on the design in the receiver as follows. When the receiver receives the raw video, it (1) decodes it through standard video codec, (2) enhances it through the VE-DNN module, and (3) plays it out. It also measures the network condition and instructs the sender which bitrate (resolution) it should use through the VEA-ABR module. In this way, we avoid the requirement for computation on the simple devices of the sender. Senders with cheap cameras are able to stream real-time video. Benefiting from recent innovations in edge computing, the receiver can be an edge server with sufficient computing power. We effectively alleviate the high bandwidth requirements for high-quality video transmission.

There are two main designs at the receiver: the video enhancement (VE-DNN) module (Sect. 4) and the bitrate adaptation (VEA-ABR) module (Sect. 5). The VE-DNN is to perform video denoising and super-resolution. The VEA-ABR module decides the quality level (bitrate). The decided bitrate (resolution) is fed back to the sender. Different from existing works, we aim to create a design that can effectively deal with two important aspects: (1) the interaction between denoising and super-resolution within VE-DNN and (2) the interaction between the VE-DNN and VEA-ABR. As for the first aspect, we should design a novel DNN model which addresses video denoising and super-resolution at the same time. As for the second aspect, we should design the VE-DNN model and the VEA-ABR algorithm for them to adopt each other. VE-DNN must support multi-granularity input multi-granularity output as required by VEA-ABR, and VEA-ABR must focus on post-DNN performance enhancement (DNN-aware) instead of being DNN-agnostic.

4   Video Enhancement DNN Design

In this section, we first state the design objectives: high quality, small inference delay, and support for multi-granularity input resolutions and multi-scale upsampling to adopt the ABR algorithm. We then introduce the model design to fulfill these objectives.

4.1   Design Objectives

The VE-DNN module is responsible to retrieve a noise-free high-resolution video stream from the noisy video stream in low resolution sent through the network with limited bandwidth. There are three major requirements for this module.

First, the output video stream should be of **high quality**. Synchronous design on video denoising and video super-resolution is needed. The direct combination of existing video denoising and video super-resolution approaches in tandem (in either order) is insufficient as the second step will amplify a small error generated from the first step, causing degraded performance in the second step. (See our experiment in Sect. 6.3 for more details.)

Second, the designed DNN should be with a **small inference delay** to satisfy the need for real-time video. Since a real-time video requires sub-400 ms total delay [13], including network delay, inference delay, and buffer delay, we should limit the inference delay to sub-200 ms to give room to other delays. The direct combination of existing denoising and super-resolution DNN models suffers from redundant computations and results in long processing times.

Third, the DNN should support **multi-granularity input** and **multi-granularity output**: i.e., input video in **multi-granularity resolutions** and upsampling in **multiple scales**. Since the network bandwidth is fluctuating, the raw video could be sent in different resolutions with different bitrates. The DNN should allow the input in different resolutions and output a high-definition video stream with a specified upscale factor to adapt to constraints at the receiver.

4.2   VE-DNN Design

To achieve the aforementioned objectives, the DNN is designed as shown in Fig. 2. It takes the corresponding input frame and two *preceding* input frames

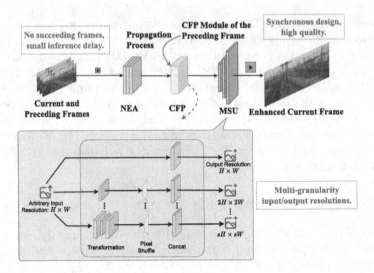

Fig. 2. Video enhancement DNN design.

in low resolution, and achieves the enhancement through an integrated pipeline. We do not input a succeeding frame as this will cause additional delay waiting for the arrival of that frame. The pipeline consists of three sub-modules: the noise-eliminated alignment (NEA) sub-module, the chronological feature propagation (CFP) sub-module, and the multi-scale upsampling (MSU) sub-module.

The NEA sub-module takes three frames in multi-granularity as input, matches time-variant frames through the guidance of similar regions, and enables the DNN to leverage the temporal redundancies among video frames to super-resolve the current frame. This frame matching and alignment process not only avoids the blurriness caused by misalignment, but also eliminates noise. Since noise in one frame is unlikely to present at the same position in an adjacent frame [4], the NEA sub-module utilizes this property to perform noise elimination by cross-checking multiple observations of the same scene. The NEA sub-module takes three input frames $\hat{I}_t$, $\hat{I}_{t-1}$, and $\hat{I}_{t-2}$, where $\hat{I}_t$ is frame to be enhanced and $\hat{I}_{t-1}$, $\hat{I}_{t-2}$ are preceding frames of $\hat{I}_t$. It outputs a noise-eliminated feature f_t.

The presence of moving objects in the video sequence often leads to the occurrence of occlusion and dis-occlusion. Some pixels are observable in one frame, but may not be observable in an adjacent frame. These temporally flickering pixels can sometimes be mistakenly recognized as noise and subsequently removed by the NEA sub-module. Therefore, we resort to propagation to address this issue. The chronological feature propagation (CFP) sub-module collects features and propagates these features to succeeding frames for reconstructions, which provides vital cues for the recovery of occluded pixels. These features contribute to both super-resolution and denoising in multiple frames so that we are able to reduce repetitive computation and achieve faster inference while obtaining a more accurate output. The CFP sub-module collects input features from both the NEA sub-module (the noise-eliminated feature f_t) and the CFP sub-module of the previous enhancement (the output of the CFP sub-module for frame $\hat{I}_{t-1}$, which is denoted f'_{t-1}). It outputs a refined feature f'_t for both the MSU sub-module, and the CFP sub-module in the succeeding reconstruction.

Due to the different resolutions of the input video, objects can be in different scales (small or large). The NEA sub-module and the CFP sub-module capture objects across various scales. Thus, we can improve the resolution of the video content at different scales using one unified pipeline. Regardless of the selection of resolution of the input video and the upscale factor, the number of data in high-dimensional features from the CFP sub-module is larger than the number of data required by the low-dimensional observations in the output video. The multi-scale upsampling (MSU) sub-module selects the data from high-dimensional features and transforms them into a noise-free frame in high resolution more precisely. The MSU sub-module consists of a set of transformation layers, a pixel-shuffle layer [24], and a concat layer. For each upscale factor s, the MSU sub-module constructs a distinct set of transformation layers, which is then fine-tuned separately. The transformation layer takes the high-dimensional feature f'_t and normalizes it to $H \times W \times s^2 C$, where $H, W,$ and C denote the height, width, and channel of the input frame, and s is the upscale factor. The pixel-shuffle layer then rearranges the elements of an $H \times W \times s^2 C$ tensor into an $sH \times sW \times C$ tensor. This tensor is then fused with the bicubic upsampled version of the noise-eliminated feature f_t using a concat layer, before constructing a noise-free high-definition frame I_t.

Given a noisy video frame in low-resolution $\hat{I}_t \in \mathbb{R}^{H \times W \times C}$ and its two preceding frames $\hat{I}_{t-1}, \hat{I}_{t-2} \in \mathbb{R}^{H \times W \times C}$, the output noise-free frame $I_t \in \mathbb{R}^{sH \times sW \times C}$ in high-definition is computed as

$$f_t = \text{NEA}(\hat{I}_t, \hat{I}_{t-1}, \hat{I}_{t-2}), \tag{1}$$

$$f'_t = \text{CFP}(f_t, f'_{t-1}), \tag{2}$$

$$I_t = \text{MSU}(f'_t, f_t). \tag{3}$$

5   VEA-ABR Algorithm

5.1   System Model for VEA-ABR Design

In this section, we propose video-enhancement-aware adaptive bitrate streaming (VEA-ABR) algorithm. Let the c_j, $j = 1, 2, \ldots, N$ denote the jth video chunk (segment). Video chunks arrive sequentially in real time, with a constant inter-arrival time C. Each video chunk contains C seconds of video. Let a_j denote the arrival time instant of the jth chunk, $a_j = C \cdot (j - 1)$. Please note that video chunks are generated in real time so that we only know c_j when it arrives. Also, we only know a chunk is the last chunk when there is no further chunk arriving, i.e., N is not known in advance.

Each chunk will be sent from the sender to the receiver via the channel first, and then it will be processed by the VE-DNN for super-resolution. The objective of VEA-ABR is to determine (1) in what resolution the c_j should be transmitted; and (2) to what resolution c_j should be super-resolved if it is received at the receiver. Different transmission delays and processing delays will be incurred for different decisions. Since we consider real-time video instead of stored video, each video chunk has a playout deadline $d_j = a_j + d$, where d is the maximum tolerable delay [15]. Each chunk must be able to be played out at d_j, otherwise, the video chunk will not be played and will be regarded as lost.

Let res_j and sres_j denote the transmitted resolution and super-resolved resolution of chuck c_j. res_j and sres_j can only be selected from a given number of available resolutions (e.g., 270p, 540p, and 1080p). Let $\mathcal{S} = \{0, 1, 2, \ldots, S\}$ denote the set of available resolutions. Without loss of generality, $\text{res}_j = 0$ denotes no transmission, and $\text{sres}_j = 0$ denotes no processing (received chunk is directly played out). Each resolution $s \in \mathcal{S}$ gives a data size. Let $D(s)$ denote the size of a video chunk with resolution s. Let $P(s_1, s_2)$ denote the processing delay of a video chunk super-resolved from s_1 to s_2 (and denoised). Please note that we allow $s_1 = s_2$, i.e., no super-resolution, meaning that it is only denoised without super-resolution. This is different from $P(s_1, 0)$, as it means that the video chunk is transmitted but not processed at all. We do not allow $s_1 > s_2 > 0$, i.e., resolution shrinks. Also, if $s_1 = 0$, s_2 can only be 0 and processing delay is 0. Let $\mathbf{S}$ denote all valid (s_1, s_2) pairs. At the receiver side, the video quality is $Q(s_1, s_2)$ if it can be played out on time. Please note that the video quality depends on both the transmitted resolution and the super-resolved resolution. For example, a video transmitted in 270p and then super-resolved (and denoised) to 1080p is not as good as a video transmitted in 540p and then super-resolved (and denoised) to 1080p. Let $Q_0(s_1) = Q(s_1, 0)$ denote the video quality of resolution s_1 without any super-resolution/denoising. We also define $Q_0(0) = Q_{\min}$. $Q_{\min}$ may not be zero as we may replay a previous chunk for this missing chuck, leading to non-zero quality. In this work, Q and Q_0 values are measured in peak signal-to-noise ratio (PSNR) in dB [11,23], which is a common metric for video quality evaluation. $Q(\cdot, \cdot)$, $D(\cdot)$, and $P(\cdot, \cdot)$ values can be known in advance, as we can measure these values based on a number of offline videos (i.e., video profiling [10,33]).

For video chunk c_j, once we determine the transmitted resolution and super-resolved resolution $\text{res}_j = s_1$ and $\text{sres}_j = s_2$, it causes transmission delay $D(\text{res}_j)/B_j$ (where B_j is the bandwidth for transmission of chunk c_j) and processing delay $P(\text{res}_j, \text{sres}_j)$. The video quality of the chunk c_j is denoted by Q_j. Q_j will be (1) $Q(\text{res}_j, \text{sres}_j)$ if video super-resolution is completed before the deadline d_j; (2) $Q_0(\text{res}_j)$ if video transmission is completed before the deadline, but the super-resolution misses the deadline; (3) $Q_{\min}$ if video transmission misses the deadline.

5.2   Quality of Experience Maximization

The objective of VEA-ABR is to maximize the overall Quality of Experience (QoE) for the whole video. Following a conventional QoE criterion [20,21,32],

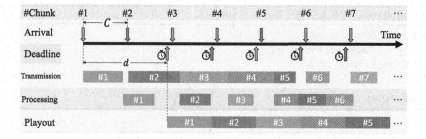

Fig. 3. Flow-shop scheduling for Real-EVE. Each chunk is transmitted, processed, and then played out. Transmission and processing must be completed by the deadline (the time it should be played out). The transmission of chunks 3–5 can only be started after the transmission of their previous trunks are completed. The processing of chunks 5–6 can only be started after the processing of their previous trunks are completed.

we characterize both the quality of each video chunk and the quality smoothness between two consecutive chunks.

$$\text{QoE} = \sum_{i=1}^{N} Q_i - \lambda \sum_{i=2}^{N} |Q_i - Q_{i-1}|, \tag{4}$$

where λ is the penalty factor and $\lambda \sum_{i=2}^{N} |Q_i - Q_{i-1}|$ is the penalty on the video unsmoothness, as the inconsistency of quality of two consecutive video chunks brings QoE penalty. Please note that unlike stored video, if a video chunk cannot be played out by its deadline, we do not wait (e.g., rebuffering) but just skip it. Otherwise, the playout time of all following video chunks will be influenced, breaking all their deadlines.

Let B_j denote the network bandwidth when we transmit chunk c_j. B_j is fluctuating, and we can predict a few B_j values in the future. We assume that at the arrival of c_j, we can predict $B_j, B_{j+1}, \ldots, B_{j+K-1}$, i.e., a window of K values. Such prediction can be well done by existing channel prediction methods [16,32].

Each video chunk is to be transmitted and then processed (super-resolved and denoised). We assume that video trunks must be transmitted and processed in order. At one time instant, we can only transmit one frame and can only process one frame. Therefore, the video chunks will pass a flow shop [6], as shown in Fig. 3. Suppose a video chunk arrives at the sender, but the sender is still sending a previous chunk, the chunk cannot be sent until the previous chunk is completed (e.g., Chunk 4 in Fig. 3). Similarly, if a video chunk is received by the receiver, but the receiver is still processing a previous chunk, the chunk cannot be processed until the previous chunk is completed (e.g., Chunk 6 in Fig. 3).

Let t_i denote the time instant that the sender completed sending chunk c_i; Let τ_i denote the time instant that the receiver completed processing chunk c_i.

Algorithm 1. VEA-ABR Sender Side.

---
1 **while** *Video is on* **do**
2   res_i =Receive_Instruct()
3   Send_Video(c_i, res_i, d_i)
4     $i := i + 1$
5 Send_Video_Status(OFF)

Algorithm 2. VEA-ABR Receiver Side Main Algorithm.

---
1 initialize $t_0 = 0$, $\tau_0 = 0$
2 **while** *Receive_Video_Status()$\neq$OFF* **do**
3   **if** *timestamp$=a_i - \Delta$* **then**
4   Predict $(B_i, B_{i+1}, \ldots, B_{i+K-1})$
5       $(res_i, sres_i) = $ VEA_MPC$(Q_{i-1}, t_{i-1}, \tau_{i-1}, B_i, B_{i+1}, \ldots, B_{i+K-1})$
6   Send_Instruct(res_i)
7     $c_i = $ Receive_Video()
8     $\tau_i = $ VE_DNN$(c_i, res_i, sres_i, d_i)$
9   Playout(c_i)

We let $t_0 = 0$ and $\tau_0 = 0$. Since chunk c_{i+1} arrives at a_{i+1}, we can derive t_{i+1} as follows.

$$t_{i+1} = \min \left(d_{i+1}, \max(a_{i+1}, t_i) + D(res_{i+1})/B_{i+1} \right), \tag{5}$$

where $\max(a_{i+1}, t_i)$ is the time instant we start to transmit c_{i+1}, and $\max(a_{i+1}, t_i) + D(res_{i+1})/B_{i+1}$ is the time instant we complete the transmission. However, if the deadline d_i is missed, we stop the transmission and this chunk is regarded as lost. Similarly, τ_{i+1} can be derived as

$$\tau_{i+1} = \min \left(d_{i+1}, \max(t_{i+1}, \tau_i) + P(res_{i+1}, sres_{i+1}) \right), \tag{6}$$

where $\max(t_{i+1}, \tau_i)$ is the time instant we start to process c_{i+1}, and $\max(t_{i+1}, \tau_i) + P(res_{i+1}, sres_{i+1})$ is the time instant we complete the processing. However, if the deadline d_i is missed, we stop the processing and the received frame is directly played out without super-resolution.

5.3   VEA-ABR Algorithm Design

The VEA-ABR Algorithm is implemented at both the sender side and the receiver side. The sender size algorithm (Alg. 1) is straightforward as the sending device may be simple. The receiver side algorithm (Alg. 2) is more complicated as the receiver in a more powerful machine is the decision maker. Alg. 2 also calls Algs. 3 and 4 respectively.

At the sender side (Alg. 1), while the video is on, the sender receives res_i instructed from the receiver and sends c_i. When the video is off, it will let the receiver know that the video is off.

Algorithm 3. VEA-MPC Algorithm.

---

1 **Function** VEA_MPC($Q_{i-1}, t_{i-1}, \tau_{i-1}, B_i, B_{i+1}, \ldots, B_{i+K-1}$):
2   maxQoE $= -\infty$
3   decision $=$ **null**
4   **for** *all valid* (s_1, s_2) **do**
5   **if** $s_1 = 0$ *and* $s_2 = 0$ **then**
6   QoE $= Q_{\min}$
7   **else**
8   QoE $=$ Predict_QoE($t_{i-1}, \tau_{i-1}, B_i, B_{i+1}, \ldots, B_{i+K-1}, s_1, s_2$)
9   **if** QoE $>$ maxQoE **then**
10   QoE $=$ maxQoE
11   decision $= (s_1, s_2)$

12   **return** decision

At the receiver side, ideally, we aim to maximize the objective (Eq. 4). However, it is not possible to know the whole video sequence and the environment condition (i.e., network bandwidth B_i) in advance. In what follows, we develop a model predictive control (MPC) based approach. Through this approach, to make the decision for chunk c_i, (1) we predict the bandwidth for the next K chunks (from i to $i + K - 1$); (2) Given the prediction, we find the best possible transmitted and super-resolved resolutions res_i and $sres_i$ to maximize the overall QoE for next K chunks; (3) We transmit c_i in res_i and super-resolve it to $sres_i$; (4) We wait for the next chunk, let $i := i + 1$, and repeat from (1). Please note that in Steps (1) and (2), even though we look ahead for K chunks and maximize the overall QoE for the K chunks, we only execute the decision on 1 chunk. This approach balances the long-term system utility (i.e., we maximize K steps) and short-term system dynamics (prediction may be inaccurate and we re-run it every step).

The overall algorithm at the receiver side is shown in Alg. 2. At time instant $a_i - \Delta$, chunk c_i is about to arrive (Line 3). Δ is a predefined safe margin, so that the instruction to the sender will arrive by time instant a_i. At this time, we predict the bandwidth in Line 4. We do not redesign a prediction mechanism but rely on existing methods, such as harmonic mean prediction scheme [25,32]. Then, we decide res_i and $sres_i$ in the function VEA-MPC (to be discussed shortly in Alg. 3). Then, the instruction res_i will be sent to the sender in Line 6. Then, c_i will be received (Line 7) and will be super-resolved from res_i to $sres_i$ (and denoised) in Line 8. Please note that we also need to declare the deadline d_i (Line 3 in Alg. 1 and Line 8 in Alg. 2) so that if the transmission or super-resolution/denoising has not been completed, it will be aborted. At the receiver, we also need to flag three states: Q_{i-1}, t_{i-1}, and τ_{i-1}. Q_{i-1} is the quality of the previous chunk, which will influence the smoothness of the video when we make a new decision on Q_i. t_{i-1} and τ_{i-1} show when the system can be freed from the previous tasks, which will influence the delay of the chunk c_i.

Algorithm 4. Predict QoE Algorithm.

---

1 **Function** `Predict_QoE`($t_{i-1}, \tau_{i-1}, B_i, B_{i+1}, \ldots, B_{i+K-1}, s_1, s_2$):

2   Initialize $D(s), \forall s, P(s_1, s_2), Q(s_1, s_2), \forall s_1, s_2$ based on profiling

3   **for** $k \leftarrow i$ **to** $i + K - 1$ **do**

4          $t_k = \max(a_k, t_{k-1}) + D(s_1)/B_k$

5   **if** $t_k \leq d_k$ **then**

6              $\tau_k = \max(t_k, \tau_{k-1}) + P(s_1, s_2)$

7   **if** $\tau_k \leq d_k$ **then**

8                  $Q_k = Q(s_1, s_2)$

9   **else**

10                  $Q_k = Q_0(s_1)$

11                  $\tau_k = d_k$

12   **else**

13              $Q_k = Q_{\min}$

14              $t_k = d_k$

15              $\tau_k = \tau_{k-1}$

16   Calculate QoE as $\text{QoE} = \sum_{k=i}^{i+K-1} Q_k - \lambda \sum_{k=i}^{i+K-1} |Q_k - Q_{k-1}|$

17   **return** QoE

Next, we discuss Alg. 3, where the VEA-MPC is called. Alg. 3 outputs the decision on res_i and sres_i, given the predicted bandwidths and states Q_{i-1}, t_{i-1}, and τ_{i-1}. In Alg. 3, for all valid decisions $(s_1, s_2) \in \mathbf{S}$, we calculate the predicted QoE (Line 8) for the next K chunks, and return the best decision that maximizes the predicted QoE (Lines 9–11).

The predicted QoE (Alg. 4) calculates the truncated QoE accumulated from c_i to c_{i+K-1}, i.e., $\sum_{j=i}^{i+K-1} Q_j - \lambda \sum_{j=i}^{i+K-1} |Q_j - Q_{j-1}|$, given the decision (s_1, s_2). For all the chunks in the window $[i, i + K - 1]$, all the chunks will be handled by the same input decision (s_1, s_2) and we calculate how much truncated QoE can be generated. Please note that the data size $D(\cdot)$, processing delay $P(\cdot, \cdot)$, and quality $Q(\cdot, \cdot)$ are given as input (Line 2). For each chunk in the window $k \in [i, i+K-1]$, we follow the transmission delay by decision s_1 in Line 4, where t_k is updated. If t_k does not miss the deadline (Line 5), we follow the super-resolution (and denoising) delay by decision s_2 (Line 6), where τ_k is updated. If the deadline is still not missed, quality $Q_k = Q(s_1, s_2)$ is generated. If super-resolution missed the deadline, quality $Q_k = Q_0(s_1)$ is generated and we do not process it. If data transmission missed the deadline, quality $Q_k = Q_{\min}$ is generated and we do not transmit it. We process all the chunks and derive $Q_i, \ldots, Q_{i+K-1}$, and calculated the truncated QoE (Line 16).

5.4 Complexity of VEA-ABR

The overall complexity of VEA-ABR is $\mathcal{O}(|\mathbf{S}| \cdot K)$, where $|\mathbf{S}|$ is the number of valid decisions and K is the size of prediction window. Typical values of $|\mathbf{S}|$ and

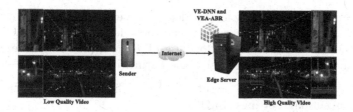

Fig. 4. Implementation setup and two enlarged video frames for visual comparison.

K are < 20 and < 10 so that the complexity of VEA-ABR is small and can be easily accomplished in the real-time environment.

6 Implementation and Experiment

6.1 System Implementation

To evaluate the proposed Real-EVE, we develop a prototype system shown in Fig. 4. We use an Oneplus 7 Pro Android phone with Sony IMX586 photosensor ($(1/2)''$ CMOS size) to act as the sender, and implement the VEA-ABR sender side algorithm using Android programming. The sender encodes the video in the bitrate specified by the edge server, and sends the video to the edge server. We employ a receiver with Intel Gold 6252 CPU and Nvidia GeForce 3090 GPU, and implement the VEA-ABR receiver side algorithm. We install Ubuntu 20.04 LTS on the edge server. The sender is physically located on the curb of a road ~ 20 km away from the lab, while the edge server and the receiver are located in the same lab. The sender accesses the Internet via a bandwidth-limited cellular network, where the signal strength is not strong enough (~ -105 dBm for 4G cellular network [7]). The sender sends captured live nighttime video with the following settings: $1/30$ s exposure time; ISO 6,400; frame rate 20 fps. We allow 4 available resolutions: bitrates in $\{0, 2.7, 4.8, 10.2\}$ Mbps, corresponding to resolutions in $\{0, 270, 540, 1080\}$p, respectively. Videos are transmitted through real-time transport protocol (RTP) [14] over UDP. We use the H.264/MPEG-4 codec throughout the prototype.

As for the VE-DNN, it is implemented using PyTorch 1.7 and trained on the DAVIS dataset [22], which contains 50 video sequences. We break the training into two parts. First, we train the NEA sub-module with mean squared error loss and adaptive moment estimation (ADAM) optimizer. The learning rate is set to 10^{-3} for the first 50 epochs then decreased by ten times every 10 epochs for the rest 30 epochs. We then end-to-end train the CFP and MSU sub-modules with Charbonnier loss, ADAM optimizer, and cosine annealing scheme. The learning rate is set to 10^{-4} for 70 epochs. Note that for each input and output resolution pair in $\{270, 540, 1080\}$p, we train the CFP and MSU sub-modules individually.

As for the VEA-ABR, it is implemented using Python 3.9. We use the peak-signal-to-noise-ratio (PSNR) [23] as the video quality metric. The inter-arrival time of the video chunk is set to 200 ms (i.e., each video chunk contains 200 ms of video), and the maximum tolerable delay d is set to 350 ms. The predefined safe margin Δ is set to 50 ms, so that the total delay is less than 400 ms [13] $(d + \Delta = 400)$. We choose a window size of $K = 4$ (800 ms). The available network bandwidth is predicted using the harmonic mean $B_i = \frac{K}{\sum_{j=i-K}^{i-1} \frac{1}{B_j}}$ for $i \in [k, k + K - 1]$ when c_k is about to arrive. The weight of penalty $\lambda = 1$, following conventional settings [16,20].

6.2   Benchmarks

We consider 9 benchmark schemes to compare with VE-DNN. We use the direct combination of two existing SOTA video denoising algorithms (DVDnet [26] and FastDVDnet [27]) and two existing SOTA video super-resolution algorithms (RRN [12] and BasicVSR++ [2]) in tandem. We swap the order of denoising and super-resolution and treat them as different benchmarks. For example, *RRN + FastDVDnet (B1)* means we run RRN first and then FastDVDnet; *FastDVDnet + RRN (B5)* means we run FastDVDnet first and then RRN. We modify the benchmarks using the same techniques as the MSU sub-module, to allow multi-granularity input and output resolutions. Thus far, we obtain 8 benchmarks denoted B1–B8 as shown in Fig. 5. We also consider RAW, which does not process the input video.

We consider 3 ABR benchmarks to compare with VEA-ABR. ABR-0 does not apply any DNN, and it leverages a traditional ABR algorithm [5] while the VE-DNN is not activated. It uses the current system bandwidth as the predicted bandwidth. ABR-1 is a DNN-aware greedy ABR algorithm that uses the same bandwidth prediction as ABR-0. It chooses the maximum possible transmission resolution (res_j) and super-resolution ($sres_j$) resolution which is expected to fit the playout deadline (d_j). ABR-2 leverages the same VEA-MPC algorithm as VEA-ABR, but is DNN-agnostic. It decides the transmission resolution using VEA-MPC, but uses a greedy algorithm to decide the super-resolution. The PredictQoE algorithm in ABR-2 is modified to take the video quality without DNN enhancement for profiling. The VE-DNN is still active, but chooses the maximum possible super-resolution.

6.3   Effectiveness of VE-DNN Module

We first validate the effectiveness of the VE-DNN module while inactivating the VEA-ABR algorithm. In this experiment, the sender sends the video in real time in 240p resolution (2.7 Mbps), and the edge server super-resolves it to 1080p (10.2 Mbps). We configure the network so that there is always enough bandwidth from the sender to the edge server (and from the edge server to the receiver). We adopt the video samples from DERF dataset [30]. Even though the videos are offline, we send them in real time to mimic real-time scenarios so as to make a fair comparison based on the same videos.

We summarize the quality and delay performance on the two-dimensional coordinate as shown in Fig. 5. Please note that the delay performance only

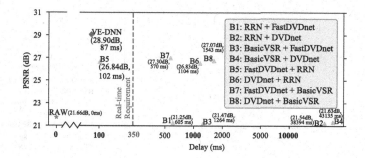

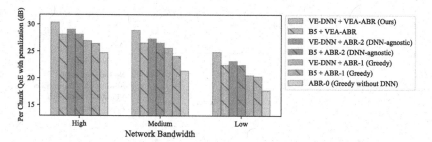

Fig. 5. Video quality vs. delay of VE-DNN and benchmarks.

Fig. 6. QoE comparison under various network bandwidth conditions.

includes the inference (processing) delay while excluding other delays as they are consistent among different benchmarks.

From Fig. 5, we observe that benchmarks B1–B4 yield even poor quality than RAW. This is because processing super-resolution before denoising causes amplified noise, which cannot be eliminated by denoising. Benchmarks B6 and B8 bring large inference delay (> 1000 ms), intolerable for real-time videos. Benchmark B7 yields a smaller delay (570 ms), but still fails to meet the real-time requirement. Compared with our VE-DNN, these benchmarks yield significantly weaker performance. The only benchmark (other than our proposed VE-DNN) acceptable by real-time video is B5. Our VE-DNN yields faster inference (87 ms versus 102 ms) and higher accuracy (28.9 dB versus 26.8 dB). This is because VE-DNN can bring more precise joint super-resolution and denoising while avoiding replicated sub-modules in super-resolution and denoising, as discussed in Sect. 4.2. Our VE-DNN is the best scheme to bring video quality gain without sacrificing much delay. To further illustrate the effectiveness of VE-DNN, a visualized video frame before and after VE-DNN is shown in Fig. 4, in which the red box regions are zoomed in. The frame after VE-DNN shows significant visual improvement in both smoothness and coherence.

6.4 Effectiveness of VEA-ABR Module

In this experiment, we run the whole prototype in realistic settings for 10 s (50 chunks) using DERF dataset. For the video enhancement DNN, we employ

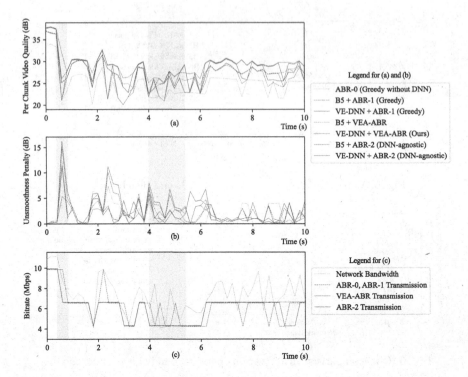

Fig. 7. Trace under the medium network bandwidth condition, where (a) shows per chunk video quality, (b) shows unsmoothness penalty, and (c) shows the bitrate. (Color figure online)

VE-DNN and B5 (the best benchmark in Sect. 6.3). For the ABR algorithm, we use ABR-1, ABR-2, and our VEA-ABR (discussed in Sect. 6.2) with the aforementioned DNNs. We also compare ABR-0, where no video enhancement DNN is applied. In total, we have $2 \times 3 + 1 = 7$ cases to compare.

We run the prototype under high, medium, and low bandwidth networks (average bandwidth being 11.3, 6.5, and 3.2 Mbps respectively) each for 10 s and record the average QoE of the seven cases in Fig. 6. Since the video QoE is $\sum_{i=1}^{N} Q_i - \lambda \sum_{i=2}^{N} |Q_i - Q_{i-1}|$ as described in Eq. 4, the QoE (with penalization) per chunk is calculated as $\frac{\sum_{i=1}^{N} Q_i - \lambda \sum_{i=2}^{N} |Q_i - Q_{i-1}|}{N}$ (shown in Fig. 6). To further identify how Real-EVE outperforms the benchmarks, we further show the trace of medium bandwidth case for the 10 s, including the video quality Q_i in Fig. 7(a), penalty $\lambda |Q_i - Q_{i-1}|$ in Fig. 7(b), and bandwidth (with the bitrate of transmitted video of different benchmarks) in Fig. 7(c) respectively.

From Fig. 6, we observe that our VE-DNN steadily outperforms benchmark B5 (VE-DNN + ABR-1 is better than B5 + ABR-1; VE-DNN + ABR-2 is better than B5 + ABR-2; and VE-DNN + VEA-ABR is better and B5 + VEA-ABR). This matches our expectation as discussed in Sect. 6.3. Also, ABR-0 is much worse compared with others as there is no video enhancement.

As for the comparison between the ABR algorithms, from Fig. 6, VEA-ABR is better than ABR-2, and ABR-2 is better than ABR-1. If we compare the results of different ABR algorithms all using VE-DNN, we have VE-DNN + VEA-ABR is better than VE-DNN + ABR-2, and VE-DNN + ABR-2 is better than VE-DNN + ABR-1.

We further track the performance under medium network bandwidth in Fig. 7 (a). When the network performance is high at the beginning (0–0.4 s), all ABR algorithms choose the highest bitrate, leading to similar performance. However, when the network bandwidth begins to fluctuate, the performance gain brought by VEA-ABR is more evident. For the video chunk arriving at $a_j = 0.6$ (pink window), ABR-0 and ABR-1 missed the transmission deadline due to the bandwidth drop, leading to the worst quality. This is because ABR-0 and ABR-1 decided on a high transmission resolution based on the current observations of the bandwidth without prediction. ABR-2 (VE-DNN + ABR-2 and B5 + ABR-2) completed transmission before the deadline, but the processing missed the deadline, leading to better quality compared with ABR-0 and ABR-1, but worse quality compared with VEA-ABR. Even if ABR-2 adopts MPC, it does not take processing delay into consideration, so the decision on the super-resolved resolution is not optimal. The VEA-ABR algorithm with bandwidth prediction and MPC is the best. It can adapt to the environment faster and suggest a more conservative transmission resolution, and super-resolve the chunk in a timely manner. B5 + VEA-ABR made the same transmission decision as VE-DNN + VEA-ABR, but B5 takes a longer time to super-resolve the chunk, and missed the deadline. This illustrates that Real-EVE is benefited from all of its key designs: bandwidth prediction and MPC to adapt to bandwidth fluctuation (VEA-ABR>ABR-0/ABR-1), video-enhancement awareness (VEA-ABR>ABR-2), and low-latency of video enhancement (VE-DNN>B5). Without even one of the key designs, the deadline could be missed and performance could be significantly decreased.

VEA-ABR + VE-DNN also improves the overall QoE through reducing the unsmoothness penalty. For example, during 4–5.5 s (yellow window) in Fig. 7. The network bandwidth is fluctuating and thus ABR-0 and ABR-1 are also quickly changing video quality to greedily utilize the bandwidth, leading to quickly changing video quality (higher penalty). ABR-2 is also MPC-based, leading to more stable transmission quality. However, ABR-2 is agnostic to super-resolution so that the video quality posterior to the quality enhancement is fluctuating fast, leading to a larger penalty. Compared with VEA-ABR + B5, since VE-DNN outperforms B5 in output video quality, VEA-ABR + VE-DNN leads to better video quality although the unsmoothness penalty is similar. In sum, VEA-ABR + VE-DNN gives the best performance through better balancing per-chunk video quality and video smoothness.

7 Conclusion and Future Directions

In this article, we propose Real-EVE, an edge computing based solution to jointly perform video denoising and video super-resolution for real-time videos.

Real-EVE comprises two key designs: (1) VE-DNN, which jointly de-noises and super-resolves real-time videos more effectively and efficiently with a small delay; and (2) VEA-ABR, which adapts sending rate in response to changing network conditions to optimize the enhanced video quality. We have implemented a real-world prototype and experiment on the performance of Real-EVE compared with a set of benchmarks. Real-EVE substantially outperforms all benchmarks, and both the VE-DNN and VEA-ABR contribute to the performance gain brought by Real-EVE.

We envision edge-assist video enhancement will open up many new research directions. One straightforward direction is to enhance the performance in terms of both delay and video quality (d and QoE will be jointly optimized). More advanced DNN models and bitrate adaption algorithms should be developed, such as transformer-based DNNs and reinforcement-learning based approaches. In addition, real-time video may have fast-changing scenes, and a dynamic DNN rather than a static DNN may further improve the video quality. Finally, if there are multiple real-time video sessions, a well-designed scale-up mechanism is desired to handle the system dynamics.

Acknowledgements. This research was supported in part by the Toronto Mobility Scheme of the University of Sydney.

References

1. Akhshabi, S., Begen, A.C., Dovrolis, C.: An experimental evaluation of rate-adaptation algorithms in adaptive streaming over HTTP. In: MMSys (2011)
2. Chan, K.C., Zhou, S., Xu, X., Loy, C.C.: BasicVSR++: improving video super-resolution with enhanced propagation and alignment. In: CVPR (2022)
3. Chen, C., Chen, Q., Xu, J., Koltun, V.: Learning to see in the dark. In: CVPR, pp. 3291–3300 (2018)
4. Chen, H., Jin, Y., Xu, K., Chen, Y., Zhu, C.: Multiframe-to-multiframe network for video denoising. IEEE (2021)
5. Dash.js: Dash.js. https://github.com/Dash-Industry-Forum/dash.js/wiki (2023)
6. Emmons, H., Vairaktarakis, G.: Flow shop scheduling: theoretical results, algorithms, and applications, vol. 182. Springer Science & Business Media (2012)
7. Engiz, B.K., Kurnaz, Ç.: Comparison of signal strengths of 2G/3G/4G services on a university campus. Int. J. Appl. Math. Electron. Comput. (Special Issue-1), 37–42 (2016)
8. Gow, R.D., et al.: A comprehensive tool for modeling CMOS image-sensor-noise performance. vol. 54, pp. 1321–1329. IEEE (2007)
9. Houdaille, R., Gouache, S.: Shaping HTTP adaptive streams for a better user experience. In: MMSys, pp. 1–9 (2012)
10. Hung, C.C., Ananthanarayanan, G., Bodik, P., Golubchik, L., Yu, M., Bahl, P., Philipose, M.: Videoedge: Processing camera streams using hierarchical clusters. In: SEC, pp. 115–131. IEEE (2018)
11. Huynh-Thu, Q., Ghanbari, M.: Scope of validity of PSNR in image/video quality assessment. Electron. Lett. **44**(13), 800–801 (2008)
12. Isobe, T., Zhu, F., Jia, X., Wang, S.: Revisiting temporal modeling for video super-resolution. In: BMVC (2020)

13. ITU-T Recommendations: One-way transmission time. https://www.itu.int/rec/T-REC-G.114-200305-I/en (2023)
14. Jacobson, V., Frederick, R., Casner, S., Schulzrinne, H.: Realtime transport protocol (RTP). https://www.ietf.org/rfc/rfc3550.txt (2014)
15. Jansen, B., Goodwin, T., Gupta, V., Kuipers, F., Zussman, G.: Performance evaluation of WebRTC-based video conferencing. SIGMETRICS 45(3), 56–68 (2018)
16. Jiang, J., Sekar, V., Zhang, H.: Improving fairness, efficiency, and stability in HTTP-based adaptive video streaming with FESTIVE. In: CoNEXT (2012)
17. Kong, X., Kong, X., et al.: Real-time mask identification for COVID-19: an edge-computing-based deep learning framework. IEEE Internet Things J. 8(21), 15929–15938 (2021)
18. Lee, R., Venieris, S.I., Lane, N.D.: Deep neural network-based enhancement for image and video streaming systems: a survey and future directions. ACM Comput. Surv. 54(8), 1–30 (2021)
19. Liu, C., Yang, H., Fu, J., Qian, X.: Learning trajectory-aware transformer for video super-resolution. In: CVPR, pp. 5687–5696 (2022)
20. Mao, H., Netravali, R., Alizadeh, M.: Neural adaptive video streaming with pensieve. In: SIGCOMM (2017)
21. Mok, R.K., Chan, E.W., Luo, X., Chang, R.K.: Inferring the QoE of HTTP video streaming from user-viewing activities. In: SIGCOMM W-MUST, pp. 31–36 (2011)
22. Perazzi, F., Pont-Tuset, J., McWilliams, B., Van Gool, L., Gross, M., Sorkine-Hornung, A.: A benchmark dataset and evaluation methodology for video object segmentation. In: CVPR (2016)
23. Setiadi, D.R.I.M.: PSNR vs SSIM: imperceptibility quality assessment for image steganography. Multimed. Tools Appl. 80(6), 8423–8444 (2021)
24. Shi, W., et al.: Real-time single image and video super-resolution using an efficient sub-pixel convolutional neural network. In: CVPR, pp. 1874–1883 (2016)
25. Sun, L., Zong, T., Liu, Y., Wang, Y., Zhu, H.: Optimal strategies for live video streaming in the low-latency regime. In: ICNP, pp. 1–4. IEEE (2019)
26. Tassano, M., Delon, J., Veit, T.: DVDnet: A fast network for deep video denoising. In: ICIP (2019)
27. Tassano, M., Delon, J., Veit, T.: FastDVDnet: towards real-time deep video denoising without flow estimation. In: CVPR (2020)
28. Viola, R., Martin, A., Zorrilla, M., Montalbán, J.: MEC proxy for efficient cache and reliable multi-CDN video distribution. In: IEEE BMSB (2018)
29. Wei, K., Fu, Y., Yang, J., Huang, H.: A physics-based noise formation model for extreme low-light raw denoising. In: CVPR (2020)
30. Xiph.org: Derf's test media collection. https://media.xiph.org/video/derf (2022)
31. Yeo, H., Jung, Y., Kim, J., Shin, J., Han, D.: Neural adaptive content-aware internet video delivery. In: USENIX OSDI, pp. 645–661 (2018)
32. Yin, X., Jindal, A., Sekar, V., Sinopoli, B.: A control-theoretic approach for dynamic adaptive video streaming over HTTP. In: SIGCOMM (2015)
33. Zhang, H., Ananthanarayanan, G., Bodik, P., Philipose, M., Bahl, P., Freedman, M.J.: Live video analytics at scale with approximation and delay-tolerance. In: USENIX NSDI (2017)
34. Zuo, X., Yang, J., Wang, M., Cui, Y.: Adaptive bitrate with user-level QoE preference for video streaming. In: INFOCOM, pp. 1279–1288. IEEE (2022)

Optimizing the Parallelism of Communication and Computation in Distributed Training Platform

Xiang Hou[✉], Yuan Yuan[✉], Sheng Ma, Rui Xu, Bo Wang, Tiejun Li, Wei Jiang, Lizhou Wu, and Jianmin Zhang

National University of Defense Technology, Changsha, China
houxiang@alumni.nudt.edu.cn,
{yuanyuan,masheng,xurui16a,bowang,tjli,jiangwei,
lizhou.wu,jmzhang}@nudt.edu.cn

Abstract. With the development of deep learning, DNN models have become more complex. Large-scale model parameters enhance the level of AI by improving the accuracy of DNN models. However, they also present more severe challenges to the hardware training platform for training a large model needs a lot of computing and memory resources, which can easily exceed the capacity of an accelerator. In addition, with the increasing demand for the accuracy of DNN models in academia and industry, the number of training iterations is also skyrocketing. In these backgrounds, more accelerators are integrated on a hierarchical platform to conduct distributed training. In distributed training platforms, the computation of the DNN model and the communication of the intermediate parameters are handled by different hardware modules, so their degree of parallelism profoundly affects the training speed. In this work, based on the widely used hierarchical Torus-Ring training platform and the Ring All-Reduce collective communication algorithm, we improve the speed of distributed training by optimizing the parallelism of communication and computation. Specifically, based on the analysis of the distributed training process, we schedule the computation and communication so that they execute simultaneously as much as possible. Finally, for data parallelism and model parallelism, we reduce the communication exposure time and the computation exposure time, respectively. Compared with the previous work, the training speed (including 5 training iterations) of the Resnet50 model and the Transformer model is increased by 23.77%–25.64% and 11.66%–12.83%.

Keywords: Distributed training simulator · Collective communication operation · Communication exposure time · Computation exposure time

---

This work is supported in part by the National Key RD Project No. 2021YFB0300300, the NSFC (62172430), the NSF of Hunan Province 2021JJ10052, the STIP of Hunan Province 2022RC3065, and the Key Laboratory of Advanced Microprocessor Chips and Systems.

Z. Tari et al. (Eds.): ICA3PP 2023, LNCS 14487, pp. 340–359, 2024.
https://doi.org/10.1007/978-981-97-0834-5_20

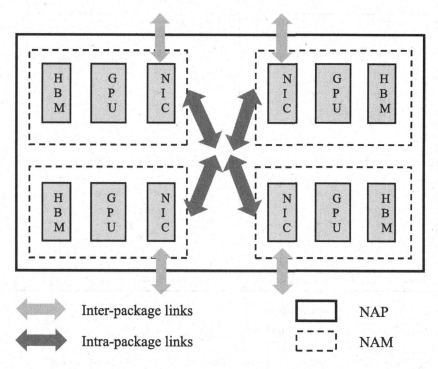

Inter-package links NAP

Intra-package links NAM

Fig. 1. Distributed DNN hardware training platform

1   Introduction

In recent years, Artificial Intelligence (AI) has profoundly affected our daily life in fields such as speech recognition [8,10,11] and image classification [3,16, 22]. Deep learning is one of the important ways to realize AI. To quickly train DNN models, researchers constantly seek to invent more effective accelerators [14,32–34], e.g. GPU, TPU. Today, the number of neural network layers has increased from the original 5 layers (LeNet5 [17]) to hundreds of layers [31]. The proliferation of model parameters makes the training process of DNN consume a lot of memory and computing resources. So using a single accelerator system for training has quite low efficiency and performance.

Using distributed DNN training platform to assign training tasks on multiple accelerators is a way to improve training speed. For example, Google's TPU system uses the 16×16 2D Torus topology to interconnect 256 TPUs to form Google Cloud TPU for training DNN models [4]. The distributed training platform divides training data and DNN models into multiple accelerators, which greatly reduces the memory and computing ability requirements of a single accelerator.

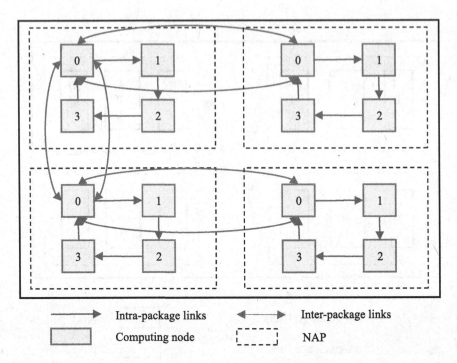

Fig. 2. The widely used hierarchical Torus-Ring topology. For inter-package links, the figure only shows the connections between all nodes labeled 0 in different NAPs.

Figure 1 shows the architecture of a typical distributed training platform, which is a hierarchical structure composed of multiple Neural Accelerator Packages (NAP) [24]. And each NAP integrates multiple Neural Accelerator Modules (NAM), each of which consists of a computing node, high-bandwidth memory, and a dedicated network interface card. Besides, the hierarchical interconnection network topology comprises the intra-package links (e.g. PCI-E, NVLink) for NAM-to-NAM communication and the inter-package links (e.g. Ethernet, InfiniBand) for NAP-to-NAP communication.

In the distributed training platform, the computation of the DNN model and the communication of the intermediate parameters are controlled by different hardware modules. Therefore, in addition to the hardware performance, the degree of parallelism of computation and communication also has a profound impact on the training speed. In this work, based on the widely used hierarchical Torus-Ring topology (Fig. 2) and the Ring All-Reduce collective communication algorithm, we make innovative scheduling of computation operations and communication operations during distributed training to improve training speed. In particular, we make the following contributions.

1. In data parallelism, we reduce the communication exposure time by overlapping the communication of the weight gradients and the computation of the activations.

2. In model parallelism, we reduce the computation exposure time by overlapping the computation of the weight gradients and the communication of the activations.
3. We simulate distributed training using five DNN workloads. And the experimental results show that, compared with the existing work, our proposed optimized design significantly improves training speed.

The paper is organized as follows. Section 2 introduces the background and some related work. In Sect. 3 and Sect. 4, we analyze the process of distributed training and propose innovative communication and computation scheduling methods. Section 5 describes the experimental method. In Sect. 6, we present the results and make some discussions. Finally, Sect. 7 concludes the paper.

2 Background and Related Work

2.1 Distributed Training

To cope with the challenges brought by the surge in model parameters and training data, researchers have begun to conduct DNN training on distributed training platforms. Distributed training refers to splitting the DNN model and/or training data among multiple accelerators to reduce the computational pressure of each accelerator.

When multiple accelerators are used to perform training tasks in parallel, there are two ways to update the parameters of the DNN models, i.e. asynchronous [5], and synchronous [6]. The asynchronous update scheme is usually used in the parameter server framework [19]. In which, the training data is distributed over worker nodes, and the server node is responsible for collecting the gradients of the worker nodes and updating their models through the reduction operation [18]. The asynchronous update scheme can generate the stale gradients [6], i.e. the model may have been updated while a worker is computing its local gradient, which reduces the training accuracy. Therefore, the current mainstream researches adopt the synchronous update scheme. That is, each node generates a local gradient through computation and then reduces the local gradient on all nodes. Finally, all nodes update their model parameters at the same time. Our work is also based on synchronous update.

In distributed training, the main parallelized strategies used include data parallelism [1], model parallelism [9], and hybrid parallelism. In data parallelism, the training data is first segmented and allocated to different computing nodes (each node has a complete DNN model) for training to generate the local gradient. Then the local gradients of multiple nodes are communicated to generate the final gradient, and finally, the weights are updated [27]. In model parallelism, the DNN model is segmented and assigned to different nodes and each node processes all training data. The segmentation of the DNN model is shown in Fig. 3.

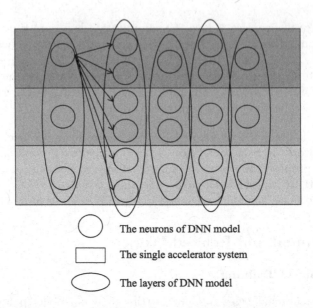

The neurons of DNN model

The single accelerator system

The layers of DNN model

Fig. 3. Model parallelism: the DNN model are split into different single accelerator system.

Due to the segmentation, communication between multiple nodes is also required to generate the final gradient and update the weights. In hybrid parallelism, all nodes are divided into different groups. And different parallelized strategies are used within or between groups.

2.2 Collective Communication Operation

To realize distributed training, the communication of three types of intermediate parameters, i.e. activations, input gradients, and weight gradients, between different nodes is needed in the three stages of training, i.e. forward pass (FP), backward propagation (BP), and weight update (WU). And the communication of the data is completed by collective communication operations [15]. Collective communication refers to multiple nodes simultaneously communicating data and performing specific operations [24]. There are three main kinds of communication in distributed training, i.e. all-gather [30], all-reduce [7], and all-to-all [28]. Figure 4 uses four nodes as an example to show the initial and final states of the three communication operations.

In the all-gather operation, each node stores its data before the communication starts. During the communication, the data of each node is transmitted to all other nodes. In the end, each node gets all the data, as shown in Fig. 4(a). In model parallelism, the communication of activations in the FP stage and the communication of input gradients in the BP stage are completed by the all-gather operation.

Node 0	Node 1	Node 2	Node 3
X0	X1	X2	X3

$\rightarrow$

Node 0	Node 1	Node 2	Node 3
X0	X0	X0	X0
X1	X1	X1	X1
X2	X2	X2	X2
X3	X3	X3	X3

(a) The all-gather operation

Node 0	Node 1	Node 2	Node 3
$X^{(0)}$	$X^{(1)}$	$X^{(2)}$	$X^{(3)}$

$\rightarrow$

Node 0	Node 1	Node 2	Node 3
$\sum_i X^{(i)}$	$\sum_i X^{(i)}$	$\sum_i X^{(i)}$	$\sum_i X^{(i)}$

(b) The all-reduce operation

Node 0	Node 1	Node 2	Node 3
$X_0^{(0)}$	$X_0^{(1)}$	$X_0^{(2)}$	$X_0^{(3)}$
$X_1^{(0)}$	$X_1^{(1)}$	$X_1^{(2)}$	$X_1^{(3)}$
$X_2^{(0)}$	$X_2^{(1)}$	$X_2^{(2)}$	$X_2^{(3)}$
$X_3^{(0)}$	$X_3^{(1)}$	$X_3^{(2)}$	$X_3^{(3)}$

$\rightarrow$

Node 0	Node 1	Node 2	Node 3
$X_0^{(0)}$	$X_1^{(0)}$	$X_2^{(0)}$	$X_3^{(0)}$
$X_0^{(1)}$	$X_1^{(1)}$	$X_2^{(1)}$	$X_3^{(1)}$
$X_0^{(2)}$	$X_1^{(2)}$	$X_2^{(2)}$	$X_3^{(2)}$
$X_0^{(3)}$	$X_1^{(3)}$	$X_2^{(3)}$	$X_3^{(3)}$

(c) The all-to-all operation

Fig. 4. The collective communication operations

The all-reduce operation is similar to the all-gather operation. The difference is that, after each node gets the data, it performs the reduction operation, as shown in Fig. 4(b). In distributed training, the all-reduce operation is usually performed in the WU stage of data parallelism. At present, the Ring All-Reduce algorithm [21] is the most popular collective communication algorithm. Figure 5 depicts the basic logical structure of the algorithm. In the algorithm, each node can only receive data from the previous node, which logically forms a ring. The algorithm can be applied to any physical topology without modification.

In terms of the all-to-all operation, each node needs to send different parts of the data to other nodes, as shown in Fig. 4(c). When using model parallelism to train the Embedding layer of DLRM [20], the all-to-all operation is indispensable in the FP and BP stages.

2.3 Related Work

Today, researchers are increasingly exploring distributed training platforms [13]. Based on the GARNET network simulator [2], Rashidi et al. [24] release a distributed training platform simulator called ASTRA-sim that can collaboratively design the collective communication algorithm and the topology. Using this simulator, they conduct in-depth research on the design of intra-package networks. Further, they connect the ASTRA-sim simulator and the NS3 simulator [26] to supply a facility that can analyze the inter-package networks [23]. To increase the utilization of the high bandwidth provided by intra-package links, Rashidi et al. [25] propose the Accelerator Collectives Engine (ACE) microarchitecture.

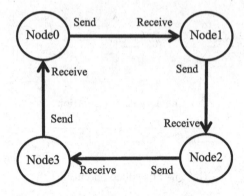

Fig. 5. The logical structure of the Ring All-Reduce algorithm

By embedding the ACE into the NIC, the distributed training speed is greatly improved.

In the distributed training platforms, computation and communication are handled by different hardware modules. In addition, under the same physical topology and collective communication algorithm, the amount of computation or communication is fixed. That is, the amount of communication or computation is independent of the execution order of communication and computation. Therefore, the degree of parallelism of computation and communication profoundly affects the training speed. However, so far, the academic research on this aspect is not sufficient. In previous work, the parallel approach used in distributed training is much the same as the DAG model described by Shi et al. [29] and the Poseidon communication architecture described by Hao et al. [35]. The DAG and the Poseidon utilize the hierarchical model structure in DNN to overlay communication and computation to improve training speed. However, as described in Sect. 3 and Sect. 4 below, these approaches still leave a lot of communication and computation exposure time that can be optimized.

3   The Parallelism of Communication and Computation in Data Parallelism

3.1   The Training Process When Using Data Parallelism

The training process when using data parallelism is conducted as follows. Firstly, the FP stage computes the activations. Then, the BP stage computes input gradients. Finally, the WU stage computes the weight gradients and updates the weights by communication. There is no collective communication during the first two stages. Yet, the WU stage uses the all-reduce collective communication operation to synchronize the computed weight gradients of all nodes.

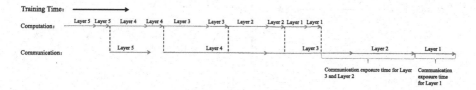

Fig. 6. The BP and WU stages for training a 5-layer DNN model in data parallelism. The blue arrows represent the BP stage, and the green arrows represent the WU stage. (Color figure online)

Since the communication of the weight gradient does not affect the computation process of the BP and WU stages, the communication of the weight gradient of any layer can be performed immediately after the computation of the weight gradients of this layer is completed. Accordingly, Shi et al. [29] propose a general Directed Acyclic Graph (DAG) model to describe the training algorithm. This model reduces training time by overlapping the computation and communication. And the Poseidon communication architecture proposed by Hao et al. [35] has a similar principle. Taking the training of a 5-layer DNN model as an example, Fig. 6 depicts the communication and computation process in the BP and WU stages.

In Fig. 6, when the computation of Layer 1 is completed, the communication of Layer 3 and Layer 2 has not been completed, which brings the communication exposure time. In addition, the communication of Layer 1 cannot be parallelized with the computation, so the communication exposure time of Layer 1 is unavoidable.

3.2   Reducing the Communication Exposure Time

In one training iteration, the above-mentioned communication exposure time is unavoidable. However, the time can be reduced if multiple training iterations are considered simultaneously. In the training process proposed by Shi et al. [29], there is no data communication when computing activations in the FP stage. If the communication and the activation computation are performed simultaneously, the communication speed can be greatly improved. Based on this insight, we optimize the training iteration of data parallelism, and the FP stage of the optimized training process is shown in Fig. 7.

In the training of the DNN model, the WU stage of any layer only needs to be completed before the computation of the activation of this layer in the next iteration. Therefore, when computing the activation of the nth layer, the weight gradient communication of the $(n+1)th$ layer in the previous iteration can be carried out. According to this idea, we move the communication exposure time

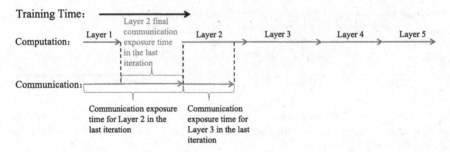

Fig. 7. The proposed training process of the 5-layer DNN model in data parallelism. The red arrows represent the FP stage. (Color figure online)

for Layer 3 and Layer 2 in the last iteration to the FP stage in this iteration. As can be observed from Fig. 7, the communication exposure time of Layer 3 is completely hidden. In addition, the communication exposure time of Layer 2 is greatly reduced and is marked as Layer 2 final communication exposure time in the last iteration, as shown in green in the figure.

To sum up, when this 5-layer DNN model is trained for 3 iterations, the exposure communication time consists of three parts, which are, (*a*) communication exposure time for Layer 1 in all 3 iterations, (*b*) Layer 2 final communication exposure time in the 2nd and 3rd iterations, (*c*) the communication exposure time for Layer 3 and Layer 2 in the 3rd iteration. Compared to the previous work, the proposed approach improves training speed by maximizing the overlap of computation and communication.

4   The Parallelism of Communication and Computation in Model Parallelism

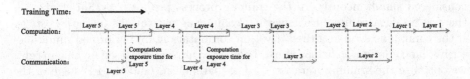

Fig. 8. The BP and WU stages for training a 5-layer DNN model in model parallelism. The blue arrows represent the BP stage, and the green arrows represent the WU stage. (Color figure online)

4.1   The Training Process When Using Model Parallelism

In model parallelism, firstly, the activations are computed layer by layer in the FP stage. After each layer is computed, the all-gather collective communication operation is executed to communicate the activations. Secondly, the BP stage computes the input gradients layer by layer. After each level is computed, the all-gather operation is performed to communicate the input gradients. Lastly, the WU phase computes the weight gradients without any communication operation.

In the BP stage, the weight gradient of the nth layer can be calculated after obtaining the input gradient of the nth layer. Moreover, the calculation can be performed synchronously with the communication of the input gradient of the nth layer. Based on these features, the current researches improve the training speed of model parallelism by overlapping the BP stage and the WU stage. Figure 8 takes a five-layer DNN model as an example to describe the BP and WU stages. In the figure, for each layer of the model, the communication of the input gradient and the computation of the weight gradient are executed in parallel.

Although this method reduces the training time, when the computation time of the weight gradient of any layer is longer than the communication time of the input gradient, there will still be a lot of computation exposure time. And the computation exposure time of the fourth and fifth layers is depicted in Fig. 8. Conversely, when the communication time is greater than the computation time, the communication exposure time is generated.

In the BP stage, the communication of the input gradient of any layer must be carried out after the computation of the input gradient, so the above communication exposure time is inevitable. For the computation exposure time, effective scheduling methods can be used to reduce it.

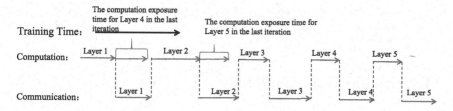

Fig. 9. The proposed training process of the 5-layer DNN model in model parallelism. The red arrows represent the FP stage. (Color figure online)

4.2   Reducing the Computation Exposure Time

In the training of DNN, the activations of each layer are calculated by multiplying the input of this layer by the weight gradients. Therefore, for any layer, the computation of the weight gradients in the previous iteration only needs to be completed before the computation of the activations in this iteration. We use this viewpoint to reduce the computation exposure time of the weight gradients. Figure 9 shows the FP stage of the proposed training process in model parallelism.

In the previous research on distributed training, when using the all-gather operation to communicate the activations, no computation occurs, which reduces the training speed. In the proposed training process, the computation exposure time of the weight gradients in the previous iteration is scheduled to the FP stage of the current iteration. In addition, as long as it can ensure that the nth layer's computation can be completed before the nth layer's activation calculations, the computation can be scheduled in parallel with any layer's communication. Comparing Fig. 8 and Fig. 9, the computation exposure time for Layer 4 and Layer 5 in the last iteration is completely hidden.

After scheduling the computation operation for model parallelism, the communication and the computation can achieve maximum parallelism in distributed training platforms. The final exposed computation includes the following parts, (a) the computation of weight gradients of the first layer, (d) the computation exposure time in the last training iteration.

Table 1. Workloads

Number	Workload	DNN layers	Parallel strategy	Iteration	Collective communication operation
1	DLRM_ HybridParallel	8	Hybrid	5	All-to-all All-reduce
2	MLP_ ModelParallel	6	Model	5	All-gather
3	MLP_HybridParallel_ Data_Model	6	Hybrid	5	All-gather All-reduce
4	Resnet50_ DataParallel	50	Data	2 or 5	All-reduce
5	Transformer_ HybridParallel	57	Hybrid	2 or 5	All-gather All-reduce All-to-all

5   Experimental Methodology

5.1   Experimental Tools

We use the ASTRA-sim [24] as the performance evaluation tool. The ASTRA-sim is a distributed DNN training platform simulator. It can simulate cycle-level communication behavior for the parallelized strategies including data parallelism, model parallelism, and hybrid parallelism. We modify it to support our proposed communication and computation scheduling methods.

Table 2. Simulation Configurations

Parameter	Value
Number of nodes within a NAP	4 or 8
Number of NAPs	4
VC/VNET	2
Buffers per VC	5000
Intra-package link width	512bits/ns
Inter-package link width	256bits/ns
Router latency	5ns
Intra-package link latency	90ns
Inter-package link latency	200ns
Flit width	2048bits
Cycle period	1ns

5.2 Workloads

The workloads used in the experiments are given in Table 1. The models in the workloads consist of two small DNN models (the model in the second and third workloads is the same) and two large DNN models. And we train them for five or two iterations. The following is a detailed description.

The DNN model in the first workload is the DLRM model [20]. The embedding layer of the model uses model parallelism due to a large number of model parameters, and the remaining 7 layers (conventional MLP layer) use data parallelism. The training of the embedding layer generates the all-to-all operation. The DNN model in the second and third workloads is a conventional DNN model with 6 layers. In the second workload, we use data parallelism. In the third workload, data parallelism is used across NAPs, and model parallelism is used across nodes within a NAP.

The DNN model used in the fourth workload is the Resnet-50 model [12], which is a deep residual network applied in the field of image recognition. We train it using data parallelism. The DNN model in the fifth workload is the Transformer model [31], which is launched by Google for natural language processing. The model contains an embedding layer, so there are all-to-all operations in training. We train it using hybrid parallelism. That is, the local dimension and the horizontal dimension use data parallelism, and the vertical dimension uses model parallelism.

5.3 The Configuration of the Hardware Platform

In the work, we apply the widely used hierarchical Torus-Ring platform and the Ring All-Reduce collective communication algorithm to explore the performance of distributed training. And we perform experiments on two sizes of training platforms with 16 and 32 processors, respectively. Table 2 shows the main simulation configurations of the distributed training platform.

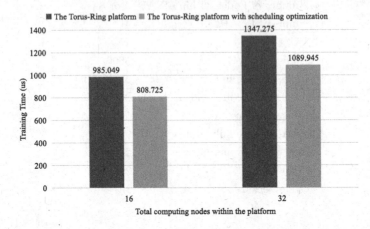

Fig. 10. The time consumed to train the DLRM_HybridParallel workload. The blue bar represent the hierarchical Torus-Ring platform. The green bar represents the platform with scheduling method optimization. (Color figure online)

6   Results and Discussion

The content of this section is arranged as follows. In Sect. 6.1, based on the training platform of the two scales, we simulate distributed training of the small DNN models. To verify the efficiency of the proposed scheduling methods, we conducted comparative experiments based on previous training methods and the proposed methods. In Sect. 6.2, we use the two large workloads to evaluate the proposed methods. Lastly, Sect. 6.3 analyzes the impact of different training iteration numbers on the acceleration effect.

6.1   The Training Time of the Three Small DNN Models

Figure 10, 11 and 12 describe the training time consumed when training the small DNN models, including 5 training iterations. The simulated training platform consists of 4 NAPs with 4 (the left of each figure) or 8 (the right of each figure) computing nodes per NAP.

We use data parallelism to train the MLP layers of the DLRM model, therefore, the all-reduce communication of weight gradient is carried out in the WU stage. In the experiment, when the computation time of any layer is greater than the communication time of the input gradient in the BP stage, we interrupt the computation when the communication is completed. Then, in the communication of the FP stage of the next iteration, we restore the previously interrupted

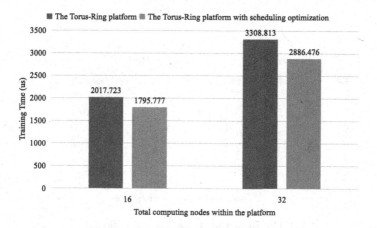

Fig. 11. The time consumed to train the MLP_ModelParallel workload. The blue bar represent the hierarchical Torus-Ring platform. The green bar represents the platform with scheduling method optimization. (Color figure online)

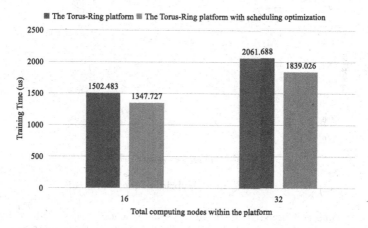

Fig. 12. The time consumed to train the MLP_HybridParallel_Data_Model workload. The blue bar represent the hierarchical Torus-Ring platform. The green bar represents the platform with scheduling method optimization. (Color figure online)

computation operation. For the second workload, when the computation time of any layer's weight gradient is greater than the communication time of the input gradient, we interrupt the computation when the communication is completed. In the FP stage of the next iteration, when the activations perform the all-gather communication operation, we continue the computation of the weight gradients. In the third workload, we use both proposed scheduling methods simultaneously.

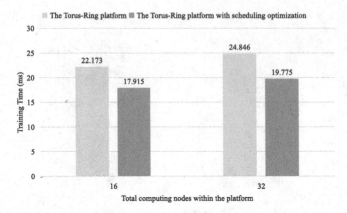

(a) The Resnet50_DataParallel workload

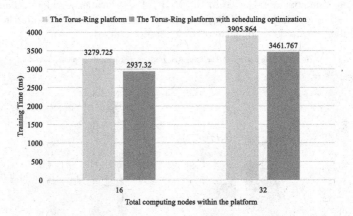

(b) The Transformer_HybridParallel workload

Fig. 13. The time consumed to train the two large workloads, including 5 training iterations. The yellow bar represent the hierarchical Torus-Ring platform, and the green bar represents the platform with scheduling method optimization. (Color figure online)

From Fig. 10, 11 and 12, it can be seen that, compared with the previous training methods, the proposed method significantly improves the distributed training speed of all three kinds of parallelism. Among them, data parallelism achieves a speedup of 1.22–1.24 times, model parallelism achieves a speedup of 1.12–1.15 times, and hybrid parallelism achieves a speedup of 1.11–1.12 times. In addition, when more computing nodes are set up in a NAP, i.e. from four to eight, the acceleration provided by our method is even more significant. For example, in Fig. 10, the speedup increases from 1.22× to 1.24×. It means that our method is more important when larger-scale distributed training platforms need to be deployed.

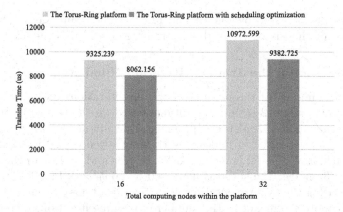

(a) The Resnet50_DataParallel workload

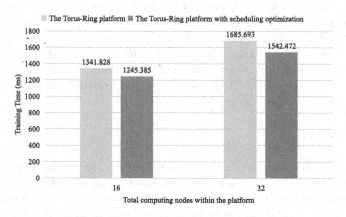

(b) The Transformer_HybridParallel workload

Fig. 14. The time consumed to train the two large workloads, including 2 training iterations. The yellow bar represent the hierarchical Torus-Ring platform, and the green bar represents the platform with scheduling method optimization. (Color figure online)

6.2   The Training Time of the Two Large DNN Models

To verify the efficiency of the proposed method in training large and complex DNN models, we simulate the distributed training of the Resnet-50 model and the Transformer model, including 5 training iterations. And the experimental results are shown in Fig. 13.

As can be seen from the figure, the training speed of the platform with proposed scheduling methods is faster compared to the hierarchical Torus-Ring platform. In the distributed training platform consisting of 16 computing nodes, the training speed of the Resnet50 model and the Transformer model is increased by 23.77% and 11.66%, respectively. In the platform consisting of 32 nodes, the training speed is increased by 25.64% and 12.83%. The results demonstrate that the proposed method is very effective in training larger DNN models.

6.3   The Impact of the Training Iteration Numbers

To analyze the impact of the number of training iterations on the acceleration effect, we also simulate 2 training iterations based on the Resnet50 model and the Transformer model. And the experimental results are shown in Fig. 14. In the platform consisting of 16 nodes, the training speed of the Resnet50 model and the Transformer model is increased by 15.67% and 7.74%. In the platform consisting of 32 nodes, the speed is increased by 16.94% and 9.29%.

By comparing the experimental results, it can be observed that the acceleration effect is more significant as the number of training iterations increases. The reasons are as follows. In the training of DNN models, the training time consumed for each iteration is the same. In the work, we reduce the training time for any iteration by scheduling communication or computation, except for the last iteration. Therefore, as the number of training iterations increases, the proportion of the total reduced communication exposure time or computation exposure time increases. From this, when increasing the number of training iterations to improve the accuracy of the DNN model, our proposed scheduling methods are more efficient in training the model.

7   Conclusion

With the continuous development of deep learning, DNN models have become more and more complex, which leads to an exponential increase in training time. To meet this challenge, training models on distributed training platforms is one of the strategies. In distributed training platforms, the computing speed of the processors and the communication speed of the interconnection links are the most important factors affecting the training speed. In addition to the hardware performance, the parallelism of communication and computation also affects the training speed. However, at present, academic research in this field is not sufficient, which results in a significant waste of time in training.

In this work, through the in-depth analysis of distributed training process, we propose novel scheduling methods to maximize the parallelism of communication and computation. Compared with the previous work, the proposed approach greatly increases the speed of distributed training by reducing the communication exposure time and the computation exposure time. And the acceleration effect becomes more significant as the number of training iterations increases. Moreover, the approach is applicable to arbitrary physical topologies and collection communication algorithms.

In future work, we will implement the proposed optimization method with the help of deep learning tools, and further study the following aspects, (a) how to efficiently interrupt/start the communication or computation in the hardware platform, and (b) the effect of the number of DNN layers and the size of each layer on performance improvement.

References

1. Abadi, M., et al.: Tensorflow: large-scale machine learning on heterogeneous distributed systems. arXiv preprint arXiv:1603.04467 (2016)
2. Agarwal, N., Krishna, T., Peh, L.S., Jha, N.K.: Garnet: a detailed on-chip network model inside a full-system simulator. In IEEE International Symposium on Performance Analysis of Systems and Software, ISPASS 2009, April 26–28, 2009, Boston, Massachusetts, USA, Proceedings (2009)
3. Akata, Z., Reed, S., Walter, D., Lee, H., Schiele, B.: Evaluation of output embeddings for fine-grained image classification. In Proceedings of the IEEE Conference on Computer Vision and Pattern Recognition, pp. 2927–2936 (2015)
4. Chao, C., Saeta, B.: Cloud tpu: codesigning architecture and infrastructure. In: Hot Chips, volume 31 (2019)
5. Chaturapruek, S., Duchi, J.C., Ré, C.: Asynchronous stochastic convex optimization: the noise is in the noise and sgd don't care. In: Advances in Neural Information Processing Systems, vol. 28 (2015)
6. Chen, J., Pan, X., Monga, R., Bengio, S., Jozefowicz, R.: Revisiting distributed synchronous sgd. arXiv preprint arXiv:1604.00981 (2016)
7. Cho, M., Finkler, U., Kung, D., Hunter, H., et al.: Blueconnect: decomposing all-reduce for deep learning on heterogeneous network hierarchy. Ibm Journal of Research and Development, **PP**(99), 1–1 (2019)
8. Chorowski, J., Bahdanau, D., Serdyuk, D., Cho, K., Bengio, Y.: Attention-based models for speech recognition. Computer ence **10**(4), 429–439 (2015)
9. Dean, J., et al.: Large scale distributed deep networks. In: Advances in Neural Information Processing Systems, 25 (2012)
10. Devlin, J., Chang, M.W., Lee, K., Toutanova, K.: Bert: Pre-training of deep bidirectional transformers for language understanding. arXiv preprint arXiv:1810.04805 (2018)
11. Graves, A., Jaitly, N., Mohamed, A.R.: Hybrid speech recognition with deep bidirectional lstm. In: Automatic Speech Recognition and Understanding (ASRU), 2013 IEEE Workshop (2013)
12. He, K., Zhang, X., Ren, S. and Sun, J.: Deep residual learning for image recognition. In: Proceedings of the IEEE Conference on Computer Vision and Pattern Recognition, pp. 770–778 (2016)
13. Hou, X., Xu, R., Ma, S., Wang, Q., Jiang, W., Lu, H.: Co-designing the topology/algorithm to accelerate distributed training. In: 2021 IEEE Intl Conf on Parallel & Distributed Processing with Applications, Big Data & Cloud Computing, Sustainable Computing & Communications, Social Computing & Networking (ISPA/BDCloud/SocialCom/SustainCom), pp. 1010–1018, 2021
14. Jouppi, N.P., Yoon, D.H., Ashcraft, M., Gottscho, M., Patterson, D.: Ten lessons from three generations shaped google's tpuv4i : Industrial product. In: 2021 ACM/IEEE 48th Annual International Symposium on Computer Architecture (ISCA) (2021)
15. Kielmann, T., Hofman, R.F., Bal, H.E., Plaat, A., Bhoedjang, R.A.. Magpie: Mpi's collective communication operations for clustered wide area systems. In: Proceedings of the seventh ACM SIGPLAN symposium on Principles and practice of parallel programming, pp. 131–140 (1999)
16. Krizhevsky, A., Sutskever, I., Hinton, G.E.: Imagenet classification with deep convolutional neural networks. In: NIPS (2012)

17. LeCun, Y., Bottou, L., Bengio, Y., Haffner, P.: Gradient-based learning applied to document recognition. Proc. IEEE **86**(11), 2278–2324 (1998)
18. Li, M., et al.: Scaling distributed machine learning with the parameter server. In: 11th USENIX Symposium on Operating Systems Design and Implementation (OSDI 14), pp. 583–598 (2014)
19. Li, M., Andersen, D.G., Smola, A.J., Yu, K.: Communication efficient distributed machine learning with the parameter server. In: Advances in Neural Information Processing Systems, **27** 19–27 (2014)
20. Naumov, M., et al.: Deep learning recommendation model for personalization and recommendation systems. arXiv preprint arXiv:1906.00091 (2019)
21. Patarasuk, P., Yuan, X.: Bandwidth optimal all-reduce algorithms for clusters of workstations. J. Parall. Distrib. Comput. **69**(2), 117–124 (2009)
22. Perez, L., Wang, J.: The effectiveness of data augmentation in image classification using deep learning. arXiv preprint arXiv:1712.04621 (2017)
23. Rashidi, S., Shurpali, P., Sridharan, S., Hassani, N., Krishna, T.: Scalable distributed training of recommendation models: An astra-sim + ns3 case-study with tcp/ip transport. In: 2020 IEEE Symposium on High-Performance Interconnects (HOTI) (2020)
24. Rashidi, S., Sridharan, S., Srinivasan, S., Krishna, T.: Astra-sim: enabling sw/hw co-design exploration for distributed dl training platforms. In: 2020 IEEE International Symposium on Performance Analysis of Systems and Software (ISPASS) (2020)
25. Rashidi, S., Sridharan, S., Srinivasan, S., Denton, M., Krishna, T.: Efficient communication acceleration for next-gen scale-up deep learning training platforms. arXiv preprint (2020)
26. Riley, G.F., Henderson, T.R.: The ns-3 network simulator. In: Wehrle, K., Güneş, M., Gross, J. (eds.) Modeling and Tools for Network Simulation, pp. 15–34. Springer Berlin Heidelberg, Berlin, Heidelberg (2010). https://doi.org/10.1007/978-3-642-12331-3_2
27. Schlkopf, B., Platt, J., Hofmann, T.: Map-reduce for machine learning on multicore. In: Advances in Neural Information Processing Systems 19: Proceedings of the 2006 Conference
28. Sepulchre, R., Paley, D.A. Leonard, N.E.: Stabilization of planar collective motion: All-to-all communication. IEEE Trans. Autom. Contr. **52**(5), 811–824 (2007)
29. Shi, S., Wang, Q., Chu, X., Li, B.: dag model of synchronous stochastic gradient descent in distributed deep learning. In: 2018 IEEE 24th International Conference on Parallel and Distributed Systems (ICPADS), pp. 425–432. IEEE (2018)
30. Träff, J.L.: Efficient all-gather communication on parallel systems with hierarchical communication structure. preparation (2003)
31. Vaswani, A.: Attention is all you need. In: Advances In Neural Information Processing Systems, 30 (2017)
32. Xu, R., Ma, S., Guo, Y., Li, D.: A survey of design and optimization for systolic array based dnn accelerators. In: ACM Computing Surveys (2023)
33. Rui, X., Ma, S., Wang, Y., Chen, X., Guo, Y.: Configurable multi-directional systolic array architecture for convolutional neural networks. ACM Trans. Architect. Code Optim. (TACO) **18**(4), 1–24 (2021)

34. Rui, X., Ma, S., Wang, Y., Guo, Y., Li, D., Qiao, Y.: Heterogeneous systolic array architecture for compact cnns hardware accelerators. IEEE Trans. Parallel Distrib. Syst. **33**(11), 2860–2871 (2021)
35. Zhang, H., et al.: Poseidon: an efficient communication architecture for distributed deep learning on {GPU} clusters. In: 2017 USENIX Annual Technical Conference (USENIX ATC 17), pp. 181–193 (2017)

FedSC: Compatible Gradient Compression for Communication-Efficient Federated Learning

Xinlei Yu, Zhipeng Gao[✉], Chen Zhao, and Zijia Mo

The State Key Laboratory of Networking and Switching Technology, Beijing University of Posts and Communications, Beijing, China
{yuxinlei518,gaozhipeng,zc_zhaochen,mozijia}@bupt.edu.cn

Abstract. Federated Learning (FL) communication costs hinge on communication frequency, device count, and per-communication-round costs. Ideally, minimizing these within the device cluster tolerance significantly curtails data traffic. Among various methods to reduce per-communication-round costs, gradient compression stands out. Gradient compression concentrates on the changes in the model rather than the model parameters. This not only prevents synchronization issues caused by varying compression levels across devices, but also incurs only minor precision loss. This makes it especially apt for distributed scenarios like FL. Existing gradient compression methods are tailored to capitalize on elements nearing zero in high-frequency settings, aiming for sparse representation and efficient encoding. Yet, in low-frequency scenarios where elements deviate from zero, this strong emphasis on sparsity results in significant compression errors. To resolve this, we propose Federated Statistical Compression (FedSC) shifts the focus from sparsity. Designed specifically for low-frequency settings, it hones in on the inherent statistical characteristics and relationships among the gradient elements. It extracts statistical information from each gradient, replacing the original gradient sent to the central server. We further introduce hierarchical estimation to improve accuracy, grouping elements in a gradient by layer label. Experimentally, FedSC outperforms most existing gradient compression algorithms for FL, converging with fewer communication bits while maintaining high model accuracy.

Keywords: Federated learning · communication-efficient · gradient compression

1 Introduction

Cloud computing [1], which forwards large amounts of data to the remote cloud for processing and training, is rapidly transitioning to the FL setting [2,3] for privacy preservation [4]. The FL system involves a central server responsible for model aggregation and multiple devices responsible for local training. The FL training process includes three steps: 1) training local model updates (model

Z. Tari et al. (Eds.): ICA3PP 2023, LNCS 14487, pp. 360–379, 2024.
https://doi.org/10.1007/978-981-97-0834-5_21

parameters or gradients) using local samples on devices, 2) sending local model updates to the central server, and 3) aggregating local models on the central server [5,6].

However, high communication costs between devices and the central server pose a critical challenge for FL due to the large number of devices attempting to send their local model updates to the central server [7,8]. With the continued development of Artificial Intelligence (AI), models on devices are becoming deeper to meet varying needs in application scenarios, such as Computer Vision (CV) [9] and Natural Language Processing (NLP) [10], which further exacerbates the communication pressure on FL [11]. Therefore, it is necessary to explore communication-efficient ways to train the global model. In the FL mechanism mentioned above, the total number of bits communicated by devices and the central server during training can be expressed as:

$$b_{total} \in \mathcal{O} \left(\underbrace{T \times r_T}_{(i)} \times \underbrace{N \times r_N}_{(ii)} \times \underbrace{|\omega|_{bits}}_{(iii)} \right) \tag{1}$$

where T represents the total number of training iterations performed by each device and r_T represents the communication frequency. We assume that δ is the number of local iterations, then $r_T = \frac{1}{\delta}$. Therefore, the multiplicative component (i) represents the communication rounds. $|\omega|_{bits}$ are the number of bits exchanged between a device and the central server in each communication round. N is the number of devices in a FL system, and r_N is the proportion of participating devices in each communication round. Thus, the high communication overhead of FL is affected by three main factors: the communication frequency (related to multiplicative component (i)), the number of participating devices (related to multiplicative component (ii)), and the per-communication-round costs (related to multiplicative component (iii)).

Most existing methods aimed at optimizing communication efficiency in FL seek to minimize one of three factors: the multiplicative component (i), by increasing the number of local iterations; the multiplicative component (ii), by applying device selection strategies; or the multiplicative component (iii), by employing gradient compression [12] or model compression [13]. However, these methods often fail to achieve sufficient compression to maximize the upper bound of FL communication data. Furthermore, the attempts to concurrently reduce all three factors have not yielded an optimal balance between high global model accuracy and minimal communication overhead. This challenge is largely tied to the issues encountered with both gradient compression and model compression.

Model compression, a method that directly manipulates model parameters using techniques such as truncating full-precision parameters or discarding insignificant ones, often results in significant precision loss. Furthermore, it may induce synchronization challenges due to inconsistent compression levels across devices. In contrast, gradient compression, which aims to compress model gradients rather than model parameters, effectively manages precision loss and doesn't incur significant additional training costs, making it more apt for various FL scenarios. This is because current methods rely on sparse compression. In high-frequency settings, almost all elements in the gradient are near zero, rendering the compression errors from sparsity negligible. However, in low-frequency scenarios where elements deviate from zero, this sparse compression leads to significant compression errors. Another challenge is predicting the maximum number of local iterations a cluster of FL devices can handle, which equates to setting the communication frequency - a task that remains intricate. Therefore, the development of gradient compression techniques that perform robustly across different communication frequency settings is crucial.

In this study, we present FedSC, a flexible gradient compression technique tailored to operate optimally across various communication frequency settings and device selection strategies in FL. To further mitigate compression errors and enhance the global model's performance, we also introduce a hierarchical estimation approach. This combination provides a potent solution, addressing major contributors to FL's high communication overhead. Our primary contributions are as follows:

- We unveil FedSC, an innovative gradient compression approach aimed at minimizing the communication costs per-communication-round costs. Distinctively, FedSC harnesses the statistical regularities present in gradients, transmitting them efficiently to the central server. The server then leverages sampling techniques to reconstruct the original element, ensuring that FedSC's effectiveness remains unaffected by communication frequency shifts.
- We propose a hierarchical estimation strategy to prevent the degradation of sampling due to the blending of distribution characteristics across layers and distortion of statistical information caused by insufficient layer elements. In this optimized workflow, gradient elements are collected as multiple vectors, each capturing statistical information that is then uploaded to the central server.
- We conduct comprehensive experiments to demonstrate that FedSC is a high-performance per-communication-round compression method.

2 Related Work

A variety of strategies have been proposed to diminish the total number of communication bits exchanged in the FL process. These strategies are succinctly outlined in Table 1. These approaches can be effectively divided into three separate categories based on the components they primarily address. These components correspond to the three multiplicative constituents (i), (ii), and (iii) encapsulated in Eq. (1).

Periodic averaging methods, which mainly focus on communication frequency, permit devices to execute multiple local iterations prior to sharing their updates [2,14,15]. This strategy significantly cuts down the communication frequency, denoted as r_T. Nevertheless, this optimization may not be completely satisfactory, as the maximum number of local iterations tolerable by each device tends to differ.

Table 1. Overview of approaches for improving communication efficiency in FL

Category of communication-efficient FL	Compression target	Multiplicative component in equation (1)	Compression idea	Shortcoming
Periodic averaging methods	Communication frequency	(i)	Performing numerous local iterations	Limitations due to variable maximum local iteration capacity per device
Device selection methods	Device count	(ii)	Choosing a subset of representative and informative devices	Trade-off between convergence speed and optimal convergence ability
Gradient compression methods	Gradients (model changes)	(iii)	Quantization of gradients	High compression errors commonly occur in low-frequency settings
			Sparsification of gradients	High compression errors commonly occur in low-frequency settings
Model compression methods	Model parameters	(iii)	Quantization of model parameters	Potential precision loss and synchronization challenges due to different quantization strategies across devices
			Distillation of model parameters	Limited knowledge transfer due to dispersed data across devices, restricting full model understanding
			Pruning of model parameters	Difficulties in identifying non-critical parameters due to partial data visibility and challenges in merging updates from differently pruned parameters

Device selection methods aim to diminish the number of devices engaged in each communication round. They strategically select a subset of devices that are both representative and informative for each training round, consequently reducing the participation ratio [16–18], r_N. However, the trade-off between convergence speed and convergence performance potentially constrains their effectiveness [19].

Lastly, per-communication-round compression methods are designed to shrink the size of model gradients or parameters transmitted during each communication round. This category encompasses techniques such as gradient quantization [20–22], gradient sparsification [23,24], model quantization [13], model distillation [25], and model pruning [26], all of which have demonstrated their potential. However, each of these methods has its shortcomings, as detailed in Table 1. Among these challenges, The fundamental challenge of gradient compression's compatibility with low communication frequency settings primarily stems from its origins in traditional distributed machine learning. In such traditional systems, devices often perform one local iteration before sending updates to a server for aggregation, resulting in a high communication frequency. To cater to the high communication frequency characteristics of traditional distributed machine learning, classical gradient compression methods often anticipate that the compressed elements will possess sparsity features conducive to secondary encoding. This expectation is based on the rule that elements in a update generated from a single iteration tend to be small and approach zero. However, in FL scenarios, the number of local iterations is often not limited to one. When traditional gradient compression methods are applied under this circumstance, with the hope to maintain the sparsity of the compressed results, substantial errors are introduced.

Deviant from conventional methods that either compress full-precision parameters or discard superfluous elements, FedSC pioneers a unique approach by substituting original model update (gradient) elements with statistical information-an underexplored strategy in the field. Unlike model compression, which often incurs substantial precision loss and synchronization challenges, FedSC circumvents these issues. Importantly, it also addresses the compatibility concerns that gradient compression often faces when paired with periodic averaging methods. Furthermore, FedSC is computationally efficient on devices, requiring only the calculation of statistical information.

3   Problem Formulation of Federated Learning

In this section, we introduce the optimization objective of FL. We consider a FL system comprised of a collection of devices $\mathcal{V} = \{1, 2, ..., N\}$ and a central server. N is the total number of devices. Each device holds a different source distribution $\mathcal{D}_i$ over instance space $\Xi = \mathcal{X} \times \mathcal{Y}$ from which it can sample training instances (x, y) [27]. Therefore, the local objective function can be defined as

$$f_i(\omega) = \mathbb{E}_{(x,y)\sim\mathcal{D}_i}[l(\omega; (x, y))] \tag{2}$$

where $\mathbb{E}[\cdot]$ represents conditional expectation. We suppose $\Omega \subseteq \mathbb{R}^d$ is the model parameter space. $\omega \in \Omega$ is a model parameter vector and $l : \Omega \times \Xi \mapsto \mathbb{R}_+$ represents loss function. That is, ω is a vector consisting of model parameters, and each element in ω represents a model parameter. Assume that $\mathcal{S}_i = \{(x_1, y_1), (x_2, y_2), ..., (x_{|\mathcal{S}_i|}, y_{|\mathcal{S}_i|})\}$ are samples that are sampled from distribution $\mathcal{D}_i$, with the size of $|\mathcal{S}_i|$. The local objective function can also be concretized as

$$f_i(\omega) = \frac{1}{|\mathcal{S}_i|} \sum_{(x,y)\in\mathcal{S}_i} l(\omega; (x, y)) \tag{3}$$

The global objective function over all samples $|\mathcal{S}| = \sum_{i=1}^{N} |\mathcal{S}_i|$ in this FL system is defned as

$$F(\omega) = \frac{\sum_{i=1}^{N} \sum_{(x,y) \in \mathcal{S}_i} l(\omega; (x,y))}{\sum_{i=1}^{N} |\mathcal{S}_i|}$$

$$= \sum_{i=1}^{N} \frac{|\mathcal{S}_i|}{|\mathcal{S}|} f_i(\omega) \tag{4}$$

The goal of FL is to train the best global model $\omega^\star$ using samples from all devices. That is, FL aims to minimize $F(\omega)$, i.e., $\omega^\star = argmin_\omega F(\omega)$.

4   Proposed Algorithm

In this section, we will first give a brief overview of the FedSC workflow. Then, we will delve into the technical details of how FedSC extracts statistical information from model updates and uses it to estimate the original elements. To provide a visual representation of this process, we include a figure (Fig. 1) outlining FedSC's workflow and implementation details. FedSC doesn't confine gradient compression to high communication frequencies, nor does it rely on sparsity-based techniques. Instead, FedSC focuses on the distribution characteristics of elements in model updates, retaining fundamental numerical values.

4.1   The Main Workflow of FedSC

The primary flow of the FedSC algorithm is encapsulated in Alg.1, which we will dissect for a more thorough understanding. To confront the issues of scalability and excessive communication overhead, only a limited subset of devices engage in each communication round for the global model training. The participants, n in total, are selected from the set of all devices $\mathcal{V}$ based on device selection methods for local training (lines 3-4). Specifically, during the k-th communication round, each participant downloads the latest global model ω^k from the central server and updates it by performing δ local Mini-Batch Gradient Descent (MBGD) iterations (lines 5-7, 13-19). Here, $g_i(\omega_i^{k,t}; \mathcal{B}_i^{k,t})$ denotes the gradient computed with a Mini-Batch $\mathcal{B}_i^{k,t}$ and learning rate $\eta_i^{k,t}$. $|\mathcal{B}_i|$ and e_i represent the Mini-Batch size and epoch size of local training, respectively, and the number of local iterations can be expressed as

$$\delta = \left\lceil \frac{|\mathcal{S}_i|}{|\mathcal{B}_i|} \right\rceil \times e_i \tag{5}$$

After local training, each participant in the k-th communication round sends the statistical information of its local model update, represented by $\mathcal{I}(\Delta\omega_i^{k+1})$, to the central server (lines 20-21). The model update mentioned here refers to the disparity in model parameters derived from δ rounds of local iterations. This could alternatively be construed as the cumulative sum of gradients acquired through δ instances of local training. Depending on the settings for communication frequency, δ will vary, which consequently brings about fluctuations in the

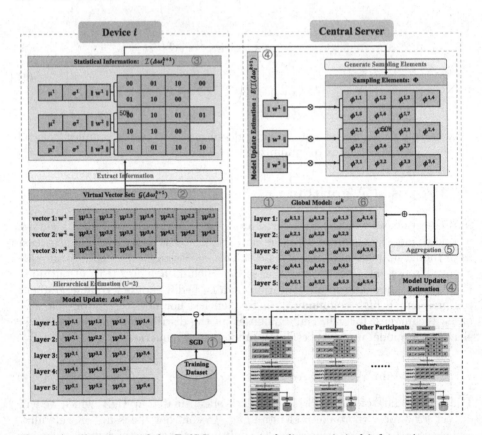

Fig. 1. An illustration of the FedSC process, including statistical information extraction and model update estimation. Depicted as a toy example for clarity. ① In the k-th communication round, every participant downloads the latest global model from the central server, updates it, and obtains a model update. ② Next, virtual vectors are created by concatenating the layers in the model update. ③ Each participant extracts statistical information from these virtual vectors and sends it to the central server. ④ The central server receives statistical information from participants and estimates the original model update through sampling. ⑤ Once the server receives statistical information from all participants and completed the estimation, it aggregates them to generate a new global model. ⑥ This process is repeated until the global model converges.

element distribution encapsulated in the model update. Our aim is to find out its statistical characteristics. The technical details of extracting statistical information, represented by $\mathcal{I}(\cdot)$, from model updates will be discussed in Sect. 4.2. Here we just need to know that this statistical information requires fewer bits to transmit than the model update. Once the server receives all statistical information from all participants, the global model is aggregated (line 8), $E(\cdot)$ represents the estimation operation using statistical information, and its process will be elaborated in Sect. 4.3.

Algorithm 1. The main workflow of FedSC

1: **function** SERVEREXECUTION ▷ Run on the server
2:    **for** each round $k = 1,2,...,T \times r_T$ **do**
3:        $n \leftarrow \max(N \times r_N, 1)$
4:        $\mathcal{V}' \leftarrow$ (select n devices to a set)
5:        **for** each device $i \in \mathcal{V}'$ **in parallel do**
6:            $\mathcal{I}(\Delta\omega_i^{k+1}) \leftarrow$ DEVICEUPDATE(i, ω^k)
7:        **end for**
8:        $\omega^{k+1} = \omega^k + \sum_{i \in \mathcal{V}'} \frac{1}{n} E(\mathcal{I}(\Delta\omega_i^{k+1}))$
9:    **end for**
10: **end function**
11:
12: **function** DEVICEUPDATE(i, ω^k) ▷ Run on device i
13:    $\omega_i^k \leftarrow \omega^k$
14:    $B \leftarrow$ (spilt $\mathcal{S}_i$ into batches of size $|\mathcal{B}_i|$)
15:    **for** each local epoch from 1 to e_i **do**
16:        **for** batch $\mathcal{B}_i \in B$ **do**
17:            $\omega_i^k \leftarrow \omega_i^k - \eta_i^k g_i(\omega_i^k, \mathcal{B}_i)$
18:        **end for**
19:    **end for**
20:    $\Delta\omega_i^{k+1} \leftarrow \omega_i^k - \omega^k$
21:    **return** $\mathcal{I}(\Delta\omega_i^{k+1})$ to server
22: **end function**

4.2 Statistical Information Extraction

In this subsection, we explain how FedSC extracts statistical information from a model update, i.e., $\mathcal{I}(\cdot)$. Before that, we need to define some concepts. We consider $\Delta\omega$ as a local model update to be uploaded from a device to the central server, where $\Delta\omega = \{W^1, W^2, ..., W^l, ..., W^L\}$. W^l in $\Delta\omega$ represents the update of the l-th layer, which for example can be a 4-dimension tensor in convolutional layer or a 2-dimension matrix in fully connected layer. For ease of description, we assume that the l-th layer in $\Delta\omega$ is a d_l-dimension vector. Thus, $\Delta\omega$ can be expressed as

$$\Delta\omega = \begin{cases} \textbf{layer1} : W^1 = \{\mathcal{W}^{1,1}, \mathcal{W}^{1,2}, ..., \mathcal{W}^{1,d_1}\}, \\ \textbf{layer2} : W^2 = \{\mathcal{W}^{2,1}, \mathcal{W}^{2,2}, ..., \mathcal{W}^{2,d_2}\}, \\ , ..., \\ \textbf{layer}l : W^l = \{\mathcal{W}^{l,1}, \mathcal{W}^{l,2}, ..., \mathcal{W}^{l,d_l}\}, \\ , ..., \\ \textbf{layerL} : W^L = \{\mathcal{W}^{L,1}, \mathcal{W}^{L,2}, ..., \mathcal{W}^{L,d_L}\} \end{cases} \tag{6}$$

The Statistical Information Extraction Operation is Comprised of Two Steps. The technical details of statistical information extraction can be better understood through the toy example illustrated in Fig. 1.

Step1. Hierarchical Estimation: The first step of the statistical information extraction operation is to create a set of virtual vectors by concatenating layers in $\Delta\omega$. The purpose of creating virtual vectors is to group elements into clusters, which makes it easier to extract information. This process of dividing the layers in $\Delta\omega$ into several virtual vectors is called hierarchical estimation and is denoted as $\mathcal{G}(\cdot)$. The outcome of $\mathcal{G}(\Delta\omega)$ is illustrated below:

$$\mathcal{G}(\Delta\omega) = \begin{cases} \textbf{vector1} : \{W^1, W^2, ..., W^{U-1}, W^U\}, \\ \textbf{vector2} : \{W^{U+1}, W^{U+2}, ..., W^{2U-1}, W^{2U}\}, \\ , ..., \\ \textbf{vector} \left\lceil \frac{l}{U} \right\rceil : \underbrace{\{..., W^l, ...\},}_{= \ U \ \text{layers}} \\ , ..., \\ \textbf{vector} \left\lceil \frac{L}{U} \right\rceil : \underbrace{\{..., W^L\}}_{\leq \ U \ \text{layers}} \end{cases} \tag{7}$$

As shown in Eq. (7), $\Delta\omega$ is divided into $\left\lceil \frac{L}{U} \right\rceil$ virtual vectors, where each vector contains U layers except for the last one. Each vector in $\mathcal{G}(\Delta\omega)$ extracts one piece of statistical information. Our method balances between concatenating all layers into one virtual vector and making each layer represent a virtual vector. The two extreme cases in our method are U = L and U = 1. When U = L, there is a significant reduction in uplink communication overhead, but the fusion of multi-layer distribution characteristics may negatively impact the model update estimation. On the other hand, when U = 1, there is no interference between the distribution characteristics of different layers, but the statistical information may become distorted if a layer contains insufficient elements. Thus, the element of U should be chosen carefully, and we will explore the effect of the element of U on FedSC performance in subsequent experiments. Next, we focus on how to extract statistical information from a vector. For the sake of clarity, we redefine $\mathcal{G}(\Delta\omega)$ as

$$\mathcal{G}(\Delta\omega) = \begin{cases} \textbf{vector1} : w^1 = \{w^{1,1}, w^{1,2}, ..., w^{1,\tau_1}\}, \\ \textbf{vector2} : w^2 = \{w^{2,1}, w^{2,2}, ..., w^{2,\tau_2}\}, \\ , ..., \\ \textbf{vectorj} : w^j = \{w^{j,1}, w^{j,2}, ..., w^{j,\tau_j}\}, \\ , ..., \end{cases} \tag{8}$$

Step2. Information Extraction: secondly, we extract statistical information from each vector in $\mathcal{G}(\Delta\omega)$. As shown in Eq. (9), each vector is assigned a distribution space that reflects its element distribution. This distribution space is represented by a tuple that includes the mean μ, standard deviation σ, and symbol code($\cdot$). The formulas for μ and σ are shown in Eq. (10) and Eq. (11)

respectively. code($\cdot$) represents the operation that encodes the signs of the elements in $\mathcal{G}(\Delta\omega)$. It's important to note that $\mathcal{G}(\Delta\omega)$ is just a set of virtual vectors that assist in extracting information and does not alter the dimension structure of layers, as illustrated in Fig. 1.

$$
\mathcal{I}(\Delta\omega) = \begin{cases}
(\mu^1, \sigma^1, \|\mathbf{w}^1\|_2, \text{code}(\frac{\mathbf{w}^1}{\|\mathbf{w}^1\|_2})), \\
(\mu^2, \sigma^2, \|\mathbf{w}^2\|_2, \text{code}(\frac{\mathbf{w}^2}{\|\mathbf{w}^2\|_2})), \\
, ..., \\
(\mu^j, \sigma^j, \|\mathbf{w}^j\|_2, \text{code}(\frac{\mathbf{w}^j}{\|\mathbf{w}^j\|_2})), \\
, ...,
\end{cases}
\tag{9}
$$

$$
\mu^j = \frac{\sum_{\gamma=1}^{\tau_j} \frac{|w^{j,\gamma}|}{\|\mathbf{w}^j\|_2}}{\tau_j} \quad j = 1, ..., \left\lceil \frac{L}{U} \right\rceil
\tag{10}
$$

$$
\sigma^j = \sqrt{\frac{\sum_{\gamma=1}^{\tau_j} \left(\frac{|w^{j,\gamma}|}{\|\mathbf{w}^j\|_2} - \mu^j \right)^2}{\tau_j - 1}} \quad j = 1, ..., \left\lceil \frac{L}{U} \right\rceil
\tag{11}
$$

The input of code($\cdot$) is a vector, and the output is also a vector. Each element in the output vector takes up 2 bits, which can encode three states. These three states represent different signs and the coding rules are as follows,

$$
\begin{cases}
00, \quad -1 < \frac{w^{j,\gamma}}{\|\mathbf{w}^j\|_2} < 0 \\
\\
01, \quad \frac{w^{j,\gamma}}{\|\mathbf{w}^j\|_2} = 0 \quad j = 1, ..., \left\lceil \frac{L}{U} \right\rceil, \quad \gamma = 1, ..., \tau_j \\
\\
10, \quad 0 < \frac{w^{j,\gamma}}{\|\mathbf{w}^j\|_2} < 1
\end{cases}
\tag{12}
$$

Finally, we estimate the cost of the uplink communication in FedSC, which is the number of bits that a device sends to the central server in each communication round after the statistical information extraction has been performed. For a local model update $\Delta\omega$ that is divided into $\lceil \frac{L}{U} \rceil$ virtual vectors, we need $2 \times \sum_{j=1}^{\lceil \frac{L}{U} \rceil} \tau_j$ bits to encode code($\Delta\omega$). The scalar $\|\mathbf{w}^j\|_2$, μ^j, and σ^j are typically represented using full precision, which we assume to be 32 bits. Thus, the amount of communication data for $\mathcal{I}(\Delta\omega)$ is given by

$$
|\mathcal{I}(\Delta\omega)| = 2 \times \sum_{j=1}^{\lceil \frac{L}{U} \rceil} \tau_j + 96 \times \left\lceil \frac{L}{U} \right\rceil
$$

$$
= 2 \times \sum_{l=1}^{L} d_l + 96 \times \left\lceil \frac{L}{U} \right\rceil
\tag{13}
$$

4.3   Model Update Estimation

The model update estimation can be considered as the inverse process of statistical information extraction. The goal of statistical information extraction is to analyze the layer distribution in $\Delta\omega$ and find distribution spaces that encompasses all its characteristics. The goal of model update estimation is to sample from these distribution spaces to estimate $\Delta\omega$. In the following, we will elaborate on the technical details of how the central server estimates the model update from $\mathcal{I}(\Delta\omega)$.

Extracting $\mathcal{I}(\Delta\omega)$ is a complex process that involves normalizing $\Delta\omega$ and performing hierarchical estimation. As a result, model update estimation is not just a matter of sampling from distribution spaces carried by $\mathcal{I}(\Delta\omega)$. Before sampling, two crucial considerations must be addressed. Firstly, it is important to determine the correspondence between the layers and the tuples in $\mathcal{I}(\Delta\omega)$. Since multiple layers share a distribution space, it is crucial to determine which space each layer should sample from. Secondly, it's essential to remember that $\mathcal{I}(\Delta\omega)$ is drawn from the normalized $\Delta\omega$, so the sampled element must be denormalized to estimate the original model update elements.

These two considerations are not difficult to resolve. For the first consideration, the correspondence between the layers and the tuples is clear. In Sect. 4.2, the layers in the model update are divided into $\lceil \frac{L}{U} \rceil$ virtual vectors based on their layer label, with U adjacent layers forming one vector. When sampling, we should also follow this arrangement. We define a set of sampling elements as Φ, and Φ should have the same layer structure as $\Delta\omega$. Since multiple layers in Φ share the same distribution space, we also need to define a virtual vector set Φ'. The purpose of this vector set is the same as $\mathcal{G}(\Delta\omega)$, which is to indicate which layers share a sample space. The relationship between elements in Φ' is specified in Eq. (14), Eq. (15), Eq. (16) and Eq. (17). For the second consideration, we simply multiply the elements in Φ by the corresponding 2-norm. The computational complexities of statistical information extraction and model update estimation in FedSC are both $\mathcal{O}(n)$, rendering the overall complexity of FedSC relatively low.

$$\Phi' = \begin{cases} \textbf{vector1} : \{\phi^{1,1}, \phi^{1,2}, ..., \phi^{1,\tau_1}\}, \\ \textbf{vector2} : \{\phi^{2,1}, \phi^{2,2}, ..., \phi^{2,\tau_2}\}, \\ , ..., \\ \textbf{vector}j : \{\phi^{j,1}, \phi^{j,2}, ..., \phi^{j,\tau_j}\}, \\ , ..., \end{cases} \tag{14}$$

$$\lim_{\tau_j \to \infty} \frac{\sum_{\gamma=1}^{\tau_j} |\phi^{j,\gamma}|}{\tau_j} = \mu^j \quad j = 1, ..., \left\lceil \frac{L}{U} \right\rceil \tag{15}$$

$$\lim_{\tau_j \to \infty} \sqrt{\frac{\sum_{\gamma=1}^{\tau_j} (|\phi^{j,\gamma}| - \mu^j)^2}{\tau_j - 1}} = \sigma^j \quad j = 1, ..., \left\lceil \frac{L}{U} \right\rceil \tag{16}$$

Table 2. Parameter settings for 1@CNN

	Layer name	Shapes
1	conv2d_1	(3, 3, 3, 96)
2	conv2d_2	(3, 3, 96, 96)
3	conv2d_3	(3, 3, 96, 96)
4	conv2d_4	(3, 3, 96, 192)
5	conv2d_5	(3, 3, 192, 192)
6	conv2d_6	(3, 3, 192, 192)
7	conv2d_7	(3, 3, 192, 192)
8	conv2d_8	(1, 1, 192, 192)
9	conv2d_9	(1, 1, 192, 10)

Table 3. Parameter settings for 2@CNN

	Layer name	Shapes
1	conv1d_1	(10, 1, 16)
2	conv1d_2	(10, 16, 32)
3	conv1d_3	(10, 32, 64)
4	conv1d_4	(15, 64, 64)
5	conv1d_5	(10, 64, 128)
6	conv1d_6	(10, 128, 128)
7	conv1d_7	(10, 128, 128)
8	dense_1	(2048, 256)
9	dense_2	(256, 128)
10	dense_3	(128, 10)

$$\begin{cases} -1 < \phi^{j,\gamma} < 0, & \text{code}(\frac{w^{j,\gamma}}{\|\mathbf{w}^j\|_2}) = 00 \\[2mm] \phi^{j,\gamma} = 0, & \text{code}(\frac{w^{j,\gamma}}{\|\mathbf{w}^j\|_2}) = 01 \\[2mm] 0 < \phi^{j,\gamma} < 1, & \text{code}(\frac{w^{j,\gamma}}{\|\mathbf{w}^j\|_2}) = 10 \end{cases} \tag{17}$$

5   Evaluation

In this section, we will provide the implementation details of FedSC and evaluate its performance in terms of global model accuracy and communication cost.

5.1   Datasets and Models

Our experiments are conducted on three benchmark datasets, MNIST [28], CIFAR-10 [29], and Digital-Speech-Recognition-Dataset(DSRD) [30], using three models, 1@CNN, 2@CNN, and 3@CNN. The parameter settings of each model are specified in Table 2, Table 3 and Table 4. It's worth noting that while these tables only highlight the key layers, in our experiments, other layers such as bias are also counted as one U. The MNIST dataset contains 60,000 training and 10,000 testing samples of 28×28 gray-scale handwritten images with ten classes. CIFAR-10 is a dataset for image classification with 60,000 photographs, divided into ten categories. DSRD is a small speech recognition dataset consisting of 741 WAV files that represent numbers from 0 to 9.

5.2   Experiment Setting

We compare our FedSC method with several state-of-the-art gradient compression methods for FL, including AdaQuantFL [31], ASTW_FedAVG [32], FedPAQ [12], and SBC [24]. Additionally, we also compare our method with the vanilla

Table 4. Parameter settings for 3@CNN

	Layer name	Shapes
1	conv2d_1	(5, 5, 1, 16)
2	conv2d_2	(5, 5, 16, 36)
3	dense_1	(1764, 128)
4	dense_2	(128, 10)

Table 5. Hyperparameters for experiments

H-param	Value		
	1@CNN	2@CNN	3@CNN
N	30	30	30
n	3	3	3
e_i	10	20	5
$|\mathcal{S}_i|$	2000	100	100
$|\mathcal{B}_i|$	30	30	30
η_i	0.001	0.01	0.01

FL method, FedAvg [2]. In our experiments, we adopt consistent FL settings, and the important hyperparameters are listed in Table 5. By default, the number of training samples and their distribution is consistent across all devices. Importantly, due to FedSC's inherent randomness, our presented results are the averages of 20 independent experiments to mitigate random fluctuations.

5.3   The Effect of the Number of Layers Within the Virtual Vector on Accuracy

In the extract information operation of FedSC, we normalize a virtual vector by calculating its 2-norm, as shown in Eq. (9). The 2-norm merges elements from multiple layers, and we then calculate the mean and standard deviation of normalized elements. The mean and standard deviation signify a distribution space that encompasses all distribution characteristics of layers within a virtual vector. However, if a virtual vector contains more layers, its 2-norm will incorporate more elements, and its mean and standard deviation will encompass more layer distribution characteristics, which can negatively affect high-performance estimation. This does not mean that the optimal value of U is 1, as having too few elements in a virtual vector can cause a distortion of the distribution space reflected by the statistics. A high-performance estimate requires a sufficient number of elements. Therefore, it is crucial to examine the impact of the number of elements in a virtual vector on the global model accuracy. In this experiment, we vary the value of U in FedSC to observe the changes in communication efficiency and accuracy. To visually show the number of elements in each virtual vector, we introduce a new indicator called "num," which refers to the number of elements in the virtual vector with the least elements among all virtual vectors. We evaluate the communication efficiency by comparing the total number of bits (b_{total}) exchanged between the central server and devices during the FL training.

The results of the experiment, presented in Table 6, are in line with what we expected. The final accuracy of the three models is unsatisfactory when the value of U is set to 1 and L. In particular, when U is set to 1, there is a significant difference in the number of elements between the virtual vectors. The limited

Table 6. Accuracy and b_{total} of FedSC with varying U after 200 communication rounds

Layers in a vector	1@CNN			2@CNN			3@CNN		
	Accuracy	b_{total}	num	Accuracy	b_{total}	num	Accuracy	b_{total}	num
U=1	84.07%	3.4902G	10	61.72%	2.6932G	10	97.71%	0.6174G	10
U=3	83.64%	3.4901G	2784	62.06%	2.6931G	208	97.79%	0.6173G	452
U=5	84.07%	3.4901G	38794	62.60%	2.6931G	5392	97.70%	0.6173G	14980
U=7	**85.00%**	**3.4901G**	**38986**	**63.84%**	**2.6930G**	**25936**	97.83%	0.6173G	1280
U=L	84.32%	3.4901G	1888138	62.96%	2.6930G	1056090	**97.93%**	**0.6173G**	**242062**

sample size for the bias layer leads to a substantial estimate error, which is the main cause of the subpar accuracy. Next, we observe the trend of the final accuracy as U changes. In 2@CNN, the final accuracy tends to increase initially and then decrease as U increases. Specifically, the accuracy gradually increases from 61.72% to 63.84%, before decreasing slightly to 62.96%. The trend for 1@CNN is similar, but slightly different for 3@CNN. This may be due to the smaller size of the 3@CNN model. Based on the experimental results, it is best to have 25K to 250K elements in a virtual vector, the optimum performance is achieved specifically when $U \geq 7$.

Finally, we examine the effect of varying U on b_{total}. Adjusting U has no noticeable impact on communication costs. The value of U must be kept above a certain threshold to maintain the number of elements in the virtual vector. This means that the expression $\lceil \frac{L}{U} \rceil$ in Eq. (13) is much smaller than $\sum_{l=1}^{L} d_l$, and altering U has a minimal effect on $\lceil \frac{L}{U} \rceil$. Consequently, hierarchical estimation has a negligible impact on FL communication costs.

5.4 Comparison with Existing Methods on Accuracy and Communication Cost

The experiment described below evaluates the communication efficiency and performance of FedSC and other communication-efficient FL methods (FedAvg, AdaQuantFL, ASTW_FedAVG, FedPAQ, and SBC) on three models (1@CNN, 2@CNN, and 3@CNN). We take into consideration three factors that influence FL communication costs as outlined in Eq. (1): communication frequency, number of participating devices, and model update size. An effective communication-efficient FL method should achieve high accuracy while keeping communication costs low, so this experiment restricts the device participation rate to a low level of 10%. To control communication frequency, the experiment adjusts δ, where $r_T = \frac{1}{\delta}$. Different δ values are set for training 1@CNN ($\delta = 670$), 2@CNN ($\delta = 80$), and 3@CNN ($\delta = 20$), allowing us to observe how model update compression of FL methods behaves under different δ. The performance and communication efficiency of the FL methods are measured using accuracy and b_{total} respectively.

Table 7 displays the accuracy and b_{total} after 200 communication rounds for three datasets trained with different communication-efficient FL methods. Among the six methods listed in Table 7, FedAvg is the only one that does

Table 7. Accuracy and b_{total} of different communication-efficient FL methods after 200 communication rounds

methods		1@CNN	2@CNN	3@CNN
FedAvg	Accuracy	85.21%	63.75%	97.83%
	b_{total}	6.5747G	5.0692G	1.1619G
ASTW_FedAVG	Accuracy	83.73%	61.83%	97.03%
	b_{total}	5.5316G	4.0928G	0.9329G
FedPAQ	Accuracy	55.22%	47.31%	96.57%
	b_{total}	3.8011G	2.9307G	0.6717G
SBC	Accuracy	80.68%	60.77%	96.99%
	b_{total}	3.5845G	2.7643G	0.6335G
AdaQuantFL	Accuracy	35.43%	45.16%	94.89%
	b_{total}	3.6030G	2.9294G	0.6527G
FedSC	Accuracy	**85.00%**	**63.84%**	**97.93%**
	b_{total}	**3.4901G**	**2.6930G**	**0.6173G**

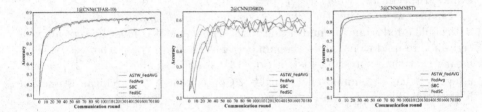

Fig. 2. The convergence speed of different communication-efficient FL methods during training of 1@CNN, 2@CNN and 3@CNN.

not compress the model update and thus can serve as a baseline for the other methods.

ASTW_FedAVG and SBC both employ a sparse compression technique, discarding non-essential elements from the model updates. While their compression methods differ, there is a trade-off between sparsification and accuracy. For instance, during the training of 1@CNN, FedAvg attains 85.21% accuracy with a communication overhead of 6.5747G, whereas SBC achieves 80.68% accuracy with a communication overhead of 3.5845G. SBC reduces communication costs by 2.9902G, however, this decrease in communication comes at the cost of a 4.53% decrease in accuracy. ASTW_FedAVG offers close-to-baseline accuracy with high communication costs. Furthermore, the accuracy difference between these two methods and the baseline is greater on 1@CNN compared to 3@CNN, which suggests that the compression error of both ASTW_FedAVG and SBC increases with the increase of δ.

Both FedPAQ and AdaQuantFL use a stochastic quantization method to compress the model updates from participating devices. Stochastic quantization is similar to sparsification and results in reduced communication costs, but at the cost of accuracy. Moreover, we find that the compression error of stochastic quantization is more sensitive to changes in δ. For example, during the training of 1@CNN, with δ set to 670, the accuracy of FedPAQ is about 30% lower than the baseline. However, when training 3@CNN with δ set to 20, the accuracy of FedPAQ is only about 1.3% lower than the baseline.

Unlike these methods, FedSC adopts a unique approach. Instead of simply truncating full-precision elements, it leverages the overall characteristics of the gradient elements as a guide to restore the original elements. This results in not only maintaining accuracy but in some instances (like in 2@CNN and 3@CNN), even surpassing the baseline accuracy. FedSC's performance is less sensitive to changes in δ, with the final accuracy remaining close to or exceeding the baseline whether δ is set to 20 or 670. This superior performance can be attributed to FedSC's focus on simulating the model update distribution, thereby encapsulating a more comprehensive understanding of the gradient elements than simple compression methods.

Next, we examine the convergence behavior of various communication-efficient FL methods. Figure 2 depicts the convergence speed of the global model when it is trained using various communication-efficient FL methods for the 1@CNN, 2@CNN, and 3@CNN datasets, respectively. We omit the convergence curves of FedPAQ and AdaQuantFL as they do not perform well in terms of accuracy. Similarly, we consider the convergence behavior of FedAvg as the baseline. The convergence speed of FedSC is not as fast as FedAvg and ASTW_FedAVG when training large models such as 1@CNN and 2@CNN. However, when training a small model like 3@CNN, FedSC converges at a speed close to the baseline. Although FedSC may not have an advantage in terms of convergence speed, its accuracy gradually surpasses the baseline as the training progresses. ASTW_FedAVG converges at about the same rate as FedAvg, but its convergence process is not always stable, particularly when training 1@CNN. In contrast, the convergence curve of SBC is very stable, but its convergence speed is the slowest. Overall, FedSC can reduce communication expenses more effectively than other methods with the same number of communication rounds. FedSC can achieve a similar level of accuracy to standard uncompressed method while using fewer communicated bits.

5.5 · Effects of Different Model Update Estimation

FedSC can be considered a class of methods because of the variety of paradigms used to estimate model updates based on statistical information. We provide a model update estimation paradigm and other paradigms can be derived from it. This experiment is an attempt to estimate model updates using other paradigms. We try to vary the constraints of Φ' mentioned in Eq. (14). These constraints can be divided into two types: the group constraint, which is defined by Eqs. (15) and (16), and the individual constraint, represented by Eq. (17). We aim to observe the impact on accuracy as we alter these two types of constraints.

Table 8. Accuracy of models after 200 communication rounds with different values of α

Model	$\alpha=1$	$\alpha=5$	$\alpha=10$	$\alpha=20$
1@CNN	82.17%	82.95%	84.76%	83.73%
3@CNN	97.82%	97.92%	97.87%	97.96%

Modifying the Equation of the Individual Factor. We modify the Eq. (17) as follows,

$$
\begin{cases}
-\mu^j - \alpha \times \sigma^j \le \phi^{j,\gamma} \le -\mu^j + \alpha \times \sigma^j, & \mathrm{code}(\frac{w^{j,\gamma}}{\|w^j\|_2}) = 00 \\[2ex]
\phi^{j,\gamma} = 0, & \mathrm{code}(\frac{w^{j,\gamma}}{\|w^j\|_2}) = 01 \\[2ex]
\mu^j - \alpha \times \sigma^j \le \phi^{j,\gamma} \le \mu^j + \alpha \times \sigma^j, & \mathrm{code}(\frac{w^{j,\gamma}}{\|w^j\|_2}) = 10
\end{cases}
\tag{18}
$$

where α is a hyperparameter. α can make a strict restriction on the variance of the elements in Φ'. That is, α restricts the elements in Φ' to be around the μ^j (or $-\mu^j$). The hyperparameter α places a strict restriction on the variance of the elements in Φ', making the elements in Φ' be close to μ^j (or $-\mu^j$). Increasing the value of α increases the gap between the elements and the mean. Table 8 displays the final accuracy of training 1@CNN and 3@CNN under different values of α. We can see that a low value of α negatively affects the final accuracy. Whether training 1@CNN or training 3@CNN, the final accuracy with $\alpha = 1$ is lower than the other α values, indicating that the distribution of elements in Φ' should not be too concentrated. Additionally, we found that small models and large models have different tolerances for element dispersion. For 1@CNN, as α increases, the final accuracy tends to increase, with the highest accuracy being achieved when the distribution is the most dispersed, as shown in Table 7. However, 3@CNN is different, with the highest accuracy at $\alpha = 20$, which is slightly higher than the one shown in Table 7. Thus, a suitable increase in the scatter of Φ can help improve accuracy.

Modifying the Equation of the Group Factor. We modify the Eq. (16) as follows,

$$
\lim_{\tau_j \to \infty} \sqrt{\frac{\sum_{\gamma=1}^{\tau_j} (|\phi^{j,\gamma}| - \mu^j)^2}{\tau_j - 1}} = \sigma^j \times \beta \quad j = 1, ..., \left\lceil \frac{L}{U} \right\rceil
\tag{19}
$$

where β is a hyperparameter. The hyperparameter β applies a "soft" constraint on the variance of the elements in Φ'. This is because the standard deviation in the group constraint is a limiting concept, which may not be able to enforce strict constraints in practice. This means that changing β can impact some elements in Φ', but not necessarily all of them. In a previous experiment, we evaluated

Table 9. Accuracy of models after 200 communication rounds with different values of β

Model	$\beta=1.25$	$\beta=1.5$	$\beta=1.75$	$\beta=2$
1@CNN	84.46%	84.51%	85.08%	84.99%
3@CNN	97.77%	97.76%	97.23%	97.3%

the effect of small variances on accuracy. Now, we analyze the impact of larger variances on accuracy. To do this, we trained 1@CNN and 3@CNN with four higher values of β and observed their accuracy, as shown in Table 9. The results suggest that the accuracy of smaller models is more sensitive to variance in elements, as the final accuracy of 1@CNN remained between 84% and 85% with increasing β, while the final accuracy of 3@CNN decreased significantly.

6 Conclusion

This paper introduced FedSC, a versatile and innovative gradient compression approach, and a hierarchical estimation strategy aimed at enhancing the performance of FL by minimizing its communication overhead. By exploiting statistical regularities within gradients and employing hierarchical estimation to reduce compression errors, FedSC proved to be a high-performance per-communication-round compression method. Comprehensive experimental results demonstrated the effectiveness of FedSC in both maintaining global model accuracy and reducing communication overhead, thereby validating its superiority over existing methods.

7 Future Work

The promising results of FedSC open opportunities for further research. A critical future focus involves integrating downstream data compression techniques into FedSC, aiming to optimize global model dispatch and reduce FL's communication overhead. This requires exploring their theoretical basis and deriving convergence proofs to ensure robust improvements. Another key direction is studying FedSC's performance in non-Independent and Identically Distributed (non-IID) scenarios, common in real-world FL implementations. Creating strategies for FedSC that consider this data diversity can enhance its real-world applicability, extending its potential impact.

Acknowledgement. This work is supported by the General Program of National Natural Science Foundation of China (62072049)

References

1. Mell, P., Grance, T., et al.: The nist definition of cloud computing (2011)
2. McMahan, B., Moore, E., Ramage, D., Hampson, S., y Arcas, B.A.: Communication-efficient learning of deep networks from decentralized data. In: Artificial intelligence and statistics, pp. 1273–1282. PMLR (2017)
3. Kairouz, P., et al.: Advances and open problems in federated learning. arXiv preprint arXiv:1912.04977 (2019)
4. Bonawitz, K., et al.: Towards federated learning at scale: system design. Proc. Mach. Learn. Syst. **1**, 374–388 (2019)
5. Konečný, J., McMahan, B., Ramage, D.: Federated optimization: Distributed optimization beyond the datacenter. arXiv preprint arXiv:1511.03575 (2015)
6. Konečný, J., McMahan, H.B., Yu, F.X., Richtárik, P., Suresh, A.T., Bacon, D.: Federated learning: strategies for improving communication efficiency. arXiv preprint arXiv:1610.05492 (2016)
7. Li, T., Sahu, A.K., Talwalkar, A., Smith, V.: Federated learning: challenges, methods, and future directions. IEEE Signal Process. Mag. **37**(3), 50–60 (2020)
8. Chen, S., Shen, C., Zhang, L., Tang, Y.: Dynamic aggregation for heterogeneous quantization in federated learning. IEEE Trans. Wireless Commun. **20**(10), 6804–6819 (2021)
9. Khan, S., Rahmani, H., Shah, S.A.A., Bennamoun, M.: A guide to convolutional neural networks for computer vision. Synthesis Lectures Comput. Vision **8**(1), 1–207 (2018)
10. Yin, W., Kann, K., Yu, M., Schütze, H.: Comparative study of cnn and rnn for natural language processing. arXiv preprint arXiv:1702.01923 (2017)
11. Sze, V., Chen, Y.H., Yang, T.J., Emer, J.S.: Efficient processing of deep neural networks: a tutorial and survey. Proc. IEEE **105**(12), 2295–2329 (2017)
12. Reisizadeh, A., Mokhtari, A., Hassani, H., Jadbabaie, A., Pedarsani, R.: Fedpaq: a communication-efficient federated learning method with periodic averaging and quantization. In: International Conference on Artificial Intelligence and Statistics, pp. 2021–2031. PMLR (2020)
13. Xu, J., Du, W., Jin, Y., He, W., Cheng, R.: Ternary compression for communication-efficient federated learning. IEEE Transactions on Neural Networks and Learning Systems (2020)
14. Wang, J., Liang, H., Joshi, G.: Overlap local-sgd: an algorithmic approach to hide communication delays in distributed sgd. In: ICASSP 2020–2020 IEEE International Conference on Acoustics, Speech and Signal Processing (ICASSP), pp. 8871–8875. IEEE (2020)
15. Haddadpour, F., Kamani, M.M., Mahdavi, M., Cadambe, V.R.: Local sgd with periodic averaging: Tighter analysis and adaptive synchronization. arXiv preprint arXiv:1910.13598 (2019)
16. Li, X., Huang, K., Yang, W., Wang, S., Zhang, Z.: On the convergence of fedavg on non-iid data. arXiv preprint arXiv:1907.02189 (2019)
17. Fraboni, Y., Vidal, R., Kameni, L., Lorenzi, M.: Clustered sampling: low-variance and improved representativity for clients selection in federated learning. In: International Conference on Machine Learning, pp. 3407–3416. PMLR (2021)
18. Cho, Y.J., Gupta, S., Joshi, G., Yağan, O.: Bandit-based communication-efficient client selection strategies for federated learning. In: 2020 54th Asilomar Conference on Signals, Systems, and Computers, pp. 1066–1069. IEEE (2020)

19. Cho, Y.J., Wang, J., Joshi, G.: Towards understanding biased client selection in federated learning. In: International Conference on Artificial Intelligence and Statistics, pp. 10351–10375. PMLR (2022)
20. Courbariaux, M., Bengio, Y., David, J.P.: Binaryconnect: training deep neural networks with binary weights during propagations. In: Advances In Neural Information Processing Systems 28 (2015)
21. Alistarh, D., Grubic, D., Li, J., Tomioka, R., Vojnovic, M.: Qsgd: Communication-efficient sgd via gradient quantization and encoding. Adv. Neural. Inf. Process. Syst. **30**, 1709–1720 (2017)
22. Shlezinger, N., Chen, M., Eldar, Y.C., Poor, H.V., Cui, S.: Uveqfed: universal vector quantization for federated learning. IEEE Trans. Signal Process. **69**, 500–514 (2020)
23. Sattler, F., Wiedemann, S., Müller, K.R., Samek, W.: Robust and communication-efficient federated learning from non-iid data. IEEE Trans. Neural Netw. Learn. Syst. **31**(9), 3400–3413 (2019)
24. Sattler, F., Wiedemann, S., Müller, K.R., Samek, W.: Sparse binary compression: Towards distributed deep learning with minimal communication. In: 2019 International Joint Conference on Neural Networks (IJCNN), pp. 1–8. IEEE (2019)
25. Li, D., Wang, J.: Fedmd: Heterogenous federated learning via model distillation. arXiv preprint arXiv:1910.03581 (2019)
26. Prakash, P., et al.: Iot device friendly and communication-efficient federated learning via joint model pruning and quantization. IEEE Internet Things J. **9**(15), 13638–13650 (2022)
27. Haddadpour, F., Mahdavi, M.: On the convergence of local descent methods in federated learning. arXiv preprint arXiv:1910.14425 (2019)
28. Deng, L.: The mnist database of handwritten digit images for machine learning research [best of the web]. IEEE Signal Process. Mag. **29**(6), 141–142 (2012)
29. Krizhevsky, A., Nair, V., Hinton, G.: The cifar-10 dataset. http://www.cs.toronto.edu/kriz/cifar.html 55(5) (2014)
30. invY: Digital speech recognition. https://github.com/AlbertYoung0112/DigitalSpeechRecognition Accessed April 13 2020
31. Jhunjhunwala, D., Gadhikar, A., Joshi, G., Eldar, Y.C.: Adaptive quantization of model updates for communication-efficient federated learning. In: ICASSP 2021–2021 IEEE International Conference on Acoustics, Speech and Signal Processing (ICASSP), pp. 3110–3114. IEEE (2021)
32. Chen, Y., Sun, X., Jin, Y.: Communication-efficient federated deep learning with layerwise asynchronous model update and temporally weighted aggregation. IEEE Trans. Neural Netw. Learn. Syst. **31**(10), 4229–4238 (2019)

Joint Video Transcoding and Representation Selection for Edge-Assisted Multi-party Video Conferencing

Fanhao Kong[1], Tuo Cao[1], Zhuzhong Qian[1(✉)], Xiaoliang Wang[1], Ming Zhao[2], Liming Wang[2], and Zhenjie Lin[2]

[1] State Key Laboratory for Novel Software Technology, Nanjing University,
Nanjing, China
{fhkong,tuocao}@smail.nju.edu.cn, {qzz,wxili}@nju.edu.cn

[2] CSG China Southern Power Grid Digital Platform Technology Company,
Shenzhen, China
zhaoming@dptc.csg.cn

Abstract. Current cloud-based multi-party video conferencing suffers from heavy workloads on media servers caused by video transcoding. Emerging edge computing can assist in offloading transcoding tasks to edge nodes. However, the resource-limited nature of edge nodes poses new challenges. First, edge nodes can real-timely transcode a video into only a subset of representations, raising the video transcoding problem of what is the set of representations each participant should transcode its video stream into. Second, since participants' downlink resources are limited, one needs to solve the representation selection problem of what representation each participant should select for receiving another participant's video. Third, the above two problems are coupled and should be optimized simultaneously. Hence, this paper studies the joint video transcoding and representation selection problem for edge-assisted multi-party video conferencing, with the aim of maximizing the overall QoE under the resource and real-time video transcoding constraints. Such a problem is formulated as a non-linear integer program and is NP-hard. To solve it, we leverage the submodular optimization technique and propose a $(1 - \frac{1}{e})$ -approximate algorithm with the polynomial computation complexity. Finally, extensive trace-driven simulations are conducted to evaluate the proposed algorithm. The results show that it outperforms the alternatives by 1.5–2.5× on average in terms of overall QoE.

Keywords: Video Transcoding · Representation Selection · Multi-Party Video Conferencing · Edge Computing

1 Introduction

Over the last decade, the widespread deployment of front-facing cameras on end devices and the rapid advancement of network technologies have provided the

Z. Tari et al. (Eds.): ICA3PP 2023, LNCS 14487, pp. 380–400, 2024.
https://doi.org/10.1007/978-981-97-0834-5_22

groundwork for multi-party video conferencing (MVC) [24]. Compared with in-person conferencing, MVC provides face-to-face multi-party interaction virtually over the network, and gains the advantages of decreased time and money costs, improved collaboration and productivity efficiency, and breaking geographical restrictions. Hence, it has been widely applied in various fields, such as remote working, online education, and geographically scattered business operations, particularly since the onset of the COVID-19 pandemic. According to the report [17], MVC market size valued at USD 25 billion in 2022 and is expected to expand at a compound annual growth rate (CAGR) of 10% from 2023 to 2032.

Current MVC applications, including Zoom, Microsoft Teams, and Google Workspace, usually apply the centralized, cloud-based architecture [24]. During a conferencing session, each participant continuously uploads its source video to media servers and receives every other participant's video from the servers. Generally, participants vary in end devices (e.g., smartphones, laptops, and desktops) and network environments (e.g., wireless or wired, poor or good). In order to provide as high as possible quality of experience (QoE), the media servers would transcode the source video into multiple videos with different representations, and forward the videos with suitable representations to adapt to users' diversities [5]. Here, a video representation is a video format defined as $\langle resolution, preset, bitrate \rangle$, and different representations usually have different video qualities, bitrates, and transcoding complexities. However, because of performing all transcoding tasks, the media servers may bear heavy computing workloads, impairing the transcoding delays and thus the QoE. Besides, they fail to scale well with the increase of participants.

Recently, edge computing has emerged as a promising way to mitigate the limitations above [?]. Specifically, it deploys computing resources (servers or devices) at the network edge and provisions computing services in close proximity to users. Then, service providers could push their services to several edge nodes to serve users, relieving the burden of the cloud. For MVC, the media servers would offload each participant's video transcoding tasks to its edge node and take charge of only selecting the suitable representation and forwarding the corresponding video. Such a distributed architecture scales well and such media servers are also called selective forward units (SFUs). However, due to the limited resources of user devices and edge nodes, edge-assisted MVC faces multiple challenges as follows:

First and foremost, video transcoding tasks are resource-consuming while edge nodes are resource-constrained. Ideally, each participant should transcode its source video into all possible representations in real time, so as to adapt to other participants' diversities. Whereas, our preliminary case study reveals that commonly-used edge nodes could real-timely transcode the source video into only a small set of representations (in Sect. 2.1). For instance, real-time video transcoding restricts the transcoding time of a video frame to be at most 50 ms while it takes over 100 ms to transcode a 2K frame into four different representations by ffmpeg [9] on Jetson TX2. Therefore, a straightforward problem is *what is the set of representations that each participant should transcode its source video into*, namely the video transcoding problem. Considering the limited computing and upload bandwidth resources of edge nodes, the video transcoding problem could be reduced from the knapsack problem and is thus hard to solve.

Second, each participant in MVC needs to receive videos from all the other participants while its download bandwidth resource is typically limited. Then, we need to decide *for each participant, what is the representation that should be selected for receiving any other participant's video*, namely the representation selection problem. Our preliminary case study indicates that users' QoE depends on not only the video itself but also user device's hardware specifications (in Sect. 2.2). For example, for the same video with 1080p resolution, users have higher QoE on smartphones (mean opinion score of 4.5) than on desktops (mean opinion score of 3.5). Thus, when selecting representations, we should consider video representations' specifications, user devices' specifications, and even participants' preferences. Furthermore, since a video representation could be selected only if we have transcoded the source video into it, representation selection decisions are coupled with video transcoding decisions. That is to say, we should jointly optimize the two decisions, which makes the problem harder.

Existing research falls insufficient for treating the aforementioned challenges. Some works [2,16,19,21,26] focus on receiver-driven adaptive bitrate control under the heterogeneity of downlink of participants, but in the context of MVC both uplink and downlink should be considered. Other works [12,18] study the selection of transcoding agents, but they fail to ensure the real-time constraints for video transcoding. MultiLive [23] models the many-to-many adaptive bitrate selection problem as a non-linear programming problem and applies Scalable Video Coding (SVC) for video encoding, but it does not consider the latency when performing video encoding tasks on edge nodes.

Therefore, in this paper, we investigate the joint video transcoding and representation selection problem for edge-assisted MVC systems. Specifically, with edge nodes' limited resources, participants' diversities in devices and network environments, and user preferences considered, we aim to maximize the overall QoE of all participants under the bandwidth resource and the real-time transcoding constraints. Such a problem is formulated as a non-linear integer program and is an NP-hard problem. To solve it, we first apply the inner-and-outer problem model to decompose it into two subproblems. The inner subproblem optimizes representation selection decisions under a given video transcoding decision, while the outer subproblem optimizes video transcoding decisions. Noting that they have the submodularity property, we propose a greedy-based algorithm to solve them one by one. The theoretic analysis shows that the proposed algorithm is a $(1 - \frac{1}{e})$-approximation algorithm and runs in polynomial time. At last, extensive trace-driven simulations are conducted and the results demonstrate the superior performance of the proposed algorithm. The main contributions of this paper are summarized as follows:

- We propose the joint video transcoding and representation selection problem for edge-assisted MVC systems. We strive to maximize the overall QoE under the bandwidth resource and the real-time transcoding constraints, which is formulated as a non-linear integer program. To the best of our knowledge, we are the first work to investigate such an issue.

- To solve the problem, we first apply the inner-and-outer problem model to decompose it into two subproblems. Noting that the subproblems have the submodularity property, we then propose a greedy-based algorithm to solve them. Theoretic analysis shows that the proposed algorithm has an approximation ratio of $(1 - \frac{1}{e})$ and runs in polynomial time.
- Extensive trace-driven simulations are conducted to evaluate the proposed algorithm. The experiment results show that it outperforms the alternative algorithms by 1.5–2.5× on average in terms of the overall QoE, meanwhile satisfying the real-time transcoding requirement.

The rest of this paper is structured as follows. Section 2 presents the motivation. Section 3 introduces the system model and problem formulation. We develop an efficient approximation algorithm in Sect. 4 and evaluate it through trace-driven simulations in Sect. 5. Finally, Sect. 6 reviews the related work, and Sect. 7 concludes this paper.

2   Motivation

In this section, we present two preliminary case studies and their main findings, which motivate us to study the joint video transcoding and representation selection problem for edge-assisted MVC.

2.1   Video Transcoding Time on Edge Nodes

Video Transcoding Parameters. Widely-used video transcoding parameters include resolution, frame rate, and preset, where resolution specifies the width and height of all frames, frame rate specifies the number of frames that are played per second, and preset specifies the preference between video quality and transcoding complexity. They together determine a video's bitrate. Generally speaking, higher resolutions and frame rates lead to better QoE. When the preset changes from fast to slow, the video quality becomes better but the transcoding complexity becomes higher, which improves the QoE but increases the transcoding time. Since the frame rates of the input and output videos are usually the same, we mainly consider to adapt the resolution and the preset.

Measurement Methodology. To assess the video transcoding time, we conduct experiments under commonly-used parameter settings on several edge nodes. The resolutions include 2K, 1080P, 720P, 540P, and 360P, the presets include fast, medium, and slow. The nodes include a server with a 1060Ti GPU, a Jetson TX2 with an NVIDIA Pascal GPU, and a laptop with an i7-6700HQ CPU. All the nodes are pre-installed with the same software and environment, where ffmpeg [9] is used for video transcoding. We take a video clip (1 s, 2K@30fps) as the source video and transcode it into multiple videos under different combinations of resolution and preset settings. The experiment results are shown in

#	Resolutions	Preset - fast			Preset - medium			Preset - slow		
		1060Ti	Jetson	Laptop	1060Ti	Jetson	Laptop	1060Ti	Jetson	Laptop
4	2K + 1080 + XXX + XXX	0.89~1.03	1.04~1.52	2.25~3.10	1.19~1.20	1.39~1.50	3.05~3.22	1.58~2.03	1.74~2.23	4.25~4.79
	1080 + XXX + XXX + XXX	0.43~0.74	0.87~1.04	1.12~2.17	0.69~0.92	1.14~1.24	1.98~2.85	0.95~1.72	1.52~1.93	3.38~4.42
3	2K + 1080 + XXX	0.35~0.63	0.83~0.95	1.43~2.32	0.68~0.73	1.10~1.12	2.20~2.53	0.84~1.03	1.41~1.77	2.45~3.02
	2K / 1080 + XXX + XXX	0.17~0.62	0.72~0.85	0.87~1.44	0.45~0.64	0.74~0.97	1.45~2.20	0.57~0.72	0.87~1.04	1.87~3.10
	XXX + XXX + XXX	0.16	0.46	0.89	0.39	0.71	0.97	0.54	0.83	1.35
2	2K + 1080 / XXX	0.35~0.47	0.45~0.75	1.14~1.57	0.46~0.55	0.78~0.92	1.62~1.97	0.53~0.75	0.89~1.11	2.07~2.46
	1080 + XXX	0.15~0.27	0.42~0.61	0.83~1.12	0.32~0.37	0.60~0.70	1.10~1.46	0.40~0.49	0.73~0.80	1.41~1.69
	XXX + XXX	0.11~0.14	0.33~0.42	0.51~0.62	0.26~0.29	0.52~0.53	0.74~0.89	0.41~0.45	0.61~0.82	0.96~1.24

Fig. 1. Video transcoding time on various edge nodes, resolutions, and presets.

Fig. 1. Since the source video is 1 s long, we regard the video transcoding that takes less than 1 s as processing in real time.

Main Findings. First, edge nodes' computing capacities are limited and they can real-timely transcode videos into only a small set of resolution-preset combinations. Second, when we append new resolution-preset combinations, improve resolutions or set slower presets, the video transcoding time increase. Thus, to achieve the best possible QoE within real-time video transcoding, we should transcode the video into a suitable set of resolution-preset combinations.

2.2 User Devices' Impact on Quality of Experience

User Devices. With the wide application of MVC, it is typical that users may hold or attend a video conference with a variety of end devices, such as smartphones, tablets, laptops, desktops, etc. The user devices are inevitably heterogeneous in terms of screen sizes and pixel densities. Intuitively, even if we play an identical video stream on user devices with different screen sizes or different pixel densities, the user-perceived QoE may differ.

Resolution / Device	2K FULLHD	1080 FULLHD	720	540	360
📱	5	4.5	4.5	4	3.5
📱	5	4.5	4	3.5	3.5
💻	4.5	4	4	3.5	3
🖥	4.5	3.5	3.5	3	2

Fig. 2. Users' MOS on various video resolutions and user end devices.

Table 1. Summary of Main Notations

Notations	Descriptions
$\mathcal{P}, \mathcal{E}, \mathcal{R}$	Set of participants, edge nodes and representations
b_r, ψ_r, ϕ_r	Bitrate, resolution and preset of representation r
r_p	Representation of the source video of participant p
e_p	Local edge node of participant p
$B_e^{\uparrow}$	Upload bandwidth capacity of edge node e
$B_p^{\downarrow}$	Download bandwidth capacity of participant p
$B_{p' \to p}$	Maximum download bandwidth that participant p reserves for p'
D	Maximum acceptable transcoding delay
$\alpha_{p' \to p}$	Preference parameter of participant p for p' regarding resolution
$\beta_{p' \to p}$	Preference parameter of participant p for p' regarding preset
$d_e(r, \mathcal{R}')$	Delay for transcoding video of $r \in \mathcal{R}$ to $\mathcal{R}' \subseteq \mathcal{R}$ on edge node e
$\Psi_p(r)$	QoE of participant p under representation r regarding resolution
$\Phi_p(r)$	QoE of participant p under representation r regarding preset
Variables	Descriptions
$x_{p,r}$	Binary variable, indicating whether the video of participant p is transcoded into representation r ($x_{p,r} = 1$) or not ($x_{p,r} = 0$)
$y_{p' \to p,r}$	Binary variable, indicating whether participant p' transmits its video to p under representation r ($y_{p' \to p,r} = 1$) or not ($y_{p' \to p,r} = 0$)

Measurement Methodology. In video streaming applications, mean opinion score (MOS) is a widely-used quantitative metric to evaluate users' experiences. The larger the MOS, the better the QoE. Furthermore, the authors in [4] have investigated the relationship among video quality, screen resolutions, and video bitrates. By analyzing the data they reveal, we obtain the users' MOS on different video resolutions and different user devices. The values of MOS range from 1 to 5, and the detailed results are shown in Fig. 2.

Main Findings. As we can see, not only video resolutions but also users' devices have a notable influence on users' QoE, especially the user devices' screen sizes. Noting that users in MVC usually use devices with various screen sizes and pixel densities, we should take user devices' specifications into consideration when optimizing video streaming to achieve better QoE.

3 System Model and Problem Formulation

In this section, we provide the system model and the problem formulation in detail. For reference ease, the main notations of this paper are listed in Table 1.

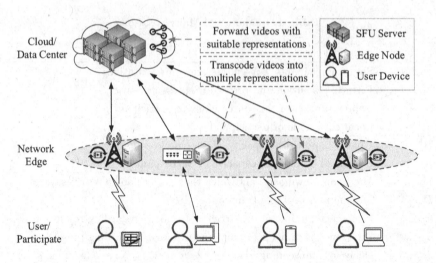

Fig. 3. Architecture of edge-assisted MVC.

3.1 System Model

System Architecture. As illustrated in Fig. 3, we consider an edge computing system that provides MVC services for multiple users. Specifically, the system consists of a cloud and a set of edge nodes, both of which are interconnected via the network. The cloud, located at the core network, hosts several SFU servers. The edge nodes, distributed at the network edge, are network access points equipped with computing devices or servers. Thus, they are able to provide computing and network capacities near the users.

Based on such infrastructure, users could use their personal devices to hold or attend video conferences. Specifically, in a multi-party video conference, each user would continuously upload its video to the edge node, then to the SFU servers, and finally to every other user in the same conference. Meanwhile, each user would also continuously download videos from every other user in the conference. Since users usually have limited and different download bandwidths, we need to real-timely transcode users' source videos into multiple representations and select the suitable representations to forward videos, so as to adapt to user diversities. To ease the computing pressure of SFU servers and the cloud, the transcoding tasks are executed at the edge nodes while selective forwarding is done by the SFU servers in the cloud. Though more than one multi-party video conferences may be held at one time in practice, we consider that we could make decisions for conferences one after another, which follows existing works [12, 23]. Therefore, we focus on one specific edge-assisted multi-party video conference.

Video Representations. We refer to the triples of resolutions, presets, and bitrates as video representations, where each representation defines a video format. For instance, the standard video representations of YouTube include (2K, medium, 16 Mbps), (1080p, medium, 8 Mbps), (720p, slow, 5 Mbps), etc. [12].

Generally speaking, with higher resolutions, slower presets, and higher bitrates, participants would perceive better QoE. In this paper, we take $\mathcal{R}$ to denote the set of all possible video representations. Besides, for any representation $r \in \mathcal{R}$, we denote its resolution, preset, and bitrate as ψ_r, ϕ_r and b_r, respectively.

Participants. We refer to the users that attend the considered conference as participants, and denote the set of participants by $\mathcal{P}$. As mentioned above, each participant would continuously upload its source video to the edge nodes and download the videos from the other participants, which consumes the upload and download bandwidth. Clearly, it uploads one video stream but downloads $|\mathcal{P}| - 1$ video streams, implying that the upload bandwidth is sufficient but the download bandwidth may be a constraint. For any participant $p \in \mathcal{P}$, we take $r_p \in \mathcal{R}$ to present the representation of its source video and take $B_p^{\downarrow}$ to present its download bandwidth capacity (in Mbps).

Edge Nodes. We take $\mathcal{E}$ to denote the set of all edge nodes. Usually, the edge nodes are heterogeneous and constrained in terms of computing and upload bandwidth capacities. Thus, for any edge node $e \in \mathcal{E}$, we take $B_e^{\uparrow}$ to denote its upload bandwidth capacity (in Mbps). Noting that edge nodes with different computing capacities have different transcoding delays, we take the function $d_e(r, \mathcal{R}')$ to denote the transcoding delays of edge node e (in milliseconds). Precisely, $d_e(r, \mathcal{R}')$ is defined as the time it takes to transcode a video of representation r into videos of representations $\mathcal{R}' \subseteq \mathcal{R}$ on edge node e. Given that video transcoding is typically done by decoding the input video into an intermediate video and re-encoding it into the output video, the function $d_e(r, \mathcal{R}')$ is an increasing function of representations r and $\mathcal{R}'$ [25]. Furthermore, we consider that participants' geographical distribution is sparse and each edge node covers only each participant at a time. Therefore, for any participant p, we refer to the edge node that covers it as its local edge node, which is denoted by e_p.

3.2 Video Transcoding and Representation Selection

Control Decisions. We take binary variables $x_{p,r} \in \{0, 1\}$ to denote the video transcoding decisions. Definitely, let $x_{p,r} = 1$ if we transcode the video of participant p into representation r, and let $x_{p,r} = 0$ otherwise. Moreover, we take binary variables $y_{p' \to p,r} \in \{0, 1\}$ to denote the representation selection decisions. Definitely, let $y_{p' \to p,r} = 1$ if we select the video of representation r to be transmitted from participant p' to p, and $y_{p' \to p,r} = 0$ otherwise. Thus, we have

$$x_{p,r} \in \{0, 1\}, \quad \forall p \in \mathcal{P}, r \in \mathcal{R}, \tag{1}$$

$$y_{p' \to p,r} \in \{0, 1\}, \quad \forall p, p' \in \mathcal{P}, r \in \mathcal{R}. \tag{2}$$

Since each participant p needs to be assigned a definite video of representation r from another participant p', we have

$$\sum_{r \in \mathcal{R}} y_{p' \to p,r} = 1, \quad \forall p, p' \in \mathcal{P}, p \neq p'. \tag{3}$$

Besides, the video of representation r must be transcoded by participant p' in order for participant p to be able to assign with it. As a result, we have

$$y_{p' \to p,r} \leq x_{p',r}, \quad \forall p, p' \in \mathcal{P}, \forall r \in \mathcal{R}. \tag{4}$$

Bandwidth Constraint. The edge nodes' upload bandwidths and the participants' download bandwidths are limited. The occupied bandwidths should be bounded by the capacities. Thus, we have the following bandwidth constraints:

$$\sum_{r \in \mathcal{R}} x_{p,r} \cdot b_r \leq B^{\uparrow}_{e_p}, \quad \forall p \in \mathcal{P}, \tag{5}$$

$$\sum_{p' \in \mathcal{P}} \sum_{r \in \mathcal{R}} y_{p' \to p,r} \cdot b_r \leq B^{\downarrow}_p, \quad \forall p \in \mathcal{P}. \tag{6}$$

Moreover, for any participant p, we require that the bitrate of the video of representation r it receives from another participant p' must be less than the downlink throughput capacity between them. Thus, we have

$$\sum_{r \in \mathcal{R}} y_{p' \to p,r} \cdot b_r \leq B_{p' \to p}, \quad \forall p, p' \in \mathcal{P}. \tag{7}$$

Real-Time Constraint. Let D be the maximum acceptable transcoding time, then the real-time video transcoding constraint is as follows: under a given video transcoding decision $\boldsymbol{x}$, we require that

$$d_{e_p}(r_p, \{r \in \mathcal{R} \mid x_{p,r} = 1\}) \leq D, \quad \forall p \in \mathcal{P}. \tag{8}$$

Participant QoE. In the MVC system, each participant's QoE depends on the qualities of the received videos from all the other participants [1] and the video quality is measured from the point of resolution and preset. Therefore, we denote by $\Psi_p(r) \geq 0$ the QoE regarding the resolution of participant p when receiving a video with representation r. Similarly, we denote by $\Phi_p(r) \geq 0$ the QoE regarding the preset of participant p when receiving a video with the representation r. Since participants may have different preferences for the two metrics and for the other participants, we introduce the weights for different metrics and participant pairs. Then, under representation selection $\boldsymbol{y}$, the QoE of participant p for participant p' could be calculated as

$$QoE_{p' \to p}(\boldsymbol{y}) = \alpha_{p' \to p} \cdot \Psi_p\Big(\sum_{r \in \mathcal{R}} r \cdot y_{p' \to p,r}\Big) + \beta_{p' \to p} \cdot \Phi_p\Big(\sum_{r \in \mathcal{R}} r \cdot y_{p' \to p,r}\Big) \tag{9}$$

where $\alpha_{p' \to p} \geq 0, \beta_{p' \to p} \geq 0$ are the QoE weights of participant p for participant p' regarding the resolution and the preset, respectively.

3.3 Problem Formulation

Focusing on the edge-assisted MVC system, we strive to optimize the video transcoding and representation selection decisions to maximize the overall QoE of all participants, subject to edge nodes' and participants' bandwidth resource constraints and the real-time transcoding constraint. Based on the models above, such a problem could be formulated as follows:

$$\mathscr{P}_1 : \max_{x,y} \sum_{p,p' \in \mathcal{P}, p' \neq p} QoE_{p' \to p}(\boldsymbol{y})$$

$$s.t. \sum_{r \in \mathcal{R}} y_{p' \to p, r} = 1, \quad \forall p, p' \in \mathcal{P}, p \neq p', \tag{1a}$$

$$y_{p \to p', r} \leq x_{p, r}, \quad \forall p, p' \in \mathcal{P}, \forall r \in \mathcal{R}, \tag{1b}$$

$$\sum_{r \in \mathcal{R}} x_{p, r} \cdot b_r \leq B_{e_p}^{\uparrow}, \quad \forall p \in \mathcal{P}, \tag{1c}$$

$$\sum_{p' \in \mathcal{P}, p' \neq p} \sum_{r \in \mathcal{R}} y_{p' \to p, r} \cdot b_r \leq B_p^{\downarrow}, \quad \forall p \in \mathcal{P}, \tag{1d}$$

$$\sum_{r \in \mathcal{R}} y_{p' \to p, r} \cdot b_r \leq B_{p' \to p}, \quad \forall p, p' \in \mathcal{P}, p \neq p', \tag{1e}$$

$$d_{e_p}(r_p, \{r \in \mathcal{R} \mid x_{p, r} = 1\}) \leq D, \quad \forall p \in \mathcal{P}, \tag{1f}$$

$$x_{p, r}, y_{p' \to p, r} \in \{0, 1\}, \quad \forall p, p' \in \mathcal{P}, r \in \mathcal{R}. \tag{1g}$$

The constraint (1a) ensures that only one representation is selected for the video streaming from one participant to another. Constraint (1b) ensures that the selected video representation must be transcoded from the source video. The following constraints, i.e., (1c)–(1e), guarantee that the occupied bandwidth resources are bounded by the capacities. The constraint (1f) is the real-time transcoding constraint, restricting that the transcoding time on any edge node is upper-bounded by the maximum acceptable transcoding delay. The last constraint (1g) specifies the domain of the video transcoding and representation selection decisions, showing that they are binary variables.

4   Algorithm Design

To solve $\mathscr{P}_1$, this section develops an efficient approximation algorithm. Specifically, we decompose $\mathscr{P}_1$ into two subproblems regarding the two decisions and then solve them one by one. Finally, we prove that the proposed algorithm is a $(1 - \frac{1}{e})$-approximation algorithm with polynomial computation complexity.

4.1 Problem Decomposition

Due to the constraint (1b), video transcoding decisions and representation selection decisions are coupled, which contributes to the first challenge of solving $\mathscr{P}_1$. Therefore, we first decompose $\mathscr{P}_1$ into two subproblems with respect to the two decisions. Considering the fact that the representation selection decisions are controlled by the video transcoding decisions, we apply the inner-and-outer problem model to decompose $\mathscr{P}_1$, where the inner problem optimizes representation selection and the outer problem optimizes video transcoding.

More specifically, the inner problem seeks the optimal representation selection decision under a given video transcoding decision $\boldsymbol{x}$. It could be formulated as

$$\mathscr{P}_2 : \max_{\boldsymbol{y}} \sum_{p,p' \in \mathcal{P}, p \neq p'} QoE_{p' \to p}(\boldsymbol{y})$$

$$s.t. \sum_{r \in \mathcal{R}} y_{p' \to p,r} = 1, \quad \forall p, p' \in \mathcal{P}, p \neq p', \tag{2a}$$

$$y_{p \to p',r} \leq x_{p,r}, \quad \forall p, p' \in \mathcal{P}, \forall r \in \mathcal{R}, \tag{2b}$$

$$\sum_{p' \in \mathcal{P}, p \neq p'} \sum_{r \in \mathcal{R}} y_{p' \to p,r} \cdot b_r \leq B_p^{\downarrow}, \quad \forall p \in \mathcal{P}, \tag{2c}$$

$$\sum_{r \in \mathcal{R}} y_{p' \to p,r} \cdot b_r \leq B_{p' \to p}, \quad \forall p, p' \in \mathcal{P}, p \neq p', \tag{2d}$$

$$y_{p' \to p,r} \in \{0,1\}, \quad \forall p, p' \in \mathcal{P}, r \in \mathcal{R}. \tag{2e}$$

For ease of presentation, we denote the solution of $\mathscr{P}_2$ under $\boldsymbol{x}$ as $\boldsymbol{y}(\boldsymbol{x})$. The outer problem regards the solution to $\mathscr{P}_2$ as subroutines or oracles, and seeks the optimal video transcoding decision. It could be formulated as

$$\mathscr{P}_3 : \max_{\boldsymbol{x}} \sum_{p,p' \in \mathcal{P}, p \neq p'} QoE_{p' \to p}(\boldsymbol{y}(\boldsymbol{x}))$$

$$s.t. \sum_{r \in \mathcal{R}} x_{p,r} \cdot b_r \leq B_{e_p}^{\uparrow}, \quad \forall p \in \mathcal{P}, \tag{3a}$$

$$d_{e_p}(r_p, \{r \in \mathcal{R} \mid x_{p,r} = 1\}) \leq D, \quad \forall p \in \mathcal{P}, \tag{3b}$$

$$x_{p,r} \in \{0,1\}, \quad \forall p \in \mathcal{P}, r \in \mathcal{R}. \tag{3c}$$

In the next subsections, we solve the inner and outer problems in sequence.

4.2   Representation Selection Under Given Video Transcoding

Since solving the outer problem relies on the solution to the inner, we begin with the inner problem, i.e., $\mathscr{P}_2$. The following lemma gives solutions to $\mathscr{P}_2$.

Lemma 1. *Given a video transcoding decision $\boldsymbol{x}$, the optimal representation selection decision for video streaming from participant p' to p is*

$$y_{p' \to p,r} = \begin{cases} 1, & r = argmax_{x_{p',r'}=1 \,\wedge\, b_{r'} \leq B_{p' \to p}} [\alpha_{p' \to p} \cdot \Psi_p(r) + \beta_{p' \to p} \cdot \Phi_p(r)], \\ 0, & otherwise. \end{cases}$$

Proof. Consider that for a given video transcoding decision $\boldsymbol{x}$, this lemma derives the representation selection decision of $\hat{\boldsymbol{y}}$. For ease of presentation, we focus on the video streaming from participant p' to p and denote the representation selected by this lemma as $\hat{r} \in \mathcal{R}$, i.e., $y_{p' \to p, \hat{r}} = 1$, and $y_{p' \to p, r} = 0, \forall r \in \mathcal{R}, r \neq \hat{r}$.

First, we prove that $\hat{\boldsymbol{y}}$ is a feasible solution to $\mathscr{P}_2$. According to this lemma, we directly have $x_{p', \hat{r}} = 1,\ b_{\hat{r}} \leq B_{p' \to p}$. Besides, in practice, different video representations usually lead to different QoEs. Thus, this lemma would select only one representation. Moreover, we usually have $\sum_{p' \in \mathcal{P}} B_{p' \to p} \leq B_p^{\downarrow}$. Then making some arrangements, we obtain $\sum_{p' \in \mathcal{P}} b_{\hat{r}} \cdot y_{p' \to p, \hat{r}} \leq \sum_{p' \in \mathcal{P}} B_{p' \to p} \leq B_p^{\downarrow}$. Summing them up, we conclude that $\hat{\boldsymbol{y}}$ satisfies all the constraints of $\mathscr{P}_2$.

Second, we prove that $\hat{\boldsymbol{y}}$ has the highest objective value of $\mathscr{P}_2$ by contradiction. Assume that there is another representation $\tilde{r} \neq \hat{r}$ that meets all of constraints of $\mathscr{P}_2$ and has higher objective value of $\mathscr{P}_2$, i.e.,

$$\alpha_{p' \to p} \cdot \Psi_p(\hat{r}) + \beta_{p' \to p} \cdot \Phi_p(\hat{r}) < \alpha_{p' \to p} \cdot \Psi_p(\tilde{r}) + \beta_{p' \to p} \cdot \Phi_p(\tilde{r}).$$

It contradicts with $\hat{r} = argmax_{x_{p', r'} = 1 \wedge b_{r'} \leq B_{p' \to p}} [\alpha_{p' \to p} \cdot \Psi_p(r) + \beta_{p' \to p} \cdot \Phi_p(r)]$. Therefore, no such $\tilde{r}$ could exist and $\hat{\boldsymbol{y}}$ has the highest objective value of $\mathscr{P}_2$.

Therefore, the derived solution of this lemma is optimal to $\mathscr{P}_2$.

Remarks. First, although rarely, it may happen that several different video representations satisfy the constraints and have the highest and the same QoE. For such cases, we just select the representation with the smallest bitrate. Second, $B_{p' \to p}$ is the upper-bound bandwidth resource for video streaming from participant p' to p, and is practically proportional to participant p's preference for p'. For example, $B_{p' \to p}$ is usually set to $\frac{(\alpha_{p' \to p} + \beta_{p' \to p}) \cdot B_p^{\downarrow}}{\sum_{p'' \in \mathcal{P}, p'' \neq p}(\alpha_{p'' \to p} + \beta_{p'' \to p})}$.

4.3 Submodular Optimization Based Video Transcoding

Now, what remains is to solve the outer problem for video transcoding, i.e., $\mathscr{P}_3$. Unfortunately, $\mathscr{P}_3$ could be reduced from the knapsack problem and is generally an NP-hard problem. It is widely believed that one can not solve an NP-hard problem optimally in polynomial time, unless $P = NP$. As a consequence, we turn to proposing an efficient algorithm with near-optimal performance.

Observing $\mathscr{P}_3$, we have two findings. The first one is that regarding video transcoding, participants' decisions have no impact on each other and they could make their decisions independently. Hence, we focus on one specific participant $p \in \mathcal{P}$ in this subsection. The second one is that the constraint (3b) shows to have the set property and we would better solve the problem from the point of set optimization. Therefore, we introduce some notations to convert $\mathscr{P}_3$ into a set optimization problem. Specifically, let the set $\mathcal{X}_p$ denote the video transcoding decision of participant p. Definitely, a representation r is in $\mathcal{X}_p$ if and only if we decide to transcode participant p's video into representation r, i.e., $\mathcal{X}_p = \{r \in \mathcal{R} | x_{p, r} = 1\}$. Still, let $\boldsymbol{y}(\mathcal{X}_p)$ denote the representation selection solution to $\mathscr{P}_2$ under $\mathcal{X}_p$. Then, for any participant $p \in \mathcal{P}$, we propose to solve

$$\mathscr{P}_4 : \max_{\mathcal{X}_p} \sum_{p' \in \mathcal{P}, p' \neq p} QoE_{p \to p'}(\boldsymbol{y}(\mathcal{X}_p))$$

$$s.t. \sum_{r \in \mathcal{X}_p} b_r \leq B_{e_p}^{\uparrow}, \tag{4a}$$

$$d_{e_p}(r_p, X_p) \leq D. \tag{4b}$$

For presentation ease, we denote $\mathscr{P}_4$'s objective function by $\Gamma_p(\mathcal{X}_p)$, which conveys the total QoE that the other participants perceive about participant p under the decision $\mathcal{X}_p$. The following lemma shows $\mathscr{P}_4$ has desirable properties.

Lemma 2. *The set function $\Gamma_p(\cdot)$ is monotone and submodular [11].*

Proof. We start with the monotonicity. Obviously, adding a new representation r into a video transcoding decision $\mathcal{X}_p$ makes the solution space of $\mathscr{P}_2$ definitely larger. Since $\mathscr{P}_2$ is a maximization problem, its objective value, as well as $\Gamma_p(\cdot)$, would increase or at least remain the same. Thus, $\Gamma_p(\cdot)$ is a monotone function.

To prove the submodularity, we introduce some notations during this proof for presentation ease. In detail, for the video streaming from participant p to p' under transcoding decision $\mathcal{X}$, we take $r_{p \to p'}^{\mathcal{X}}$ and $QoE_{p \to p'}(r_{p \to p'}^{\mathcal{X}})$ to denote the selected representation and the corresponding QoE, respectively. That is,

$$QoE_{p \to p'}(y(\mathcal{X})) = \alpha_{p \to p'} \cdot \Psi_p(r_{p \to p'}^{\mathcal{X}}) + \beta_{p \to p'} \cdot \Phi_p(r_{p \to p'}^{\mathcal{X}}) = QoE_{p \to p'}(r_{p \to p'}^{\mathcal{X}}).$$

Moreover, consider any two subsets $\mathcal{X}, \mathcal{Y} \subseteq \mathcal{R}$ and any representation $r \in \mathcal{R}$ such that $\mathcal{X} \subset \mathcal{Y}, r \notin \mathcal{Y}$. Then, when participant p changes its video transcoding decision from $\mathcal{X}$ to $\mathcal{X} \cup \{r\}$, the representation selection decision for some other participants may accordingly change from $r_{p \to p'}^{\mathcal{X}}$ to r. We denote these participants by $\mathcal{P}_{\mathcal{X}}^r$. Then, the total QoE deviation could be calculated as

$$\Gamma_p(\mathcal{X} \cup \{r\}) - \Gamma_p(\mathcal{X}) = \sum_{p' \in \mathcal{P}_{\mathcal{X}}^r} QoE_{p \to p'}(r) - QoE_{p \to p'}(r_{p \to p'}^{\mathcal{X}}),$$

$$\Gamma_p(\mathcal{Y} \cup \{r\}) - \Gamma_p(\mathcal{Y}) = \sum_{p' \in \mathcal{P}_{\mathcal{Y}}^r} QoE_{p \to p'}(r) - QoE_{p \to p'}(r_{p \to p'}^{\mathcal{Y}}).$$

First, since $\mathscr{P}_2$ is a maximization problem and $\mathcal{X}$ is a subset of $\mathcal{Y}$, we have $QoE_{p \to p'}(r_{p \to p'}^{\mathcal{X}}) \leq QoE_{p \to p'}(r_{p \to p'}^{\mathcal{Y}})$. Second, for any participant $p' \in \mathcal{P}_{\mathcal{Y}}^r$, we have $QoE_{p \to p'}(r_{p \to p'}^{\mathcal{X}}) \leq QoE_{p \to p'}(r_{p \to p'}^{\mathcal{Y}}) < QoE_{p \to p'}(r)$, leading to $p' \in \mathcal{P}_{\mathcal{X}}^r$. However, a participant p' in $\mathcal{P}_{\mathcal{X}}^r$ is not necessarily in $\mathcal{P}_{\mathcal{Y}}^r$. Therefore, $\mathcal{P}_{\mathcal{Y}}^r$ is indeed a subset of $\mathcal{P}_{\mathcal{X}}^r$, i.e., $\mathcal{P}_{\mathcal{Y}}^r \subseteq \mathcal{P}_{\mathcal{X}}^r$. Combining the two yields

$$\Gamma_p(\mathcal{X} \cup \{r\}) - \Gamma_p(\mathcal{X}) \geq \Gamma_p(\mathcal{Y} \cup \{r\}) - \Gamma_p(\mathcal{Y}).$$

As a consequence, we conclude that $\Gamma_p(\cdot)$ is a submodular function.

Now that $\mathscr{P}_4$ is proved to be maximizing a monotone and submodular function, we propose a greedy-based algorithm to solve it. It works as follows (also shown in Algorithm 1). Every participant $p \in \mathcal{P}$ makes its decisions independently (Line 1). It first initializes its video transcoding decision $\mathcal{X}_p$ to empty and a auxiliary variable B to zero (Line 2), where B presents the currently used bandwidth resource. Then it iteratively seeks the unused representation that satisfies the constraints of $\mathcal{P}_4$ and has the highest margin performance gain (Line 4). If such representation dose not exist, it stops seeking and regards the current decision as its final decision (Line 6). On the contrary, if such representation exists, it updates its decision and the auxiliary variable (Line 8). When all participants stops seeking, the algorithm converts the set decisions into our original decisions (Line 11) and outputs them (Line 12).

Algorithm 1. The proposed algorithm

Input: Problem parameters, i.e, $\mathcal{P}, \mathcal{R}, \mathcal{E}, b_r, \psi_r, \phi_r, r_p, e_p, B_e^\uparrow, B_p^\downarrow, B_{p' \to p}, D$, etc.
Output: Video transcoding decision $\boldsymbol{x}$, representation selection decision $\boldsymbol{y}$
1: **for** every participant $p \in \mathcal{P}$ **do**
2:     $\mathcal{X}_p \leftarrow \emptyset, B \leftarrow 0$;
3:     **while** true **do**
4:         $r^* \leftarrow argmax_{r \in \mathcal{R} \backslash \mathcal{X}_p, B + b_r \leq B_{e_p}^\uparrow, d_{e_p}(r_p, \mathcal{X}_p \cup \{r\}) \leq D} \; \Gamma_p(\mathcal{X}_p \cup \{r\}) - \Gamma_p(\mathcal{X}_p)$;
5:         **if** r^* is null **then**
6:             break;
7:         **end if**
8:         $\mathcal{X}_p \leftarrow \mathcal{X}_p \cup \{r^*\}, B \leftarrow B + b_{r^*}$;
9:     **end while**
10: **end for**
11: Convert $\{\mathcal{X}_p | p \in \mathcal{P}\}$ into $\boldsymbol{x}$, obtain $\boldsymbol{y}$ according to Lemma 1 ;
12: **return** $\boldsymbol{x}, \boldsymbol{y}$.

Theorem 1. *The proposed algorithm is a $(1 - \frac{1}{e})$-approximation algorithm for the proposed problem $\mathcal{P}_1$. Its computation complexity is $O(|\mathcal{P}|^2 |\mathcal{R}|^2)$.*

Proof. According to Lemma 2, $\mathcal{P}_4$ is maximizing a monotone and submodular function. Besides, according to [13], the greedy algorithm for such problem is a $(1 - \frac{1}{e})$-approximation algorithm. Thus, the proposed algorithm is a $(1 - \frac{1}{e})$-approximation algorithm for $\mathcal{P}_4$. Moreover, since participants' making decisions independently has no impact of the total QoE, it is also a $(1 - \frac{1}{e})$-approximation algorithm for $\mathcal{P}_3$. Finally, the decomposition method is lossless and Lemma 1 gives the optimal representation selection decision under any given video transcoding decisions. As a consequence, the proposed algorithm is a $(1 - \frac{1}{e})$-approximation algorithm for $\mathcal{P}_1$.

As for the computation complexity, the algorithm involves $|\mathcal{P}|$ participants. For each participant, it conducts at most $|\mathcal{R}|$ iterations. Furthermore, for each iteration, it traverses the representation set $\mathcal{R}$ and the participant set $\mathcal{P}$. Therefore, its total computation complexity is $O(|\mathcal{P}|^2 |\mathcal{R}|^2)$.

5   Evaluation

In this section, we conduct extensive simulations based on real-world network traces to evaluate the performance of our proposed algorithm. The simulation results show that our proposed algorithm improves the overall QoE against other baseline algorithms by 1.5–2.5× on average, meanwhile meeting the real-time requirement. The details are discussed in the following.

5.1   Evaluation Settings

Dataset: We evaluate the proposed algorithm using two real throughput traces datasets: Commercial Dataset [23] and Belgium 4G/LTE Dataset [22]. The former dataset consists of uplink and downlink throughput data from three servers located in different geographical locations over 72 h. And the latter dataset contains network throughput measurements in 4G networks along several paths in and around the city of Ghent, Belgium.

Settings: All optional resolutions in our simulation are 2K, 1080p, 720p, 540p, 360p, and 270p. And all optional presets are veryfast, faster, fast, medium, slow, slower, veryslow supported in ffmpeg [9]. We set the maximum transcoding latency D as 30 ms, i.e., at least 30 frames can be processed within one second. For the QoE model, we set $\alpha_{p' \to p} = 1, \beta_{p' \to p} = 1$ for any participant pair $p, p' \in \mathcal{P}$. Moreover, we mainly refer to the resolution $\Psi_p(r)$ and preset $\Phi_p(r)$ influences of QoE studied by Cermak *et al.* [4] and [10]. Due to the dynamic of network bandwidth in the system, we divide the whole time span into some time slots. And then, at the beginning of each time slot, we execute our proposed algorithm to decide the set of video stream representations to send and the video stream representations to receive for each participant. For video transcoding time $d_e(r, \mathcal{R}')$, we can conduct experiments to obtain them in advance, which will be read into memory as configuration before the algorithm runs.

Baselines: The main performance metrics we focus on are transcoding latency and overall QoE in the evaluation. We compare our proposed algorithm with two baseline algorithms. The one is the fixed bitrate algorithm, namely Non-adaptive. It prioritizes the network bandwidth constraint, while the QoE for other participants is ignored. The other one is the K-means algorithm. It first aggregates other participants into several clusters mainly based on their network bandwidth. And then, it selects the set of video streams to be generated based on the clustering centers. Considering the uplink throughput constraint of each participant, we set the number of clusters used in the K-means algorithm to 2.

5.2   Evaluation Results

Figure 4 and Fig. 5 present the simulation results of our proposed algorithm against baseline algorithms on each dataset. We repeat each evaluation ten times and take the average as the simulation result. The Non-adaptive algorithm always transcodes the original video stream into the lowest quality video

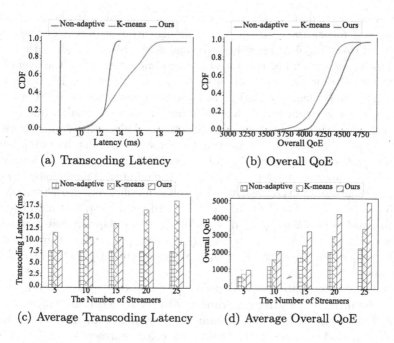

(a) Transcoding Latency

(b) Overall QoE

(c) Average Transcoding Latency

(d) Average Overall QoE

Fig. 4. Simulation results on the Commercial network dataset.

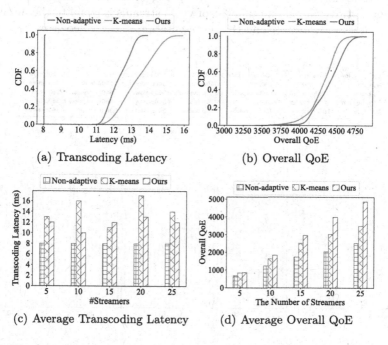

(a) Transcoding Latency

(b) Overall QoE

(c) Average Transcoding Latency

(d) Average Overall QoE

Fig. 5. Simulation results on the Belgium 4G/LTE network dataset.

stream for all participants, regardless of network dynamics. We mainly compare our proposed algorithm with the K-means algorithm in terms of transcoding latency and overall QoE. Figure 4(a), (b) and 5(a), (b) clearly show that i) The Non-adaptive algorithm has the lowest transcoding latency, but the overall QoE is the worst. ii) Our proposed algorithm not only achieves higher overall QoE compared to the K-means algorithm, but also has lower transcoding latency. This is because our algorithm can dynamically adjust both the quality and quantity of video streams at the same time. iii) Our proposed algorithm can improve QoE as much as possible, meanwhile meeting the requirement of transcoding latency. Then, we evaluate the performance of our algorithm with different numbers of streamers. In the current MVC applications, the number of participants in a session is in the order of tens or less. We change the number of video streamers from 5 to 25. As shown in Fig. 4(c), (d) and 5(c), (d), our proposed algorithm still maintains stable performance and outperforms Non-adaptive and K-means up to 2.5× and 1.5× respectively. In addition, our proposed algorithm can always meet the transcoding latency requirement regardless of the growth of the number of streamers.

To further investigate the superiority of our proposed algorithm, we conducted a detailed comparative experiment with the K-means algorithm. Figure 6 and Fig. 7 show the set of video streams that a participant needs to transcode out over a period of time. Note that, each color represents a different video stream. The line on top of each figure represents the overall QoE obtained from the corresponding set of video streams. In Fig. 6, it can be observed that for each time slot, the maximum number of video streams is equal to the number of cluster centers in K-means, and the representation of the video streams changes

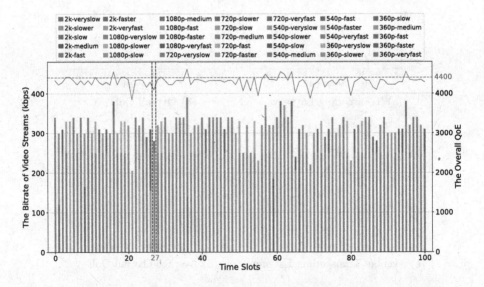

Fig. 6. The set of video streams over time slots for the K-means algorithm.

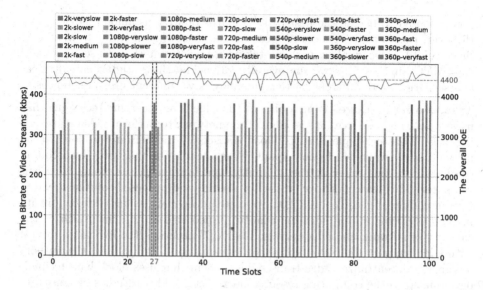

Fig. 7. The set of video streams over time slots for our proposed algorithm.

dynamically. However, in our proposed algorithm, as shown in Fig. 7, not only the representation of the video stream changes dynamically within each time slot, but also the number of video streams adjusts accordingly. For instance, at the 27th slot, the K-means algorithm decided on the set of video streams of {2k-fast, 360p-fast} with the corresponding overall QoE of 4094. However, our proposed algorithm decided on the set of video streams of {1080p-fast, 720p-slower, 540p-slow} with the corresponding overall QoE of 4570.

6   Related Work

This section reviews the most related works from two categories and highlights the differences between their studies and ours.

Multiparty Video Conferencing. Previously, some studies present the centralized architecture for MVC systems. Airlift [8] was the first to propose a cloud-based architecture for video conferencing, which utilizes the resources of the cloud's bandwidth to enhance the overall conferencing experience. Hajiesmaili *et al.* [12] cast a joint problem of user-to-agent assignment and transcoding-agent selection, and devised an adaptive parallel algorithm to find a close-to-optimal solution. They mainly focus on the rich computing resources of cloud servers to improve the overall QoE. In recent years, decentralized architecture is becoming popular in MVC. Celerity [6] examines the problem of video conferencing in P2P architecture using the network utility maximization framework. In [14], the authors aim to maximize the received video quality under uplink-downlink capacity constraints. In vSkyConf [25], a fully decentralized, efficient algorithm was proposed to decide the best delivery paths of streams among participants. However, these studies generally focus on the network constraint to maximize the received video quality, without taking into account the transcoding cost. And MVC aims to improve the overall QoE, while these studies cannot meet this scenario where different participants may have different user preferences.

Transcoding at Edge. Transcoding at the edge is an emerging technique that uses computing resources at the edge of the network to perform video transcoding tasks. This technique aims to offload transcoding tasks from the cloud to the network edge, where each participant is responsible for their own transcoding computation, thus relieving the computational burden on the cloud. In recent years, several researchers have conducted studies and proposed solutions related to transcoding at the edge. Shi [20] proposed a deep learning-based fuzzy bitrate matching scheme at the edge for adaptive video streaming, which utilizes the capacity of network and edge servers. Dogga [7] proposed the live video service with edge support for transcoding and transmission. And in [15], the authors designed a decentralized edge transcoding system that uses blockchain technology to manage the computing resources at the edge. These studies demonstrate the potential of transcoding at the edge, while for video conference, the tolerable end-to-end delay is less than 400 ms [12], which requires to design an efficient many-to-many adaptive bitrate algorithm.

7   Conclusion

In this paper, we study the joint video transcoding and representation selection problem for edge-assisted multi-party video conferencing systems, with the aim of maximizing the overall QoE under the bandwidth resource constraints and the real-time transcoding constraint. We first decompose the problem into two subproblems and then solve them by developing a greedy-based algorithm. Theoretical analysis shows that the proposed algorithm is a $(1 - \frac{1}{e})$-approximation algorithm and runs in polynomial time. We evaluate its performance through extensive trace-driven simulations and various scenario settings. The results show that our proposed algorithm outperforms other alternative algorithms, with 1.5–2.5× improvement of the overall QoE on average.

Acknowledgments. This work is partially supported by National Science Foundation of China, under grant No. 61832005; China University Industry Research Innovation Foundation, under grant No. 2021FNA04005.

References

1. Ahmed, A., Shafiq, Z., Bedi, H., Khakpour, A.: Suffering from buffering? Detecting QoE impairments in live video streams. In: 2017 IEEE 25th International Conference on Network Protocols (ICNP), pp. 1–10. IEEE (2017)
2. Akhtar, Z., et al.: Oboe: auto-tuning video ABR algorithms to network conditions. In: Proceedings of the 2018 Conference of the ACM Special Interest Group on Data Communication, pp. 44–58 (2018)

3. Block, P., et al.: Social network-based distancing strategies to flatten the Covid-19 curve in a post-lockdown world. Nat. Hum. Behav. **4**(6), 588–596 (2020). https://doi.org/10.1038/s41562-020-0898-6

4. Cermak, G., Pinson, M., Wolf, S.: The relationship among video quality, screen resolution, and bit rate. IEEE Trans. Broadcast. **57**(2), 258–262 (2011)

5. Chang, H., Varvello, M., Hao, F., Mukherjee, S.: Can you see me now? A measurement study of Zoom, Webex, and Meet. In: Proceedings of the 21st ACM Internet Measurement Conference, pp. 216–228 (2021)

6. Chen, X., Chen, M., Li, B., Zhao, Y., Wu, Y., Li, J.: Celerity: a low-delay multiparty conferencing solution. In: Proceedings of the 19th ACM International Conference on Multimedia, pp. 493–502 (2011)

7. Dogga, P., Chakraborty, S., Mitra, S., Netravali, R.: Edge-based transcoding for adaptive live video streaming. In: HotEdge (2019)

8. Feng, Y., Li, B., Li, B.: Airlift: video conferencing as a cloud service using inter-datacenter networks. In: 2012 20th IEEE International Conference on Network Protocols (ICNP), pp. 1–11. IEEE (2012)

9. FFmpeg (2022). https://ffmpeg.org/. Accessed Jan 2022

10. Preset parameter in H.264 (2022). https://trac.ffmpeg.org/wiki/Encode/H.264. Accessed Jan 2022

11. Fujishige, S.: Submodular Functions and Optimization. Elsevier (2005)

12. Hajiesmaili, M.H., Mak, L.T., Wang, Z., Wu, C., Chen, M., Khonsari, A.: Cost-effective low-delay design for multiparty cloud video conferencing. IEEE Trans. Multimedia **19**(12), 2760–2774 (2017)

13. Krause, A., Golovin, D.: Submodular function maximization. Tractability **3**, 71–104 (2014)

14. Kurdoglu, E., Liu, Y., Wang, Y.: Dealing with user heterogeneity in P2P multiparty video conferencing: layered distribution versus partitioned simulcast. IEEE Trans. Multimedia **18**(1), 90–101 (2015)

15. Liu, Y., Yu, F.R., Li, X., Ji, H., Leung, V.C.: Decentralized resource allocation for video transcoding and delivery in blockchain-based system with mobile edge computing. IEEE Trans. Veh. Technol. **68**(11), 11169–11185 (2019)

16. Mao, H., Netravali, R., Alizadeh, M.: Neural adaptive video streaming with pensieve. In: Proceedings of the Conference of the ACM Special Interest Group on Data Communication, pp. 197–210 (2017)

17. Video conferencing market (2023). http://www.gminsights.com/industry-analysis/video-conferencing-market. Accessed 15 May 2023

18. Ooi, W.T., van Renesse, R.: Distributing media transformation over multiple media gateways. In: Proceedings of the Ninth ACM International Conference on Multimedia, pp. 159–168 (2001)

19. Sengupta, S., Ganguly, N., Chakraborty, S., De, P.: HotDASH: hotspot aware adaptive video streaming using deep reinforcement learning. In: 2018 IEEE 26th International Conference on Network Protocols (ICNP), pp. 165–175. IEEE (2018)

20. Shi, W., et al.: Learning-based fuzzy bitrate matching at the edge for adaptive video streaming. In: Proceedings of the ACM Web Conference 2022, pp. 3289–3297 (2022)

21. Spiteri, K., Urgaonkar, R., Sitaraman, R.K.: Bola: Near-optimal bitrate adaptation for online videos. IEEE/ACM Trans. Netw. **28**(4), 1698–1711 (2020)

22. Van Der Hooft, J., et al.: HTTP/2-based adaptive streaming of HEVC video over 4G/LTE networks. IEEE Commun. Lett. **20**(11), 2177–2180 (2016)

23. Wang, Z., et al.: MultiLive: adaptive bitrate control for low-delay multi-party interactive live streaming. IEEE/ACM Trans. Netw. **30**(2), 923–938 (2021)

24. Wei, J., Bojja Venkatakrishnan, S.: DecVi: adaptive video conferencing on open peer-to-peer networks. In: 24th International Conference on Distributed Computing and Networking, pp. 336–341 (2023)
25. Wu, Y., Wu, C., Li, B., Lau, F.C.: vSkyConf: cloud-assisted multi-party mobile video conferencing. In: Proceedings of the Second ACM SIGCOMM Workshop on Mobile Cloud Computing, pp. 33–38 (2013)
26. Yadav, P.K., Shafiei, A., Ooi, W.T.: QUETRA: a queuing theory approach to dash rate adaptation. In: Proceedings of the 25th ACM International Conference on Multimedia, pp. 1130–1138 (2017)

Performance Comparison of Distributed DNN Training on Optical Versus Electrical Interconnect Systems

Fei Dai[1,2]([✉]) [ID], Yawen Chen[1] [ID], Zhiyi Huang[1] [ID], Haibo Zhang[1] [ID],
and Hui Tian[3] [ID]

[1] School of Computing, University of Otago, Dunedin, New Zealand
travis.dai@otago.ac.nz
[2] School of Computing, Eastern Institute of Technology | Te Pūkenga, Hawke's Bay,
Napier, New Zealand
[3] School of Information and Communication Technology, Griffith University,
Brisbane, Australia

Abstract. Parallel and distributed Deep Neural Network (DNN) training have become integral in data centers, significantly reducing DNN
training time. The interconnection type among nodes and the chosen
all-reduce algorithm critically impact this speed-up. This paper examines the efficiency differences in distributed DNN training across optical
and electrical interconnect systems using various all-reduce algorithms.
We first explore the Ring and Recursive Doubling (RD) all-reduce algorithms in both systems, followed by formulating a communication cost
model for these algorithms. Performance comparison is then carried out
via extensive experiments. Our results reveal that, in 1024-node systems, the Ring algorithm outperforms the RD algorithm in optical and
electrical interconnects when data transfer exceeds 64 MB and 1024 MB,
respectively. We also find that both Ring and RD algorithms in optical
interconnect systems reduce average communication time by around 75%
compared to electrical interconnect systems across four different DNNs.
Interestingly, the communication time of the RD algorithm, but not the
Ring algorithm, reduces as the number of wavelengths increase in optical
interconnects. These findings provide valuable insights into DNN training
optimization across various interconnect systems and lay the groundwork
for future related research.

Keywords: Performance comparison · Distributed DNN training ·
Electrical interconnect · Optical interconnect · Performance
comparison · Parallel computation

1 Introduction

Deep Neural Networks (DNNs), fundamental algorithms in Deep Learning (DL),
have found widespread applications in domains such as image classification, language translation, and speech recognition [1]. However, training DNN models

© The Author(s), under exclusive license to Springer Nature Singapore Pte Ltd. 2024
Z. Tari et al. (Eds.): ICA3PP 2023, LNCS 14487, pp. 401–418, 2024.
https://doi.org/10.1007/978-981-97-0834-5_23

with large datasets can be a painstakingly long process, often taking days or even weeks, which is not practical in most cases. As a response, numerous parallel and distributed DNN training methods have emerged. Among these, data parallelism is widely adopted: each worker trains the DNN using its local dataset and iteratively exchanges model parameters (e.g., gradients) with other workers. Stochastic Gradient Descent (SGD), the leading method for DNN training, invokes intense communications for the all-reduce operation in distributed DL training [14]. However, with an escalating number of workers, network communication traffic balloons, necessitating higher bandwidth.

Electrical interconnect systems for distributed DNN training are advancing rapidly. Yet, their bandwidth improvements lag behind the progression of computing units [15]. High communication overheads from factors like crosstalk, dielectric loss, and switching noise render electrical interconnect a critical obstacle to enhancing overall system performance [16]. When communication overhead outweighs the gains from parallel computation, training performance deteriorates. Optical interconnect emerges as a promising alternative [23], offering high bandwidth, low power costs, and minimal latency. Leveraging wavelength division multiplexing (WDM), optical interconnect allows data transmission through a waveguide using different wavelengths. These benefits position optical interconnects as efficient performers in intensive device-to-device communications and can thus accelerate data-parallel distributed DNN training.

Nonetheless, optical interconnect also carries overheads such as optical-to-electrical (OE)/electrical-to-optical (EO) conversion costs, insertion loss from light transmission through the waveguide, and configuration delay of the microring resonator (MRR). These factors, along with communication patterns governed by the all-reduce algorithm, influence the communication time in data-parallel distributed DNN training. To date, no studies have compared the communication performance of all-reduce algorithms between optical and electrical interconnect systems in data-parallel distributed DNN training.

Thus, this paper investigates the comparison of all-reduce algorithms for distributed DNN training efficiency on optical and electrical interconnect systems under varying configurations. We aim to answer the following research questions: *1) How does the performance of the Ring and Recursive Doubling (RD) all-reduce algorithms differ between optical and electrical interconnect systems with varying sizes of transferred data? 2) What improvements can optical interconnect systems bring to distributed DNN training compared with electrical systems? 3) Are existing all-reduce algorithms utilizing optical resources in the optical interconnect system optimally for distributed DNN training?* The key contributions of this paper are summarized as follows:

1. We first analyze Ring and RD all-reduce algorithms and summarize their communication step and the amount of communication traffic in each step. Then, we formulate the communication costs of these two algorithms on the optical and electrical interconnect systems for data-parallel distributed DNN training.

2. We conduct extensive simulations to compare the communication performance of those two all-reduce algorithms between the optical interconnect system and the electrical interconnect system using two experiment settings. Our results indicate: a) The Ring all-reduce algorithm outperforms the RD algorithm when the transferred data is more than 64 MB and 1024 MB respectively in the 1024 nodes optical and electrical interconnect systems. b) The optical interconnect system demonstrates superior scalability compared to its electrical counterpart. Additionally, the Ring and RD all-reduce algorithms on the optical interconnect system reduce the average communication time by 74.74% and 75.35% for four DNNs compared with the electrical interconnect system. c) In the optical interconnect system, the Ring all-reduce algorithm's communication time is not impacted by the number of wavelengths. In contrast, the RD algorithm's communication time decreases with an increasing number of wavelengths.

The remaining part of the paper proceeds as follows: Sect. 2 introduces the related works. Section 3 provides the necessary background on optical and electrical interconnect system and data-parallel distributed DNN training. Section 4 first illustrates the all-reduce algorithms, then presents the formulation of communication costs for the two all-reduce algorithms in optical and electrical interconnect systems. Section 5 outlines the experimental setup and compares the performance of the two all-reduce algorithms across optical and electrical interconnect systems. Finally, Sect. 6 concludes the paper.

2   Related Work

An immense amount of research has been conducted to expedite distributed DNN training in electrical interconnect systems. For instance, Plink [2] employs two-level hierarchical aggregations to balance traffic and curtail communication delay. Blink [3], designed specifically for the DGX-2 GPU, uses multiple directed spanning trees to optimize link utilization. Poseidon, a robust communication framework for distributed DL on GPUs, has been devised to address the difficulties associated with scaling out due to the necessity for regular network synchronization [19]. Other studies have focused on enhancing all-reduce performance by reducing network contention [4], bolstering ring all-reduce performance for large-scale clusters [5], and decomposing all-reduce on heterogeneous network hierarchies [6]. A growing trend also suggests integrating optical interconnect systems for superior performance. For instance, a hybrid electrical-optical switch architecture proposed in [7] achieved a 10% improvement in training time. Sip-ML, another work utilizing optical interconnect for distributed machine learning training, showed a 1.3–9.1× improvement in DNN model training time through simulations [8]. RAMP is an all-optical network architecture that greatly improves energy efficiency and training time for recommendation systems such as Megatron and DLRM [21].

Certain studies have compared the performance between optical and electrical interconnect systems. The study in [9] using the FFT benchmark found over

70% gain in speedup and efficiency in optical over electrical systems. Another study [10] compared power dissipation, data rate, and loss between the two systems, and simulations revealed higher bandwidth, lower channel loss, and reduced power consumption in optical interconnects. The study referenced in [20] compared the performance of multi-layer perceptron training between electrical and optical network-on-chips. Experimental results demonstrate that the MLP training time for ONoC surpasses ENoC in both training duration and energy efficiency. However, these studies did not focus on comparisons specific to distributed DNN training scenarios. To the best of our knowledge, this paper is the first to compare the performance of optical and electrical interconnect systems in the context of data-parallel distributed DNN training using different all-reduce algorithms.

3   Background

3.1   Optical and Electrical Interconnect Systems

This section first elucidates the major distinctions, benefits, and drawbacks of optical and electrical interconnect systems. It then discusses the particular architectures of these systems used in our study. The fundamental difference between optical and electrical interconnect systems lies in the transmission technology they employ to facilitate communication between hosts or nodes. In electrical interconnect systems, all nodes are connected via electronic interconnect in a multilevel electronic hierarchy, such as a fat tree. Conversely, optical interconnect systems utilize arrays of optical switches or Sip interfaces to interconnect nodes [8]. Communication in electrical interconnect systems transpires through hierarchical electrical switches. Electrical packets from the source traverse through electrical links and switches to reach the destination. In contrast, transmission among nodes in optical interconnect systems leverages different wavelengths for parallel communication via waveguides using WDM technology. The benefits of optical communication include low transmission delay (10 ms for optical switch, 25 μs for MRRs Sip interface), low power consumption (essentially independent of distance), and high bandwidth (up to 40 Gb/s per wavelength, with 64 wavelengths per waveguide) [8]. However, one significant disadvantage of optical interconnect systems is the need for numerous optical components, thereby incurring configuration overhead. Compared to optical interconnect systems, electrical interconnect systems offer lower bandwidth (25 Gb/s) and require a much more complex structure for switch interconnect as the cluster scales.

Figure 1 and Fig. 2 present an overview of the electrical and optical interconnect systems respectively, as used in distributed DNN training. The architecture of the electrical interconnect comprises two levels, where hosts directly connect to level 1 electrical switches. These switches' uplinks connect to level 2 electrical switches, while their downlinks connect to hosts. Each electrical switch features 36 ports. Figure 2 illustrates the architecture of the optical interconnect system, which aligns with the system proposed in [8] and is based on the Sip interface.

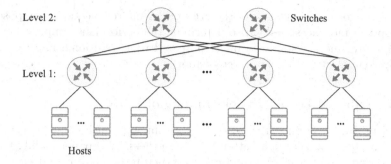

Fig. 1. Electrical interconnect architecture.

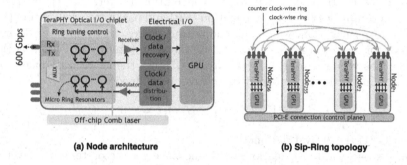

(a) Node architecture   **(b) Sip-Ring topology**

Fig. 2. Optical interconnect architecture: (a) Node architecture, (b) Sip-Ring topology [8].

Figure 2(a) shows the composition of nodes leveraging TeraPHY silicon photonics technology. Each node incorporates four optical interfaces, each containing 64 micro-ring resonators (MRRs) to select and forward any subset of 64 wavelengths. On the transmit side (Tx), an off-chip comb laser generates light steered into the node via a fiber coupler towards an array of MRRs. These modulate the accelerator's transmitting data at 40 Gbps per wavelength. Conversely, the receive side (Rx) features a second array of MRRs that select wavelengths targeted to the accelerator and pass through the remaining wavelengths. Figure 2(b) depicts nodes interconnected with their adjacent nodes in a ring topology (only two rings for two directions are displayed for clarity). In the data plane, traffic transmits across four single-mode fiber rings, two in each direction (clockwise and counter-clockwise). Routing and wavelength assignment (RWA) configuration is performed in the control plane, where wavelengths can be dynamically placed around the fiber ring. We assume that each host in both electrical and optical interconnect systems uses a GPU as its computing device, and all GPUs across hosts are homogeneous. The details of system parameters are described in Sect. 5.

The primary motivations for comparing these two interconnect architectures are twofold: (1) to evaluate the distributed DNN training performance of existing all-reduce algorithms on state-of-the-art optical interconnections composed of MRR optical interfaces; and (2) to determine the performance improvement an optical interconnect system can offer compared to a traditional multi-tier, fat-tree topology electrical interconnect system.

3.2   Distributed Data-Parallel Training of DNN

A DNN typically contains L layers ($L \geq 2$), among which common types are fully connected, convolutional, attention, and others. This paper mainly focuses on DNNs with fully connected and convolutional layers. Given that each DNN layer can be a fully connected or convolutional layer, the neurons in layer ℓ receive vectors or tensors as input. We first elucidate the scenario for a fully connected layer, then extend it to the convolutional layer. The DNN training process encompasses forward propagation (FP) and backward propagation (BP). In forward propagation, we denote the output vector/tensor in layer ℓ (or the input vector/tensor of layer $\ell - 1$) as $Z^{(\ell)}$, the weight matrix at layer ℓ as $W^{(\ell)}$, and the bias vector/tensor in layer ℓ as $B^{(\ell)}$. Thus, the forward propagation of DNN training with n_ℓ neurons at layer ℓ can be defined as:

$$Z^{(\ell)} = \phi(W^{(\ell)} Z^{(\ell-1)} + B^{(\ell)}), \ell = 2, 3, ..., L. \tag{1}$$

where $\phi(*)$ is the activation function.

During backward propagation, we define the error vector/tensor in layer ℓ as E_ℓ. We can then express this as:

$$E^{(\ell)} = (E^{(\ell+1)} (W^{(\ell)})^T) \phi'(Z^{(\ell)}), \tag{2}$$

where $\phi'(*)$ is the derivative function of $\phi(*)$.

Using the error vector/tensor, the gradient of the weight $\Delta W^{(\ell)}$ can be calculated as:

$$\Delta W^{(\ell)} = (Z^{(\ell)})^T E^{(\ell+1)}. \tag{3}$$

Upon obtaining the gradient, weights are updated as follows:

$$W^{(\ell)} = W^{(\ell)} - \sigma \Delta W^{(\ell)}, \tag{4}$$

where σ is the learning rate. Given that a convolutional layer can also be transformed into a matrix-matrix multiplication by the im2col function [11], albeit with different matrix operand dimensions, the training of the convolutional layer can be represented by the above equations.

In data-parallel distributed DNN training, training is parallelized by partitioning the input data across a batch size of b. The whole training model is replicated, such that each worker maintains a local copy but processes different training batches. In the weight update process of BP, all-reduce algorithms synchronize the gradients among the workers. Assuming n nodes in the cluster system, the update process in data-parallel distributed DNN training becomes:

$$W^{(\ell)} = W^{(\ell)} - \sigma \Delta W^{(\ell)}$$

$$= W^{(\ell)} - \frac{\sigma(\Delta W_1^{(\ell)} + \Delta W_2^{(\ell)} + \dots + \Delta W_n^{(\ell)})}{n}. \tag{5}$$

4   Performance Model

4.1   All-Reduce Algorithm

During the data-parallel distributed DNN training process (specifically, the weight update process of BP), all nodes in the interconnect systems need to receive data from all other nodes using all-reduce algorithms. Consequently, the communication pattern and communication traffic are dependent on the specific all-reduce algorithms used. Given the distinct characteristics of electrical and optical interconnect systems, the communication steps and sizes of transferred data can vary across different all-reduce algorithms, leading to different training times. Hence, we consider two well-known communication schemes for performing all-reduce operations in a system with n nodes and d as the size of transferred data to be reduced:

- **Ring all-reduce:** The communication in Ring all-reduce consists of two stages - reduce and broadcast stage. Ring all-reduce first reduce-scatters data of size $\frac{d}{n}$ on all nodes in $n-1$ steps during the reduce stage. Then, all nodes conduct all-gather operations to collect reduced data of size $\frac{d}{n}$ from other nodes until every node obtains reduced data of size d. This process also takes $n-1$ steps. Therefore, the total number of steps required by the Ring all-reduce is $2(n-1)$ steps. Figure 3(a) illustrates how all-reduce operations are conducted by the Ring all-reduce algorithm in a 4-node optical interconnect system. The Ring all-reduce communications comprise two stages, each with three steps. Since reduction operations are conducted every step during the reduce stage and all-gather operations are conducted every step during the broadcast stage, the transferred data in each communication step is $\frac{d}{4}$. During each communication step, each node sends data to its adjacent node using one wavelength.
- **Recursive Doubling (RD) all-reduce:** RD with pairwise exchange enables logarithmic scaling of $O(log_2 n)$. During each communication step, each node in the group sends data of size d to its corresponding peer appropriate for the current step. After nodes receive the data, reduction operations are executed in parallel. The overall all-reduce operation among all nodes is built up by recursively conducting a series of smaller all-reduce operations over orthogonal sub-groups of nodes until the entire target group has been reduced. Figure 3(b) demonstrates how the all-reduce communications are conducted by RD in a 4-node optical interconnect system. In the first step, nodes (nodes 1 and 2) in group one and nodes (nodes 3 and 4) in group two exchange data using two wavelengths. In the second step, nodes 1 and 3 form a new group, and nodes 2 and 4 form another group. Then, nodes in the groups exchange

the reduced data within the group. According to the communication pattern of RD, the number of wavelengths requirement doubles in every step, and the size of transferred data in each step is d.

4.2   Communication Cost

Communication Step. The communication time in distributed DNN training is influenced by the number of communication steps and the amount of communication traffic in each step, which are determined by the all-reduce algorithm employed in the interconnect systems. Hence, we examine the number of communication steps involved in the Ring and RD all-reduce algorithms, as well as the amount of communication traffic generated in each communication step for both optical and electrical interconnect systems.

In optical interconnect systems, the number of communication steps in the all-reduce algorithm depends on the communication patterns and the available number of wavelengths. In Ring all-reduce, nodes simultaneously send data of size $\frac{d}{n}$ to adjacent nodes, taking $2(n-1)$ steps. For RD all-reduce, nodes send data of size d to their corresponding pair in each step, with multiple wavelengths

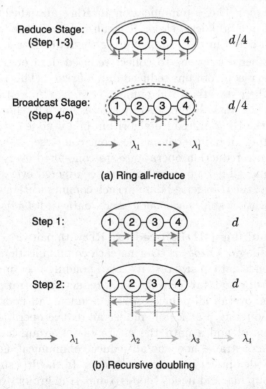

Fig. 3. Illustration of Ring and RD all-reduce operations in a 4-node optical interconnect system.

Table 1. Comparison of the number of communication steps for Ring and RD all-reduce algorithms in optical and electrical interconnect systems.

Algorithms	Network	Communication steps
Ring	Optical	$2(n-1)$
	Electrical	$2(n-1)$
RD	Optical	$\lceil log_2 n \rceil, (n \leq 2w)$ $2\lceil log_2 n \rceil - \lceil log_2 2w \rceil, (n > 2w)$
	Electrical	$\lceil log_2 n \rceil$

used concurrently to prevent conflict. When the required number of wavelengths exceeds the available number in the system, additional communication steps are needed. For instance, if the available number of wavelengths is 64, a system with 128 nodes would need $\lceil log_2 128 \rceil = 7$ steps to complete the all-reduce operations, while a system with 1024 nodes would need $2\lceil log_2 1024 \rceil - 7 = 13$ steps to do so. Contrarily, in an electrical interconnect system, which is based on packet switching, the number of communication steps for Ring and RD all-reduce are $2(n-1)$ and $\lceil log_2 n \rceil$, respectively. We summarize the communication steps for the corresponding interconnect systems in Table 1, where the number of available wavelengths in the optical interconnect system is denoted as w.

Communication Time. This subsubsection explores the communication time in data-parallel distributed DNN training, defined as the time required for all nodes to exchange data of size d with every other node. Given our previous assessment of the number of communication steps and the size of transferred data, we can now establish the communication time for both the Ring and RD all-reduce algorithms in the optical and electrical interconnect systems.

In an optical interconnect system, let a denote the reconfiguration time of the MRR and delay of O/E/O conversion, and β_O represent the bandwidth per wavelength. Based on Table 1, the communication time of Ring all-reduce in a system with n nodes for data-parallel distributed DNN training can be calculated by

$$T_{O-Ring} = 2(n-1)\left(\frac{d\eta}{n\beta_O} + a\right), \tag{6}$$

where η represents the storage size of one parameter.

Similarly, the communication time of RD all-reduce in an optical interconnect system with n nodes can be calculated as

$$T_{O-RD} = \theta\left(\frac{d\eta}{\beta_O} + a\right), \tag{7}$$

where θ is the number of communication steps when using the RD all-reduce algorithm.

We also formulate the communication time of the two all-reduce algorithms in an electrical interconnect system for data-parallel distributed DNN training.

We use α to denote the link latency and β_E to denote the network bandwidth. According to Table 1, the communication cost of Ring all-reduce in a n nodes electrical interconnect system can be calculated as

$$T_{E-Ring} = 2(n-1)\left(\alpha + \frac{d\eta}{n\beta_E}\right). \tag{8}$$

Likewise, the communication cost of RD all-reduce in an electrical interconnect system with n nodes can be calculated as

$$T_{E-RD} = \lceil log_2 n \rceil \left(\alpha + \frac{d\eta}{\beta_E}\right). \tag{9}$$

5   Evaluation

This section details the methodologies used for simulating data-parallel distributed DNN training across both optical and electrical interconnect systems. In our approach, we developed a custom optical interconnect simulator in Python, aimed at illustrating the behavior of both Ring and RD all-reduce algorithms. The accuracy and reliability of this simulator were ensured through meticulous modeling and disciplined software development techniques. Given the lack of real-world optical interconnect hardware for direct validation, we implemented additional safeguards. Specifically, we cross-verified the simulator's outputs with analytical and mathematical models to assure both accuracy and consistency. Though not tested against real-world configurations, our simulator is grounded in widely-accepted models of optical interconnects and machine learning frameworks. This simulated environment provides a thorough understanding of the relevant algorithms within the context of optical communications. For the electrical interconnect system, we replicated a conventional data center cluster environment. These typically employ a multi-layer Fat-tree topology and utilize electrical packet switches, as cited in [18]. This emulation was realized using SimGrid framework's version 3.3 discrete-event simulator, which is well-known for its effectiveness in modeling Fat-tree electrical interconnect systems [12]. It is worth noting that we used float32 as the data type for computations, and GPUs served as accelerators in both simulation scenarios.

To bolster the validity of our simulations, we used four widely-recognized DNN benchmarks: AlexNet, VGG16, ResNet50, and GoogLeNet. These models were strategically chosen for their compatibility with GPU memory constraints, a critical prerequisite for distributed data-parallel training since each GPU must house a full copy of the DNN model. To ensure a diverse set of test conditions, we executed these benchmarks using both the MNIST and ImageNet datasets, as cited in [11]. To precisely profile computation times, memory footprints, and data sizes for backpropagation, among other variables, we utilized the TensorFlow profiler [17]. Our test system for profiling was equipped with an Intel i7-6700K 4.0 GHz CPU, 64 GB RAM, and a GeForce GTX TITAN XP GPU. This process yielded invaluable insights, such as optimal batch sizes and data transfer volumes for each of the four DNN models under varying numbers of GPUs. Notably, the profiler also verified the uniformity of gradient quantities-excluding the input and output layers-across multiple datasets within the same batch during distributed data-parallel training.

In terms of network-level simulations, we employed SimGrid to model the communications behavior of two key all-reduce algorithms in the electrical interconnect system designated for distributed DNN training. Table 2 outlines the simulated platform settings along with their associated parameters, all of which adhere to SimGrid's default configurations. Given that SimGrid does not natively support real DNN applications, we devised a specialized benchmark to simulate data-parallel distributed DNN training, as detailed in Algorithm 1. For simulations involving the optical interconnect system, we relied on parameters detailed in Table 2, which were sourced from references [8,22]. These parameters were implemented in our custom optical interconnect simulator. To ensure a rigorous performance evaluation, we benchmarked both the optical and electrical interconnect systems by comparing the efficiencies of the Ring and RD all-reduce algorithms. This comparison was conducted using both SimGrid and our in-house optical simulator. Estimates of communication time for these setups were derived through numerical computation, factoring in the number of parameters as calculated in both SimGrid and our custom simulator.

Algorithm 1: Pseudo-code of the benchmark

   input : $N \leftarrow$ the number of nodes
              $M \leftarrow$ the number of parameters
              $I \leftarrow$ the number of iterations
   output: Communication time of benchmark C

1   $t_1 \leftarrow$ MPI_Wtime()
2   Allocate buffers with $M \times$ sizeof(float) bytes
3   Initialize the value of send buf
4   **for** $i \leftarrow 1$ **to** I **do**
5   **for** $i \leftarrow 1$ **to** M **do**
6 | MPI_Allreduce(send_buf, rec_buf, M, MPI_FLOAT, ...)
7   **end**
8   **end**
9   $C \leftarrow$ MPI_Wtime() - t_1

5.1   Experiment Setup

Our experimental framework consists of two distinct sets of tests. The first set aims to investigate the impact of varying data transfer sizes on the Ring and RD all-reduce algorithms within both the electrical and optical interconnect systems. In the second set, we focus on the scalability of the two all-reduce algorithms in optical and electrical interconnect systems. We manipulate the number of nodes within our in-house simulator and SimGrid while maintaining the data transfer sizes of four representative DNN models. Additionally, we scrutinize the influence of varying wavelengths in the optical interconnect system, using a configuration of 1024 nodes. It is important to note that our evaluation of communication time is predicated upon one epoch of training time, given the repetitive nature of the training process.

Table 2. Parameters of simulated architecture.

Interconnect	Parameters setup
Electrical network	Two-level fat-tree topology,
	Router bandwidth: 25 Gbps,
	Router delay: 25 μs,
	Packet size: 128 bytes,
	Shortest-path routing
Optical network	Ring topology with two directions,
	64 wavelengths per wavegiude,
	MRRs reconfiguration delay: 25 μs,
	O/E/O conversion delay: 1 cycle/flit,
	Packet size: 128 bytes,
	Flit size: 32 bytes,
	40 Gbps/wavelength

5.2   Simulation Results

Results of the First Set. We compared the performance of the Ring and RD all-reduce algorithms within the electrical and optical interconnect systems by deploying these algorithms in our in-house simulator and SimGrid. This configuration consisted of 1024 nodes and used transferred data sizes of 16 MB, 64 MB, 256 MB, 1024 MB, 4096 MB, and 16384 MB. For the optical interconnect system, we assumed 64 wavelengths.

Figure 4 illustrates the performance of the Ring and RD all-reduce algorithms within optical and electrical interconnect systems consisting of 1024 nodes, with incrementally increasing data transfer sizes. A zoomed-in segment of the line charts that depict the performance of these algorithms on optical and electrical interconnect systems is presented in the inset of Fig. 4. From this inset, we can observe that the Ring all-reduce algorithm surpasses the RD all-reduce algorithm when data transfer exceeds 64 MB within the optical interconnect system. This can be attributed to the trade-off between data transfer sizes and the number of communication steps in the Ring and RD all-reduce algorithms, with 64 MB acting as a pivot point in the optical interconnect system. However, this pivot point differs within the electrical interconnect system. As shown in the inset of Fig. 4, the communication cost of the Ring all-reduce algorithm remains higher than the RD all-reduce algorithm until the transferred data exceeds 1024 MB.

This is due to the Ring all-reduce algorithm's larger number of communication steps and low bandwidth utilization in the electrical link when the data transfer is small, while the RD all-reduce algorithm's smaller number of communication steps and high bandwidth utilization of the electrical link.

Comparing the two interconnect systems, the optical interconnect system outperforms the electrical interconnect system for both the Ring and RD all-reduce algorithms. This superior performance of the optical interconnect system can be attributed to its higher bandwidth, lower latency, and smaller number of communication steps.

Results of the Second Set. To evaluate the scalability of the two all-reduce algorithms in both interconnect systems, we scaled the nodes (128, 256, 512, and 1024) within the electrical and optical interconnect systems, employing transferred data sizes from four conventional DNN models. The batch sizes for AlexNet, VGG16, GoogLeNet, and ResNet50 were set at 512, 48, 64, and 1024, respectively, which are sizes that maximize GPU memory usage for all-reduce algorithms. All results in Fig. 5 are normalized by dividing by the result of O-RD in GoogLeNet.

Figure 5 presents a comparison of the communication times of the Ring and RD all-reduce algorithms in the electrical and optical interconnect systems, with different DNN models across various scales. From Fig. 5(a) to (d), it is apparent that: 1) the communication time of both all-reduce algorithms in the optical and electrical interconnect systems increases with the number of nodes, and 2) the communication time for both all-reduce algorithms is lower in the optical interconnect system compared to the electrical interconnect system. We observe the most significant change in communication time with the RD all-reduce algorithm in the optical interconnect system. The communication time for RD remains low and fairly constant with increasing nodes for DNNs with small data transfer sizes, such as ResNet50 and GoogLeNet. However, for DNNs with large data transfers, such as AlexNet and VGG16, the communication time for RD is higher than that of Ring all-reduce. This is because the communication time for RD grows proportionally with the data transfer size, as described in Eq. (7). Contrarily, in the electrical interconnect system, RD always outperforms Ring all-reduce due to RD's higher link utilization and relatively constant communication steps with the increase in nodes. Comparing the Ring all-reduce algorithm in optical and electrical interconnect systems, O-Ring significantly reduces communication time compared with E-Ring. This performance gain for the Ring all-reduce algorithm in the optical interconnect system is due to the system's higher bandwidth and lower latency. Overall, applying the RD and Ring all-reduce algorithms in the optical interconnect system reduces communication time on average by 74.74% and 75.35%, respectively, compared to the electrical interconnect system.

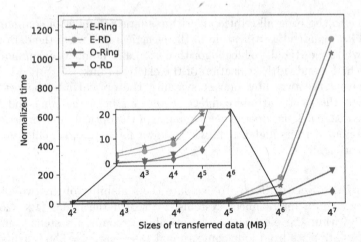

Fig. 4. Performance of the algorithms in the 1024 nodes optical and electrical interconnect system using different amount of transferred data.

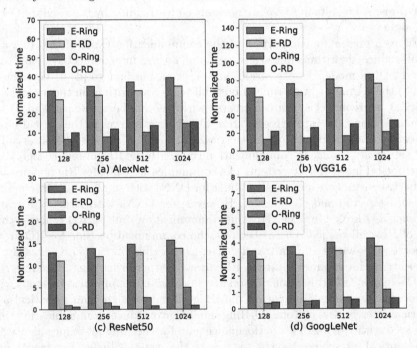

Fig. 5. The performance comparison of two all-reduce algorithms in electrical and optical interconnect systems.

Furthermore, we evaluated the impact of wavelength on the performance of the two all-reduce algorithms in the 1024 node optical interconnect system, varying the number of wavelengths (8, 16, 32, 64, 128) with the data sizes from the four traditional DNN models. Figure 6 displays the performances of the two all-reduce algorithms using different numbers of wavelengths for the four DNNs. As the number of wavelengths increases, the communication time of RD decreases while Ring all-reduce remains constant. This suggests that the communication cost of RD is related to the number of wavelengths, while Ring is not.

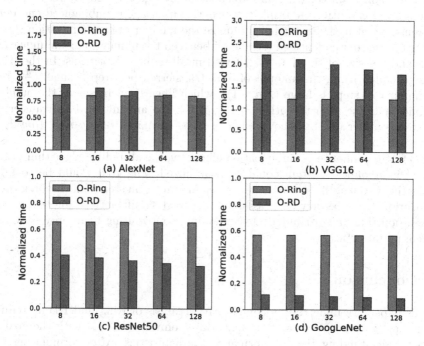

Fig. 6. The wavelength impact on Ring and RD all-reduce in the 1024 nodes optical interconnect systems.

From the above simulation results, we can draw the following conclusions: 1) In the 1024-node optical and electrical interconnect systems, Ring all-reduce outperforms RD when the data transfer size exceeds 64 MB and 1024 MB, respectively. 2) The optical interconnect system exhibits superior scalability over the electrical interconnect system. Both Ring and RD all-reduce in the optical interconnect system reduce average communication time by 74.74% and 75.35%, respectively, compared to the electrical interconnect system across all four DNNs. 3) In the optical interconnect system, the communication time of Ring is unaffected by the number of wavelengths, while the communication time of RD decreases as the number of wavelengths increases.

5.3 Discussion

In light of our findings, we believe that a hybrid approach to implementing all-reduce algorithms could significantly improve distributed DNN training performance in both electrical and optical interconnect systems. This approach would allow us to tailor the algorithm's behavior to the specific characteristics of the system and the size of the data being transferred.

In the context of optical interconnect systems, we observed that neither the Ring nor the RD all-reduce algorithms fully exploit the available wavelengths at each step. For example, Ring all-reduce uses only one wavelength per step, and although RD can utilize multiple wavelengths, its usage is unbalanced and does not reuse wavelengths to further minimize the number of communication steps.

Furthermore, we identified a trade-off between the number of communication steps and the size of the data transferred in all-reduce algorithms. In the Ring all-reduce algorithm, the amount of data transferred per step is small ($\frac{d}{n}$), but the number of steps is large ($2(n-1)$), which linearly increases as the number of nodes increases. On the other hand, in RD, the amount of data transferred per step is larger (d), but the number of steps is smaller ($2\lceil log_2 n \rceil - \lceil log_2 2w \rceil$, where $n > 2w$).

To further enhance communication efficiency, the all-reduce algorithm in the optical interconnect system could be customized. The goal would be to fully utilize the wavelengths available at each communication step while keeping communication traffic as low as possible. This could significantly reduce the communication time in distributed DNN training, contributing to improved overall system performance.

6 Conclusion

In this paper, we undertook a comparative analysis of Ring and RD all-reduce algorithms within optical and electrical interconnect systems, with the goal of better understanding their performance characteristics. After formulating the communication cost of the two all-reduce algorithms, we carried out extensive simulations to further compare their performance under various system settings. Our findings can be summarized as follows: 1) The Ring all-reduce algorithm demonstrated superior performance over RD when transferring data exceeding 64 MB and 1024 MB in optical and electrical interconnect systems with 1024 nodes, respectively. 2) The optical interconnect system proved to have better scalability than the electrical interconnect system. Furthermore, both Ring and RD all-reduce operations within the optical interconnect system experienced an average communication time reduction of 74.74% and 75.35% compared to the electrical interconnect system across four distinct DNNs. 3) Within the optical interconnect system, the communication time of the Ring algorithm did not show a dependency on the number of wavelengths. However, the communication time of the RD algorithm could be reduced with an increase in the number of wavelengths. Our study offers a foundation for future research to focus on optimizing all-reduce algorithms for distributed DNN training to further reduce communication time in optical interconnect systems. Moreover, exploring the performance differences of these algorithms in different topologies could offer additional insights to enhance system performance in distributed DL applications.

Acknowledgements. We thank the reviewers for taking the time and effort necessary to review the manuscript. Besides, we acknowledge using New Zealand eScience Infrastructure (NeSI) high-performance computing facilities as part of this research (Project code: uoo03633).

References

1. Khan, A.R., Kashif, M., Jhaveri, R.H., Raut, R., Saba, T., Bahaj, S.A.: Deep learning for intrusion detection and security of Internet of Things (IoT): current analysis, challenges, and possible solutions. Secur. Commun. Netw. **2022**, 1–13 (2022)
2. Luo, L., West, P., Nelson, J., Krishnamurthy, A., Ceze, L.: PLink: discovering and exploiting locality for accelerated distributed training on the public cloud. Proc. Mach. Learn. Syst. **2**, 82–97 (2020)
3. Wang, G., Venkataraman, S., Phanishayee, A., Devanur, N., Thelin, J., Stoica, I.: Blink: Fast and generic collectives for distributed ML. Proc. Mach. Learn. Syst. **2**, 172–186 (2020)
4. Yuichiro, U., Yokota, R.: Exhaustive study of hierarchical allreduce patterns for large messages between GPUs. In: 2019 19th IEEE/ACM International Symposium on Cluster, Cloud and Grid Computing (CCGRID), pp. 430–439 (2019)
5. Jiang, Y., Gu, H., Lu, Y., Yu, X.: 2D-HRA: two-dimensional hierarchical ring-based all-reduce algorithm in large-scale distributed machine learning. IEEE Access **8**, 183488–183494 (2020)
6. Cho, M., Finkler, U., Serrano, M., Kung, D., Hunter, H.: BlueConnect: decomposing all-reduce for deep learning on heterogeneous network hierarchy. IBM J. Res. Dev. **63**(6), 1:1–1:11 (2019)
7. Nguyen, T.T., Takano, R.: On the feasibility of hybrid electrical/optical switch architecture for large-scale training of distributed deep learning. In: 2019 IEEE/ACM Workshop on Photonics-Optics Technology Oriented Networking, Information and Computing Systems (PHOTONICS), pp. 7–14 (2019)
8. Khani, M., et al.: SIP-ML: high-bandwidth optical network interconnects for machine learning training. In: Proceedings of the 2021 ACM SIGCOMM 2021 Conference, pp. 657–675 (2021)
9. Gu, R., Qiao, Y., Ji, Y.: Optical or electrical interconnects: quantitative comparison from parallel computing performance view. In: 2008 IEEE Global Telecommunications Conference, IEEE GLOBECOM 2008, pp. 1–5 (2008)
10. Shin, J., Seo, C.S., Chellappa, A., Brooke, M., Chatterjee, A., Jokerst, N.M.: Comparison of electrical and optical interconnect. In: IEEE Electronic Components and Technology Conference, pp. 1067–1072 (1999)
11. Wei, J., et al.: Analyzing the impact of soft errors in VGG networks implemented on GPUs. Microelectron. Reliab. **110**, 113648 (2020)
12. Casanova, H., Legrand, A., Quinson, M.: SimGrid: a generic framework for large-scale distributed experiments. In: Tenth IEEE International Conference on Computer Modeling and Simulation, UKSim2008, pp. 126–131 (2008)
13. Alotaibi, S.D., et al.: Deep Neural Network - based intrusion detection system through PCA. Math. Prob. Eng. **2022**, 1–9 (2022)
14. Huang, J., Majumder, P., Kim, S., Muzahid, A., Yum, K.H., Kim, E.J.: Communication algorithm-architecture co-design for distributed deep learning. In: 2021 ACM/IEEE 48th Annual International Symposium on Computer Architecture (ISCA), pp. 181–194. IEEE (2021)

15. Ghobadi, M.: Emerging optical interconnects for AI systems. In: IEEE 2022 Optical Fiber Communications Conference and Exhibition (OFC), pp. 1–3 (2022)

16. Dai, F., Chen, Y., Huang, Z., Zhang, H., Zhang, F.: Efficient all-reduce for distributed DNN training in optical interconnect systems. In: Proceedings of the 28th ACM SIGPLAN Annual Symposium on Principles and Practice of Parallel Programming, pp. 422–424 (2023)

17. TensorFlow: Optimize TensorFlow performance using the Profiler (n.d.). https://www.tensorflow.org/guide/profiler. Accessed 2 Sept 2023

18. Wang, W., et al.: TopoOpt: co-optimizing network topology and parallelization strategy for distributed training jobs. In: 20th USENIX Symposium on Networked Systems Design and Implementation, NSDI 2023, pp. 739–767 (2023)

19. Zhang, H., et al.: Poseidon: an efficient communication architecture for distributed deep learning on GPU clusters. In: 2017 USENIX Annual Technical Conference, USENIX ATC 2017, pp. 181–193 (2017)

20. Dai, F., Chen, Y., Huang, Z., Zhang, H., Zhang, H., Xia, C.: Comparing the performance of multi-layer perceptron training on electrical and optical network-on-chips. J. Supercomput. **79**(10), 10725–10746 (2023)

21. Ottino, A., Benjamin, J., Zervas, G.: RAMP: a flat nanosecond optical network and MPI operations for distributed deep learning systems. Opt. Switching Netw. **51**, 100761 (2023)

22. Dai, F., Chen, Y., Zhang, H., Huang, Z.: Accelerating fully connected neural network on optical network-on-chip (ONoC). arXiv preprint arXiv:2109.14878 (2021)

23. Xia, C., Chen, Y., Zhang, H., Zhang, H., Dai, F., Wu, J.: Efficient neural network accelerators with optical computing and communication. Comput. Sci. Inf. Syst. **20**(1), 513–535 (2023)

Dynamic Path Planning Based on Traffic Flow Prediction and Traffic Light Status

Weiyang Chen[1,2] , Bingyi Liu[1,3]([✉]) , Weizhen Han[1,2] , Gaolei Li[4] , and Bin Song[1]

[1] School of Computer Science and Artificial Intelligence, Wuhan University of Technology, Wuhan 430070, China
{weiyangchen,byliu,hanweizhen,SongBinCS}@whut.edu.cn
[2] Sanya Science and Education Innovation Park, Wuhan University of Technology, Sanya 572000, China
[3] Chongqing Research Institute, Wuhan University of Technology, Chongqing 401135, China
[4] School of Electronics, Information and Electrical Engineering, Shanghai Jiao Tong University, Shanghai 200240, China
Gaolei_Li@sjtu.edu.cn

Abstract. Traffic flow prediction and path planning are crucial components of effective intelligent transportation systems research. The intelligent transportation system can optimize vehicle driving routes by utilizing predicted traffic flow data for each road segment and considering the periodic changes in traffic light patterns at intersections. However, most studies on traffic flow prediction have overlooked the frequency domain information of traffic flow sequences, resulting in a lack of effective modeling of this vital frequency domain information. Furthermore, existing path planning approaches only consider factors such as traffic density and road length in decision-making, neglecting the influence of traffic light status on vehicle travel time. We propose a traffic flow prediction model called mWDN-LSTM-ARIMA to address these issues, incorporating frequency feature extraction and residual testing. Additionally, we present a path planning method that leverages the traffic flow predictions from mWDN-LSTM-ARIMA and takes into account the periodic transformation law of traffic lights at urban intersections. Our experimental results validate the effectiveness of the proposed approach in reducing the average travel time and waiting count of vehicles.

Keywords: Deep Reinforcement Learning · Traffic Flow Prediction · Path Planning · Traffic Light Status

This work was supported by National Natural Science Foundation of China (No. 62272357), Key Research and Development Program of Hubei (No. 2022BAA052), Key Research and Development Program of Hainan (No. ZDYF2021GXJS014), Science Foundation of Chongqing of China (cstc2021jcyj-msxm4262), and Research Project of Chongqing Research Institute of Wuhan University of Technology (ZD2021-04, ZL2021-05).

Z. Tari et al. (Eds.): ICA3PP 2023, LNCS 14487, pp. 419–438, 2024.
https://doi.org/10.1007/978-981-97-0834-5_24

1   Introduction

With the acceleration of urbanization and the increasing population and number of vehicles, the construction of urban road infrastructure is progressing relatively slowly, failing to keep up with the rapid growth of urban car ownership. The substantial rise in the number of vehicles has led to significant challenges for the urban transportation system, including safety issues and traffic congestion [1]. According to data provided by Baidu Maps, in the third quarter of 2020, traffic in five cities in China experienced severe congestion, while traffic in another 14 cities had moderate congestion. Congestion leads to wastage of fuel, time, and environmental impact, incurring significant costs. The U.S. Traffic Congestion Index indicates that the annual congestion costs in certain American cities might amount to billions of dollars. Therefore, improving urban traffic congestion and enhancing traffic safety and environmental health are crucial for urban development and social progress.

The integration of mobile edge computing (MEC) with path planning in intelligent transportation systems enhances system performance by enabling distributed and collaborative path planning [2–4]. By leveraging edge computing resources, MEC facilitates real-time information exchange and coordination, leading to parallel path planning and collective optimization. Moreover, deep learning has become increasingly popular in traffic flow sequence processing and analysis, resulting in the application of various types of deep neural network models that have produced satisfactory results. After Fu et al. [5] first applied Gated Recurrent Unit (GRU) to traffic flow recognition and showed better prediction than Long Short-Term Memory (LSTM), GRU emerged in traffic prediction. However, most models, including GRU, ignore the frequency domain information of traffic flow sequences and still lack effective modeling of the significant frequency domain information.

In addition, single traffic flow prediction models, such as Locally Weighted Regression (LWR), Autoregressive Integrated Moving Average (ARIMA), and Kalman Filter (KF), suffer from incomplete data feature extraction and low prediction accuracy. While each traffic flow prediction model has benefits and drawbacks, integrating various models might result in complementary benefits and enhance prediction accuracy. A CNN-LSTM multilayer hybrid model was suggested by Du et al. [6] for enhancing vehicle flow prediction. The LSTM models traffic data's temporal and regular features, whereas the Convolutional Neural Network (CNN) model describes spatial dependencies relating to local projections. Furthermore, in urban road conditions, the existing path planning scheme only considers factors such as traffic density, road length, and road conditions in the decision-making process, ignoring the impact of traffic light status on vehicle travel time.

Therefore, based on historical city traffic flow data, we propose mWDN-LSTM-ARIMA, a traffic flow prediction model designed for urban road conditions. This model serves as the foundation for an improved path planning approach that involves adjusting traffic lights' periodic shifts at intersections, thus enhancing travel efficiency. The path planning method optimizes vehicle routes by leveraging anticipated traffic flow data per road section and incorporating the cyclic transformation pattern of intersection traffic lights. Additionally, to expedite training, the neural network of the Proximal Policy Optimization (PPO) algorithm is enhanced with added convolutional layers.

In summary, the main contributions of this paper are as follows:

- We propose an mWDN-LSTM-ARIMA traffic flow prediction model. ARIMA predicts the residual series of mWDN-LSTM, and the prediction results are combined with the mWDN-LSTM model for the final traffic flow prediction results.
- We present a deep reinforcement learning-based path planning method that considers the waiting time for traffic lights in the reward function and the state of all traffic lights on the road in the observation design of the agent.
- For the spatial feature extraction of the road network model and the data processing of the road network weight matrix, We improve the existing deep reinforcement learning algorithms by adding convolutional layers to the original neural network of PPO.
- We conduct extensive simulation experiments to validate the proposed method. The results demonstrate that the proposed path planning method outperforms the compared algorithm in terms of reducing travel time, both in traffic congestion and smooth traffic flow periods, and achieving better algorithm convergence.

2 Related Work

2.1 Traffic Flow Prediction Study

Research on traffic flow prediction has yielded significant results and continues to evolve. Initially, traffic flow prediction was mainly based on statistical analysis, and as research progressed, numerous nonlinear theoretical models were proposed. One such model is ARIMA, a well-known parametric time series forecasting model. It was proposed by Box and Jenkins and has been widely used for traffic flow prediction [7]. Various nonlinear theoretical models, including the wavelet transform model, have been proposed in traffic flow prediction research. One of the key advantages of these models is their ability to capture nonlinear patterns in traffic flow data, enabling more accurate predictions and improved traffic management strategies. Sun et al. [8] proposed a method that integrates Fourier analysis with wavelet denoising techniques. A two-layer fast Fourier transform-based traffic flow prediction method was developed by utilizing two discrete wavelet transforms with different thresholds to decompose the raw data into high-frequency noise and identify low-frequency traffic flow trends.

Deep learning has gained immense popularity in traffic flow prediction, giving rise to a variety of neural network-based models. These models excel in learning intricate patterns and extracting valuable insights from complex, high-dimensional data [9]. By leveraging such capabilities, these models have significantly improved the accuracy of predicting vehicular traffic flow. They can effectively capture and utilize the spatial-temporal dependencies present in multi-source heterogeneous datasets, leading to more precise predictions.

The traditional LSTM model employs the BackPropagation Through Time (BPTT) algorithm for network parameter optimization, which has high complexity and can quickly converge to a locally optimal solution. To address this, Xu and Guo [10] researched an LSTM model with GWO optimization parameters, preventing the model from falling into a local optimum solution.

However, due to the variation in climate, population density, and traffic conditions worldwide, the traffic flow data generated also has considerable differences. Therefore, a single traffic flow prediction model may only apply to some traffic flow datasets. Owing to their capability of combining various models and overcoming the limitations of individual models, combined traffic flow prediction models have been widely used in traffic flow prediction studies. Hou et al. [11] presented an adaptive hybrid model for short-term traffic flow prediction. They first used the ARIMA method and non-linear Wavelet Neural Network (WNN) to predict traffic flow. Then, the outputs of the two independent models were analyzed and combined using fuzzy logic. The weighted results were considered as the final predicted traffic volume of the hybrid model. The results demonstrate that the hybrid model outperforms the two individual models in short-term traffic flow prediction, regardless of stable or fluctuating conditions. In [12], two different combined prediction models were introduced based on the Seasonal Autoregressive Integrated Moving Average (SARIMA) model and the Non-Autoregressive (NAR) model. Additionally, the study conducted further investigations on the performance of the combined prediction model specifically applied to traffic flow during the epidemic period.

2.2   Path Planning Based on Deep Reinforcement Learning

Traditional path planning algorithms typically require an accurate environment model to be constructed beforehand, which can be inefficient for complex environments. In contrast, deep reinforcement learning algorithms can interact with the environment directly to obtain samples and learn the optimal policy without requiring a pre-constructed model of the environment.

As intelligent technology innovates rapidly, path planning development is constantly improving. Deep reinforcement learning has been critical in developing intelligent path-planning technology. For example, Koh et al. [13] introduced a deep reinforcement learning-based vehicle path planning method for real-time interaction between vehicles and complex urban traffic environments. They improved the traditional Deep Q-Network (DQN) algorithm by adopting the Double-DQN algorithm to solve the over-valuation problem. Peng et al. [14] combined deep reinforcement learning with dynamic planning methods to propose a path planning model that generates multiple selectable paths. The method generates multiple selectable paths based on the Q-value, which equalizes the distribution of traffic in the road network and alleviates problems such as congestion drift. In [13], a path-planning method for self-driving vehicles based on reinforcement learning and deep inverse reinforcement learning was proposed, which incorporates the driver's driving style as the learning objective and leverages reinforcement learning and inverse reinforcement learning to achieve the desired driving behavior.

MEC plays a pivotal role in enhancing the efficiency and effectiveness of path planning by bringing computational resources closer to the network's edge, thereby reducing latency in the path planning process [15,16]. Miao et al. [17] recommended using unmanned aerial vehicles (UAVs) as MEC nodes in the sky, introducing a multi-UAV-assisted MEC offloading algorithm for coordinating with mobile devices. This algorithm involves global and local path planning, managed by a ground station and onboard computers.

3   System Model

We present a path planning strategy based on traffic flow forecast and light status in this section. The following sections will present detailed descriptions of the traffic flow prediction model, traffic light timing relationship model, road network model, and path planning model based on deep reinforcement learning.

3.1   Traffic Flow Prediction Model

Real-life traffic flow sequences contain trending patterns of increasing traffic flow due to continuous urban development and cyclical patterns of daily, weekly, and monthly cycles. In addition, the traffic flow sequence is also affected by various random disturbances. The complexity of traffic flow sequences increases due to the superposition of various factors, making traffic flow prediction challenging. The multi-level wavelet decomposition technique can overcome this challenge by extracting multi-scale features from the traffic flow sequence by decomposing it into sub-series of different frequencies. The high-frequency sub-series provides detailed information about local changes, whereas the low-frequency sub-series provides information about traffic flow trends and overall pattern changes.

The Multilevel Wavelet Decomposition Network (mWDN) can decompose the original traffic flow sequences. In the traffic flow sequence decomposition, the low-frequency and high-frequency subsequences of layer $i + 1$ need to be further decomposed at layer i. The low-frequency subsequence and the high-frequency subsequence at layer $i + 1$ are described as Eq. 1 and Eq. 2, respectively.

$$x^l(i+1) = \sigma \left(W^l(i+1)x^l(i) + b^l(i+1) \right) \tag{1}$$

$$x^h(i+1) = \sigma \left(W^h(i+1)x^h(i) + b^h(i+1) \right) \tag{2}$$

where σ represents the sigmoid activation function and $b^l(i+1)$, $b^h(i+1)$ stand for the deviation vector initialized close to 0 at layer $i+1$. $W^l(i+1)$ and $W^h(i+1)$ represent the low-pass filter matrix and high-pass filter matrix, respectively.

The high-frequency subsequences obtained from each layer of mWDN and the subsequences obtained from the last layer of decomposition are denoted as $X(N) = \{x^h(1), x^h(2), \ldots, x^h(i), \ldots, x^h(N), x^l(N)\}$, $x^h(i)$ represents the high-frequency component of the i-th layer. Extracting features from these sub-series, LSTM is utilized, where $X(N)$ is used as the input of $N+1$ independent LSTM subnets, and each LSTM subnet predicts the future state of a subsequence. Ultimately, the results of LSTM predictions are combined using a fully connected neural network to produce the preliminary prediction results Y_1.

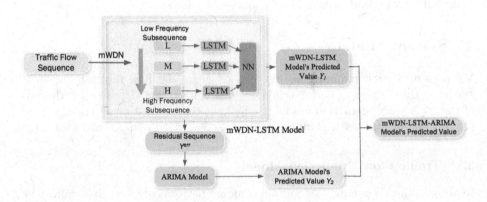

Fig. 1. Diagram of mWDN-LSTM-ARIMA model

Despite the high prediction accuracy of mWDN-LSTM, the residual sequences still exhibit autocorrelation, indicating the presence of unextracted traffic flow features. The residuals are analyzed using ARIMA to further optimize the prediction results of mWDN-LSTM. The mWDN-LSTM-ARIMA model can effectively capture the multi-scale features of traffic flow sequences and accurately predict the traffic flow in complex urban road networks.

The mWDN-LSTM-ARIMA model's structure is depicted in Fig. 1. It mainly consists of the mWDN-LSTM neural network and the ARIMA model. The prediction process of the mWDN-LSTM-ARIMA model is as follows. Firstly, the traffic flow sequences are decomposed into low-frequency and high-frequency sub-sequences by the mWDN. Then, the LSTM extracts the sub-series features and predicts each subsequent state. Subsequently, the preliminary prediction Y_1 results are obtained by integrating the results of LSTM predictions through a fully connected neural network. Next, the residuals of the preliminary prediction results Y^{err} are analyzed using ARIMA to obtain the prediction result Y_2 if Y^{err} is not white noise. Finally, the optimized prediction results of traffic flow are obtained by adding the prediction result of residuals Y_2 to the preliminary results Y_1. Take note that if Y_1 is white noise, then the Y_1 is the mwDN-LSTM-ARIMA model prediction result.

3.2 Traffic Light Timing Relationship Model and Road Network Model

In the simulation of a traffic scenario, the traffic signal is partitioned into four phases, which are referred to as Phase A, B, C, and D. During Phase A, vehicles are permitted to proceed straight or turn right in the north-south direction. In Phase B, vehicles are permitted to turn left in the north-south direction. Phase C allows vehicles to proceed straight and turn right in the east-west direction, while in Phase D, vehicles are allowed to turn left in the east-west direction. The traffic lights operate in a cycle consisting of four phases, forming a pattern that repeats over time. This pattern follows a specific sequence of Phase A, B, C, D, and then back to Phase A, as depicted in Fig. 2.

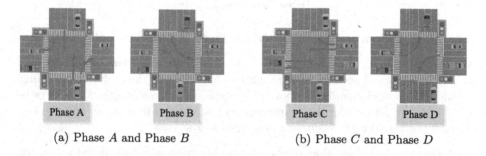

Phase A	Phase B	Phase C	Phase D

(a) Phase A and Phase B (b) Phase C and Phase D

Fig. 2. Traffic light phase diagram

The traffic light state represents the set of phases that the traffic lights at each intersection in the entire road network at the current moment, which can be denoted as $S_{tra} = \{ph_1, ph_2, \cdots, ph_n\}$, where S_{tra} denotes the state of traffic light and ph_i represents the phase of the i_{th} traffic light.

The mWDN-LSTM-ARIMA model is used to predict the average speed of vehicles on road segments. The predicted results are combined into a road average speed matrix V, which is defined as an $m \times m \times k$ tensor:

$$
V = \left\{
\begin{matrix}
v_{11}^1, & v_{12}^2, & v_{13}^3, & \cdots, & v_{1\,m}^k \\
v_{21}^1, & v_{22}^2, & v_{23}^3, & \cdots, & v_{2\,m}^k \\
v_{31}^1, & v_{32}^2, & v_{33,}^3 & \cdots, & v_{3\,m}^k \\
\vdots & \vdots & \vdots & \vdots & \vdots \\
v_{m1}^1, & v_{m2}^2, & v_{m3}^3, & \cdots, & v_{mm}^k
\end{matrix}
\right\}
\tag{3}
$$

where m is the number of nodes in the road network, k denotes the time horizon, and v_{ij}^t represents the predicted average travel speed of the road segment from node i to node j at time t. The length of the road segment is a fixed value. For example, the length between node i and node j can be expressed as l_{ij}. The estimated value of the travel time of the road segment between node i and node j at time t is t_{ij}^t, and it is calculated as Eq. 4:

$$t_{ij}^t = \frac{l_{ij}}{v_{ij}^t}. \tag{4}$$

In the weight matrix, we set the weight value of each road segment as the estimated value of travel time t_{ij}^t in time period t. The weight matrix W can be calculated based on the average driving speed prediction matrix V and the fixed length of the road segment:

$$W = \begin{Bmatrix} t_{11}^t, & t_{12}^t, & t_{13}^t, & \cdots, & t_{1m}^t \\ t_{21}^t, & t_{22}^t, & t_{23}^t & \cdots, & t_{2m}^t \\ t_{31}^t, & t_{32}^t, & t_{33}^t, & \cdots, & t_{3m}^t \\ \vdots & \vdots & \vdots & \vdots & \vdots \\ t_{m1}^t, & t_{m2}^t, & t_{m3}^t, & \cdots, & t_{mm}^t \end{Bmatrix} \tag{5}$$

3.3   Path Planning Model Based on Deep Reinforcement Learning

We propose a path planning model that combines traffic flow prediction and traffic light status based on mWDN-LSTM-ARIMA and the traffic light timing relationship model mentioned above. The decision process of agents at intersections is described as a Markov game and can be defined by a tuple, and the key elements of the Markov Game are described below:

- **Agent:** Each vehicle that sends a path-plan request is considered an agent, and VANET is considered as the environment.
- **State:** The state s of an agent is the current traffic status, which includes the current agent position, the vehicle's destination, the weight matrix of the road network W, and the state of the traffic lights in the whole road network. $State$ can be defined as $[\text{pos}_v, \text{des}_v, W, S_{tra}]$.
- **Action:** The action a refers to the road that the agent can choose. The dimension of the action space is determined by the number of roads that are connected to the current road, which in this case is 3, corresponding to the options of turning left, turning right, and going straight.
- **Reward:** The agent aims to minimize its travel and waiting times, which are used to calculate the reward r. The reward r is defined as the negative sum of the travel time and waiting time at traffic lights incurred by the agent while executing its path plan. Deep reinforcement learning aims for long-term rewards, and therefore, cumulative discounted rewards are used to update the network parameters. The formula for cumulative discounted rewards is as follows:

$$R_t = \sum_{k=0}^{T-t} \gamma^k r_{t+k} \tag{6}$$

where R_t is the cumulative discount reward, and r_{t+k} is the reward of step $t + k$. The discount factor γ is a value between 0 and 1 that determines the weight given to future rewards in the agent's decision-making process.

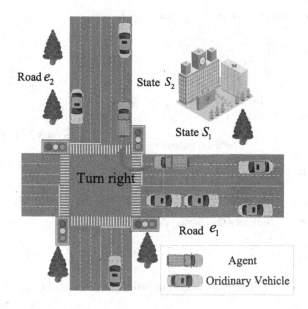

Fig. 3. Schematic diagram of intelligent body state transfer process

When the agent is in state s_1 in Fig. 3, it takes action a_1 based on the observed information and transitions to state s_2. Then, it repeats the process of selecting actions and updating the neural network parameters until it converges into a path-planning strategy that obtains the shortest path.

4   Deep Learning Algorithm Design and Model Training

4.1   Deep Reinforcement Learning Algorithm Design

We selected the Proximal Policy Optimization (PPO) algorithm for offline learning and improved the algorithm to handle the observed information of the agents better. The improved PPO algorithm for path planning with traffic flow prediction and light status involves modifying the neural network to incorporate features of the road network weights matrix W. A convolutional layer is added to the original neural network to quickly and accurately extract weight features and spatial features between road segments. The convolutional kernel in the convolutional neural network convolves the input at different positions, reusing the parameters of the convolutional kernel several times, thereby significantly reducing the number of parameters that need to be learned. The structure diagram of the improved neural network is illustrated in Fig. 4.

The input to the neural network is *State*, which consists of the weight matrix W, the traffic light state S_{tra}, the current position of the agent pos_v, and the destination des_v. The weight matrix W is fed into two convolutional layers and one fully connected layer for feature extraction. The output of the convolutional

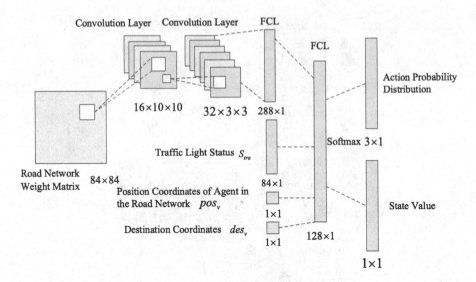

Fig. 4. Improved neural network structure diagram.

layers is concatenated with S_{tra}, pos_v, and des_v, and then passed through fully connected layers to obtain the policy *Policy* and the state value *Value*.

The output *Policy* represents the probability distribution of each action that the agent can take in the current state, and its dimension is consistent with the number of available actions. *Value* is used to evaluate the advantages and disadvantages of different policies in a given state and to select the best policy accordingly. *Value* indicates the expected reward that the agent can obtain by taking different actions in the state.

4.2   Model Training Process

The neural network takes in the agent's observation and outputs *Policy* and *Value*. When the agent reaches state s, it first outputs *Policy* and *Value* based on the observation of the current state s. As shown in Fig. 5, the agent then selects and executes an action using a sampling function, after which it enters the next state, and the environment provides the corresponding reward r. In one step of the improved PPO algorithm, the data collected by the sampler on-policy is denoted as (s, a, r) and a set of data $(s_1, a_1, r_1), (s_2, a_2, r_2), \cdots, (s_n, a_n, r_n)$ is collected for each episode of the sampler during the process of exploring the optimal path in the road network. Here, i represents the number of steps in a round, and the state, action, and reward of step i are denoted as s_i, a_i and r_i, respectively.

The collected data is deposited into the experience pool replay buffer and the advantage function $\hat{A}_t$ is calculated to evaluate the current policy's goodness or badness. The evaluator *Critic* generates *Value* based on the data sampled by the sampler and updates its network parameters according to the advantage function $\hat{A}_t$ and $Value$.

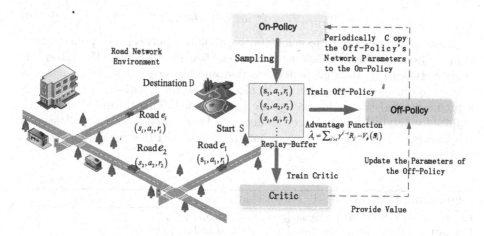

Fig. 5. Diagram of model training process

The strategy distribution of the sampler is denoted as $p_{\theta'}(s_t, a_t)$, the strategy distribution of the learner as $p_\theta(s_t, a_t)$, and the difference between the advantage function $\hat{A}_t$ and the state value $Value$ as $A^{\theta'}(s_t, a_t)$. Then, the objective function of the path planning strategy can be formulated as Eq. 7:

$$J^{\theta'}(\theta) = E_{(s_t, a_t) \sim \pi_{\theta'}} \left[\frac{p_\theta(a_t \mid s_t)}{p_{\theta'}(a_t \mid s_t)} A^{\theta'}(s_t, a_t) \right] \tag{7}$$

where θ and θ' denote the parameters of the learner and the sampler, respectively. s_t and a_t represent the state and action at time step t, and $\pi_{\theta'}$ represents the sampling policy. The first term in the formula is a ratio that measures the advantage of the current policy relative to the sampling policy. $A^{\theta'}(s_t, a_t)$ is the advantage function of the state-action pair (s_t, a_t) and reflects the size of the advantage of choosing that action in the current state relative to choosing an action in the sampling policy. The optimization objective of this objective function is to maximize the expected reward.

The gradient of the objective function $J^{\theta'}(\theta)$ yields for updating the parameters of the learner:

$$\nabla J^{\theta'}(\theta) = E_{(s_t, a_t) \sim \pi_{\theta'}} \left[\frac{p_\theta(s_t, a_t)}{p_{\theta'}(s_t, a_t)} A^{\theta'}(s_t, a_t) \nabla \log p_\theta(a_t^n \mid s_t^n) \right] \tag{8}$$

The parameters θ of the learner's network are updated using Eq. 8 while the parameters θ' of the sampler's network are fixed for a certain period and then gradually updated to replicate θ.

The similarity between a and b is also ensured in the PPO algorithm by adding Kullback-Leibler (KL) divergence to the objective function $J^{\theta'}(\theta)$, as shown in Eq. 9:

$$J_{PPO}^{\theta'}(\theta) = J^{\theta'}(\theta) - \beta KL(\theta, \theta') \tag{9}$$

Equation 9 is employed to calculate the similarity between two policies, with $KL(\theta, \theta')$ quantifying the difference between their distributions. The KL divergence between the updated policy function and the original policy function is computed during each iteration of the algorithm. If the KL divergence surpasses a predetermined threshold, no update is performed. This threshold should be adjusted individually to balance the disparity between old and new policies and the learning pace. The pseudo-code for the improved PPO algorithm is shown in Algorithm 1:

Algorithm 1. Improved PPO algorithm

Input: Observations of the agent $[\text{pos}_v, \text{des}_v, W, S_{tra}]$.
Output: Action probability distribution $Policy$ and state value $Value$

1: **for** $k = 0, 1, 2 \ldots$ **do**
2:     Excute T-step policy parameters π_θ;
3:     The data collected by the sampler on-policy is recorded as (S_t, A_t, R_t);
4:     Estimate the advantage function $\hat{A}_t = \sum_{t' > t} \gamma^{t'-t} R_t - V_\phi(S_t)$;
5:     $\pi_{old} \leftarrow \pi_\theta$;
6:     **for** $m \in [1, \ldots, M]$ **do**
7:         $J_{PPO}(\theta) = \sum_{t=1}^{T} \frac{\pi_\theta(A_t|S_t)}{\pi_{old}(A_t|S_t)} \hat{A}_t - \lambda \hat{E}_t [D_{KL} (\pi_{old} (\bullet \mid S_t) \| \pi_\theta (\bullet \mid S_t))]$
8:         Update the $Policy$ function parameters θ based on $J_{PPO}(\theta)$;
9:     **end for**
10:    **for** $b \in [1, \ldots, B]$ **do**
11:        $L(\phi) = -\sum_{t=1}^{T} \left(\sum_{t' > t} \gamma^{t'-t} R_{t'} - V_\phi(S_t) \right)^2$;
12:        Update the $Value$ function parameters based on $L(\phi)$;
13:    **end for**
14:    Calculate $d = \lambda \hat{E}_t \lceil D_{KL} (\pi_{old} (\bullet \mid S_t) \| \pi_\theta (\bullet \mid S_t)) \rceil$
15:    **if** $d < d_t/a$ **then**
16:        $\lambda \leftarrow \lambda/b$
17:    **else if** $d > d_t/a$ **then**
18:        $\lambda \leftarrow \lambda \times b$
19: **end for**

---

5   Experiment

5.1   Experiment Setting

We utilized the traffic flow dataset of Futian District, Shenzhen, provided by the Shenzhen government data open platform, which contains details about the road network structure in the area and traffic flow information for each road segment, such as road segment ID, length, and the average speed of vehicles during each time slice. Traffic flow data is feature-reduced on SUMO (Simulation of Urban Mobility), an open-source transportation simulation software used to model the operation of cities and transportation systems. Table 1 lists the simulation parameters employed in the experiment, and the structure of the road network in the area of the experimental scenario is shown in Fig. 6:

Table 1. Simulation parameters

Parameter	Value
Scenario	Futian District, Shenzhen City
Area size	$4 \times 3.9 \, \text{km}^2$
Number of intersections	84
Number of road segments	208
Signal duration	[20,30,40,50,60,70,80] s
Maximum vehicle speed	70 km/h

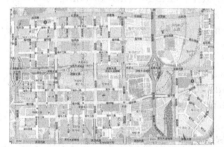 

(a) The real map of Futian District   (b) Abstraction map of Futian District

Fig. 6. Generative diagram of the road network

In the experiment, the vehicle that requires route planning starts from outside the road network and exits the network after reaching its destination. The starting points are defined as set $S = (s_1, s_2, \cdots, s_n)$, and the ending points are defined as set $D = (d_1, d_2, \cdots, d_n)$. The starting and ending points are combined to form matrix $U = S^T D$, where S^T denotes the transpose of S.

$$U = \left\{ \begin{matrix} u_{11}, & u_{12}, & u_{13}, & \cdots, & u_{1n} \\ u_{21}, & u_{22}, & u_{23}, & \cdots, & u_{21} \\ u_{31}, & u_{32}, & u_{33}, & \cdots, & u_{3n} \\ \vdots & \vdots & \vdots & \vdots & \vdots \\ u_{n1}, & u_{n2}, & u_{n3}, & \cdots, & u_{nn} \end{matrix} \right\} \tag{10}$$

where u_{ij} denotes the path planning task whose starting point is s_i and ending point is d_j.

5.2   Evaluation Indexes and Compared Methods

Evaluation Indexes. To assess the effectiveness of our path planning approach in comparison to the conventional path planning method without considering the traffic light state, we consider the following metrics:

- Average Travel Time: Average Travel Time means the time it takes for a vehicle to travel from various starting points to different destinations.
- Average Waiting Count: Average Waiting Count represents the number of times a vehicle slows down to a speed of less than 0.1 m/s while on its route.

Algorithm for Comparing Convergence of Algorithms. To demonstrate the convergence performance of the improved PPO algorithm in complex large road networks, we compare the improved PPO algorithm with the other deep reinforcement learning algorithms. The comparison methods compared with are described as follows:

1. Deep Q-Network (DQN): DQN [18] uses a neural network to map states to Q-values, i.e., the desired value of the output action given a state. The algorithm learns the optimal policy by training the neural network to obtain the optimal result.
2. Double Deep Q-Network (DDQN): DDQN [19] uses two separate deep neural networks, the primary, and the target networks, to approximate the action-value function. The target network receives less frequent updates compared to the primary network, and its parameters are periodically copied from the primary network.
3. Dueling DQN: Dueling DQN [20] separates the state-value function and action advantage function representations, which allows the agent to learn the value of a particular state independently of the actions that can be taken. This separation is used to estimate Q-values more accurately.
4. Actor-Critic: Actor-Critic [21] consists of two parts: *Actor* and *Critic*. *Actor* learns the system's action strategy and maximizes the cumulative reward by continuously trying to act; *Critic* learns the system's value function and maximizes the cumulative reward by continuously evaluating good and bad states.

5. Advantage Actor-Critic (A2C): A2C [22] is a variation of the Actor-Critic that aims to reduce the variance of the policy gradient update. It achieves this by computing the advantage function of the action, which measures the difference between the expected value of the action and the estimated value of the state.

Path Planning Algorithm for Comparing Travel Times. To assess the performance of our path planning method in reducing average travel time and waiting count under varying traffic light durations, we compare it with other path planning methods that do not consider traffic light states. The following methods are used for comparison:

1. Dynamic-Dijkstra: Dijkstra's algorithm is a classical graph search algorithm used to solve the vehicle path problem. The road network is represented as a graph where nodes are intersections, edges are roads, and the weight of each edge is the weight matrix W.
2. Dynamic-A^*: Dynamic-A^* is a dynamic version of the A^* algorithm for finding the shortest path between two points. It adjusts the heuristic function based on new information, allowing for efficient pathfinding in dynamic environments.
3. Multi-Route Dynamic Programming Model(MRDP): MDPM employs dynamic planning with deep reinforcement learning to tackle the urban route planning problem. It generates multiple paths based on Q values but solely observes road weights in reinforcement learning agents.

5.3   Analysis of Experimental Results

Algorithm Convergence Analysis. Figure 7 illustrates the convergence of reward for various deep reinforcement learning algorithms, with the DQN algorithm exhibiting the slowest convergence speed and the improved PPO algorithm displaying the fastest. This can be attributed to two reasons: firstly, the neural network of the A2C algorithm is enhanced to expedite the extraction of spatial features of the road network and the processing of data; secondly, the PPO algorithm uses offline learning by separating sampling and training.

The online learning-based deep reinforcement learning algorithm's intelligence can only use each batch of sampled data once as they interact with the environment. It must be constantly sampled when calculating strategy gradients and updating parameters, resulting in an extensive training volume and slow convergence rate. Offline learning-based deep reinforcement learning algorithms, on the other hand, use a model that separates sampling and training, where the sampler is responsible for interacting with the environment to collect state, action, reward, and other data, and the learner uses the data collected by the sampler to train its policy. The PPO algorithm uses offline learning, and the neural network parameters θ can be updated multiple times using the same batch of data to improve the efficiency of updates and speed up the convergence.

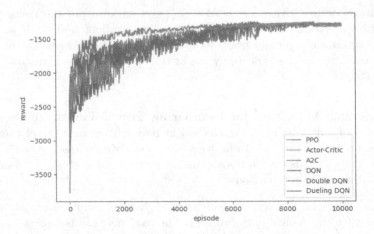

Fig. 7. Comparison of the convergence of the reward function of each model

Travel Time Analysis. The mWDN-LSTM-ARIMA model can be used to predict the average speed of each road segment in three-time intervals: 0–5 min, 5–10 min, and 10–15 min in the future. To determine which time interval is more suitable for calculating the road network weights in real traffic scenarios, we conducted feature reduction on SUMO traffic flow data during the morning rush hour from 8:00 to 9:00. Starting from when a vehicle enters a new road segment, we used the mWDN-LSTM-ARIMA model to predict the average speed of each road segment in the next 5 and 10 min. Based on the predicted results of different time intervals, we calculated the weights of the road network. The evaluation results of various indicators are shown in Fig. 8 and Fig. 9, indicating that compared with 5–10 min, selecting the predicted speed values in the 0–5 min can reduce the average travel time and average waiting count.

Figure 10 illustrates the experimental results of mWDN-LSTM-ARIMA using data from the noon period of 1:00 to 2:00. As shown in Fig. 8 and Fig. 10, the average travel time and the average waiting count increase during the noon hour compared to the morning peak hour. However, our proposed method still significantly improves performance, reducing the average travel time by 3.41% and the average number of stops by 5.24% with a traffic light duration of the 20 s, compared to the MRDP approach that does not consider traffic light conditions. These results demonstrate the effectiveness of our proposed protocol, which outperforms other route planning methods in terms of average travel time reduction and waiting count reduction in both morning peak congestion and midday traffic flow, indicating its good generalizability.

Our approach leads to a reduction of 3.41% in the average travel time and a decrease of 5.24% in the average number of stops compared to the MRDP scheme without considering the traffic light condition during the noon hour with a traffic light duration of the 20 s. These results demonstrate the effectiveness of our proposed protocol, which outperforms other route planning methods in terms

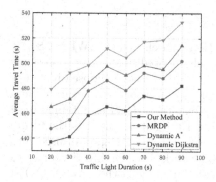
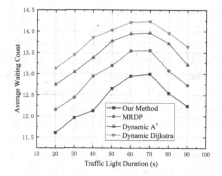

Fig. 8. Results during the morning peak and when the predicted time interval is 0–5 min

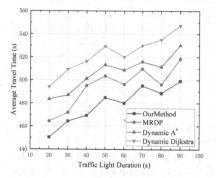
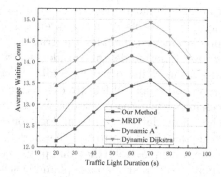

Fig. 9. Results during the morning peak and when the predicted time interval is 5–10 min

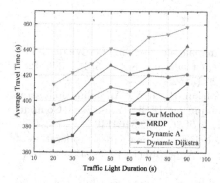
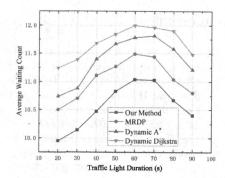

Fig. 10. Results for the midday period and the predicted time interval is 0–5 min

of average travel time reduction and waiting count reduction in both morning peak congestion and midday traffic flow, indicating its good generalizability.

6 Conclusion

We propose a combined traffic flow prediction model, called mWDN-LSTM-ARIMA, which employs a multi-step approach to predict each road section's traffic flow speed accurately. First, the time series data is decomposed into sub-series of different frequencies using mWDN, which allows for more accurate identification of fluctuation changes in the data. Next, the sub-series obtained from mWDN are used as separate models in LSTM. Each subnet predicts the future state of one sub-series, and the prediction results of all subnets are fused into a whole using a fully connected layer neural network. Finally, ARIMA analyzes the residual sequences and optimizes the prediction value of the mWDN-LSTM model. This multi-step approach allows for more accurate and robust traffic flow predictions.

A path planning method based on deep reinforcement learning was proposed to reduce vehicle travel time and improve travel efficiency by incorporating traffic light patterns at urban intersections. The method utilizes traffic flow speed prediction and includes the traffic light state in the agent's observation. The reward function design considers the waiting time for the traffic light. The PPO algorithm is improved by adding a convolutional layer to the original neural network for spatial feature extraction of the road network model and data processing of the road network weight matrix. The convolution kernel convolves the input data at different locations, reducing the number of parameters to be learned and improving the model's generalization ability. The experimental results demonstrate that our proposed path planning method performs better than the comparison path planning scheme in decreasing the average travel time and average waiting count during morning peak congestion and midday traffic flow.

References

1. Liu, B., et al.: A region-based collaborative management scheme for dynamic clustering in green VANET. IEEE Trans. Green Commun. Netw. **6**(3), 1276–1287 (2022). https://doi.org/10.1109/TGCN.2022.3158525
2. Shao, X., Hasegawa, G., Dong, M., Liu, Z., Masui, H., Ji, Y.: An online orchestration mechanism for general-purpose edge computing. IEEE Trans. Serv. Comput. **16**(2), 927–940 (2023). https://doi.org/10.1109/TSC.2022.3164149
3. Liu, B., et al.: A novel framework for message dissemination with consideration of destination prediction in VFC. Neural Comput. Appl. **35**(17), 12389–12399 (2023). https://doi.org/10.1007/s00521-021-05754-9
4. Liu, B., et al.: Collaborative intelligence enabled routing in green IoV: a grid and vehicle density prediction-based protocol. IEEE Trans. Green Commun. Netw. **7**(2), 1012–1022 (2023). https://doi.org/10.1109/TGCN.2022.3188026
5. Fu, R., Zhang, Z., Li, L.: Using LSTM and GRU neural network methods for traffic flow prediction. In: 2016 31st Youth Academic Annual Conference of Chinese Association of Automation (YAC), pp. 324–328. IEEE (2016). https://doi.org/10.1109/yac.2016.7804912

6. Du, S., Li, T., Gong, X., Yang, Y., Horng, S.J.: Traffic flow forecasting based on hybrid deep learning framework. In: 2017 12th International Conference on Intelligent Systems and Knowledge Engineering (ISKE), pp. 1–6. IEEE (2017). https://doi.org/10.1109/iske.2017.8258813

7. Box, G.E., Jenkins, G.M., Reinsel, G.C., Ljung, G.M.: Time Series Analysis: Forecasting and Control. Wiley, Hoboken (2015)

8. Sun, P., AlJeri, N., Boukerche, A.: A fast vehicular traffic flow prediction scheme based on fourier and wavelet analysis. In: 2018 IEEE Global Communications Conference (GLOBECOM). pp. 1–6. IEEE (2018). https://doi.org/10.1109/glocom.2018.8647731

9. Zhang, D., Kabuka, M.R.: Combining weather condition data to predict traffic flow: a GRU-based deep learning approach. IET Intel. Transp. Syst. **12**(7), 578–585 (2018). https://doi.org/10.1049/iet-its.2017.0313

10. Guo, S., Lin, Y., Feng, N., Song, C., Wan, H.: Attention based spatial-temporal graph convolutional networks for traffic flow forecasting. In: Proceedings of the AAAI Conference on Artificial Intelligence, vol. 33, pp. 922–929 (2019). https://doi.org/10.1609/aaai.v33i01.3301922

11. Hou, Q., Leng, J., Ma, G., Liu, W., Cheng, Y.: An adaptive hybrid model for short-term urban traffic flow prediction. Phys. A **527**, 121065 (2019). https://doi.org/10.1016/j.physa.2019.121065

12. Wang, Y., Jia, R., Dai, F., Ye, Y.: Traffic flow prediction method based on seasonal characteristics and SARIMA-NAR model. Appl. Sci. **12**(4), 2190 (2022). https://doi.org/10.3390/app12042190

13. You, C., Lu, J., Filev, D., Tsiotras, P.: Advanced planning for autonomous vehicles using reinforcement learning and deep inverse reinforcement learning. Robot. Auton. Syst. **114**, 1–18 (2019). https://doi.org/10.1016/j.robot.2019.01.003

14. Peng, N., Xi, Y., Rao, J., Ma, X., Ren, F.: Urban multiple route planning model using dynamic programming in reinforcement learning. IEEE Trans. Intell. Transp. Syst. **23**(7), 8037–8047 (2021). https://doi.org/10.1109/tits.2021.3075221

15. Meng, X., Shao, X., Masui, H., Lu, W.: Intelligent predicting method for optimizing remote loading efficiency in edge service migration. Mob. Netw. Appl. **27**, 2218–2231 (2022)

16. Huang, Y., Zhang, H., Shao, X., Li, X., Ji, H.: RoofSplit: an edge computing framework with heterogeneous nodes collaboration considering optimal CNN model splitting. Futur. Gener. Comput. Syst. **140**, 79–90 (2023)

17. Miao, Y., Hwang, K., Wu, D., Hao, Y., Chen, M.: Drone swarm path planning for mobile edge computing in industrial internet of things. IEEE Trans. Industr. Inf. **19**(5), 6836–6848 (2023). https://doi.org/10.1109/TII.2022.3196392

18. Fan, J., Wang, Z., Xie, Y., Yang, Z.: A theoretical analysis of deep Q-learning. In: Learning for Dynamics and Control, pp. 486–489. PMLR (2020)

19. Zhang, W., Gai, J., Zhang, Z., Tang, L., Liao, Q., Ding, Y.: Double-DQN based path smoothing and tracking control method for robotic vehicle navigation. Comput. Electron. Agric. **166**, 104985 (2019). https://doi.org/10.1016/j.compag.2019.104985

20. Sewak, M.: Deep Q Network (DQN), double DQN, and dueling DQN. In: Deep Reinforcement Learning, pp. 95–108. Springer, Singapore (2019). https://doi.org/10.1007/978-981-13-8285-7_8

21. Zanette, A., Wainwright, M.J., Brunskill, E.: Provable benefits of actor-critic methods for offline reinforcement learning. In: Advances in Neural Information Processing Systems, vol. 34, pp. 13626–13640 (2021)
22. Liu, G., Li, X., Sun, M., Li, P.: An advantage actor-critic algorithm with confidence exploration for open information extraction. In: Proceedings of the 2020 SIAM International Conference on Data Mining, pp. 217–225. SIAM (2020). https://doi.org/10.1137/1.9781611976236.25

A Time Series Data Compression Co-processor Based on RISC-V Custom Instructions

Peiran Du[ID] and Zhaohui Cai(✉)[ID]

The School of Computer Science, Wuhan University, Wuhan 430072, China
{2019300003090,zhcai}@whu.edu.cn

Abstract. With the widespread use of IoT devices, the issue of high data transaction pressure introduces a new set of challenges. To overcome the conflict of limited network resources and the growing requirement of transforming generated data from IoT devices, academia has attempted to bring compression algorithms into IoT devices. These research efforts have resulted in a series of compression algorithms for time series data. Meanwhile, with the research on the RISC-V instruction set, its custom instructions provide a platform for implementing DSA. In this thesis, we propose a set of custom extended instructions to compute the compression of time series float data based on the Nuclei open source RISC-V core Hummingbird E203, using the Nuclei Instruction Co-unit Extension (NICE) interface. The compression co-processor uses asynchronous memory access and is independent of the core's main pipeline. Cores with the co-processor demonstrate over 95% acceleration compared to the software solution, achieving compression rates of around 3–10 when working on 32-bit float data with more than 10,000 input points. Furthermore, the implementation requires only a minimal amount of hardware resources.

Keywords: co-processor · time series data · compress · RISC-V

1 Instruction

During the widespread use of IoT devices, they have become increasingly important in our daily lives and production processes. With the massive deployment of IoT devices, a significant amount of data is collected, including a substantial portion of time series data [13]. However, these data are typically used infrequently, making storage and transfer costs a crucial consideration. Consequently, compression techniques are commonly employed at the edge nodes of IoT devices.

Nevertheless, edge computing devices face strict limitations in terms of power, cost, and other resources. Therefore, optimizing compression algorithms for these platforms has become a crucial solution [8]. Furthermore, utilizing domain-specific architectures (DSAs) instead of general-purpose computing devices can offer superior performance at lower power consumption in specific domains [10], aligning with our requirement for adequate computational performance within limited power constraints.

Z. Tari et al. (Eds.): ICA3PP 2023, LNCS 14487, pp. 439–454, 2024.
https://doi.org/10.1007/978-981-97-0834-5_25

With the increasing attention on open source Instruction Set Architectures (ISAs), the RISC-V instruction set has emerged as a foundational platform for embedded devices, featuring an extendable ISA structure [2]. Building upon existing research, we propose a set of extended instructions and a co-processor based on the RISC-V ISA, designed to significantly accelerate the compression process of time series data at a minimal cost.

In the context of compression, the gateway assumes the role of a Fog node [4]. Our co-processor can be utilized as an edge or fog gateway, enabling efficient compression algorithms tailored to the specific characteristics of the input data [5]. As depicted in Fig. 1, our data exhibits slow variation and conforms well to a log distribution. Therefore, we primarily focus on achieving high compression rates for this specialized data. Additionally, our real-time requirements necessitate immediate output of compressed data, without the need to wait for all the data to be generated.

Considering these constraints and taking into account the performance in a hardware implementation, we find that the Delta-of-Delta method utilized in the Gorilla Database [11] is the most efficient. Other candidates, such as LZW, which is not designed for time series data and lacks compatibility with our real-time usage, or Zig-Zag, which only demonstrates efficiency when compressing data with a significant number of zero bits, fail to meet all our requirements.

The algorithm employed in the Gorilla Database exhibits exceptional efficiency in compressing our continuous time series data. Over a considerable period, the compressed size can be reduced to 1.37 bytes per entry of 4 bytes, as illustrated in Fig. 2. This represents a significant advantage of 'the chosen algorithm:

- It performs exceptionally well on time series data, particularly those characterized by slow variations.
- It is capable of generating results in real-time during input, thus meeting our requirement for low-latency compression.

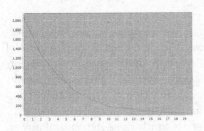
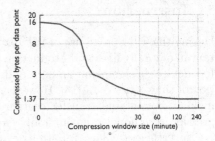

Fig. 1. Data Varying Steps   **Fig. 2.** Compress Window Size

On the other hand, we have chosen the RISC-V instruction set for our co-processor, which means it can only operate in conjunction with a RISC-V

System-on-Chip (SoC). The RISC-V ISA offers several advantages, as outlined below:

- It provides reserved space for implementing Domain-Specific Architecture (DSA).
- It is extensively utilized in IoT devices, which is the primary environment for our co-processor.
- The RISC-V ISA is relatively simple, resulting in significantly lower static power consumption. This enables fog devices utilizing RISC-V instructions to operate efficiently under strict power and cooling constraints.

We have selected the Hummingbird E203 SoC provided by Nuclei for our implementation. The E203 SoC is an embedded system-on-chip specifically designed with a 2-stage pipeline to optimize power consumption, as well as varying length long-pipelines for complex executing units. Our co-processor can be seamlessly integrated as an execution unit, as illustrated in Sect. 3.2. Additionally, similar to many open-source RISC-V SoCs, the Hummingbird E203 features a special NICE (Nuclei Instruction Co-unit Extension) interface that allows for the embedding of custom execution units into the processor's main pipeline. Consequently, our co-processor has the potential to serve as a standard extension module for various devices utilizing the Hummingbird E203.

Overall, our research focuses on the development of an embedded compression co-processor specifically designed for time series data. The contributions of our work can be summarized as follows:

- We have developed a RISC-V extension scheme for compression at the application level. By sending the start address and data length, the co-processor can independently access memory, perform compression, and return the result. The entire process, apart from the initial start and result retrieval, remains transparent to the CPU.
- The co-processor incorporates several caches, enabling rapid retrieval of results during each visit. Typically, the result can be obtained in just 2 cycles, significantly reducing the overall execution time.
- We have implemented and tested a prototype of our co-processor on an FPGA platform utilizing the SoC provided by Hummingbird E203. Additionally, we have evaluated the performance of our application using the extended RISC-V instructions, uncovering the relationship between the compression rate and the pace of data variation.

The structure of this thesis is organized as follows: Sect. 2 outlines the project's motivation in terms of cost and performance considerations. Section 3 provides a comprehensive introduction to the prior knowledge of the RISC-V ISA, E203 SOC, and the Gorilla Algorithm. In Sect. 4, we present the current research on RISC-V co-processors and instruction extensions, along with a survey of hardware implementation approaches for Gorilla. The experimental results and evaluation are presented in Sect. 5. Finally, Sect. 6 outlines future work and concludes the thesis.

2 Motivation

2.1 Continues Data Access Latency

In our compression algorithm, we consider the input data as a continuous block in memory, which can be efficiently pre-fetched to the caches. However, in the software implementation, this capability is not available. The cache process remains transparent to the software, and memory access is synchronous, leading to unstable latency during the fetch process. To overcome these limitations, we employ multiple caches to store pre-fetched data points. This enables the co-processor to operate independently of the CPU, thereby accelerating the compression algorithm and maximizing the utilization of memory bandwidth.

2.2 Expensive XOR and Reassemble Operations

When employing software implementations on our edge nodes, the resulting latency fails to meet our requirements. On average, it takes over 300 CPU cycles to process a single data point (see Table 6). Consequently, an embedded device operating at 100MHz using such an algorithm can only achieve a bandwidth of 10Mbps. This performance is significantly inadequate, as we require a speed improvement of approximately tenfold.

Upon analysis, we observed that the majority of cycles (over 60%) are consumed by the computation of leading and trailing trivial zeroes, which can be effectively pipelined in hardware implementations. Therefore, the most efficient approach for us is to leverage hardware solutions, such as FPGA, to implement our algorithms. This can significantly enhance the processing efficiency and alleviate the burden of these tedious computations.

3 Preliminary

In this section, we will introduce the technologies utilized in our research and provide the necessary background information.

3.1 RISC-V ISA

RISC-V is an open-source instruction set architecture (ISA) initially proposed by the University of California, Berkeley in 2010 [1]. It offers a more simplified and energy-efficient alternative to IA32/IA64 ISA. Additionally, RISC-V is distinguished by its open-source nature, refined design, and extensibility, setting it apart from other RISC ISAs such as ARM or OpenRISC. Notably, it provides reserved coding space for custom extensions, particularly for domain-specific architectures (DSA). Many system-on-chip (SoC) implementations of RISC-V feature an extendable interface for custom instruction extensions, which is widely adopted in embedded devices.

For instance, Western Digital utilizes RISC-V for storage applications, while Intel recognizes its ecosystem potential for industry stakeholders [9]. Consequently, RISC-V offers high usability and extensibility, making it well-suited for specific application domains.

3.2   Hummingbird E203

The E203 is a widely adopted open-source RISC-V hardware description language (HDL) implementation developed by Nuclei Systems Technology. It serves as a 32-bit RISC-V system-on-chip (SoC) that supports the RV32IMAC standard extensions. As depicted in Fig. 3, the E203 features a 2-stage pipeline architecture with a variable-length Long-Pipes execution unit. This design comprises a Fetch stage (IFU) in the first stage, followed by Decode (EXU), Execute (EXU), Write Back (WB), and Memory Access (LSU) stages. These design considerations ensure the suitability of the E203 for low-power applications.

Similar to the Rocket Chip's RoCC interface [3], Nuclei has developed the Nuclei Instruction Co-unit Extension (NICE) interface to facilitate the integration of custom extended co-processors. As illustrated in Fig. 4, the co-processor functions as an execution unit within the pipeline, interacting seamlessly with the NICE interface. It can receive requests, send responses, and access the LSU directly through the channel provided by NICE.

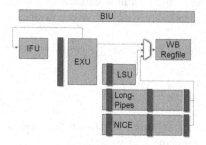

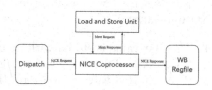

Fig. 3. E203 Micro Architecture   **Fig. 4.** NICE Interface

3.3   Gorilla Database Algorithm [11]

Facebook has successfully discovered an efficient compression and storage method for time series data, particularly focusing on continuous floating-point data. This achievement is demonstrated in the Gorilla Database, where their compression algorithm achieves an impressive compression ratio of 1.37 bytes per data point (down from 16 bytes) for specific datasets. The compression algorithm encompasses both time stamp compression and value compression. However, in our application, we only require compression of the values. Consequently, the following sections will introduce the value compression algorithm employed in the Gorilla Database.

Table 1. XOR with previous values

Series	Float expression	XOR result
1	0x40280000	/
2	0x40380000	0x00100000
3	0x402e0000	0x00160000
4	0x40280000	0x00600000
5	0x40418000	0x00698000

The compression process consists of two steps: XOR and encoding. We specifically limit the input data points to 32-bit floating-point values. Similar to the approaches discussed in [7,12], we observe that our data points exhibit minimal variations compared to their neighboring data points. In light of this observation, we have found XOR to be highly effective in compressing float values, as it disregards the unchanged bits. In the first XOR step, each data point is XORed with its preceding point, except for the first point, resulting in a new value referred to as the XORed value. As depicted in Table 1, XORed values often feature multiple leading and trailing zeroes.

Subsequently, we employ the following encoding scheme to compress the XORed values:

1. The first value is stored with no compression.
2. If the XORed value is zero, a single '0' bit is stored.
3. If the XORed value is non-zero, the number of leading and trailing zeroes in the XORed value is calculated. Then, the following steps are taken:
   a. If the number of leading and trailing zeroes is equal to or greater than the previous value, a single control bit '0' is stored, followed by the meaningful XORed value (i.e., the non-zero interval of the previous XORed value).
   b. Otherwise, the encoding proceeds as follows: first, a single control bit '1' is stored, followed by the length of the number of leading zeroes in the next 5 bits. Next, the length of the meaningful bits of the XORed value is stored in the next 6 bits. Finally, the meaningful bits of the XORed value are stored.

3.4 SpinalHDL

SpinalHDL is a high-level hardware description language that allows generating Verilog or VHDL code from Scala. Compared to Verilog or VHDL, SpinalHDL offers improved readability, maintainability, and extensibility. Thanks to the functional features and object-oriented design of Scala, SpinalHDL provides enhanced expressive capabilities, leading to increased efficiency and correctness in hardware design.

Moreover, components designed with SpinalHDL can seamlessly integrate with other non-SpinalHDL parts within the project. Several factors influenced our decision to choose SpinalHDL:

- The SpinalHDL compiler offers various optimizations, resulting in improved performance and reduced area utilization.
- SpinalHDL enhances readability and maintainability, facilitating the design of complex hardware systems.
- SpinalHDL is a Scala library, allowing us to leverage all the features of Scala in our hardware design process.

The capabilities of SpinalHDL align perfectly with our requirement for designing quick prototypes in our project.

4 Design

In this section, we present the implementation and micro architecture of our Gorilla compression co-processor. Following a similar approach to [11], we limit the input data to 32-bit floating-point values and output data to a 32-bit stream. Consequently, our co-processor can process an average of 2 to 4 data points per cycle, achieving a compression rate ranging from 0.1 to 0.5. Notably, our co-processor operates asynchronously with the CPU, delivering an average of one 32-bit result per cycle.

Table 2. RISC-V base opcode map

inst 4:2 inst 6:5	000	001	010	011	100	101	110	111(>32 b)
00	LOAD	LOAD-FP	custom-0	MISC-MEM	OP-IMM	AUIPC	OP-IMM-32	48 b
01	STORE	STORE-FP	custom-1	AMO	OP	LUI	OP-32	64 b
10	MADD	MSUB	NMSUB	NMADD	OP-FP	reserved	custom-2/rv128	48 b
11	BRANCH	JALR	reserved	JAL	SYSTEM	reserved	custom-3/rv128	>=80 b

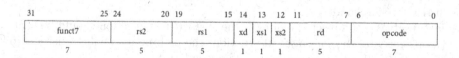

31		25 24		20 19		15 14	13	12 11		7 6		0
	funct7		rs2		rs1		xd	xs1	xs2	rd		opcode
	7		5		5		1	1	1	5		7

Fig. 5. RISC-V Instruction Encode Format

4.1   Instruction Design

As depicted in Table 2, the RISC-V architecture has reserved dedicated coding space for extension purposes. Each instruction can encode two source registers (rs1, rs2) and a destination register (rd), along with their corresponding enable bits (xs1, xs2, xd), as illustrated in Fig. 5. Consequently, the address and length of the data to be compressed need to be passed through registers, and the initial processing should be completed using an R-type instruction. The compression result can then be returned using the rd field of the same R-type instruction. The following provides a detailed description of the instruction design.

Table 3. Custom Instructions

	inst.init	inst.reset	inset.run
func7 1:0	00	10	01
rs1	*Read address for data input*	(Disable)	(Disable)
rs2	*Data length (by byte)*	(Disable)	(Disable)
rd	(Disable)	(Disable)	*Output data*

As shown in Table 3, we introduce three types of custom instructions, each serving a specific purpose. The usage of these instructions is described below:

- *inst.init*: To facilitate asynchronous calculation, we have opted for an efficient method of transmitting memory addresses to the co-processor through registers. The co-processor can directly access memory via the memory access channel provided by E203. The *inst.init* instruction requires two input registers, providing the co-processor with the start address and the length of the input data. Upon receiving the *inst.init* instruction, the co-processor transitions into the *init.working* state.
- *inst.run*: To support real-time applications, the co-processor needs to promptly return each result. Each invocation of the *inst.run* instruction causes the co-processor to return the next compressed bit string stored in the destination register (rd). It is important to note that when using *inst.run*, the co-processor accesses memory, so the data to be compressed must be stored in RAM beforehand.
- *inst.reset*: When a reset is required, the *inst.reset* instruction should be sent. Upon receiving the *inst.reset* instruction, the co-processor will transition to the **stat.IDLE** state, regardless of its current state.

Figure 6 illustrates the finite-state machine representing the working process of the co-processor. Initially, without any instructions, the co-processor is in the **IDLE** state, denoted by a state code of 000. Upon receiving the *inst.init* instruction, the co-processor transitions to the **init.working** state, indicating the initialization phase. After several cycles, the co-processor enters the **init.ready** state, indicating that the initialization process is complete, and it is waiting for

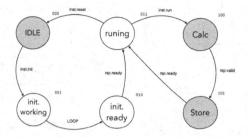

Fig. 6. Finite-State machine for working process of the co-processor

the NICE interface to permit return. If the co-processor receives the *rsp.ready* signal from the NICE interface, it transitions to the **running** state. In this state, the co-processor returns the compressed data in each cycle. If the co-processor receives the *inst.reset* instruction, it transitions back to the **IDLE** state, regardless of its current state.

The custom instruction is only necessary for transitioning from the **IDLE** state to the **init.working** state, performing calculations in the **init.working** state, and transitioning from the **running** state to the **IDLE** state. The remaining processes are handled by the co-processor itself, without requiring the assistance of the CPU. It is important to note that this figure provides an overview of the co-processor's entire process, which includes various internal processes of the co-processor.

4.2 Architecture Overview

Our compression co-processor consists of four main components (as depicted in Fig. 7): the control unit, load and store unit (LSU), cache, and core unit. When a custom instruction is received, it is forwarded to the Ctrl unit via the NICE interface. The Ctrl unit maintains the finite-state machine shown in Fig. 6 and transmits control signals or data to other components. The LSU, when required, retrieves data points from RAM through the NICE memory interface. This process occurs asynchronously, separate from the main pipeline of the CPU. The Ctrl unit is responsible for managing the address and length offsets. Additionally, we have designed a 16-line input cache to expedite the data loading process for continuous data points. This cache enhances CPU utilization when the co-processor operates based on the cached data.

The core unit receives data points from the cache and stores the results in the L-Cache and buffer (as illustrated in Fig. 8). The calculation of the first data point result typically requires 5–15 cycles, while subsequent data points can be processed within 2 cycles in most scenarios, considering the pipeline's bandwidth.

The **Core Unit** (as depicted in Fig. 8) serves as the primary execution component for compression. It comprises four main parts: the calculation unit (CU),

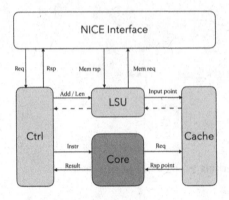

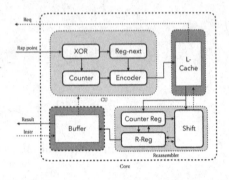

Fig. 7. Structure of the Co-processor

Fig. 8. Compress co-processor inner structure

the local cache (L-Cache), the reassembler, and the output buffer. The Core Unit performs XOR operations with the previous value to generate a 32-bit stream output, which may include multiple leading and trailing trivial zeroes. These zeroes need to be either dropped or re-encoded in the CU's encoder, resulting in a 64-bit stream output. This step takes into account the alignment requirements of the hardware wire and pads the output with zeros as necessary. The reassembler then removes the trivial zeroes and splices the stream back into a 32-bit format.

The **Encoder** is responsible for re-encoding the 32-bit stream output from the CU. It eliminates the leading and trailing trivial zeroes and applies the Gorilla compression algorithm. The Encoder produces a 64-bit stream output, which may contain trailing trivial padding zeroes. The inner working process of the Encoder is as follows:

- Firstly, the input data is separately shifted left and right by 5 times and then XORed with the original data. This result is used to calculate the number of leading and trailing zeroes. This process takes 6 cycles but can be pipelined with 1 data input per cycle, contributing to the co-processor's performance improvement compared to software implementation. The process is as follows:
- Let's suppose the input data is represented by *input*, which is a 32-bit stream. We have two variables, *leading* and *trailing*, both represented by 5-bit binary numbers. In this example, we will focus on *leading*, but the calculation of *trailing* follows a similar approach. Only when the high 16 bits of *input* are all zero, *leading[0]* will be unset; otherwise, it will be set. To calculate *leading[1]*, we need the result of *leading[0]* to determine the offset. If *leading[0]* is set, we test *input[32:24]*; otherwise, we test *input[16:8]*. If these 8 bits are all zero, then *leading[1]* will be unset. Similarly, other bits of *leading* and *trailing* are calculated in a similar manner. By repeating this process for 6 cycles, we

obtain the values of *leading* and *trailing*. This process utilizes a shift register, allowing it to be pipelined to increase the overall bandwidth.
- After obtaining the number of leading and trailing zeroes, they are compared with the previous values. The non-trivial bits are re-encoded based on the comparison results, and padding zeroes are added at the end.

Reassembler component is responsible for dropping the trailing trivial zeroes and splicing the remaining bits into a 32-bit stream. To address the speed mismatch between the Calculation Unit (CU) and the Reassembler, we have designed an L-Cache. Additionally, the Buffer Cache is available to store a 32-bit result, supporting background asynchronous calculation.

The Reassembler effectively eliminates the trailing trivial zeroes and splices the bits into their correct positions. It comprises a shift register, a bit counter, and a result cache. The Reassembler can process the shift and bitwise operations on a single data point and insert it into the appropriate position in the result cache. This process typically consumes multiple clock cycles, as the shift operation cannot be pipelined. Consequently, the Reassembler requires several cycles to complete the output for one data point.

To optimize the performance, we have incorporated the L-Cache and the Output Buffer. The L-Cache can store up to 16 lines, while the Output Buffer can store up to 4 lines. These components enable the co-processor to continue its operations even when the CPU is busy or the Load and Store Unit (LSU) is occupied with other tasks. This design ensures that the Reassembler process and the waiting time for writing back can be synchronized, maximizing the utilization of the data bandwidth.

Total Working Process of the co-processor can be divided into three steps: Init, Calc, and Reset. The calculation process begins with the Init step, which requires the data point to be stored at a specific RAM address. To initiate this process, the *Inst.init* instruction is called, and the Load and Store Unit (LSU) initializes the start address and length accordingly. The LSU loads the data point whenever the Cache is empty and the NICE interface is ready. Subsequently, the data is stored in the Cache, awaiting compression.

During the Calc step, the Calculation Unit (CU) pipeline can XOR and Encode a data point per cycle when the L-Cache in the Core is not filled. Consequently, the L-Cache is typically full. Once the Buffer in the Core is empty, indicating that the result has been read, the Reassembler begins its operation to generate the next 32-bit result. Depending on the number of leading zeroes in the L-Cache output, the Reassembler can complete the calculation in 3–16 cycles. On average, the compression of one data point consumes 2.3 cycles. When the current address of the LSU reaches the end address, the co-processor sends a specific result of 32 bits, all set, which would never occur in the calculated process.

Compared to software implementation, our co-processor achieves a significant performance improvement. This is due to allowing a wider pipeline width and the efficient design of the Encoder and Reassembler. As a result, the process of counting the leading and trailing zeroes can be completed in two cycles, and the Reassemble process can be performed asynchronously.

4.3 Software Design

Here we provide a sample program that demonstrates the usage of the co-processor at the software level. The working process consists of three stages: Init, Calc, and Reset. To ensure that the working process meets our expectations, we need to check the return value of each custom instruction. In real-time applications, it is necessary to return a 32-bit stream each time the *inst.run* instruction is called. Algorithm 1 illustrates how to use our custom instructions for compression.

Before executing the algorithm, we need to store the data to be compressed in a specific area in RAM and keep track of the starting address and length. Upon calling the algorithm, we first verify the validity of the input length (which should be greater than zero). Then, we initialize the co-processor using the address and length by making the initial call to *inst.init*. Next, we introduce a counting variable i to indicate the offset. The program enters a loop, where we check whether $rAddr[i]$ is set to its maximum value, indicating the end of the algorithm. If not, we increment the counting variable i by 4, pointing to the next address for storage. The loop continues until the calculation is completed, signified by receiving the value 0xFFFF_FFFF. At that point, the loop is terminated, and the counting variable represents the length of the resulting compressed data, which is returned.

5 Implementation and Evaluation

Here, we present the implementation and evaluation of our prototype co-processor. Our work demonstrates the advantages of using custom instructions compared to the software method.

5.1 Simulation

Based on the E203 architecture, we implemented a prototype of our co-processor on the Artix-7 FPGA platform (Nuclei DDR 200T). The RISC-V tool chains were used to compile custom instructions using inn's pseudo-instructions.

$$.insn \ r \ opcode, \ func3, \ func7, \ rd, \ rs1, \ rs2 \tag{1}$$

Based on our software design, we conducted experiments using sample programs to verify the functions and performance of our co-processor. We established a RISC-V compiled environment under Linux, operating at a frequency of 50MHz. Additionally, we utilized the clock cycle counter CSR to measure the cycles of the co-processor.

In our experimental environment, we selected the Artix-7 FPGA platform (Nuclei DDR 200T) operating at a frequency of 50MHz. The input data consisted

Data: input data address, input data length(by bytes)

Result: the compressed data will be returned each time *inst.run()* is called

function Compress (address, length, rAddr) âĘŠ rLen **then**

//*judge whether the input data length is valid*

if *length* >= *0* **then**

    //*custom instruction inst.init*

    inst.init(address. length) ;

    //*calculate the offset*

    i = 0;

    //*the custom instruction returns **0xFFFF_FFFF** to show the end of the calculation*

    while *rAddr[i]* != *0xFFFF_FFFF* **do**

        //**custom instruction** *inst.run*

        rAddr[i] = inst.run() ;

        //*address offset increase 4*

        i += 4 ;

    end

    //*custom instruction inst.reset*

    inst.reset()

end

return i ;

end

Algorithm 1: sample code for using the co-processor

of continuously varying floating-point values. Figure 9 illustrates a subset of 1000 data points with a varying step of 8. This dataset aligns well with our deployment environment. The distribution of the varying step, i.e., the difference between consecutive data points, is shown in Fig. 10. Most of the data points exhibit slow variation and can be approximated by a curve, capturing the essential information from the sensors.

On the input data set depicted in Fig. 9, we achieved an average compression rate of 0.24, with an average compression cycle count of 11 per data point. This represents a 10-20 times improvement compared to the software solution.

Furthermore, we conducted tests to examine the relationship between performance and the input data characteristics. By varying the length of the input data from 100 data points to 16000 data points and adjusting the varying step between 2 and 16, we explored the impact of initialization on performance and compression rate. To eliminate the influence of different memory implementations, we set the memory latency to 0. Detailed results are presented in Tables 4 and 5, demonstrating the relationship between average cycles and compression rate for different input data varying steps and lengths.

In summary, we observed that as the input data length increased, the compression rate improved, and the number of cycles required to compress data points decreased. When the data varying rate was lower and the input data length was longer, the co-processor exhibited better performance and achieved higher compression rates.

Our design, featuring parallel LSU access, significantly improves the latency of accessing linear data, such as data arrays. In our implementation, the co-processor required 6 cycles for calculation, but due to the pipeline design, this process was reduced to 2 cycles. The parallel LSU ensures that the co-processor can operate on cached data without utilizing CPU cycles. Consequently, in most scenarios, the co-processor can promptly return results, requiring only 1 cycle in addition to the memory access latency. Our prototype achieved compression rates ranging from 0.11 to 0.342, depending on the input data characteristics. The co-processor spent 8 to 15 cycles on each data point, with an average cycle count of 11.

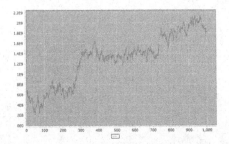

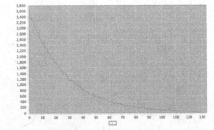

Fig. 9. Data Slice of Sample Input

Fig. 10. Data Varying step Distribution

Our co-processor implementation consumes fewer than 2000 LUT logic gates and utilizes some on-chip RAM resources. This resource utilization is considered acceptable for most embedded environments. In comparison to the software solution, our co-processor offers a significant advantage in terms of resource consumption and power consumption.

Table 4. Average cycle (cycles per input point)

len\step	2	3	4	5	6	8	10	12	16
100	2.2	2.42	2.89	3.13	2.99	3.39	3.14	3.39	3.74
500	1.772	1.962	2.222	2.426	2.544	2.578	2.778	2.826	3.092
1000	1.683	1.978	2.171	2.289	2.361	2.559	2.680	2.679	2.908
2000	1.713	2.035	2.196	2.317	2.431	2.526	2.632	2.753	2.934
4000	1.635	1.916	2.137	2.277	2.369	2.525	2.623	2.763	2.906
8000	1.650	1.893	2.103	2.222	2.340	2.472	2.584	2.685	2.872
16000	1.653	1.890	2.079	2.228	2.328	2.457	2,574	2.666	2.839

Table 5. Average compression rate(output length/input length)

len\step	2	3	4	5	6	8	10	12	16
100	0.14	0.19	0.28	0.30	0.30	0.35	0.32	0.36	0.42
500	0.12	0.17	0.21	0.25	0.27	0.29	0.32	0.33	0.37
1000	0.10	0.17	0.21	0.24	0.25	0.28	0.30	0.31	0.35
2000	0.10	0.17	0.21	0.24	0.26	0.28	0.30	0.32	0.36
4000	0.10	0.16	0.20	0.23	0.25	0.28	0.30	0.32	0.35
8000	0.10	0.16	0.21	0.23	0.25	0.28	0.30	0.32	0.35
16000	0.10	0.16	0.20	0.23	0.25	0.28	0.30	0.32	0.35

Table 6. On Board Result of sample data compress

Data points	Software Solution (100 MHz)		Proposed (100 MHz)	
	Cycle/point	Throughput (Mbps)	Cycle/point	Throughput (Mbps)
tsgas	10.396	307.8	26.2	122.137
tswesad	10.322	310.0	25.2	126.984

5.2 Benchmark

Benchmarking with standard datasets is an important method to evaluate the overall performance of the co-processor. In our study, we selected tsgas and tswesad single precision datasets as our input, as mentioned in [6]. The benchmarking tests were conducted on the Artix-7 FPGA development board. During the test, we considered the impact of memory latency and the CPU's data writing back on the co-processor's performance. As a result, the co-processor was not able to achieve its maximum performance potential. To ensure a fair comparison, we set the CPU clock frequency to 100 MHz, which is the same as the software solution running on the same platform. Table 6 presents the performance comparison between our co-processor and the software solutions implementing the same algorithm. Our co-processor demonstrated a speedup of 12–13 times compared to the software implementation.

6   Conclusion and Future Work

This thesis introduces the implementation of a Gorilla-Based Compress co-processor that leverages the NICE or ROCC extension channel, provided by E203 and Rocket cores, to enable independent memory access and calculation for achieving high performance in the Compress core. Additionally, we propose a set of compression instructions based on RISC-V custom extensions, along with a software programming interface. Our approach surpasses existing software solutions in terms of efficiency.

To validate the feasibility of the co-processor in a real IoT environment, we have developed a prototype on an FPGA platform. The results demonstrate that our co-processor can effectively meet the real-time requirements while maintaining efficiency.

In our future work, we plan to expand our efforts by developing a non time-series data compression co-processor to address the needs of multimedia data transport. This extension will cater to the compression and transport requirements of streaming media data.

References

1. Waterman, A., Krste Asanovic, R.V.F.: The RISC-V Instruction Set Manual, Volume I: User-Level ISA, Document Version 20191213 (2019)
2. Asanovic, K.P., Eecs, D.A.: Instruction Sets Should be Free: The Case for RISC-V
3. Asanovic, K., et al.: The Rocket Chip Generator (UCB/EECS-2016-17) (2016)
4. Bonomi, F., Milito, R.A., Zhu, J., Addepalli, S.: Fog computing and its role in the internet of things. In: MCC 2012 (2012)
5. Gia, T.N., Qingqing, L., Queralta, J.P., Tenhunen, H., Zou, Z., Westerlund, T.: Lossless compression techniques in edge computing for mission-critical applications in the IoT. In: 2019 Twelfth International Conference on Mobile Computing and Ubiquitous Network (ICMU), pp. 1–2 (2019)
6. Knorr, F., Thoman, P., Fahringer, T.: Datasets for benchmarking floating-point compressors (2020). https://doi.org/10.48550/ARXIV.2011.02849
7. Lindstrom, P., Isenburg, M.: Fast and efficient compression of floating-point data. IEEE Trans. Visual Comput. Graphics 12(5), 1245–1250 (2006)
8. Mochizuki, S., Komuro, N.: Power saving method using compressed sensing technique for IoT-based time-series environment monitoring system. In: 2021 IEEE International Conference on Consumer Electronics-Taiwan (ICCE-TW), pp. 1–2 (2021)
9. Nicholas, G.S., Gui, Y., Saqib, F.: A survey and analysis on SOC platform security in arm, intel and RISC-V architecture. In: 2020 IEEE 63rd International Midwest Symposium on Circuits and Systems (MWSCAS), pp. 718–721 (2020)
10. Patterson, D.: 50 years of computer architecture: from the mainframe CPU to the domain-specific TPU and the open RISC-V instruction set. In: 2018 IEEE International Solid - State Circuits Conference - (ISSCC), pp. 27–31 (2018)
11. Pelkonen, T., et al.: Gorilla: a fast, scalable, in-memory time series database. Proc. VLDB Endowm. 8(12), 1816–1827 (2015). https://doi.org/10.14778/2824032.2824078
12. Ratanaworabhan, P., Ke, J., Burtscher, M.: Fast lossless compression of scientific floating-point data. In: Data Compression Conference (DCC 2006), pp. 133–142 (2006)
13. Xu, X., Huang, S., Chen, Y., Browny, K., Halilovicy, I., Lu, W.: Tsaaas: time series analytics as a service on IoT. In: 2014 IEEE International Conference on Web Services, pp. 249–256 (2014)

MSIN: An Efficient Multi-head Self-attention Framework for Inertial Navigation

Gaotao Shi[1,2(✉)], Bingjia Pan[1,2], and Yuzhi Ni[3]

[1] College of Intelligence and Computing, Tianjin University, Tianjin, China
{shgt,pbjxxs}@tju.edu.cn
[2] Tianjin Key Laboratory of Advanced Networking (TANK), Tianjin, China
[3] Tianjin International Engineering Institute, Tianjin University, Tianjin, China
cs_niyuzhi@tju.edu.cn

Abstract. Inertial Measurement Unit (IMU) makes an outstanding contribution to indoor inertial navigation in the era of ubiquitous computing, as it is widely integrated into portable devices. Many prominent works have been proposed by taking gyroscope and accelerometer readings as input to estimate the velocity and orientation. However, most of them focus on the local features of IMU (i.e., single sensor temporal feature or local spatial feature), eventually leading to drift on the trajectory. In this paper, we propose a robust model to mitigate the problem of jitters and drifts in trajectory prediction by exploiting the spatial dependence in accelerometer and gyroscope readings, as well as the contextual relation in motion terms through in-depth analyses of IMU readings. In particular, we design a framework (MSIN) to fuse the local spatial dependence of multiple sensors and incorporate the local spatial and global temporal features by using the multi-head self-attention mechanism. We have conducted extensive experiments on two public datasets and the results show that MSIN achieves a significant improvement in RTE (Relative Trajectory Error) performance, with improvements of up to 6.14% and 15.19% over state-of-the-art methods for RoNIN-Unseen and RIDI-Unseen, respectively.

Keywords: Inertial Navigation · Multi-head self-attention · IMU · IoT

1 Introduction

The convergence of ubiquitous computing, Internet of Things, and mobile computing has paved the way for a new generation of location-based services that are more accurate and personalized than ever before. The existing methods for location and tracking are mainly based on pre-established infrastructure, such as GPS satellite [6] and WiFi access points [37]. However, both have the inherent limitation that GPS signal is unstable indoors, WiFi is constrained by the

This work was supported by the National Natural Science Foundation of China (NSFC) under Grant No. 61872266.

range of deployments and its performance suffers from the impact of multi-path indoors. Inertial Measurement Units (IMU) is a sophisticated sensor that is widely integrated into smartphones and other wearable devices (i.e., smartphones, smartwatches, earphones, and etc.)) [12,30]. The popularity of IMU in various mobile devices provides a supplemental locating approach for ubiquitous services where the GPS or WiFi cannot work [18,39]. Inertial Navigation [34] is a significant majority of location and tracking, which applies a self-contained navigation technique to track the position and orientation of an object relative to a known starting point, orientation, and velocity. And IMU sensors (i.e., accelerometer, gyroscope, and magnetometer) are used in inertial navigation to estimate the positions and orientations of a moving subject based on a sequence of IMU measurements.

Traditional inertial navigation algorithms mainly obtain the position information by double integration of sensor data, but the existence of bias noise and measurement errors in the data makes the traditional algorithms suffer from serious trajectory drift during the localization process [21,24]. To solve this problem, many works have proposed inertial navigation algorithms based on prior knowledge [21,27], but the existence of constraints makes the traditional algorithms difficult to be extended. Due to the rapid development of artificial intelligence technologies, data-driven neural networks breakthrough that limitations and gain significant performance improvements in inference. RIDI [36] is the first to integrate machine learning techniques into the field of inertial navigation, aiming at cascading velocity regression models to correct sensor errors. IONET [4], IDOL [2], and TLIO [25] improved positioning accuracy in various degrees by using neural networks. However, most existing works do not sufficiently extract the global dependencies among the tokens of IMU data. To the best of our knowledge, the best-performing model RoNIN [14] only uses the local features extracted by the ResNet [13] to accomplish trajectory prediction.

To explore the potential of IMU data, we conduct comprehensive analyses of IMU characteristics, and then two key findings have been obtained from IMU data distributions and context patterns. First, data pattern differs distinctively when behavior, the position of the device, and other conditions change. And the data pattern generally lasts for a long time due to the continuity of activities. Second, we find that the gyroscope and accelerometer data have strong spatial dependence. Evident fluctuations in gyroscope readings, which will cause jitter in predicted trajectories, can be mitigated by introducing steadier accelerometer readings. In summary, the key to exploring the potential of IMU lies in learning the general representations from the long-term contextual relationships and spatial dependencies of multiple sensors.

Motivated by the aforementioned findings, we introduce Transformers consisting of convolutional neural network (CNN) and multi-head attention mechanism [33] to process IMU data and propose the Multi-head Self-attention architecture for Inertial Navigation (MSIN). To fuse multi-sensor data and extend the feature dimensions, we design CNN and deep residual network modules to enhance spatial feature representation. Inspired by the utilization of the multi-head attention mechanism in BERT [7], we renovate the multi-head mechanism to learn the correlation between representation subspaces and then obtain generalizable representations of long-term temporal features. To demonstrate the effectiveness of MSIN, we conduct numerous experiments within two open datasets (RIDI [36] and RoNIN [14]) and the results show that MSIN achieves better performance than previous works. The positioning error is 1.34 m on a trajectory of 172 m in the RoNIN-Unseen dataset and 1.64 m on a trajectory of 75 m in the RIDI-Unseen dataset. The contributions are summarized as follows:

- We conducted numerous analyses of IMU data and obtained two key observations that are reflected in the spatial dependence of multiple sensors and long-term contextual temporal relations.
- Based on two key observations of IMU data, this paper devises the MSIN based on the multi-head attention mechanism, which combines Transformers to learn the correlation between long-term tokens of IMU. Our model effectively captures the high-level semantic information of IMU data by integrating spatial features with contextual relations to further improve the localization accuracy of inertial navigation.
- MSIN achieves the excellent performance of inertial navigation on RIDI and RoNIN datasets. Extensive ablations and insightful case studies are given.

The rest of this paper is organized as follows. Section 2 reviews the related works about inertial navigation. Section 3 shows the motivation for our work. Section 4 introduces MSIN workflow along with the design details. The experiment details and evaluation results are discussed in Sect. 5. Section 6 concludes this paper.

2 Related Work

IMU data has been widely utilized in various applications, such as human activity recognition [17], human-computer interaction [26], localization and tracking [14, 21], and etc. Here, we only present a concise summary of the existing research on positioning methods using IMU data.

2.1 Conventional Methods

Research on inertial navigation and position technology began in the late 20th century. Levi and Judd [21] proposed the Pedestrian Dead Reckoning (PDR) algorithm in 1996, which calculates the pedestrian's step length and step number based on accelerometer data, and then solves the pedestrian motion based on gyroscope data. Nilsson et al. [27] proposed ZUPT, a zero velocity correction algorithm. ZUPT is an inertial navigation and heading projection system based on shoe-strapped inertial sensors, which reduces equipment costs while improving short-range positioning accuracy. The proposal of ZUPT greatly facilitated the traditional inertial navigation algorithms. In combination with ZUPT, Guo, and M. Uradzinski [11] proposed an algorithm with fused accelerometers, gyroscopes, and magnetometers after initial alignment to reduce the error in gait by estimating and compensating the drift error at each step. And Ilyas Muhammad [16] utilized ZUPT to reduce the drift and inertial sensor errors by shoe-tied inertial sensors, the algorithm calculates the true heading angle by using magnetic anisotropy detection in the heading estimation process and correcting the compensation with Extended Kalman Filter (EKF) [9] at each update step.

Acceptable results are shown with PDR, ZUPT, and their derivatives in terms of localization accuracy, but the limitations are also obvious. It is difficult for PDR to adapt to the changes in pedestrian stride length, while ZUPT has strict requirements on how the device is worn. Some works [10,24] reduce the impact of sensor errors on positioning results by means of cumulative error cancellation algorithms but are struggling with cumulative errors in remote navigation solutions.

2.2   Deep Learning Based Methods

With the booming development of deep learning recently and its remarkable effect in the field of CV [8] and NLP [3,7], many works adopt deep neural networks to extract more effective features from IMU data. As the first attempt, Yan et al. [36] proposed RIDI, a neural network based on regression linear velocity. RIDI has similar positioning errors as Visual inertial navigation and outperforms previous conventional algorithms, demonstrating the effectiveness of neural networks on IMU processing. To cure the inertial system drift, IONeT [4] breaks the cycle of continuous error propagation, defines inertial navigation as a sequential learning problem for the first time, and learns the position transformation of polar coordinates within the independent window by two-layer LSTM [15] learning. The remarkable effect of IONeT's velocity fitting has inspired many works, Chen et al. [5] proposed Motiontransformer, an IONet variant based on adversarial neural networks and domain adaptation techniques, which extracts domain-invariant features of the original sequence from an arbitrary domain. Herath et al. [14] proposed the RoNIN model in 2020 and designed R-ResNet, R-LSTM, and R-TCN to accomplish the velocity regression task on independent time windows, and then obtain the position information by one integration. The R-ResNet deepens the depth of the model through the residual structure and surpasses other models according to our knowledge. Nevertheless, none of these methods take into account the global dependencies.

Recently, Transformers [33] have shown great power in long-range temporal dependencies. The CTIN [29] model as the first Transformer-based model proposed by Rao et al. employs a ResNet-based encoder to capture spatial information from IMU data, as well as the Transformer for contextual features modeling. However, CTIN has a large number of model parameters and a high training cost, making it impossible to be placed in practice. Xu et al. proposed LIMU-BERT [35], adapting BERT to IMU representation. However, LIMU-BERT mainly focuses on the representation of IMU, therefore does not probe deeply into inertial navigation.

Unlike the previous usage of Transformers, MSIN renovates the multi-head mechanism into inertial navigation based on the observation of IMU features. However, most previous works [14] on IMU data are aiming to extract the local contextual relationship rather than combinate the local and global which is inspired by other surrogate tasks [19, 28, 38] from images and text. LIMU-BERT has considered this problem, but it is not optimized for inertial navigation. This paper provides an in-depth analysis of inertial navigation tasks based on IMU data, which emphasizes the combination of local spatial relationships and global temporal features to enhance the robustness of the performance in different datasets. However, due to the limitation of the length of the article, clustered streaming data processing is not discussed further in this paper, which has been well explored in [22, 23]

3 Motivation

To discover what IMU readings are saying, we have carefully studied the characteristics of the IMU and obtained the following key observations, which will guide the design of MSIN. Notably, we discarded the analysis of magnetometers, which are susceptible to environmental influences. Figure 1 provides three traces of IMU (accelerometer, gyroscope) readings of various activities with different placements.

Spatial Distribution Characteristics. In Fig. 1(b), gyroscope readings are more volatile than accelerometer readings since the gyroscope is more sensitive to movement [31]. And by introducing accelerometer readings, the fluctuations of the gyroscope readings can be mitigated to improve the stability of orientation prediction and avoid jitters in the predicted trajectory. Therefore, if only gyroscope readings are used as input data for deep neural networks, it will inevitably cause the problem of data homogenization. In other words, the cross-use of multiple sensors can provide more semantic information and improve the overall performance of the system [35]. Therefore, data fusion of multiple IMU sensors is indispensable for representation learning to extract spatial features.

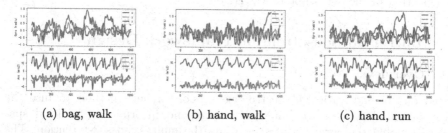

| (a) bag, walk | (b) hand, walk | (c) hand, run |

Fig. 1. IMU measurements of daily activities with different device placements.

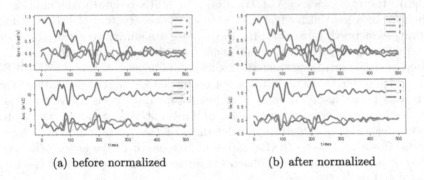

(a) before normalized   (b) after normalized

Fig. 2. Sensor data distribution before and after normalization

Temporal Characteristics. Observing the IMU readings when the devices were placed in the bag in Fig. 1(a) and in hand in Fig. 1(b), we find that the readings of IMU in hand change more frequently. This is likely because the orientation of the phone changes as the hands move. There is a clear difference in the frequency and amplitude of IMU readings in Fig. 1(b) and Fig. 1(c). More intense motion, more volatile IMU readings. What's more, the periodic patterns are apparently existed in Fig. 1 due to the continuity of speed in activities. In a nutshell, temporal relations in local (i.e. periodical patterns) and global (i.e. patterns of different activities) play a critical role in IMU data usage.

It is clear that the range of IMU readings varies widely for different sensors. In Fig. 2(a), the range of gyroscope readings is $[-0.5, 1.5]$, but the accelerometer readings range is $[5, 15]$. In order to get a good adaptation to the model, we adopted a simple but effective normalization method on accelerometer readings to narrow the difference in the value range of sensors, which can be formalized as follows:

$$acc_i = \frac{acc_i}{9.8m/s^2}, i \in \{x, y, z\} \tag{1}$$

where acc_i is the measurements in the i-axis of the accelerometer. After the linear normalization, the reading range of the accelerometer is essentially the same as that of the gyroscope in Fig. 2(b).

4   Multi-head Self-attention Framework Design

4.1   Overview

Based on the motivation in Sect. 3, we propose a new method to fuse the local spatial dependence of multiple sensors and incorporate the local spatial and global temporal features by using the multi-head self-attention. The framework of the Multi-head Self-attention mechanism for inertial navigation (MSIN) is presented in Fig. 3 which consists of three major components including *spatial feature extractor*, *transformer encoder*, and *velocity prediction*. The

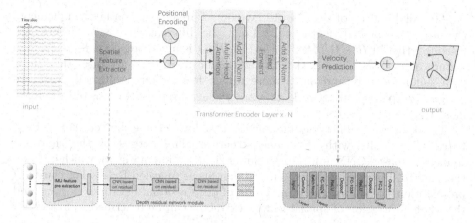

Fig. 3. The framework of MSIN

MSIN attempts to regress the velocity with two key modules and then predict the trajectory for inertial navigation. We first make the data normalization before feeding it into subsequent modules. In order to explore the spatial dependence of multi-sensor, we design the spatial feature extractor that fuses accelerometer and gyroscope data by CNN and residual connection and outputs spatial tokens (the striped rectangles at the bottom of Fig. 3). The transformer encoder takes spatial tokens as input and learns the long-term temporal features to further extract the correlation of tokens. In the end, the velocity module regresses the depth feature into two-dimensional velocity vectors by convolutional units and fully connected layers in supervised training.

4.2   Spatial Feature Extractor

Note that RNN-based models only learn features on one-dimensional time-series data, which will ignore the characteristic distribution of the gyroscope and accelerometer existing in space simultaneously. Inspired by the popularity of CNN in image feature extraction, we proposed the spatial feature extractor based on CNN and residual connection. The normalized IMU sequences are divided into multiple fixed windows $X \in \mathbf{R}^{N \times L \times S_{dim}}$, where N is the batch size, L is the length of the fixed window and S_{dim} defaults to 6, representing the accelerometer and gyroscope for a total of six axes of sensor data. For instance, the details of an IMU sample $x_i \in X$ can be expressed as $x_i = [gyro_x, gyro_y, gyro_z, acc_x, acc_y, acc_z]$. To explore the potential of limited IMU features, we extend the feature dimensionality and fuse multiple IMU sensors by projecting the data X into a higher space by:

$$H = Proj(X) \tag{2}$$

The combination of the Conv1d layer and fully-connection layer implements the projection of IMU features. By Eq. 2, the different dimension features are fused together in $H \in \mathbf{R}^{N \times L \times H_{dim}}$, which H_{dim} is much larger than the original feature dimension. Compared with the original feature X, H fuses the sensor data simultaneously and has a more robust feature representation. Next, we designed deep residual networks to extract general representations of IMU readings.

The deep residual network module is stacked with three shortcut-connection-based CNN units with the same structure. The details of the shortcut-connection-based CNN unit (i.e., the lower right section in Fig. 3), which includes the Conv1d, Batch Normalization, and non-linear function ReLU. For clarity of presentation, we simply the formulation of deep residual networks as f_exc, and the whole procession of spatial feature extractor can be described as follows:

$$token^i = f_{exc}(Proj(X^i)) \tag{3}$$

where $token^i$ is a $l \times H_{dim}$ matrix and contains the spatial features of the i-th IMU windows X^i. Additionally, l is a small integer relative to the original window length L. In conclusion, the spatial feature extractor obtains high-level spatial representations across multiple sensors (the striped rectangles in Fig. 3) and reduces the cost of subsequent encoder training.

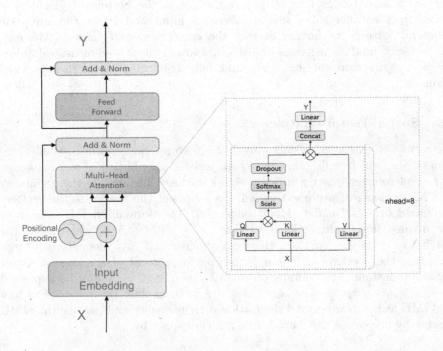

Fig. 4. The workflows of the spatial feature extractor

4.3   Transformers Encoder

As illustrated in Fig. 4, MSIN adapts the Transformers with the multi-head mechanism to extract global dependencies of temporal from the output of the spatial feature extractor. We briefly formulate the process of the encoder as follows:

$$E = f_{enc}(token^i + PE(i)) \tag{4}$$

where E is a $l \times H_{dim}$ matrix, and $PE(\cdot)$ is the positional encoding function [33], which is added to maintain a sequential relation before tokens are fed into the encoder. All parameters in $PE(\cdot)$ are trainable to dynamically map the order index into a vector with the length of H_{dim}. The transformer encoder then takes tokens after positional encoding as input and repeats R_{nums} times before the final representations got. The transformer encoder involves two residual components, which can be formulated as:

$$T^r = LayerNorm(FFN(Z^{r-1}) + Z^{r-1}) \tag{5}$$

$$Z^{r-1} = LayerNorm(MultiAttention(P^{r-1}) + P^{r-1}) \tag{6}$$

$$P^0 = token_i + PE(i) \tag{7}$$

where $r \in [1, \cdots, R_{num}]$ denotes the output of the r-th Transformer encoder. The $MultiAttention(\cdot)$ is the multi-head attention layer with A_{num} heads to extract the characterization from different sequence positions of different tokens in parallel. It is worth noting that A_{nums} in the experiment requires a complete division of H_{dim}. The $FFC(\cdot)$ is essentially two fully-connected layers with ReLU. Finally, we get the final representations $T^{R_{num}}$ of an IMU subsequence X^i.

4.4   Velocity Regression Module

The aim of the Velocity Regression Module is to obtain a 2-D velocity vector with the representations generated by MSIN. Figure 5 shows the structure of this module, which essentially consists of convolutional units, fully connected layers, and non-linear activation functions. The processing of the Velocity Regression Module can be expressed as follows:

$$D = FC_2(Pred(B)) \tag{8}$$

$$B = BatchNorm(Conv(T)) \tag{9}$$

where $Pred(\cdot)$ denotes the multilayer perceptron (MLP) with ReLU and Dropout layer [32]. And $FC_2(\cdot)$ is a fully-connected layer with input dimension H_{dim} and output 2 respectively. The $D \in \mathbf{R}^{N \times 2}$ indicates the final 2D velocity vector.

An integration and interpolation operation is required for the regressed velocity vector to implement the inertial navigation location. We first integrate the

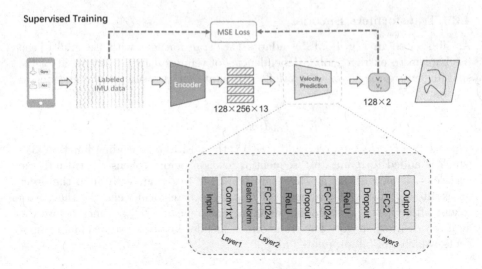

Fig. 5. The workflows of the spatial feature extractor

velocity between windows to get the displacement change between windows. Second, the trajectory information of a specific step is obtained by accumulating the displacement change between windows after introducing the initial position. Finally, the position information of the whole sequence is obtained by one-dimensional linear interpolation.

5   Performance Evaluation

5.1   Datasets and Experiment Settings

We extensively evaluated the generalizability of MSIN on two well-known datasets and performed extensive ablation experiments. In this subsection, we will introduce the datasets and experiment settings.

RIDI [36]. It contains measurements of IMU sensors and cameras across multiple human subjects and multiple device placements. More specifically, the IMU sequences are collected by Lenovo Phab2 Pro, yielding the data at 200 Hz. A Visual Inertial Odometry system is used on the Google Tango phone to record the ground truth trajectories. The test set is divided into two parts, one in which subjects have appeared in the training set, called Seen test set. And the other is called the Unseen test set where subjects do not appear in the training set. In our experiment, we use 50 trajectories in RIDI for training and select 23 and 6 sequences from the Seen and Unseen test sets respectively for evaluation.

RoNIN [14]. It is one of the largest and most credible inertial navigation datasets. The human subject is attached to an Asus Zenfone AR with a harness and handles Galaxy S9 and Pixel 2XL freely to collect IMU data to simulate real human behavior. The RoNIN includes over 276 long and complex sequences in 3 buildings collected by 20 human subjects, and the IMU data is sampled at 200 Hz. The training set consists of 68 sequences, and the validation set consists of 16 sequences. The test set is also divided into the Seen and Unseen. The Seen test set consists of 32 sequences with an average length of 420m and the Unseen includes 32 sequences with an average length of 455m.

Setting Details. MSIN and other baseline models are implemented with Python and PyTorch. We trained all models on a single NVIDIA Geforce 3070 GPU with 8G memory. Duration training, the batch size is fixed to 128 and the ADAM optimizer [20] is used with a starting learning rate of $1e^{-4}$. In practice, H_{dim} is empirically set as 128. The window size L and the number of heads A_{num} are set as 200 and 4 respectively according to our ablation experiments. We choose the mean square error (MSE) as the loss function of MSIN, which can be defined as follows:

$$loss = \frac{1}{m} \sum_{i=1}^{m} (pred_i - targ_i)^2 \tag{10}$$

where $pred_i$ and $targ_i$ are the prediction velocity and ground truth of the i-th window, respectively.

5.2 Baselines

We evaluate our framework by comparing it with three existing models, which are the best ones in inertial navigation as far as we know.

R-LSTM. It stacks unidirectional LSTM to enrich the input feature by concatenating the output of a bilinear layer. R-LSTM belongs to the Sequence to Sequence model, which takes the IMU data of the accelerometer and gyroscope as input to regress a 2D vector for each frame.

R-TCN. This model improves the drawbacks of the LSTM with sequential inputs by enabling parallel processing through the use of a fully convolutional network [1]. It guarantees that the output at a given moment is only relevant to the current moment and earlier information.

R-ResNet. As we know, R-ResNet takes the ResNet-18 architecture [13] to extract sensor data features at adjacent moments. It is trained to regress a velocity vector and then integrates every five frames to estimate motion trajectories. R-ResNet performs better in the test set owing to the deeper network by residual connection.

5.3 Metrics

For an inertial navigation system, the common metrics include the absolute trajectory error (ATE) and relative trajectory error (RTE). ATE intuitively reflects the accuracy of the algorithm and the global consistency of the trajectory by calculating the root mean square error (RMSE) of the estimated trajectory and ground truth for the entire sequence. The other metric RTE is applied to evaluate the amount of drift in the system trajectory. RTE describes the accuracy of the difference between two frames of poses within a fixed time difference, which is equivalent to the error of a direct odometer measurement. ATE and RTE can be formulated as Eqs. 11 and 12.

$$ATE = \sqrt{\frac{1}{n} \sum \|pos(t) - \hat{pos}(t)\|^2} \tag{11}$$

$$RTE = \frac{1}{n} \sum ((pos(t + \delta t) - pos(t)) - (pos(\hat{t} + \delta t) - \hat{pos}(t))) \tag{12}$$

where $pos(t)$ and $\hat{pos}(t)$ are the position coordinates of the actual trajectory and the predicted trajectory at time t, respectively. And δt means a small time period.

Table 1. Position evaluation. We compare MSIN with four competing methods: R-ResNet, R-LSTM, R-TCN, and Pedestrian Dead Reckoning (PDR) on two public datasets: the RIDI dataset and the RoNIN dataset.

Dataset	Test subjects	Metric	PDR	R-ResNet	R-LSTM	R-TCN	MSIN
RoNIN	Seen	ATE	29.04	3.86	4.01	4.62	**3.56**
		RTE	21.76	2.81	2.66	2.86	**2.65**
	Unseen	ATE	27.48	5.36	6.32	5.75	**5.26**
		RTE	23.07	4.56	4.53	4.32	**4.28**
RIDI	Seen	ATE	32.87	1.56	1.62	1.56	**1.43**
		RTE	38.12	1.58	1.72	1.66	**1.34**
	Unseen	ATE	31.25	1.63	1.75	1.69	**1.55**
		RTE	37.64	1.97	2.13	2.13	**1.84**

5.4 Comparisons

In the experiment, we investigate the performance of MSIN and other baseline models mentioned in Sect. 5.2. The results are shown in Table 1, from which we can see that all models driven by data obtain better performance than the conventional one (PDR), and our proposed model MSIN outperforms the other baseline models in all cases on both ATE and RTE. The lower localization error in the RIDI test set is due to its short trajectory length with an average length

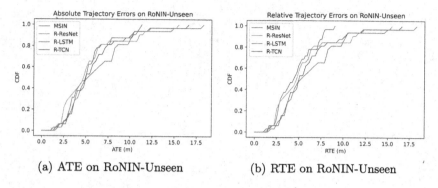

(a) ATE on RoNIN-Unseen   (b) RTE on RoNIN-Unseen

Fig. 6. Performance comparison of position errors on the RoNIN-Unseen.

of 103 m while 420 m in RoNIN. We observe that localization errors in Seen are generally lower than that in Unseen, which verifies the contribution of user behaviors in training.

To illustrate the effectiveness and robustness of MSIN, we focus on analyzing the Unseen results in the table. R-LSTM and R-TCN performance fluctuate widely, for instance, R-LSTM can get 1.75 m on ATE in RIDI but 6.32 m on RoNIN. This is partly because R-LSTM and R-TCN have difficulty in fusing multi-sensor data and make errors in orientation prediction. The deeper CNN-based network layers and residual connections make the performance of R-ResNet generally outperforms R-TCN and R-LSTM. However, R-ResNet suffers in complex trajectories from lacking temporal modeling (i.e. RTE on RoNIN-Unseen). Our proposed MSIN goes beyond the representation learned by ResNet and RNNs, and models more effective informative associations. The results illustrated in Table 1 show that data fusion and long-term temporal modeling are beneficial for efficient trajectory prediction.

In addition, we plot the CDF curve in Fig. 6 for a complete comparison. As illustrated in Fig. 6(c) and Fig. 6(d), the steep trend of the curve for the MSIN clearly indicates that our model is superior to that of the other baseline models in all trajectories. The ATE of MSIN finally reaches 11.2m on the RoNIN-Unseen, which is much shorter than other models. The RTE result of MSIN is even less than 10m which is an unprecedented achievement.

More specifically, we visualized a trajectory a051_3 of the RoNIN-Unseen in Fig. 7. There are significant drifts and jitters in the fitting process of R-LSTM and R-TCN observed in Fig. 7(c) and Fig. 7(d). The lack of correction for multi-sensor information results in a large amount of jitter in model predictions, which aligns with our previous conclusion on the performance of these two models. Figure 7(a) and Fig. 7(b) depict that the overall results of the MSIN and R-ResNet trajectory fit are similar, but the R-ResNet drift seriously in the upper part of the trajectory. The drawback of R-RetNet is that only local relationships are learned, while features cannot be preserved for long periods of time and the cumulative error finally leaded to a severe drift. For the upper part of the

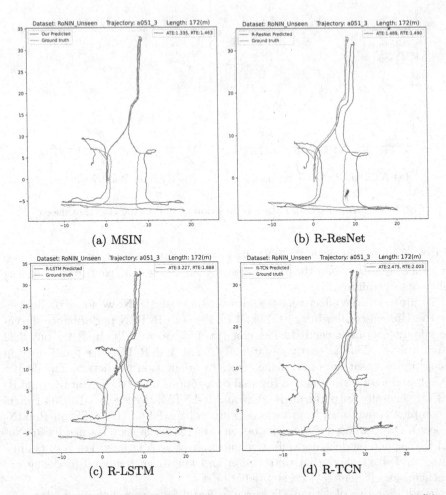

Fig. 7. Visualization of the RoNIN $a051\_3$ trajectory results of the ground truth (blue), our model, and other three baseline models (orange). (Color figure online)

trajectory, R-LSTM and R-TCN both surpass R-ResNet due to the preservation of the long-term features. As for ATE and RTE of four models in $a051_3$, MSIN has only accumulated 1.34m and 1.46m respectively for 172m, which are much lower than others, and the result confirms the superiority of MSIN for IMU data-based inertial navigation.

To demonstrate the stability of different datasets, we conduct the same experiments on RIDI-Unseen and visualized *shali_bag*1. The trajectories of each model are shown in Fig. 8. The results also show that MSIN outperforms others and only accumulated 1.64 m and 1.01 m in ATE and RTE respectively. In summary, the performance gains of the MSIN are remarkable thanks to the combination of spatial features and global temporal features by MSIN. Furthermore, the results

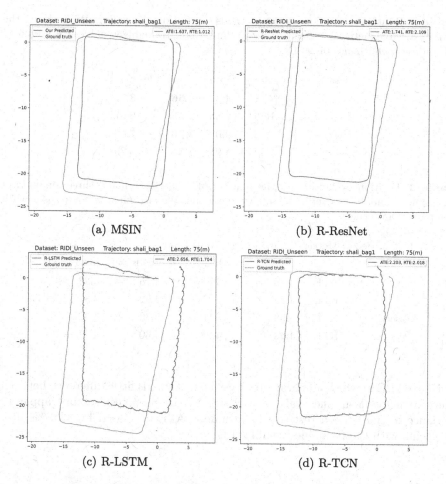

Fig. 8. Visualization of the RIDI *shali_bag*1 trajectory results of the ground truth (blue), our model, and other three baseline models (orange). (Color figure online)

in the RoNIN-Unseen and RIDI-Unseen demonstrate the generalization capability of MSIN.

5.5   Ablation

In this section, we further investigate the impact of each component in our model. The position encoding of each token in the IMU sequence significantly impacts the performance of our model. Therefore, we compare the different ways of PE by ATE and RTE in RoNIN. As Table 2 shows, the ATE of MSIN without PE achieves 4.43m and 9.03m in the Seen and Unseen test sets, which are much larger than MSIN with PE. This finding strongly suggests that MSIN requires location information to target learning of global dependencies of IMU sequences. The performance of MSIN with trainable variables PE generally outperforms the one

Table 2. Position evaluation on the methods of positional encoding. We compare the impact of three methods of positional encoding: None, trainable variables PE and fixed PE on the performance of MSIN.

Test subjects	Metric	None	Learnable	Fixed
Seen	ATE	4.34	**3.56**	3.71
	RTE	2.83	**2.65**	2.67
Unseen	ATE	9.03	**5.26**	7.86
	RTE	4.91	**4.28**	5.29

Table 3. Position evaluation on the number of heads. We choose three amounts of heads: 4, 8, and 16 as candidates for the best one. We also make a comparison with the pure Transformer.

Test subjects	Metric	Transformer	Multi-head		
			heads = 4	heads = 8	heads = 16
Seen	ATE	5.36	3.56	3.75	5.23
	RTE	4.29	2.65	2.62	3.37
Unseen	ATE	5.58	5.26	5.33	5.42
	RTE	4.43	4.28	4.30	4.31

with fixed PE in which all parameters are fixed. There is little difference between the two methods in the evaluation results of the Seen test set but apparent variance in the Unseen test set, for example, ATE decreased by 33.07% after replacing with the trainable variables PE.

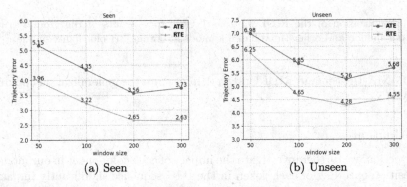

(a) Seen   (b) Unseen

Fig. 9. Trajectory Error of MSIN with different window size in RoNIN-Seen and RoNIN-Unseen

Here we focus on the impact of window size on performance. If the window is too small, the spatial feature extraction may not take advantage of the convolution operation, and the later encoder cannot fully capture the dependencies among tokens due to only a small number of tokens. On the contrary, large windows increase the training cost as well as lead to weight imbalance. As Fig. 9 depicts that ATE and RTE generally continue to decline as window size increases. More specifically, there is a slight performance improvement in RTE when the window size increases to 300 but also increase the ATE. Consequently, the best performance is obtained for our model when the window size is 200.

Finally, We investigate the effect of the number of heads in the multi-head attention mechanism and make a comparison with the pure Transformer. As shown in Table 3, The most suitable number of heads is 4, while all multi-head cases surpass the pure Transformer in both the Seen and Unseen test sets. These ablation experiments verify that each of the modules we designed is effective.

MSIN is designed to be computationally lightweight, minimizing the demand for computational resources during training. It can be trained efficiently with flexible training requirements and parameterization settings. Additionally, our future plans involve optimizing the model further and adapting it for mobile devices, ensuring its seamless performance on resource-constrained platforms.

6 Conclusion

This paper studies the problem of inertial navigation with IMU sequences. Unlikely previous works, we focus on the combination of the long-term temporal (to reduce the overall trajetory drifts) and local spatial (to avoid jitters) features according to the findings of the analysis of IMU data. And we proposed the MSIN base on the CNN and Transformer for the purposes of the extraction of local spatial features and integration of long-term dependencies. For inertial navigation, we designed a Velocity Module to regress the 2D velocity vector and then located the position by introducing the accumulation of the amount of displacement change between windows. Further, the extensive evaluation of MSIN has been conducted on two public datasets to demonstrate the generalization of MSIN. Our experiments show that the proposed MSIN outperforms other competing models in ATE and RTE.

References

1. Bai, S., Kolter, J.Z., Koltun, V.: An empirical evaluation of generic convolutional and recurrent networks for sequence modeling. arXiv preprint arXiv:1803.01271 (2018)
2. Brossard, M., Barrau, A., Bonnabel, S.: Ai-imu dead-reckoning. IEEE Trans. Intell. Veh. 5(4), 585–595 (2020)
3. Brown, T., et al.: Language models are few-shot learners. Adv. Neural. Inf. Process. Syst. 33, 1877–1901 (2020)
4. Chen, C., Lu, X., Markham, A., Trigoni, N.: Ionet: learning to cure the curse of drift in inertial odometry. Proc. AAAI Conf. Artif. Intell. 32, 6468–6476 (2018)
5. Chen, C., et al.: Motiontransformer: transferring neural inertial tracking between domains. Proc. AAAI Conf. Artif. Intell. 33, 8009–8016 (2019)

6. Cummins, C., Orr, R., O'Connor, H., West, C.: Global positioning systems (GPS) and microtechnology sensors in team sports: a systematic review. Sports Med. **43**, 1025–1042 (2013)
7. Devlin, J., Chang, M.W., Lee, K., Toutanova, K.: Bert: pre-training of deep bidirectional transformers for language understanding. arXiv preprint arXiv:1810.04805 (2018)
8. Dosovitskiy, A., et al.: An image is worth 16x16 words: transformers for image recognition at scale. arXiv preprint arXiv:2010.11929 (2020)
9. Einicke, G.A., White, L.B.: Robust extended kalman filtering. IEEE Trans. Signal Process. **47**(9), 2596–2599 (1999)
10. Gao, Z., Li, Q., Li, C., Liu, N.: Iekf-swcs method for pedestrian self-navigation and location. J. Syst. Simulat. **27**(9), 1944–1950 (2015)
11. Guo, H., Uradziński, M., Yin, H., Yu, M.: Indoor positioning based on foot-mounted imu. Bull. Polish Acad. Sci. Tech. Sci. **63**(3), 629–634 (2015)
12. Han, C., Zhang, L., Tang, Y., Huang, W., Min, F., He, J.: Human activity recognition using wearable sensors by heterogeneous convolutional neural networks. Expert Syst. Appl. **198**, 116764 (2022)
13. He, K., Zhang, X., Ren, S., Sun, J.: Deep residual learning for image recognition. In: Proceedings of the IEEE Conference on Computer Vision and Pattern Recognition, pp. 770–778 (2016)
14. Herath, S., Yan, H., Furukawa, Y.: Ronin: robust neural inertial navigation in the wild: benchmark, evaluations, and new methods. In: 2020 IEEE International Conference on Robotics and Automation (ICRA), pp. 3146–3152. IEEE (2020)
15. Hochreiter, S., Schmidhuber, J.: Long short-term memory. Neural Comput. **9**(8), 1735–1780 (1997)
16. Ilyas, M., Cho, K., Baeg, S.H., Park, S.: Drift reduction in pedestrian navigation system by exploiting motion constraints and magnetic field. Sensors **16**(9), 1455 (2016)
17. Jiang, W., Yin, Z.: Human activity recognition using wearable sensors by deep convolutional neural networks. In: Proceedings of the 23rd ACM International Conference on Multimedia, pp. 1307–1310 (2015)
18. Jiang, Y., Li, Z., Wang, J.: Ptrack: enhancing the applicability of pedestrian tracking with wearables. IEEE Trans. Mob. Comput. **18**(2), 431–443 (2018)
19. Joshi, M., Chen, D., Liu, Y., Weld, D.S., Zettlemoyer, L., Levy, O.: Spanbert: improving pre-training by representing and predicting spans. Trans. Assoc. Comput. Linguist. **8**, 64–77 (2020)
20. Kingma, D.P., Ba, J.: Adam: a method for stochastic optimization. arXiv preprint arXiv:1412.6980 (2014)
21. Levi, R.W., Judd, T.: Dead reckoning navigational system using accelerometer to measure foot impacts. US Patent 5,583,776 (1996)
22. Li, W., Liu, D., Chen, K., Li, K., Qi, H.: Hone: mitigating stragglers in distributed stream processing with tuple scheduling. IEEE Trans. Parall. Distrib. Syst. **32**(8), 2021–2034 (2021)
23. Li, W., et al.: Efficient coflow transmission for distributed stream processing. In: IEEE Conference on Computer Communications (IEEE INFOCOM 2020), pp. 1319–1328. IEEE (2020)
24. Liu, H., Li, Q.: 12-dimensional zero velocity state updating intelligent algorithm for pedestrian dead reckoning. J. Syst. Simulat. **30**(11), 4387 (2012)
25. Liu, W., et al.: Tlio: tight learned inertial odometry. IEEE Robot. Automat. Lett. **5**(4), 5653–5660 (2020)

26. Liu, Y., Li, Z., Liu, Z., Wu, K.: Real-time arm skeleton tracking and gesture inference tolerant to missing wearable sensors. In: Proceedings of the 17th Annual International Conference on Mobile Systems, Applications, and Services, pp. 287–299 (2019)
27. Nilsson, J.O., Skog, I., Händel, P., Hari, K.: Foot-mounted ins for everybody-an open-source embedded implementation. In: Proceedings of the 2012 IEEE/ION Position, Location and Navigation Symposium, pp. 140–145. IEEE (2012)
28. Pathak, D., Agrawal, P., Efros, A.A., Darrell, T.: Curiosity-driven exploration by self-supervised prediction. In: International Conference on Machine Learning, pp. 2778–2787. PMLR (2017)
29. Rao, B., Kazemi, E., Ding, Y., Shila, D.M., Tucker, F.M., Wang, L.: CTIN: robust contextual transformer network for inertial navigation. Proc. AAAI Conf. Artif. Intell. **36**, 5413–5421 (2022)
30. Saha, S.S., Sandha, S.S., Garcia, L.A., Srivastava, M.: Tinyodom: hardware-aware efficient neural inertial navigation. Proc. ACM Interact. Mobile Wearable Ubiquit. Technol. **6**(2), 1–32 (2022)
31. Shoaib, M., Bosch, S., Incel, O.D., Scholten, H., Havinga, P.J.: Fusion of smartphone motion sensors for physical activity recognition. Sensors **14**(6), 10146–10176 (2014)
32. Srivastava, N., Hinton, G., Krizhevsky, A., Sutskever, I., Salakhutdinov, R.: Dropout: a simple way to prevent neural networks from overfitting. J. Mach. Learn. Res. **15**(1), 1929–1958 (2014)
33. Vaswani, A., et al.: Attention is all you need. Adv. Neural. Inf. Process. Syst. **30**, 6000–6010 (2017)
34. Woodman, O.J.: An Introduction to Inertial Navigation. University of Cambridge, Computer Laboratory, Tech. Rep. (2007)
35. Xu, H., Zhou, P., Tan, R., Li, M., Shen, G.: Limu-bert: unleashing the potential of unlabeled data for IMU sensing applications. In: Proceedings of the 19th ACM Conference on Embedded Networked Sensor Systems, pp. 220–233 (2021)
36. Yan, H., Shan, Q., Furukawa, Y.: RIDI: robust IMU double integration. In: Ferrari, V., Hebert, M., Sminchisescu, C., Weiss, Y. (eds.) ECCV 2018. LNCS, vol. 11217, pp. 641–656. Springer, Cham (2018). https://doi.org/10.1007/978-3-030-01261-8_38
37. Yang, C., Shao, H.R.: Wifi-based indoor positioning. IEEE Commun. Mag. **53**(3), 150–157 (2015)
38. Yang, S., Quan, Z., Nie, M., Yang, W.: Transpose: keypoint localization via transformer. In: Proceedings of the IEEE/CVF International Conference on Computer Vision, pp. 11802–11812 (2021)
39. Yao, S., Hu, S., Zhao, Y., Zhang, A., Abdelzaher, T.: Deepsense: a unified deep learning framework for time-series mobile sensing data processing. In: Proceedings of the 26th International Conference on World Wide Web, pp. 351–360 (2017)

Absorb: Deadlock Resolution for 2.5D Modular Chiplet Based Systems

Yi Yang[✉], Tiejun Li, Yi Dai, Bo Wang, Sheng Ma, and Yanqiang Sun[✉]

National University of Defense Technology, Changsha, China
{yang_yi,tjli,daiyi,bowang,masheng,yq_sun}@nudt.edu.cn

Abstract. With Moore's Law slowing down, the development of SoCs has encountered a bottleneck. Integrating more functional units and larger on-chip storage leads to a dramatic increase in chip area, resulting in lower chip yields and higher costs. Most researches and industry products began to seek to use advanced connection and packaging technologies to decompose the raw chip into multiple smaller, higher yield, and more cost-effective chiplets, and then packet them. Interposer-based 2.5D integration, as an emerging packaging technology, is widely used in chiplet-based systems. However, even if both the interposer and chiplets are deadlock-free, deadlock dependency cycles across them may still occur after integration. To address these problems, this paper proposes a deadlock resolution called *Absorb* for 2.5D integrated chiplet systems, which is different from deadlock avoidance and deadlock recovery. By regularly absorbing inter-chiplet packets, global deadlock freedom is achieved, and no extra VCs are for deadlock resolution. Our proposed Absorb maintains the modularity of each chiplet and imposes no restrictions on the routing algorithm. Our evaluations show that compared with the previously proposed deadlock-free designs in 2.5D-chiplet systems, Absorb provides an average performance improvement of about 7.5%, and the area overhead is less than 6%.

Keywords: Network-on-Chip · deadlock · 2.5D-chiplet system

1 Introduction

In recent years, the cost of systems-on-chip (SoCs) has increased significantly as chip functions have become more complex. First of all, completing the design of all functional units of the chip under the most advanced technology greatly increases the design cost; secondly, more functional units and larger on-chip storage will lead to an increase in chip area, which in turn will result in a decrease in chip yield. These factors lead to higher manufacturing costs for chip. The poor scalability, low yield, and high manufacturing cost of traditional 2D SoCs are known as motivations for 2.5D-chiplet systems. This concept decomposes a traditional monolithic SoC into several smaller chiplets, each of which is cheaper to develop, easier to reuse across multiple products, and implemented using the

© The Author(s), under exclusive license to Springer Nature Singapore Pte Ltd. 2024
Z. Tari et al. (Eds.): ICA3PP 2023, LNCS 14487, pp. 474–487, 2024.
https://doi.org/10.1007/978-981-97-0834-5_27

most suitable technology. Decomposing the entire SoC into multiple chiplets for independent production and then integrating them together brings new challenges. The design challenge for the network-on-chip (NoCs) is mainly reflected in the new deadlock problem caused by inter-chiplet traffic [12]. As shown in Fig. 1, even if the traditional NoC deadlock-free method is adopted to achieve no deadlock in interposer and each chiplet, the inter-chiplet deadlock may still occur after integrating each independently designed chiplet through the interposer. We use Fig. 1 as the baseline architecture, but the approach proposed in this paper can be applied to all 2.5D-chiplet systems.

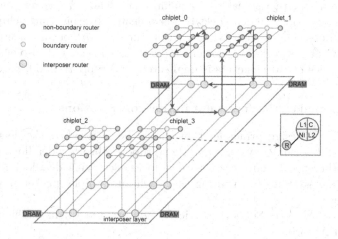

Fig. 1. Baseline architecture with inter-chiplet deadlock crosses two chiplets.

A deadlock occurs when there is a cyclic resource dependence in the network. The existence of deadlock will seriously endanger the correctness of the system on chip. NoC, the most classic interconnect architecture for on-chip communication systems in SoCs, already has many methods to deal with interconnect deadlock [3–9]. Although these methods achieve deadlock free, they are not suitable for the 2.5D-chiplet architecture.

With the widespread use of 2.5D-chiplet technology, for the inter-chiplet deadlock, some methods have been proposed, and they can be divided into two categories: deadlock avoidance and deadlock recovery. The main methods of deadlock avoidance are: MTR places turn restrictions on the boundary router to build a global acyclic CDG [12]; RC separates inter-chiplet and intra-chiplet traffic by adding RC_buffer (RCB) to the boundary router and restricting inter-chiplet traffic to be injected only after obtaining RCB space through the permission network [13]; DeFT uses VNet allocation to isolate chiplet to interposer and interposer to chiplet traffic, thus avoiding the creation of inter-chiplet dependent cycle [14]. The main methods of deadlock recovery is that UPP determines the existence of a deadlock by using a simple timeout mechanism, and then adds an additional synchronization protocol to coordinate the routers to complete the deadlock recovery operation [15].

Existing deadlock-free designs for 2.5D-chiplets all suffer from the problems of limiting network performance and high overhead: Deadlock avoidance method (MTR, RC, DeFT) while limit the packet forwarding, thus affecting the performance of most packet transports; deadlock recovery method (UPP) requires additional deadlock detection and recovery mechanisms, and the false positives and lags of deadlock detection can all have a significant impact on network performance.

In summary, for the deadlock problem in Fig. 1, we propose a solution - Absorb, which is neither a deadlock avoidance nor deadlock recovery mechanism. Absorb removes potential deadlocks by periodically absorbing inter-chiplet packets without detecting or identifying whether the deadlock occurs or not. Absorb ensures that the network does not crash due to deadlocks, but also ensures that the performance of most packets is unaffected, and supports modular design for each chiplet.

This paper makes the following major contributions:

- We delve the design principles that should be satisfied by the deadlock-free design in 2.5D-chiplet systems and then propose Absorb, a deadlock resolution that neither avoids nor reacts to deadlocks, achieving deadlock-freedom in 2.5D-chiplet systems by ensuring that each inter-chiplet packet can reach its destination.
- We proved the correctness theoretically and provide a lightweight implementation of Absorb.
- We fully evaluate Absorb by using a network simulator and provide system performance results using full system simulations. Moreover, we compare Absorb with the prior solutions in terms of performance and area.

2   Motivation

There are many traditional topology-independence solutions to the NoC deadlock problem. These solutions can be applied to irregular topologies formed by chiplet architectures, but they break the modularity of the chiplet design. In recent years, many deadlock-free approaches have been proposed to support the modular design of 2.5D-chiplet architectures. However, these approaches cost too much for the very rare event of deadlock. Therefore, in this section, we present the design motivation for this paper from these two aspects.

The main reason why traditional NoC methods cannot be directly applied in 2.5D-chiplets is the need for global information. Instead, chiplet technology benefits from the decomposition of a single SoC into different, independently designed and implemented chiplets that support integration using advanced packaging techniques. So chiplets should support design in a modular way without overall system information so that they can be reused in different SoCs. The main requirements that a modular design should specifically meet are topological modularity, routing modularity, and VC modularity [15].

Topological Modularity: The topology adopted by each chiplet can be designed independently without considering the topology of other modules in the system.

Routing Modularity: The routing algorithm adopted for each chiplet's internal routing can be designed independently.

VC Modularity: Each chiplet can independently design the number of VCs needed for each port based on its overhead budget, load size, etc., so it needs to ensure that there is no global minimum limit on the number of VCs.

In a real system, the formation of a cyclic buffer dependency which causes a real deadlock in on-chip or off-chip networks is very rare, and it is not cost-effective to spend a high price to solve it. To demonstrate that deadlock is a very rare event, DRAIN [11] has conducted experiments on the probability of deadlock by running PARSEC with fully adaptive routing on an 8 × 8 mesh topology, the results are shown in Fig. 2. It can be seen that deadlock does not occur until the removed link is raised to 16 under 4VCs, and in the absence of link removal, no deadlock occurs for all loads in both 1VC and 4VCs configurations.

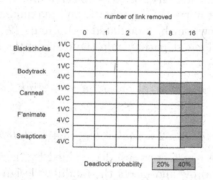

Fig. 2. Probability of deadlock under PARSEC workloads.

Although the real deadlock is very rare in practice, once a deadlock occurs, it will have a huge impact on the entire system. Therefore, in order to ensure the correctness of the system, the deadlock problem must be solved. In summary, our design goal is to solve the rare deadlock problem with low overhead, and support each chiplet to be designed in a modular way.

3 Absorb

In this section, we will introduce the new solution proposed in this paper, Absorb, from both theoretical and implementation aspects.

3.1   Theory

For the formation of the inter-chiplet deadlock, it is easy to see that there must be a packet forwarded upward in this deadlock cycle, and the destination of this packet must be located in the reachable range of this boundary router. This deadlock can be solved by ejecting the upward packets within the deadlock loop. To be specific, the core of the Absorb is that the boundary routers periodically absorb upward packets which could be the blocked packets in the deadly loop, and delivers them along its path to the destination NI.

Definitions:

base_vc: The Absorb operation is performed only within this VC.

absorb_pkt: Packets that need to be submitted to the destination NI for the Absorb operation.

back_pkt: Packets that are backtracked in the Absorb operation.

pre_absorb: The period needed to pause the VA before the Absorb operation to stabilize the in-flight traffic.

absorb_threshold: Absorb operation occurs every absorb_threshold cycles.

absorb_window: The number of cycles required for the Absorb operation.

pre_absorb, absorb_threshold, and absorb_window are all pre-configured for the router. The calculation formulas of the parameters are given below. As presented by formula (1), the length of pre_absorb is relevant with the router latency, link latency, number of ports, VCs per port, and maximum packet length. The length of absorb_window can be calculated by formula (2) with the maximum packet length and the longest path within the reach of the boundary router.

$$pre\_absorb = ((packet\_length - 1) \times router\_latency + link\_latency) \times VCs \times ports \tag{1}$$

$$absorb\_window = 2 \times packet\_length + path\_length - 1 \tag{2}$$

The values of these parameters are determined by the integration architecture, which are configurable and meets the modular design requirements.

Assumption 1: VCT flow control is adopted, hence each VC has to buffer the whole packet.

Note: Absorb requires the boundary router to periodically absorb the packets forwarded upward from the interposer to the destined chiplet. In order not to generate truncated packets, the VC needs to have enough space to buffer the whole packets and use VCT flow control. Under WH flow control, there is no guarantee that the packet is completely buffered in the VC after the pre_absorb stage. The next successive delivery and backtrack operations will generate packet truncation. To support Absorb with wormhole routers, this requires adding package truncation support, similar to prior works in deflection routers.

Assumption 2: The destination of the packet forwarded by boundary router to the chiplet is in the reachable range of this boundary router, and there is no intersection of the chiplet router corresponding to each boundary router.

Note: In a modular chiplet-based system, the routing algorithm needs to handle the transmission of three types of packets: internal packets from the chiplet and the interposer, packets transmitted from the chiplet to the interposer, and packets transmitted from the interposer to the chiplet. Absorb has no restrictions on the first two types of packets. For the third type of packets, Absorb requires that the chiplet router corresponding to each boundary router has no intersection, which ensures that the links do not conflict during the Absorb operation.

Assumption 3: absorb_pkt can be ejected to the destination NI.

Note: In NoC, when the packet is routed to its destination, it will be popped to the NI ejection queue. In extreme cases, when the traffic is extremely intense and exceeds the processing capability of the processing element (PE, e.g., core, cache, directory), the packets will be blocked at the destination NI, thereby the ejection queue could be full. Absorb uses the pre_absorb stage to suspend the router from forwarding new packets. In this stage, NI does not receive new packets, but only submits to the PE the packets already buffered in the ejection queue, and at the end of the pre_absorb stage, NI will have enough space to store the absorb_pkt.

Assumption 4: All routers are bi-directional links, and the link inputs can be buffered to any input buffer.

Note: To achieve the delivery-backtracking operation, Absorb requires the router to be a bidirectional link and the emptied buffer to be used to buffer the back_pkt, which requires the input of the link can be store in any input buffer. In this regard, we presents an adoptable router design that uses a bus structure to control back_pkt to be buffered in the corresponding input buffer.

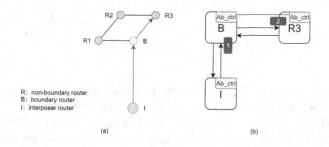

(a)   (b)

Fig. 3. Partial expanded view of the inter-chiplet deadlock.

Absorb Operation: Figure 3(a) shows a part of the structure in Fig. 1. We expand the deadlock resource dependency to get Fig. 3(b). The packet in the input buffer of B's Up-inport requests the input buffer of R3's Horizontal-inport, and the input buffer in R3 is occupied. When the absorb_threshold is reached, pre_absorb is performed to stop the VC allocation of routers in the network and stabilize in-flight-flit, so at the end of pre_absorb, all packets in the network are fully buffered in the VC. Next, enter absorb_window, assuming that the P1's path is B-R3-R2, and all the base_vc required along the path have been occupied by the existing packet, denoted by P1, P2 and P3 respectively, the head of the packet is denoted by H1, H2 and H3, and the body of the packet is denoted by B1, B2 and B3, then the whole process is shown in Fig. 4.

Step1: Ab_ctrl of B reads the routing information of P1 in the Up-inport and enables its next-hop router R3. B delivers H1 to R3, and R3 transmits H2 back to B via the horizontal inport of the enabled direction, as shown in Fig. 4(b).

Step2: After R3 receives H1, its Ab_ctrl determines that its next hop is R2, enables R2, R3 delivers H1 to R2, and R2 returns H3 to R3, while B1 and B2 are exchanged between R3 and B, as shown in Fig. 4(c).

Step3: After R2 receives H1, it determines that P1 has reached the destination, and can eject it to NI.

Step4: After the ejection of P1 is completed, P2 and P3 will return to their original positions at the same time.

Step5: Restart the VC allocation of routers in the network and exit absorb_window.

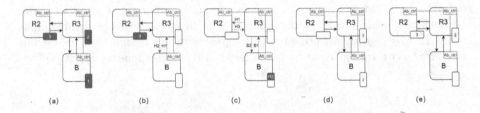

Fig. 4. Example of Absorb operation.

The above delivery-backtracking-return process only operates within base_vc, and the inter-chiplet packets can be transmitted without deadlock by periodically performing Absorb operation in the network. In addition, since the interposer and chiplets are deadlock-free and livelock-free, the global deadlock freedom and livelock freedom can be achieved eventually.

3.2   Modifications to the Chiplet Routers

In order to implement the periodic Absorb operation, the chiplet router needs to be adjusted. Backtracking is performed by in-place packet switching between buffers of neighboring routers, Fig. 5 gives the conceptual idea [10]. Unlike software switching, which requires additional temporary storage, in-place switching is conceptually identical to circular shift registers in hardware. Intuitively,

Absorb allows any blocked packet to be guaranteed to advance to its destination through the switching mechanism, regardless of the network congestion.

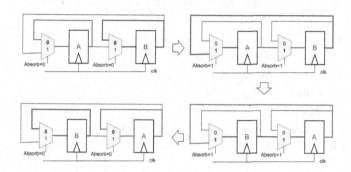

Fig. 5. Inplace packet switches.

Except the in-place packet switching unit, we need to add a control unit called Ab_ctrl. Ab_ctrl mainly contains three timers, which are pre_absorb, absorb_window and absorb_threshold. Ab_ctrl needs to process signals from neighboring routers and collaborates with neighboring routers. Moreover, Absorb router needs to utilize the bus structure to realize that input packets from any inport can be buffered to any input buffer. By adding the structures mentioned above, Absorb operation can be triggered periodly. When the absorb_threshold is reached, the chiplet router enter the pre_absorb and stop VA. After pre_absorb, the chiplet routers enter the absorb_window and perform the Absorb operation as shown in Fig. 4. In this stage, chiplet routers use in-place packet switching and bus structures achieve the delivery-backtracking operation. When the absorb_window ends, Ab_ctrl restarts VA and the network return to the normal state until the next absorb_threshold reached.

4   Evaluation

In this section, a comprehensive evaluation of Absorb is performed using the Gem5 [2] simulator with the Garnet [1] embedded in it. The chiplet architecture used is consistent with the baseline shown in Fig. 1. In addition, Absorb is compared with two other deadlock-free approaches for chiplet-based modular systems, specifically RC and DeFT. The routing algorithm used in both chiplets and interposer is XY routing. The binding of non-boundary router and boundary router in Absorb follows the principle of shortest distance. RC's permission network also follows the principle of shortest distance, RCB set to 1-packet-size, other configuration parameters refer to Table 1.

Table 1. Simulation configurations

Full system simulation configurations	
Cores	x86 in-order cores,1 GHZ
L1 I cache	32 KB per core
L1 D cache	64 KB per core
L2 cache	2 MB per core
Cache coherence	MOESI
Real applications	SPLASH-2X
Network configurations	
Router/link latency	1 cycle
Link width	128 bits
Flow control	VCT
Packet size	data packet: 5 flits, control packet: 1 flit
Absorb_threshold	1000, 5000, 10000
Topology	Baseline architecture: 1 4 * 4 mesh interposer, 4 4 * 4 mesh chiplets
VCs	3 VNets, with 1 or 2 or 4 VCs per VNet, each VC has 5 flit-sized buffers
Synthetic simulation	uniform_random, shuffle, transpose, bit_complement, bit_reverse, bit_rotation

4.1   Synthetic Traffic

Figure 6 shows the latency comparisons of Absorb, RC, and DeFT under six synthetic traffic patterns with 2VCs per VNet. The bottleneck of RC is mainly in the atomicity allocation of RCB. When a chiplet router has to inject an inter-chiplet packet, it needs to send an RCB request to its corresponding boundary router. If the RCB has enough free buffer, the packet can be injected into the network, otherwise the packet will wait at the NI, blocking subsequent packets from being injected into the network, until an RCB answer is obtained. This causes high queuing delay, thus resulting in high overall packet latency; DeFT requires a minimum number of 2VCs to achieve deadlock freedom, because DeFT physically isolates the formation of deadlock dependency cycle by restricting the VNet allocation of inter-chiplet packets. The VNet allocation restriction makes it more difficult to establish the forwarding conditions for inter-chiplet packets, thus blocking them in the VC and affecting other packet forwarding.

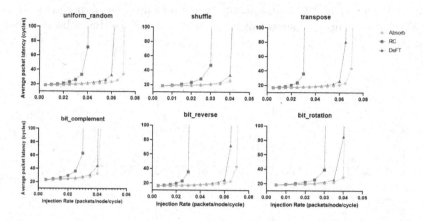

Fig. 6. Latency comparison under six synthetic traffic patterns as the injection rate increases by 0.005. Each VNet has 2 VCs. absorb_threshold = 5000.

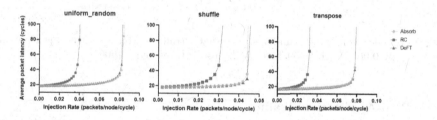

Fig. 7. Latency comparison under three synthetic traffic patterns as the injection rate increases by 0.002. Each VNet has 4 VCs. absorb_threshold = 10000.

Figure 7 shows the results of 4VCs per VNet under three traffic patterns: uniform_random, shuffle, and transpose. The trends of latency in the rest of traffic patterns are similar. With the increase of VCs per VNet, the forwarding condition of inter-chiplet packets in the DeFT is easier to be satisfied, so its latency saturation point moves back at lighter injection rate. The pre_absorb period of the Absorb is performed for each VC per inport, so the overhead of each pre_absorb increases. However, the probability of deadlock is much lower under 4VCs compared to 1VC, and deadlock becomes a much less probable event. Therefore, Absorb can increase the absorb_threshold, so that the overhead of pre_absorb is relatively lower, and the full performance runtime (time = absorb_threshold-pre_absorb-absorb_window) in the network remains longer, thus a comparable latency performance to DeFT is also achieved. RC is limited by the RCB size, so the latency performance does not improve as the number of VCs increases.

To fit better with the increased VC number, RC needs to increase the RCB size, and the NI separation injection queue need to be used, which causes huge hardware overhead. Therefore, in order to adapt the RCB set to 1-packet-size, also to ensure fairness and reflect the advantages of RC and Absorb without the minimum number of VC requirements, we supplement the experiment of 1VC per VNet, the results are shown in Fig. 8, and you can see that Absorb is better than RC.

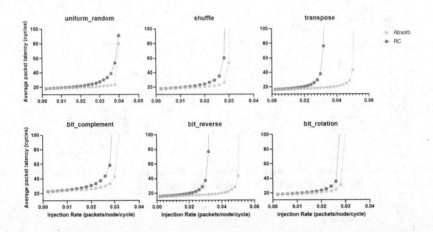

Fig. 8. Latency comparison under six synthetic traffic patterns as the injection rate increases by 0.002. Each VNet has 1 VC. absorb_threshold = 1000.

4.2   Real-Application Traffic

Figure 9(a) shows the normalized runtime results evaluated by using Gem5 simulator to run SPLASH-2X [16]. The results are normalized to those of RC and the configurations are shown in Table 1. By observing the results, we can see that Absorb and DeFT have significantly improved compared to RC. As VCs per VNet is increased to 4, the bottleneck of DeFT will be relatively weaker, so the performance of DeFT is also excellent. But the runtime performance of Absorb is equal to DeFT under most loads or slightly better, with significant improvements under raytrace and volrend. In summary, Absorb has an average 13% improvement in runtime compared to RC and an average 2% improvement compared to DeFT.

4.3   Area and Power

The results of the normalized link power comparison are shown in Fig. 9(b). The extra link power overhead of RC mainly comes from the request and answer of permission, DeFT has no explicit extra link power overhead, and Absorb mainly refers to the back_pkt backtrack and recovery operations. From the results, we can observe that DeFT and Absorb have optimized link power overhead compared to RC, and Absorb is almost the same as DeFT, which shows that the link power overhead of Absorb is very considerable.

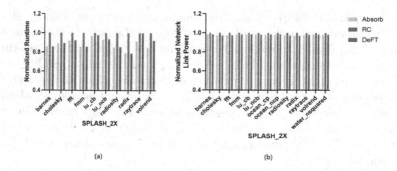

Fig. 9. (a) Normalized runtime comparison. (b) Normalized link power comparison. Full system simulations with 4 VCs per VNet. absorb_threshold = 10000.

Figure 10 shows the comparison of the hardware overhead where the network frequency is 1 GHz. Synthesizing using the Synopsys Design Compiler. The baseline router has an area of 121206 μm^2 with 5 ports. The hardware overhead on boundary routers and non-boundary routers in RC are considered respectively which denoted by RC-bound and RC-nonbound. The results in Fig. 10 are normalized to the baseline router. RC-bound has the largest area overhead due to the RCB. RC-nonbound needs to add additional control logic to send RCB requests and accept RCB answers. The overhead of DeFT includes the logic for VN-assignment algorithm and look-up tables to store data for fault-tolerant VL selection. Absorb imposes less than 6% hardware overhead and the overhead comes due to the bus structure and timers. In summary, Absorb achieves performance and flexibility benefits with a suitable hardware overhead.

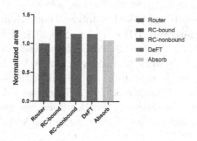

Fig. 10. Hardware overhead comparison.

5   Conclusion

This paper proposes a novel solution Absorb to the inter-chiplet deadlock in 2.5D-chiplet systems. This approach does not detect or identify a deadlock might occur, and just periodically absorbs the inter-chiplet packets to solve the potential deadlock. The Absorb operation is specifically implemented as three-stage using delivery-backtracking-return. Each boundary router periodically absorbs the upward forwarding packet and submits it along the path to its destination, achieving global deadlock freedom by ensuring the arrival of the inter-chiplet packet. Absorb is the first deadlock-free design on a 2.5D-chiplet modular architecture not based on avoidance or recovery mechanisms, which allows packets to be inserted and submitted to the destination without adding additional buffer. Our simulation results show that Absorb achieves an average runtime improvement of 7.5% compared to other state-of-the-art designs, and its area overhead is less than 6%. These results demonstrate the promising application of Absorb in 2.5D-chiplet systems.

References

1. Agarwal, N., Krishna, T., Peh, L.S., et al.: GARNET: a detailed on-chip network model inside a full-system simulator. In: 2009 IEEE International Symposium on Performance Analysis of Systems and Software, pp. 33–42. IEEE (2009)
2. Binkert, N., Beckmann, B., Black, G., et al.: The gem5 simulator. ACM SIGARCH Comput. Archit. News **39**(2), 1–7 (2011)
3. Dally, W.J., Seitz, C.L.: Deadlock-free message routing in multiprocessor interconnection networks. IEEE Trans. Comput. **C-36**(5), 547–553 (1988)
4. Duato, J.: A new theory of deadlock-free adaptive routing in wormhole networks. IEEE Trans. Parallel Distrib. Syst. **4**(12), 1320–1331 (1993)
5. Duato, J., Pinkston, T.M.: A general theory for deadlock-free adaptive routing using a mixed set of resources. IEEE Trans. Parallel Distrib. Syst. **12**(12), 1219–1235 (2001)
6. Fallin, C., Craik, C., Mutlu, O.: CHIPPER: a low-complexity bufferless deflection router. In: 2011 IEEE 17th International Symposium on High Performance Computer Architecture, pp. 144–155. IEEE (2011)
7. Ramrakhyani, A., Krishna, T.: Static bubble: a framework for deadlock-free irregular on-chip topologies. In: 2017 IEEE International Symposium on High Performance Computer Architecture (HPCA), pp. 253–264. IEEE (2017)
8. Chen, L., Wang, R., Pinkston, T.M.: Critical bubble scheme: an efficient implementation of globally aware network flow control. In: 2011 IEEE International Parallel & Distributed Processing Symposium, pp. 592–603. IEEE (2011)
9. Ramrakhyani, A., Gratz, P.V., Krishna, T.: Synchronized progress in interconnection networks (SPIN): a new theory for deadlock freedom. In: 2018 ACM/IEEE 45th Annual International Symposium on Computer Architecture (ISCA), pp. 699–711. IEEE (2018)
10. Parasar, M., Jerger, N.E., Gratz, P.V., et al.: SWAP: synchronized weaving of adjacent packets for network deadlock resolution. In: Proceedings of the 52nd Annual IEEE/ACM International Symposium on Microarchitecture, pp. 873–885 (2019)
11. Parasar, M., Farrokhbakht, H., Jerger, N.E., et al.: DRAIN: deadlock removal for arbitrary irregular networks. In: 2020 IEEE International Symposium on High Performance Computer Architecture (HPCA), pp. 447–460. IEEE (2020)
12. Yin, J., Lin, Z., Kayiran, O., et al.: Modular routing design for chiplet-based systems. In: 2018 ACM/IEEE 45th Annual International Symposium on Computer Architecture (ISCA), pp. 726–738. IEEE (2018)

13. Majumder, P., Kim, S., Huang, J., et al.: Remote control: a simple deadlock avoidance scheme for modular systems-on-chip. IEEE Trans. Comput. **70**(11), 1928–1941 (2020)
14. Taheri, E., Pasricha, S., Nikdast, M.: DeFT: a deadlock-free and fault-tolerant routing algorithm for 2.5 D Chiplet Networks. In: 2022 Design, Automation & Test in Europe Conference & Exhibition (DATE), pp. 1047–1052. IEEE (2022)
15. Wu, Y., Wang, L., Wang, X., et al.: Upward packet popup for deadlock freedom in modular chiplet-based systems. In: 2022 IEEE International Symposium on High-Performance Computer Architecture (HPCA), pp. 986–1000. IEEE (2022)
16. Zhan, X., Bao, Y., Bienia, C., et al.: PARSEC3. 0: a multicore benchmark suite with network stacks and SPLASH-2X. ACM SIGARCH Comput. Archit. News **44**(5), 1–16 (2017)

A Multi-server Authentication Scheme Based on Fuzzy Extractor

Wang Cheng[ID], Lin You[ID], and Gengran Hu[✉][ID]

School of Cyberspace, Hangzhou Dianzi University, Hangzhou 310018, China
grhu@hdu.edu.cn

Abstract. Biometrics have been widely used in various authentication occasions, especially in the Internet field. However, the security of biometrics has attracted a lot of attention. Once the biometric template used for authentication is leaked, the user's private data will face a great threat. As a key generation technology, fuzzy extractor can not only solve the problem of biometric template leaking, but can also realize the generation and regeneration of the secret keys. In previous research, we proposed an improved fuzzy extractor scheme (**PrrFE**). In this paper, a practical fuzzy extractor is applied to generate user's Bio-ID for authentication along with a multi-server authentication scheme based on **PrrFE**. In this scheme, the users can use their own biometrics to generate a Bio-Key and then achieve authentication through the generated Bio-ID on more than one server. In our scheme, the biometric data and Bio-Key do not need to be stored anywhere. We have performed security analyses on our scheme. Our two experiments with the database FVC2002 and the NIST 302d have shown that the genuine acceptance rate (GAR) of real users is 99.16% in the optimal case. And it has a acceptable network delay.

Keywords: Biometrics · Multi-Servers · Bio-Key · Fuzzy Extractors · Authentication

1 Introduction

Now more and more people have been enjoying the conveniences brought by various internet application services. We know that the first step to enjoying these applications is to complete users' identity authentications. Currently, the most common method of securing a user's access through internet is by using a password, a physical device, or a biometric such as a fingerprint or facial image [8]. However, the first two methods have some disadvantages. One of these is that a password can be easily forgotten, and the other is that a physical device has to be carried around. Nowadays, biometrics have received widespread attention and been applied for identification to access to various internet systems.

This work was supported by the National Natural Science Foundation of China (No. 61772166) and the Key Program of the Natural Science Foundation of Zhejiang Province of China (No. LZ17F020002).

Zahid et al. [6] proposed a biometric authentication system based on DNA. This system realizes authentication by converting DNA information collected from the human body into DNA sequence IDs and saving them in a safe database. In 2022, Anveshini et al. [1] proposed a scheme to use fingerprint authentication at ATMs. In this method, the user's fingerprint is converted into a unique string and stored in the EC_2 database. Raunak et al. [11] used an Near-Infrared (NIR) camera to capture finger images, and extract finger vein images from them through filtering operations. The user's finger vein image was saved in the database for authentication.

However, due to the non-renewability of biometrics, it means that once a user's biometrics are leaked or stolen, it is difficult to use the biometrics again, which puts forward high requirements for the safe preservation of biometrics. It is vital and full of challenging to securely store the biometric data used in the user identification and authentication system on an internet server [7].

The emergence of fuzzy extractor solves this problem very well. Fuzzy extractor is a biometric key generation scheme, which combines biometric technology and cryptography, and it makes use of both of cryptographical advantage and biometric advantage [14]. Fuzzy extractors can handle biometric uncertainty and generate secure keys.

In 2004, Dodis et al. [5] first proposed a theory called fuzzy extractor, whose goal is to extract strong passwords from some noisy information such as biometrics. The fuzzy extractor is composed of $FE = (Gen, Rep)$ with two algorithms, that is, the generating algorithm (shorted as Gen) and the reproducing algorithm (shorted as Rep). The generating algorithm Gen use an inputted string $\mathbf{w}$ (one sampling of the noisy random sources) to output a string R and a public helper string Pub. Meanwhile, another sampling from the former noise random sources $\mathbf{w}'$ and the stored public helper String Pub will used in the reproducing algorithm Rep. When the $\mathbf{w}$ and $\mathbf{w}'$ are similar enough, the output string R' of Rep is the same as the R of Gen. The regenerable string R can be a secret key and applied to cryptographical algorithms, and the publicly available string Pub will not demonstrate any information about the original input $\mathbf{w}$. However, they only declared this innovation but did not provide an implementation example.

The fuzzy extractor proposed by Dodis had two pivotal algorithms, that is, a stronger extractor and a secure sketch (shorted as SS). The stronger extractor is aimed to generate a uniformly distributed long string as a secure password from the sampled noise information. The function of algorithm secure sketch SS is to correct the error between two samples, which is usually implemented by linear error-correcting codes.

1.1   Related Works

In 2017, Li et al. [9] assumed that the sampled source information was encoded into a one-dimensional biometric feature vector $X = \{x_1, \ldots, x_n\}$. They defined a number line to express vector elements of the biometric information and applied the Chebyshev distance in their fuzzy extractor. Though their fuzzy extractor based identification protocol runs in around 110 ms, but their experiment test were based on their assumed simulated biometric data instead of the biometric data extracted from real biometrics. In 2020, Fuller et al. [3] proposed computational security of fuzzy extractors. Their proposal solved the gaps between known negative results, existential constructions, and polynomial-time constructions. Wen et al. [13] constructed a reusable fuzzy extractor from the Learning With Errors (LWE) assumption. Their scheme does not require a specific input source.

As we can see in Table 1, from our referred works on fuzzy extractors above, the authors does not provide practical constructions or implementations for their fuzzy extractors on some biometrics or other noisy sources [3,13]. While some other works, such as the one presented in [9], they only employed fictitious biometric data for their simulations.

Table 1. Summary of related fuzzy extractors

Scheme	[3]	[13]	[10]	[2]	[4]	[12]	PrrFE [17]
Robustness	✓	✓	×	×	✓	✓	✓
Reusability	✓	✓	×	×	✓	×	✓
Biometrics	×	×	ECG	Fingerprint	×	×	Fingerprint
Implement	×	×	✓	✓	×	✓	✓

1.2   Our Contributions

In our following work, we focus on the gap between fuzzy extractors and some approaches for biometric identification. Our work is dedicated to pushing the instantiation of fuzzy extractors. Fuzzy extractors provide capabilities for privacy preservation of biometrics. Our research will be based on our proposed practical reusable and robust fuzzy extractor (shorted as **PrrFE**) [17]. With its reusability, it is possible to generate multiple independent keys based on one biometrics. Then a multi-server authorization scheme is proposed. In this scheme, one user can generate independent distributed keys on multiple servers by one biometrics, and the biometrics does not need to be stored on any servers.

The subsequent chapters of this paper are scheduled as follows: In Sect. 2, some preliminaries related our work is introduced. In Sect. 3, we show our main construction of proposed authorization scheme. The security analysis of our scheme is provided in Sect. 4. In subsequent section, we will show our experiment simulation based on fingerprints. The conclusion remarks are given in the last section.

2   Preliminaries

In this section, We examine basic essentials of fuzzy extractors and associated theories which will be used in the following discussions. We also provide a description of the notations used in this paper.

In Table 2, we provide some necessary notations to characterize the construction of our proposed schemes.

Table 2. Notions used in this paper.

SS	the secure sketch algorithm
SS-Gen/SS-Rec	the generating or recovery algorithm of SS
Gen/Rep	the generating or reproducing algorithm of fuzzy extractor
B/B'	the biometrics for a key generation or recovery
R	the key generated by the fuzzy extractor
Pub	the public helper string generated by the fuzzy extractor
SE_j	server j
$Cert_j$	server j's certificate
$SEpk_j$	server j's public key/encryption by the public key
$SEsk_j$	server j's secret key/decryption by the secret key
CA_{pk}	verify signature by CA's public key
$\parallel$	connection string
DB_{in}/DB_{out}	database entry or fetch data

2.1   Fuzzy Extractor

Secure Sketch. A secure sketch [8] is a component of a fuzzy extractor, and it includes two algorithms, denoted as SS-Gen and SS-Rec, respectively, described in Fig. 1.

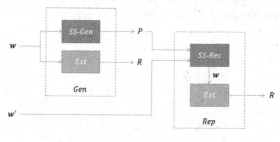

if $dis(\mathbf{w}, \mathbf{w}') \leq t,\ R' = R$

Fig. 1. Construction of a Fuzzy Extractor.

Fuzzy Extractor. There are two main algorithms in fuzzy extractor: a generating algorithm (Gen) and a reproducing algorithm (Rep), which are described in detail as following Fig. 1:

Properties of Fuzzy Extractors. Correctness [3]: The correctness requires that when the distance between the two sampled samples $\mathbf{w}$ and $\mathbf{w}'$ is close enough, that is, when the threshold t is set, if condition $dis(\mathbf{w}, \mathbf{w}') \leq \mathbf{t}$ is satisfied, $R' = R$ will be true. Therefore, we can consider that R can be accurately reproduced.

Security [3]: The condition for security is that R must be uniformly random if the random source has sufficient entropy.

Robustness [16]: $Gen\,(\mathbf{w}) \rightarrow (R, Pub)$, For all $Pub' \neq Pub$, $Rep(\mathbf{w}', Pub') \rightarrow \perp$.

Reusability [15]: For the multiple samples $\mathbf{w}_0, \mathbf{w}_1, \ldots, \mathbf{w}_\rho$ from a noisy source, the Gen algorithm generates corresponding $(r_0, p_0), (r_1, p_1), \ldots, (r_\rho, p_\rho)$. Its security requirement is that for $i \neq j$, $p_i \neq p_j$, $Rep\,(\mathbf{w}_0, p_j) \rightarrow r_j$ are known by the adversary. And among the remaining r_i, there is still some pseudorandomness.

2.2   The Practical Reusable and Robust Fuzzy Extractor (PrrFE)

The **PrrFE** scheme [17] is composed of two algorithms which are called generating algorithm Gen and reproducing algorithm Rep, respectively, where the Gen algorithm is $Gen(\mathbf{B}) \rightarrow (R, Pub)$, and the Rep algorithm is $Rep(\mathbf{B}', Pub) \rightarrow (R)$. Please read [17] for more details.

3   Construction of Multi-server Authentication Scheme Based on PrrFE

We suppose that there is a server sequence $SE = (SE_1, \ldots, SE_s)$, user i's username ID_i and his biometric is $\mathbf{B}_i$. The generation algorithm of the fuzzy extractor is Gen, and the regeneration algorithm is Rep. Through this scheme, users can use multiple samples of the same biometric $\mathbf{B}_i$ to generate independent keys among multiple servers for authentication.

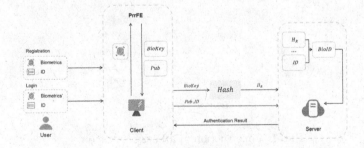

Fig. 2. Our Multi-Server Authentication Scheme

A brief flowchart of our scheme is shown in Fig. 2. The scheme has two stages: user registration and user login.

3.1 User Registration

There are two sub-steps in the user registration phase. Step one is *Bio-Key* generation based on **PrrFE** and user's biometrics. Step two is *Bio-ID* generation based on generated *Bio-Key* and user's identity. Here we simply choose the username as the user identity. In practice, the user ID card or social security number can also be selected. The user registration stage process is shown in Fig. 3.

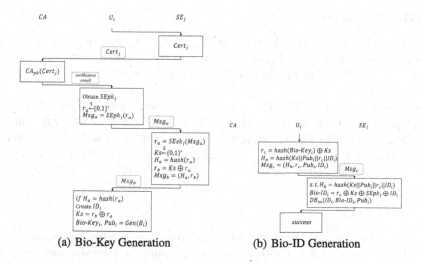

(a) Bio-Key Generation   (b) Bio-ID Generation

Fig. 3. Process of User registration

Bio-Key Generation

1. **Server initialization**: User i initiates a registration request to some server SE_j by his client, SE_j agrees to the request and returns its own digital certificate $Cert_j$ (such as a X.509 certificate). Usually a digital certificate will include the server's public key ($SEpk_j$) and the supported encryption algorithms.
2. **Client initialization**: After the client receive $Cert_j$, the client will verify the legitimacy of the certificate through the certification authority (CA). If the certificate is legal, then the client generates a random numbers r_a and encrypts it with the server's public key. Finally the client generates a message Msg_a including the encrypted result and sends Msg_a to the server.
3. **Server verification**: When the server receives Msg_a, it decrypts it with its own private key ($SEsk_j$). If Msg_a has not been tampered with during transmission, r_a will be obtained. After that, the server randomly generates the session key K_s, and XORs K_s with the hash value H_a of r_a to obtain r_b. Finally the server returns $Msg_b = (H_a, r_b)$ to the client.

4. **User Bio-Key generation**: After the client receives Msg_b, it computes the hash value of r_a locally, and compares the value with the received H_a. If the value is the same, it is believed that the Msg_b was not intercepted or tampered with by a third party during the transmission process. The client computes $Ks = r_b \oplus r_a$ at the same time. The user inputs his own username ID_i a and biometrics $\mathbf{B}_i$ on the local biometrics acquisition device. The client runs the fuzzy extractor generation algorithm $\mathrm{Gen}(\mathbf{B}_i)$ and gets a key $(Bio\text{-}Key_i)$ and Pub_i.

Bio-ID Generation

1. **User registration**: In this step, the client calculates $r_c = hash(Bio\text{-}Key_i) \oplus Ks$ and $H_b = hash(Ks||Pub_i||r_c||ID_i)$. The client then includes the parameters H_b, r_c, Pub_i and ID_i in the Msg_c and sends it to the server, and finally destroys $Bio\text{-}Key_i$.
2. **Server enroll and Bio-ID generation**: The server receives Msg_c, then computes the hash value together with the received r_c, Pub_i, ID_i and the value Ks stored locally. If the calculated hash value is the same as the received H_b, then the server generates $Bio\text{-}ID_i = r_c \oplus Ks \oplus SEpk_j \oplus ID_i$. Finally, the server saves ID_i, $Bio\text{-}ID_i$ and Pub_i in the database.

3.2 User Login

In the user login stage, the user first needs to sample the biometrics again and input the his identity. The $Bio\text{-}Key$ is recovered from the user's biometrics by our **PrrFE**. Finally, the $Bio\text{-}ID'$ is regenerated on the server through the $Bio\text{-}Key$, and the $Bio\text{-}ID'$ is compared with the original $Bio\text{-}ID$ to authenticate the user. The user login stage process is shown in Fig. 4.

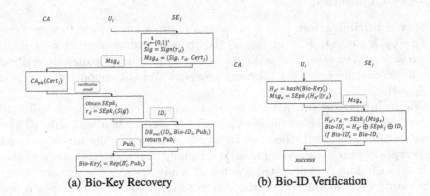

(a) Bio-Key Recovery   (b) Bio-ID Verification

Fig. 4. Process of User Login

Bio-Key Recovery

1. **Initialization**: The user initiates a login request to the server SE_j through the client. SE_j agrees to the request and returns its own digital certificate $Cert_j$ to the client. At the same time, the server selects a nonce r_d, uses its own private key to calculate the signature $Sig = SEsk_j(r_d)$, and returns $Msg_d = (Sig, r_d, Cert_j)$ to the client.
2. **Bio-Key recovery**: When the client receives Msg_d, it first needs to verify the certificate's legitimacy. If the certificate is valid, then the signature Sig included in the Msg_d is verified by the public key on the certificate. Finally, the username ID_i is sent to the server SE_j. The server searches the database for the corresponding record and returns the public helper string Pub_i. The user sample the same biometric again and calculates $Bio\text{-}Key_i' = Rep(\mathbf{B}_i, Pub_i)$

Bio-ID Verification

1. **User login**: After $Bio\text{-}Key$ recovery, the client calculates $H_{R'} = hash(Bio\text{-}Key_i')$. Finally, the client encrypts $H_{R'}$ and r_d with the public key of server, and send Msg_e containing the encrypted result the to the server.
2. **Server authentication**: The server receives Msg_e from the user and calculate $(H_{R'}, r_d) = SEsk_j(Msg_e)$, $Bio\text{-}ID_i' = H_{R'} \oplus SEpk_j \oplus ID_i$. If $Bio\text{-}ID_i' = Bio\text{-}ID_i$, the user is authenticated.

4 Security Analyses

We have proved the security of our proposed fuzzy extractor in [17]. This security means that an attacker cannot recover the user's biometrics and the generated keys from the public data (Pub). In this section, we show our authentication scheme based on the fuzzy extractor can resist the following common attacks.

4.1 Offline Password Guessing

In the user registration stage, the attacker can only use the equation $r_c = hash(R_i) \oplus Ks$ to guess the password. In the whole communication process, Ks never appears in the form of plaintext, and so the attacker cannot know $hash(R_i)$ at the same time. In the user login phase, the message containing the key is $Msg_e = SEpk_j(H_{R'}, r_d)$, which is encrypted and transmitted with the server's public key, so only the corresponding server SE_j can use its private key to decrypt and obtain $H_{R'}$. And since R_i is also never transmitted in plaintext, an adversary who correctly guesses $hash(R_i)$ does not exist in polynomial time due to the collision resistance of the hash function. Therefore, offline password guessing attacks is invalid in our protocol.

4.2   Man-in-the-Middle Attack

In the registration phase, suppose that an attacker intercepts the message containing the user's key R_i, and attempts to replace the key R_i in it with his own key R_A. The formula contained R_i is $r_c = hash(R_i) \oplus Ks$, and Ks is an unknown parameter x_1 to the attacker. If the attacker forges $r'_c = hash(R_A) \oplus x_1$ and $H'_b = hash(x_1||Pub_i||r'_c||ID_i)$, send $Msg'_c = (H'_b, r'_c, Pub_i, ID_i)$ to the server. When the server is verifying, the server has Ks and calculates $hash(Ks||Pub_i||r'_c||ID_i)$ at the same time. Obviously, the attacker forged message is $H'_b = hash(x_1||Pub_i||r'_c||ID_i) \neq hash(Ks||Pub_i||r'_c||ID_i)$, so the attacker cannot successfully forge a request message that used when he registers.

4.3   Replay Attack

During the login phase, an attacker hopes to achieve a replay attack by intercepting and retaining the user's login message. The valuable message for the attacker is also $Msg_e = SEpk_j(H_{R'}, r_d)$. If the attacker sends the message to the server SE_j again, although r_d is a public parameter, and r_d is also a nonce, that is, r_d is a totally different number during each login process. Therefore, Msg_e is an outdated message after being used once. Therefore, this protocol can resist replay attacks.

4.4   Server Phishing Attack

When a user tries to obtain authorization on server SE_j, if a malicious server SE_* wants to impersonate the real server SE_j, the first way is to use a forged certificate $Cert_*$. Apparently the certificate cannot be verified by the CA. The second way is that the malicious server SE_* sends the certificate of SE_j to the user to deceive the user's trust. However, in the login phase, the server needs to sign the random number r_a with the private key first and send the signature to the user with the signature Sig included in Msg_d. However, SE_* cannot obtain the private key of the origin server, so it cannot forge a correct signature. Therefore, the spoofing server fails.

4.5   Database Off-Base Attack

Suppose that the server's database is attacked, so that all (ID_i, K_i, Pub_i) stored in it are leaked. The attacker hopes to use the leaked K_i to log in to user U_i's account. In the last process of login, the message sent by the client to the server is $Msg_e = SEpk_j(H_{R'}, r_d)$, but $K_i = hash(R_i) \oplus SEpk_j$. If the private key is not leaked, the attacker cannot know $hash(R_i)$, and so he cannot login user U_i's account.

5 Experiments

To evaluate the performance of proposed scheme, we implement the multi-server authentication protocol described in Sect. 3. In order to better simulate the real network environment, the simulation is performed between two hosts (one as a client and one as a server). The client runs on Windows11 21H2, has 8 GB of memory, an Intel i5-8265U CPU, and a network bandwidth of 100 Mbps. The Server runs on Windows Server 2012 R2, has 2 GB of memory, an Intel Platinum 8255C CPU, and a network bandwidth of 6 Mbps. The server environment is Apache 2.4.35 with MySQL 5.7.23.

We use two databases for our experiments. The first is FVC2002 DB2 Set A. There are total 100 fingers in the database, and there are 8 different live scan fingerprint images for each finger. The image is $296px \times 560px$ in size. In the registration phase, we select Image 1 from total 8 fingerprint images of one finger as the fingerprint image, and in the login phase, we select Image 2 as the fingerprint image. The second is NIST Special Database 302. We choose one of the sub-databases, SD302d, as our experimental object. In this database, a total of 500 fingerprint data are collected, and we select 2 sampled images for each fingerprint. Among them, we also set Image 1 as the image in the registration phase, and Image 2 is used togetheras the image in the login phase. As a result, a total of 600 sets of experiments will be performed based on above two databases. Some sample fingerprint images of above two database is shown in Fig. 5.

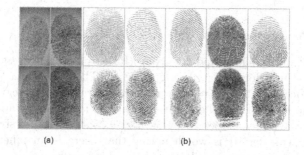

(a)   (b)

Fig. 5. (a) FVC2002 DB2. (b) NIST Special Database 302d.

5.1 Accuracy Performance Analysis

The first test shows us the genuine acceptance rate (GAR) of the real users and the false acceptance rate (FAR) of an attacker who can break our system in the login phase. In the test, we extracted 15 and 20 minutiae for per fingerprint, respectively. And in each case, 4 different modes of biometric verification thresholds are provided. The first is the strict mode, which requires the biometrics (fingerprint minutiae) recovered by the fuzzy extractor to be more than 93% accurate during the login phase. The second is the hard mode, which requires a value of 70%, 53% in the normal mode and 47% in the easy mode. In testing

FAR, for each user, we randomly select fingerprints from the used fingerprint database (except the fingerprint of the current user) to attack our scheme.

As the result shown in Table 3, when the number of the sampled minutiae is $n = 20$ and the security mode is 'Normal', the GAR is equal to 98.16% and the FAR is equal to 1.50%. Too strict authentication mode may cause real users to be rejected. An overly easy authentication mode can also lead to attacks by illegal users. Hence, in the practical application environments, we must consider the trade-off between our security and efficiency.

Table 3. GAR and FAR of our authentication protocol with 15 and 20 minutiae.

Users	key length	number of minutiae	Mode	Verified real users	GAR %	Verified illegal users	FAR %
600	128	15	Strict	104	17.33	0	0.00
600	128	15	Hard	507	84.50	4	0.67
600	128	15	Normal	583	97.16	5	0.83
600	128	15	Easy	591	98.50	7	1.16
600	128	20	Strict	64	10.67	0	0.00
600	128	20	Hard	477	79.50	5	0.83
600	128	20	Normal	589	98.16	9	1.50
600	128	20	Easy	594	99.16	11	1.83

5.2   Time Performance Analysis

We then test the average round-trip time (RTT) of each process during user registration and user login (based on 600 users). There are three stages in the registration process: certificate request, registration request and registration data transfer. In the login phase, there are also three phases: server authentication, the transfer of the public helper string of fuzzy extractor (helper string transfer) and login data transfer. It is worth noting that there is an inherent network delay of 12 ms between the local host and the cloud server.

As we can see, the registration data transfer and helper string transfer take the most time during user registration and login. This is because the two phases transmit the public helper string Pub between the server and the client.

Later, a new test is conducted. In this test, we save the Pub in the user's local storage. Although this will cause a certain amount of local storage overhead (usually 5–7 MB), it can improve the latency between the user and the server. The results of the above two tests are shown in Fig. 6.

Comparing the two ways of saving Pub, we find that saving the Pub locally greatly improves network latency (from an average of 363 ms to 94 ms when user login). Thanks to the properties of the fuzzy extractor, this also does not affect the security of the system. The comparison results are shown in Fig. 6.

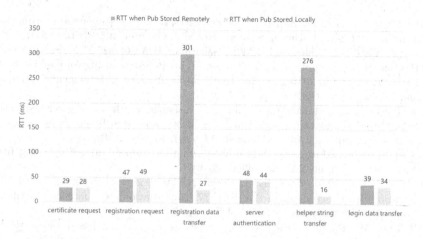

Fig. 6. Comparison of RTT at Each Stage of Above Two Tests.

6 Conclusion

A new authentication scheme based on our **PrrFE** is proposed in this paper. Based on this scheme, the users do not need to memorize keys, and at the same time, it further protects the privacy data of users. In this authentication scheme, the users are authenticated by their biometrics, but the biometrics do not need to be stored locally or in the server in any form. Our experiments show that the optimal GAR of our scheme is 99.16%, the average registration network delay is 104 ms and the login network delay is 94 ms.

We'll concentrate on discovering more applications for identity authentication in more environments such as blockchains or IoT systems in our upcoming work. At the same time, we will study how to further improve our authentication efficiency.

References

1. Anveshini, D., Revathi, V., Eswari, A., Mounika, P., Meghana, K., Aparna, D.: Pattern recognition based fingerprint authentication for ATM system. In: Proceedings of the 2022 International Conference on Electronics and Renewable Systems (ICEARS), pp. 1708–1713 (2022)
2. Arakala, A., Jeffers, J., Horadam, K.J.: Fuzzy extractors for minutiae-based fingerprint authentication. In: Lee, S.-W., Li, S.Z. (eds.) ICB 2007. LNCS, vol. 4642, pp. 760–769. Springer, Heidelberg (2007). https://doi.org/10.1007/978-3-540-74549-5_80
3. Benjamin, F., Meng, X., Leonid, R.: Computational fuzzy extractors. Inf. Comput. **275**(104602), 1–20 (2020)
4. Apon, D., Cho, C., Eldefrawy, K., Katz, J.: Efficient, reusable fuzzy extractors from LWE. In: Dolev, S., Lodha, S. (eds.) CSCML 2017. LNCS, vol. 10332, pp. 1–18. Springer, Cham (2017). https://doi.org/10.1007/978-3-319-60080-2_1

5. Dodis, Y., Reyzin, L., Smith, A.: Fuzzy extractors: how to generate strong keys from biometrics and other noisy data. SIAM J. Comput. **38**, 97–139 (2004)
6. Zahid, A.Z.G., Al-Kharsan, I.H.M.S., Bakarman, H.A., Ghazi, M.F., Salman, H.A., Hasoon, F.N.: Biometric authentication security system using human DNA. In: Proceedings of the 2019 First International Conference of Intelligent Computing and Engineering (ICOICE), pp. 1–7 (2019)
7. Javier, G., Arun, R., Marta, G., Julian, F., Javier, O.: Iris image reconstruction from binary templates: an efficient probabilistic approach based on genetic algorithms. Comput. Vis. Image Underst. **117**(10), 1512–1525 (2013)
8. Katarzyna, K., Marek, O.: Multiply information coding and hiding using fuzzy vault. Soft. Comput. **23**, 4357–4366 (2019)
9. Li, N., Guo, F., Mu, Y., Willy, S., Surya, N.: Fuzzy extractors for biometric identification. In: Proceedings of the IEEE 37th International Conference Distributed Computing Systems, Atlanta, GA, USA, pp. 667–677 (2017)
10. Rakesh, M., Parthasarathy, V.: A secure fuzzy extractor based biometric key authentication scheme for body sensor network in internet of medical things. Comput. Commun. **153**, 545–552 (2020)
11. Raunak, B., Lakshmi, S.K., Kishan, K.N., Prakhar, B., Spoorthi, B.A., Sandesh, B.J.: Finger vein authentication system. In: Proceedings of the 2021 International Conference on Computer Communication and Informatics (ICCCI), pp. 1–7 (2021)
12. Shahriar, E., Siavash, B.S.: Lightweight fuzzy extractor based on LPN for device and biometric authentication in IoT. IEEE Internet Things J. **8**, 10706–10713 (2021)
13. Wen, Y., Liu, S.: Reusable fuzzy extractor from LWE. In: Proceedings of the Australasian Conference on Information Security and Privacy, pp. 13–27 (2018)
14. Wu, L., Xiao, P., Yuan, S., Jiang, S., Chen, C.: A fuzzy vault scheme for ordered biometrics. J. Commun. **6**(9), 682–690 (2011)
15. Xavier, B.: Reusable cryptographic fuzzy extractors. In: Proceedings of the 11th ACM Conference Computer and Communications Security, Washington, DC, USA, vol. 2004, pp. 82–91 (2004)
16. Xavier, B., Dodis, Y., Jonathan, K., Rafail, O., Adam, S.: Secure remote authentication using biometric data. In: Proceedings of the International Conference Theory and Applications of Cryptographic Techniques, Berlin, Heidelberg, Germany, vol. 3494, pp. 147–163 (2005)
17. You, L., Cheng, W., Hu, G.: Construction and implementation of practical reusable and robust fuzzy extractors for fingerprint. IACR Cryptology ePrint Archive, p. 1102 (2021). https://eprint.iacr.org/archive/2021/1102/1630416188.pdf

Author Index

B

Bao, Wei 320

C

Cai, Zhaohui 439
Cao, Mei 263
Cao, Tuo 380
Cao, Yuan 17, 30
Chai, Baobao 97
Chen, JunQi 204
Chen, Na 30
Chen, Sheng 17, 30
Chen, Si 150
Chen, Weiyang 419
Chen, Yawen 401
Chen, Yifei 224
Cheng, Wang 488
Cheng, Yuxia 224

D

Dai, Chenglong 150
Dai, Fei 401
Dai, Yi 474
Dong, Yipeng 130
Du, Peiran 439

F

Fang, Minhui 224
Feng, Tao 187

G

Gao, Jiaheng 109
Gao, Xianming 187
Gao, Yaru 30
Gao, Zhipeng 360
Ge, Binbin 81
Ge, Liming 320
Guan, Xinjie 243
Guo, Xiaojun 1

H

Han, Weizhen 419
Hou, Xiang 340
Hu, Gengran 488
Hu, Xiaofei 49
Hu, Yitao 109
Huang, Fanding 224
Huang, Zhiyi 401

J

Jiang, Haifeng 170
Jiang, Shanqing 187
Jiang, Wei 340
Jin, Shuyuan 301

K

Kong, Fanhao 380

L

Li, Gaolei 419
Li, Guanghui 150
Li, Jialun 61
Li, Songchen 224
Li, Tiejun 340, 474
Li, Wei 150
Li, ZhiKe 204
Lin, Jiacheng 30
Lin, Jiaqi 187
Lin, Zhenjie 380
Liu, Bingyi 419
Liu, Junwei 17
Liu, Liang 81
Liu, Suhui 97
Liu, Zhong 283
Lu, HuaCheng 204
Lu, Jianbo 263
Lu, Jintian 301
Lu, Wei 97

M
Ma, Sheng   340, 474
Mo, Xuan   61
Mo, Zijia   360

N
Ni, Yuzhi   455

P
Pan, Bingjia   455
Pan, Guanyan   224
Peng, Wenbing   224

Q
Qi, Junjie   81
Qian, Chengzhi   17
Qian, Zhuzhong   380

S
Shang, Jingjie   170
Shang, Xinzheng   17
Shi, Gaotao   455
Shu, Lang   224
Song, Bin   419
Sun, Jiakun   301
Sun, Yanqiang   474

T
Tang, Chaogang   170
Tang, Yin   243
Tian, Hui   401

W
Wan, Xili   243
Wang, Bei   224
Wang, Bo   340, 474
Wang, Dan   81
Wang, Liming   380
Wang, Shuhao   170
Wang, Xiaoliang   380
Wang, Yabo   301

Wang, Yong   204
Wang, Yue   49
Wu, Dechong   61
Wu, Huaming   170
Wu, Lizhou   340
Wu, Weigang   61
Wu, Zhihao   1

X
Xiao, Xin   283
Xu, Lin   130
Xu, Rui   340

Y
Yan, Biwei   97
Yang, Yi   474
You, Lin   488
Yu, Jiguo   97
Yu, Nanxiang   263
Yu, Xinlei   360
Yuan, Dong   320
Yuan, Yuan   340

Z
Zhang, DaHuan   204
Zhang, Haibo   401
Zhang, Jianmin   340
Zhang, Lei   130
Zhang, Shoujun   170
Zhang, Tingting   263
Zhao, Chen   360
Zhao, Mengying   263
Zhao, Ming   380
Zhao, Qinglin   150
Zhao, Xiaohu   49
Zhao, Zehui   81
Zhou, Bing Bing   320
Zhou, Nanxin   187
Zhou, Tao   49
Zhu, Aichun   243

Printed in the United States
by Baker & Taylor Publisher Services